AF541624

POLLUTION
THE UGLY FACE OF ENVIRONMENT

POLLUTION
THE UGLY FACE OF ENVIRONMENT

By
Dr. Syed Aftab. Iqbal
M.Sc. , Ph.D., FICS
FICC, FIAEM, MNASc.
(Chief Editor & Co-author

DISCOVERY PUBLISHING HOUSE PVT. LTD.
NEW DELHI-110 002

Published by:
Tilak Wasan
DISCOVERY PUBLISHING HOUSE PVT. LTD.
4383/4B, Ansari Road, Darya Ganj
New Delhi-110 002 (India)
Phone : +91-11-23279245, 43596064-65
Fax : +91-11-23253475
E-mail : discoverypublishinghouse@gmail.com
namitwasan9@gmail.com
sales@discoverypublishinggroup.com
web : www.discoverypublishinggroup.com

Reprinted: **2019**

First Edition: **2011**

ISBN: 978-81-8356-810-4

Pollution: ***The Ugly Face of Environment***

Printed at:
Infinity Imaging Systems
Delhi

Shivraj Singh Chouhan
Chief Minister

Government of Madhya Pradesh
BHOPAL – 462 004
S.No. - 2036/September 16, 2010

Message

I am happy to know that book Pollution: The Ugly Face of Environment is being penned by you.

It is relevant topic, Human follies, irresponsible behaviour and loss of respect for the mother nature have collectively endangered our environment. Today, when environmental problems are aggravating, humans express a deep sense of remorse about their misdeeds. Awareness towards causes and possible remedies is imperative. Involvement of voluntary organizations can make a difference.

I hope the book will be used as an instrument of behaviour change.

Regards

(Shivraj Singh Chouhan)

Preface

Environment is the sum of all social cultural, biological, physical and chemical factors which constitute the surroundings of human beings who are both creator and molders of their environment. Environment is never constant it is continuously changing slowly by nature, rapidly or drastically by all living beings, specially man.

Since last few decades global environment is passing through serious changes and challenges, viz. pollution explosion leading to scarcity of resources, drought floods, earthquakes, natural or man made disasters have seriously affected the life on our planet.

Depletion of Ozone layer causes skin and eye diseases and even cancer through the U.V. light, global warming had endangered a number of islands and cities due to rapid rise in sea level, Water and air pollution is rapidly increasing the sick society, indiscriminate use of pesticides causing a threat to the whole global society, precious animal and plant varieties are going to extinct.

Air pollutants (toxic gases, SPM, toxic metal ions and smog) and noise pollution cause headache, fatigue and deep anxiety to everyone. Thousands of people died as a result of disasters like Chernobyl, Minimata and London somg etc. The victims of MIC gas in Bhopal are still suffering from various ailments and is a worst example of chemical disasters in the world.

The situation became alarming when the most developed countries like USA, U.K. China, Canada, Australia etc. has take a "U" turn in Copenhagen conference and totally refused to cut short the emissions of green house gases, neither have offered any financial support to developing countries.

Now in such circumstances it has become the moral and social responsibility of every one especially, scientist, academicians, teachers, intellectuals and students to cure and to nourish our earth by way of taking immediate steps in following directions:

- Exploration of plant resources and animals, their identification, conservation, multiplication and maintenance.
- Biodiversity conservation which has direct consumptive value in food, agriculture, medicine and industry.
- Dissipating knowledge of environment awareness and Environment Impact Assessment which will be vital in making appropriate scientific decisions at the time of disasters.
- Abatement of pollution, realizing the trend of pollution in various environment media like air, water and soil, Steps which include stringent regulations, developing environmental standards, control of vehicular pollution, spatial environmental planning including industrial estates.
- Environmental research programmers, which specially deal with problems related at air, water and soil pollution, development of suitable cost effective technologies for abatement of pollution, development of eco-friendly biological and other interventions for prevention of pollution and development of strategic/technologies/instruments etc. for control of pollution.

- This environmental publications will encourage students, teachers and young scientist through constructive editing and rapid publications on various thrust areas of environment so as the make each and every person well aware about the current responsibilities of environment.

In the same direction a unique effort is made by way compilation of a book entitled **Pollution: The Ugly Face of Environment** in which senior professors of various universities of India and abroad have contributed their best articles on each and every aspect of environment. The authors have tried their level best to furnish recent developments in the field of environment to make this book a comprehensive, lucid and systematic for better understanding of the subject. The book is equally useful for the students of U.G. and P.G. studies as well as for common literate persons.

Dr. S.A. Iqbal
(Chief Editor)

Contents

CHAPTER 1

Atmosphere

Dr. S.A. Iqbal[1]

Atmosphere is the protective blanket of gases surrounding the earth, which sustains life on earth and saves it from the hostile environment of outer space. It absorbs most of the cosmic rays from outer space and a major portion of the electromagnetic radiations from the sun.

The atmosphere plays a key role in maintaining the heat balance of the earth, through absorption of I.R. radiation emitted by the sun and re-emitted from the earth.

The major components of atmosphere are nitrogen, oxygen while the minor components are argon, carbon dioxide and some traces of gases. The atmosphere is the sources of oxygen (essential for life on earth) and carbon dioxide (essential for plant photosynthesis). It also supplies nitrogen to nitrogen fixing bacteria and ammonia manufacturing plants utilize to yield chemically bound nitrogen essential for life.

Unfortunately, with progress in science and technology man has been dumping waste materials into the atmosphere, which are posing a problem for survival of mankind itself on earth.

VARIOUS REGIONS OF ATMOSPHERE

The atmosphere may be broadly divided into four regions as shown in table. It extends upto 500 km with temperature varying from a minimum of -2°C to maximum of 1200°C.

Region	Altitude range (km)	Temperature (°C)	Chemical species
Troposphere	0–11	15 to –56	N_2,O_2,CO_2,H_2O
Stratosphere	11-50	–56 to –2	O_3
Mesosphere	50–85	–2 to –92	O_2^+,NO^+
Thermosphere	85–500	–92 to 1200	O_2^+,O,NO^+

Troposphere

Troposphere consists 70% of the mass of the atmosphere. Air is uniform, with respect to density and temperature. Density decreases exponentially with increasing altitude. In respect of composition troposphere is more or less homogenous in the absence of air pollution. The temperature in troposphere falls off uniformly with increase in altitude. The air near the ground level is heated by radiation from the earth. The cold layer (–56°C) at the top of the troposphere is called stratosphere, which makes temperature inversion, i.e. transition form negative to positive lapse rate (slope of temperature-altitude curve).

[1]Department of Chemistry, Saifia Science College, (Barkatullah University), Bhopal (India)

Stratosphere

The stratosphere is the quiescent layer having a positive lapse rate. The temperature increases with increase in altitude with a maximum of –2°C at the upper limit of stratosphere. Ozone in this region absorbs U.V. radiation and raises the temperature causing a positive lapse rate.

$$O_3 + hv\ (220 - 330\ mm) \rightarrow O_2 + O$$

Ozone plays an important role in the stratosphere, it acts as protective shield for life on earth from the injurious effects of sun's U.V. rays.

In case of slow mixing in stratosphere, the residence times of molecules or particles, in this region are quite long. If the pollutants can some how reach or injected into the stratosphere, they pose long term global hazards compared to their impact in the much denser troposphere.

Mesosphere

The mesosphere shows negative lapse rate, *i.e.* temperature, falls with increase in altitude. This is due to low levels of U.V. absorbing species, particularly ozone.

Thermsphere

Thermosphere immediately above the mesophere, the temperature rises once again, giving a positive lapse rate (maximum temperature 1200°C). Here the atmospheric gases, particularly oxygen and nitric oxide, split into atoms and also undergo ionization after absorption of solar radiation in the far U.V. region.

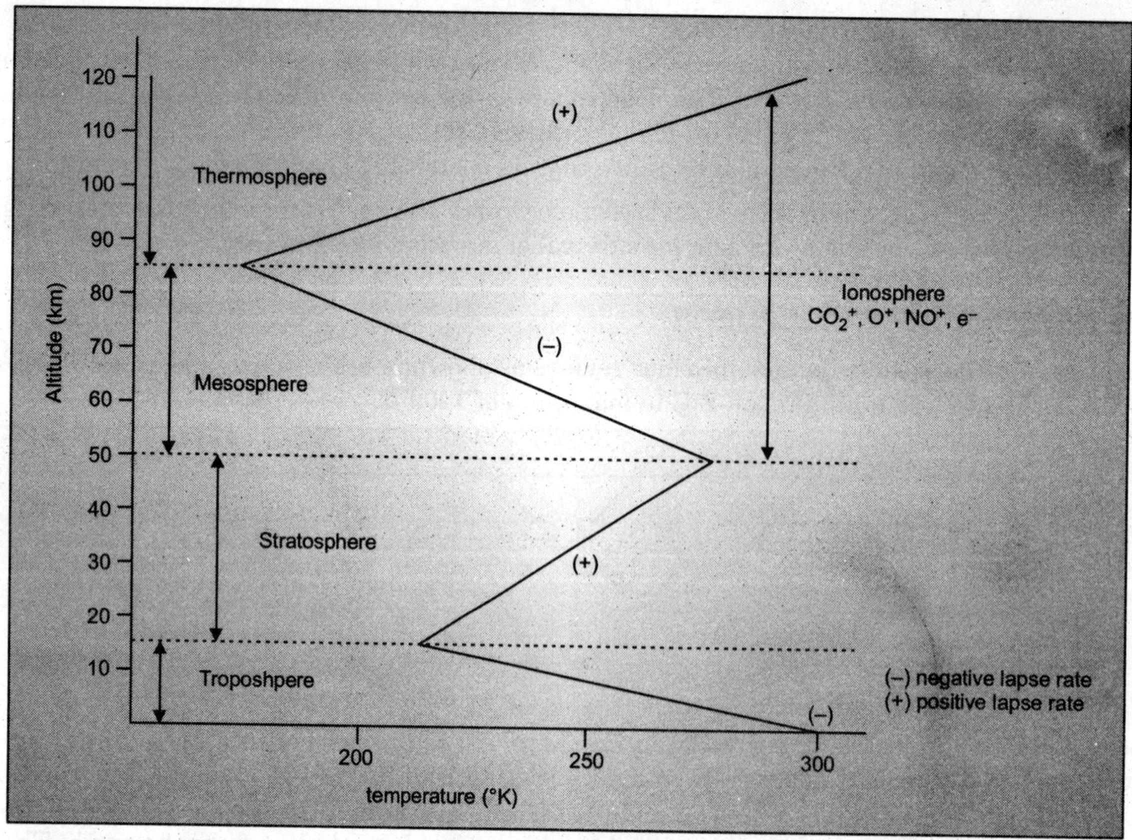

AIR POLLUTION

The atmosphere has broadly been classified into three categories - major, minor and trace.

Major components are - Nitrogen, Oxygen, water vapour. Minor components are - Argon, carbon dioxide

Trace components are - Neon, Helium, Methane, Krypton, Ozone, Nitrous oxide, Hydrogen, Xenon, Sulphur dioxide, NO_2, NH_3, CO, Iodine.

Air is never found clean in nature, due to natural and man-made pollution. Gases such as CO, SO_2 and H_2 are continually released into the atmosphere through natural activities *e.g.* volcanic activity, vegetation decay and forest fires. In addition to these "natural pollutants" there are man-made pollutants -gases, mists and particulate aerosols - resulting from the chemical and biological processes used by man.

Air pollutants disturb the dynamic equilibrium in the atmosphere and therefore affect man and his environment.

Sources of Air Pollution

There are five primary pollutants which together contribute more than 90 % of global air pollution. These are-

1. Carbon mono-oxide
2. Nitrogen oxides
3. Hydrocarbons
4. Sulphur oxides
5. Particulates

Carbon mono-oxide [CO] : It is a colourless, odourless and tasteless gas. It is insoluble in water and is 96.5% as heavy as air. The basic chemical reactions yielding CO are-

1. Incomplete combustion of fuel or carbon containing compounds.

$$2C + O_2 \longrightarrow 2CO$$

2. Reactions between CO_2 and carbon containing materials at elevated temperature in industrial processes.

   e.g. In blast furnaces $\quad CO_2 \leftrightarrow C + 2CO$

3. Dissociation of CO_2 at high temperature

$$CO_2 \leftrightarrow CO + O$$

Sources of Carbon mono-oxide [CO] : Natural processes, *e.g.* Volcanic action, natural gas emission, electrical discharge during storms, seed germination, and marsh gas production contribute in a small measure to CO in atmosphere. The significant contribution is from human activities. The annual emission on a global scale is 350 million tonnes (human source 275 and natural 75 million tonnes), of which USA alone release more than 100 million tonnes of CO into atmosphere.

(A) Transportation contributes about 84% of CO includes motor vehiclese (59.2%), aircraft (2.4%) and railroads (0.1%)

(B) Next in magnitude are miscellaneous sources-16.9 % the main components are forest (7.2%) and agricultural burning (8.3%). Agricultural burning of forest debris, crop residues, brush, weeds and other vegetation.

(C) Industrial process- mainly iron and steel industries and petroleum and paper industry, constitute the third largest contributor of CO (9.6%) to the air. the annual input of CO into the atmosphere by human activities is expected to double its concentration in the surrounding atmosphere every 5 years.

Control of Carbon mono-oxide Pollution

The considerable efforts for control of CO pollution are mainly directed at automobiles because transportation sources are responsible for 74% of all CO emissions and gasoline fed internal combustion engines are primarily accountable for it.

It is well known that the internal combustion engines emissions consist of a mixture of CO, NO_X, hydrocarbons and particulate matter.

It follows that a low air-fuel ratio reduces NO_X emissions remarkably but increases CO and hydrocarbon emissions.

Extensive investigations have seen and are being made for control of automotive emissions along the following lines:-

1. Modification of internal combustion engines to reduce the amounts of pollutions formed during fuel combustion.
2. Development of exhaust system reactors which will complete the combustion process and change potential pollutants into more acceptable material.
3. Development of substitute fuel for gasoline which will yield low concentration of pollutants upon combustion.
4. Development of pollution free power sources to replace the internal combustion engine.

Use of catalytic concreters into two stages helps in eliminating pollutants trom exhaust gases before they are discharged into the atmosphere. In the first converter (NO_X) is reduced to (N_2 + NH_3) in the presence of finally divided Pt, as catalyst and the reducing gases, CO and hydrocarbons. The production of NH_3 is kept at minimum under carefully controlled conditions. In the second converter air is introduced to provide an oxidizing atmosphere for complete oxidation of hydrocarbon and CO into H_2O and CO_2 in the presence of finally divided Pt catalyst.

[Tetraethyl lead, $Pb(C_2H_5)_4$ is used in gasoline in internal combustion engine to reduce knocking. Gasoline containing Pb $(C_2H_5)_4$ is the major source of Pb pollution in the environment. Unleaded gasoline (i.e. lead free gasoline) has recently come into vogue in USA and to a lesser intent in Europe. It is slightly costly than leaded gasoline, but considered necessary to curb Pb pollution in the environment.]

The third possible approach to CO pollution is a substitute fuel for gasoline. Natural gas in both compressed (CNG) and liquefied (LNG) forms has been used as fuel. Although it is pollution free fuel but there are problems of steady supply and economic storage. Alcoholism is other substitute but their combustion products, aldehydes are eye irritants.

The fourth possible solution to automotive emission problem is alternate power sources e.g. streams, electric and gas turbine engines. But none of these are economically viable as compared to gasoline

NITROGEN OXIDES (NOx)

NOx represent composite atmospheric gases, nitric oxide (NO) and nitrogen dioxide (NO_2).

NO is a colorless gas, odorless but NO_2 has a reddish brown colour and pungent suffocating odour.

$$N_2 + O_2 \xrightarrow{1210-1765°C} 2NO$$
$$NO + O_3 \longrightarrow NO_2 + O_2$$

The formation of NO is favoured at high temperature, normally attained during many combustion processes involving air. Rapid cooling of combustion products prevents dissociation of NO.

The 2nd reaction is also favoured at temperatures about 1100°C, but the amount of NO_2 formed is usually less than 0.5%.

Sources: Natural bacterial action discharges about 5×10^8 tonnes of NOx, all over the world (mainly in the form of NO). Whereas man made sources annually release 5×10^7 tonnes of NOx.

The distribution from natural resources is more or less uniform but from man made sources caries depending upon rural/urban areas. In urban atmosphere, NOx is 10-100 times greater than in rural areas. The major man made source is combustion of coal, oil, natural gas and gasoline from power plants, industries, domestic and commercial heating plants etc.

Control : The use of catalytic converters for control of auto motive emissions provides for removal of (NOx) in first stage as described earlier.

Power plants emit about 50-1000ppm of (NO)x. such emissions can be reduced by 90% by a two stage combustion process.

1. The fuel (coal/oil gas) is fired at a relatively high temperature with a sub-stoichiometric amount of air; say 90-95% of the stoichiometric requirement. This yield of NO is limited in the absence of excess of ozone.
2. Fuel burnt out is completed at a relatively low temperature in excess air, under this condition NO is not formed.

A possible approach to (NO)x removal form stack gas is the chemical sorption process using H_2SO_4 solution or alkaline solution containing $Ca(OH)_2$ and $Mg(OH)_2$

HYDROCARBONS AND PHOTOCHEMICAL SMOG

Hydrocarbons are chemicals compounds having only C & H . Methane occurs naturally and is the principal constituent of the fuel known as natural gas. It is colourless, odourless (the odour of LPG is due to the sulphur compound added to it so that humans can detect it). Ethylene and propane are also gases.

Sources- Biological decomposition of organic matter, seepage from natural gas and oil fields and volatile emissions from plants has been major causes for the release of hydrocarbons such as methane, ethylene and aniline.

Incomplete combustion of fuels, motor vehicle exhaust, petroleum refineries, agricultural burring, motor fuel marketing, manufacture of explosives and cracking of natural gas in petrochemical plants have been the human activities sources that emits hydrocarbons.

Natural sources, particularly trees, emit large quantities of hydrocarbons in the atmosphere CH_4 is major naturally occurring hydrocarbon. It is produced in considerable quantities by bacteria in the anaerobic decomposition of organic matter in water, sediments and soil.

$$2(CH_2O)_2 \xrightarrow{\text{bacteria}} CO_2 + CH_4$$

Domesticated animals contribute 5 million tonnes of CH_4 to atmosphere each year. CH_4 has a mean residence time of about 3-7 years in the atmosphere.

Human activities contribute about 15% of HC emitted to atmosphere each year. Automobiles are the major sources in this respect. An approximate break up of global HC emissions (about 570 x 10^6 tonnes per year) is given below:

Petroleum 55%, coal 3.3%, wood 2.2%, refuse burning and incinerators 28.3%, solvent evaporation 11.3%. Due to HC, maximum damage to plants and human beings occurs.

Hydrocarbons are removed off from the atmosphere by several chemical and photochemical reactions. They are thermodynamically unstable towards oxidations and tend to be oxidized through a series of steps. The end products are CO_2, solid organic particulate matter that settles from the atmosphere or water soluble products, *e.g.* acids, aldehydes which are washed by rain.

Control : (of hydrocarbons and photochemical pollutant) O_3 and PAN(Peroxy acetyl nitrate) are secondary pollutants, so their control ultimately depends on the control of their primary precursors, hydrocarbons and nitrogen oxides. The control of (NO)x has been described earlier and that of hydrocarbon in connection with the control of auto exhaust emissions.

PHOTOCHEMICAL SMOG/OXIDANTS

Sometimes, in the presence of sunlight, atomic oxygen from the photochemical reduction of NO_2 also reacts with a number of reactive hydrocarbons (such as CH_4, C_2H_6, $C_6H_5CH_3$ etc. all of which originate from burning of fossil fuels or directly from plants) to form reactive intermediates called free radical. These radicals then take part in series of reactions to form still more radicals that combine with O_2, HC and NO. As a result NO_2 is regenerated and nitric oxide disappears, O_3 accumulates and a number of secondary pollutants are formed such as formaldehyde, aldehydes and peroxy acetyl nitrate (PAN) ($C_2H_3O_5N$). All of these collectively form photochemical smog.

Photochemical smog is oxidizing smog having a high concentration of oxidants. It is characterized by brown, hazy fumes which irritate eye and lungs, lead to the cracking of rubber and extensive damage to plant life.

The probable mechanism of smog formation is:-

1. Reactive HC (those with C=C groups) from auto exhaust interact with O_3 to form a HC free radical $RCH_2°$.
2. $RCH_2°$ rapidly reacts with O_2 to form another free radical $RCH_2O_2°$.
3. RCH_2O_2 reacts with NO to produce NO_2 and the free radical $RCH_2O°$.
4. This new free radical next interacts with O_2 to yield a stable aldehyde, RCHO and hydroperoxyl radial $HO_2°$.
5. $HO_2°$ then reacts with another molecule of NO to give NO_2 and HO°.
6. HO° is extremely reactive and rapidly reacts with a stable HC RCH_3 to yield H_2O and regenerate the HC free radical. $RCH_2°$ thereby completing cycle. One complete cycle yields 2 molecules of NO_2, one molecule of aldehyde RCHO, and regenerates free radical $RCH_2°$ to start all over again. Very soon there is a rapid built up of smog products.
7. The aldehyde RCHO may initiate another route by interacting with HO° radical, leading to the formation of an acyl radical RC°=O, peroxylacyl radical $RCOO_2°$ (by reaction with O_2) and finally peroxyl acyl nitrate PAN (by reaction with NO_2). PAN is one of the most potent eye irritants found in smog.

Smog forming reactions

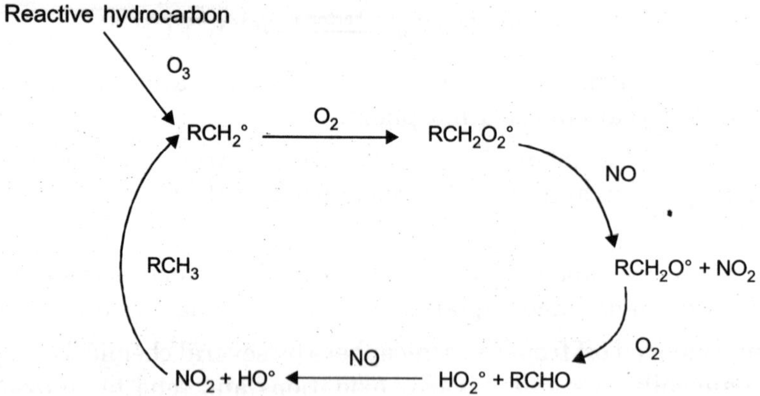

Route to PAN formation :

$$RCHO \xrightarrow{HO°} RC°{=}O \xrightarrow{O_2} RC({=}O)OO° \xrightarrow{NO_2} RC({=}O)OONO_2$$

acyl radical Persuyl acyl radical [PAN]

SULPHUR DIOXIDE [SO_2]

SO_2 is a colourless, non flammable with a pungent odour. It is produced from the combustion of any sulphur bearing material. SO_2 is always accompanied by little a SO_3. The mixture is denoted by SOx.

$$S + O_2 \rightarrow SO_2$$

$$2SO_2 + O_2 \leftrightarrow 2SO_3$$

Under normal humid conditions of the atm., SO_3 invariably reacts with water vapour to form droplets of H_2SO_4.

$$SO_3 + H_2O \longrightarrow H_2SO_4$$

Sources : Natural processes *e.g.* volcanoes provide 67% of the SOx pollution which is evenly distributed all over the globe. Man made sources contribute about 33% of SOx pollution which is, however, localized in some urban areas. Among man made sources, fuel combustion (coal) in stationary sources accounts for 74%, industries 22% and transportation 2% of the total SOx emission. In other words coal fired power stations are the major culprit in this respect, followed by industrial plants.

Sulphate aerosols in urban air are generally smaller than 2μ so they can easily penetrate the innermost passages of the lungs (pulmonary regions) of humans and cause severe respiratory troubles, particularly among older people (70 + years of age).

Control : There are four possible approaches to the removal and control of SOx emissions:

1. Removal of SOx from the fuel gases.
2. Removal of sulphur from the fuel before burning
3. Use of low sulphur fuels
4. Substitutions of other energy sources for fuel combustion

Power plants, the major man made (anthropogenic sources) of SOx pollution are normally built with tall smokestacks to disperse the plume over a wide area. This reduces the local problem but creates problems in areas far away from the power plants.

SOx from flue gases can be conveniently eliminated by using chemical scrubbers. The flue stack gases are passed through slurry of limestone, CaCOs, which absorbs SO_2 quite efficiently.

$$2CaCO_3 + 2SO_2 \longrightarrow 2CaSO_4 + 2CO_2$$

This method appears economical, but is not favored by power plant officials since the primary drawback is that $CaSO_4$ in large amounts poses a waste disposal problem.

An alternative process is based on a reaction between HSO_3- ions (from SO_2) and citrate ions. The flue gas is cooled to 50°C or lower and freed from particulates and traces of H_2SO_4, it is then led into an absorption tower and brought into contact with a solution containing citrate ions (hb Cit.-)

$$SO_2 + H_2O \longrightarrow HSO_3- + H^+$$

$$HSO_3- + H_2{\cdot}cit \rightleftharpoons (HSO_3.H_2Cit)^{-2}$$

The solution is next fed into a closed vessel into which H_2S is bubbled.

$$(HSO_3.H_2Cit.)^{-2} + H^+ + H_2S \rightleftharpoons 2S + H_2\ Cit.^- + 3H_2O$$

Sulphur precipitates out and is later melted and removed from the solution. Part of sulphur is converted to H_2S and circulated to above system. By this method about 99% SO_2 is removed from the flue gas and the recovered sulphur is 99.99% pure.

The removal of sulphur from fuels before burning needs a variety of approaches. Organically bound sulphur in coal can be removed with difficulty through process like carbonization, liquefaction or gassification.

Alternate energy sources such as hydroelectric plants and nuclear power plants are still not economically viable.

PARTICULATES

Small, solid particles and liquid droplets are collectively termed particulates. These are present in atmosphere in fairly large numbers and sometimes pose a serious air pollution problem.

The most important physical property is size. Particulate range in size from a diameter of 0.0002 μ (about the size of a small molecule) to a diameter of 500μ (1 μ = 10^{-6} m) with life time varying from a few seconds to several months.

The particulates possess large surface areas in general and hence present good sites forsorption of various organic and inorganic matters. Their optical properties viz. ability to scatter light and reduce visibility is important in determining their effects on solar radiation. Atmospheric particles can scatter and absorb sunlight, thus reducing visibility. The visual range is approximately inversely proportion to concentration of particulates.

Inorganic Particulate Matter

1. Metal oxides are produced whenever fuels containing metals are burnt. Thus Fe_3O_4 (particulate) is formed during the combustion of pyrite containing coal:

$$3FeS_2 + 8O_2 \longrightarrow Fe_3O_4 + 6SO_2$$

2. Important particulate matter originating from the combustion of leaded gasoline are lead halides. $Pb(C_2H_5)_4$ (TEL) in leaded gasoline reacts with O_2 and halogenated scavengers (dichloroethane and dibromoethane) to yield lead halides which are volatile and emerge through the exhaust system, but which condenses to form particles

$$Pb(C_2H_5)_4 + O_2 + C_2H_4Cl_2 + C_2H_4Br_2 \longrightarrow CO_2 + H_2O + PbCl_2 + PbBrCl + PbBr_2$$

   This is the basis of Pb pollution in the atmosphere.

3. Aerosol mists appear whenever H_2SO_4 droplets are formed

$$2SO_2 + O_2 + 2H_2O \longrightarrow 2H_2SO_4$$

   In the presence of basic air pollutants, such as NH_3 or CaO, salts are formed:

$$\underset{\text{(droplet)}}{H_2SO_4} + \underset{\text{(gas)}}{2NH_3} \longrightarrow \underset{\text{(droplet)}}{(NH_4)_2SO_4}$$

$$\underset{\text{(droplet)}}{H_2SO_4} + \underset{\text{(particle)}}{CaO} \longrightarrow \underset{\text{(droplet)}}{CaSO_4 + H_2O}$$

   At low humidity, H_2O is lost from these droplets and a solid aerosol results.

4. Fly ash is another type of particulate resulting from the combustion of high – ash fossil fuels. Smaller ash particles enter furnace flues and emerge from the stack, in the absence of collector devices, and enter atmosphere.

Organic Particulate Matter

Polycyclic aromatic hydrocarbon (PAN) one important components of organic particulate matter because of their carcinogenic nature.

Some typical PAN compounds are :

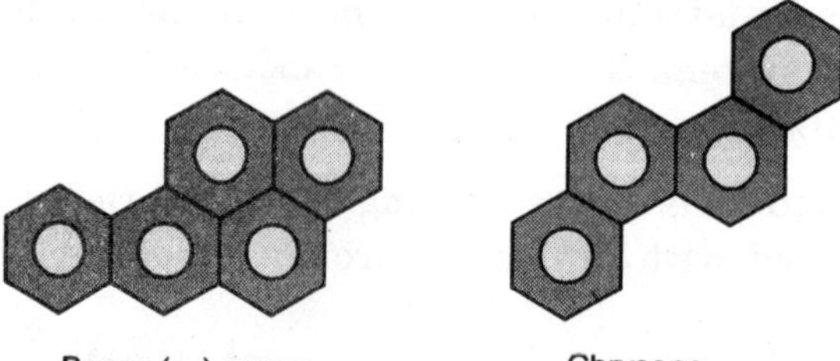

Benzo (α-) pyrene   Chrysene

These are mostly found in solid phase.

Toxic organic micropollutants

Globally it is considered that VOC natural emissions (metabolism and decomposition of plants) are 5 times higher than the emissions generated by human activity.

Anthropogenic emissions are generated by road traffic, fuel burns, industrial processes, solvents evaporation.

Oxidant smog - mechanism of ozone generation

Mechanisms of O_3 generation are different as the reactions occur in troposphere or stratosphere.

In stratosphere, O_3 is generated by O_2 photolysis in presence of UV radiations with $\lambda < 240$ nm:

$$O_2 \xrightarrow{hv} O\cdot + O\cdot \qquad \text{and} \qquad O_2 + O\cdot \rightarrow O_3$$

In presence of UV radiations with $\lambda < 310$ nm O_3 decomposes:

$$2O_3 \xrightarrow{hv} 3O_2$$

In troposphere O_3 is generated by the same reaction:

$$O_2 + O\cdot \rightarrow O_3$$

As solar radiations with X < 240 nm can not reach the troposphere, oxygen atom can be generated only by NO_2 photolysis:

$$NO_2 \rightarrow NO + O$$

Reaction rate depends on light intensity; the reaction does not occur during night. NO_2 photolysis does not quantitatively generate O_3; O_3 reacts with NO and generates NO_2. Thus, O_3 generation competes with oxidation of NO to NO_2. This competition implies generation of radicals with high oxygen content (hydroperoxy and organic peroxy radicals).

Organic compounds and CO are important sources of peroxyl radicals. Life time of organic compounds is short because they are involved in radical reactions in which hydroxyl radicals take part. These reactions are combustions at low temperatures; in these conditions CH_2, H_2 and CO are oxidized to CO_2 and H_2O.

The presence of hydrocarbons and nitric oxides in the air increases the availability of free radicals. This mechanism of reaction allows the evaluation of maxim number of O_3 molecules which are generated by a certain compound oxidation. Theoretically, each carbon atom in saturate hydrocarbons generates three molecules of O_3. For substituted organic compounds (alcohols) carbon atom which is linked to OH is partially oxidized and it is less implied in O_3 generation (1).

The high reactivity of O_3 molecule makes possible its consumption in reactions with other chemical influence O_3 stability which can persist in the atmospheric from several hours to several days.

Tropospheric O_3 tends to concentrate in high populated urban regions in which important amounts of VOC, O_3 precursors, accumulate.

Other chemical species are involved in photochemical decomposition of organic compounds. As ihe organic molecule is attacked by a free radical, the radical reactions cycle generates O_3 only if NO concentration is 10 – 40 ppt at least. During these reactions, decomposition of peracetyl nitrate is a source of NO_x.

Nitric oxides are not consumed during elementary photochemical processes which generate O_3. Although, in photochemical systems, the lifetime of nitric oxides is limited because the hydroxyl radicals react very fast with NO_2 and generate $HONO_2$.

If NO_x concentration is higher than the organic compounds concentration, hydroxyl radicals can be consumed stopping O_3 generation. During a photochemical episode, 1 – 3 ozone molecules are generated for each NO molecule. In unpolluted troposphere (free troposphere), O_3 generation is elevated (10 – 100 O_3 molecules for one NO_x molecule). Although, in free atmosphere O_3 generation is limited by NO_x decreased availability (in non-urbane continental regions NO_x concentrations vary between 50 and 1000 ppt; in free troposphere it varies between 10 and 100 ppt) (3, 4).

O_3 stability in atmosphere

The diminution of O_3 concentration in atmosphere occurs by its decomposition and deposition on the ground, at the surface of seas and oceans, plants. Plants are the main receiver for O_3 which destroys organic molecules in vegetal tissue, especially during day and the intensive vegetation period. Ozone decomposition by photolysis generates O_2 and oxygen atoms which can restart the cycle with generation of new O_3 molecules which can react with water vapors generating hydroxyl radicals or they can react with other molecular species.

Oxidant smog - mechanism of peroxyacyl nitrates generation

In the atmosphere peroxyacyl nitrates are not generated as they are; they are generated *in situ* by photochemical reactions having NO_x and VOC as precursors.

Depending on organic radical, peroxyacyl nitrates can be: peroxy acetyl nitrates (PAN): $CH_3CO(O)ONO_2$; peroxy propionyl nitrates (PPN): $CH_3CH_2CO(O)ONO_2$; peroxy *n*-butyl nitrates (PnBN): $CH_3CH_2CH_2CO(O)ONO_2$ etc, Among these, PAN plays an important in atmospheric chemistry.

The reactions of PAN formation are based on generation of acetyl radicals by radiation of some VOC (hydrocarbons, alcohols, aldehydes). For example:

$$CH_3-CHO\cdot + HO \rightarrow H_2O + CH_3-CO\cdot \quad (1)$$

$$CH_3-CO\cdot + O_2 \rightarrow CH_3-C(=O)-O-O\cdot \quad (2)$$

$$CH_3-C(=O)-O-O\cdot + NO_2 \rightarrow CH_3-C(=O)-O-O-NO_2 \quad (3)$$

In addition to the reaction with NO_2, peracetyl radical can also react with NO generating NO_2 and CH_3 radicals. These species generate formaldehyde by oxidation.

$$CH_3-C(=O)-O-O\cdot + NO \rightarrow NO_2 + CO_2 + CH_3\cdot \quad (4)$$

$$\left.\begin{array}{ll} CH_3\cdot + O_2 \rightarrow CH_3-O-O\cdot & (5a) \\ CH_3-O-O\cdot + NO \rightarrow NO_2 - CH_3-O\cdot & (5a) \\ CH_3-O-O\cdot \rightarrow NO\cdot - O-\dot{O} + CHO & (5a) \end{array}\right\} \quad (5)$$

PAN decomposition occurs only in the presence of NO. The rate of PAN decomposition increases fast with the temperature. In the atmosphere PAN concentration depends on temperature, NO_2/NO ratio (competition between reaction 3 and 4 and VOC availability and reactivity VOC can generate acetyl radicals). Therefore, the relationship between PAN in the air and the emission of its precursors (VOC and NO_x) is not linear and, in the same time dependent on O_3, VOC and NO_x concentrations.

Acyl radicals having a higher number of carbon atoms (propionyl-, *n*-buthyryl-) generates PPN or PnBN, and not PAN. Ethanol as direct precursor can be oxidized to acetaldehyde.

PPN/PAN ratio can be an index of the impact of using ethanol as an additive for vehicles fuels compared to using gasoline. Both PPN day variation and ratio PPN/O_3 are similar to those for PAN.

PAN stability in the atmosphere

As it has already been shown, PAN stability in the atmosphere is limited (35 minutes). The daytime evolution of oxidant smog composition is expressed by PAN/O_3 ratio. The ratio decreases with increasing temperatures. This confirms PAN thermal decomposition mechanism (1). This ratio has a maximum value during night (the temperature diminution promotes PAN stability) and a minimum value at non (mid-day), when temperature is higher. A secondary maximum is noticed in the morning when hydroxyl radicals availability is increased (a more intensive road traffic) promoting PAN generation.

Exposure to PAN

PAN impact upon human health, plants and environment in general justifies specialists' interest for the study of this compound. The conversion of human exposure expressed by ppb (μg/person/day). 1 ppb PAN = 4, 95 (μg/m^1 la 25°C and 1 atm. and for the 23 m^3 of air which are daily represent 114 (μg/person/ day inhaled. The inhaled doze would be 570 μg/person/day if PAN concentration in inhaled air is 5 ppb.

REFERENCES

1. *Adelman Z.E.* A reevaluation of the carbon bond-IV photochemical mechanism. Master of Science Thesis – Department of Environmental Sciences and Engineering, School of Public Health , University of North Carolina at Chapel Hill, 1999.
2. *Atkinson R.D.L., Baulch R.A., Cox R.F., Hampson Jr. J.A., Kerr M. J., Rossi and Troe J.* Kinetic and photochemical data for atmosphere chemistry, organic species: Supplement VII. *J. Phys. Chem.* Ref. Data. 1999, 28: 191-393.
3. *Cocheo V.* Polluting agents and sources of urban air pollution. ann Ist Super Sanita 2000; 36(3): 267-74.
4. *Caffinery J.S., Marley N.A.,* Atmosphereic chemistry and air pollution. *Scientific World Journal* 2003; 7(3): 199-234.
5. *Lareto F.* Emission of isoprenoids by plants: their role in atmospheric chemistry, response to the environment, and biochemical pathways. *J., Environ Pathol Toxicol Oncol.* 1997; 16(2-3): 119-24.

CHAPTER 2

The Biogeosphere

Dr. S.A. Iqbal[1] and Dr. H.C. Kataria[2]

Thee Biogeophysical Environment

The nature of the biosphere and its relationship to living organisms are determined in large measure by the physical characteristics of the earth. The earth is the fifth largest planet in the solar system, having a diameter of slightly less than 12,800 km. Its mass would be *5.52* times as great as an equal volume of water. This is called its *density.* The surface of the his slightly irregular, although this would be hard to discern an observer in space.

Lithosphere

The land part of the earth is called the *lithosphere,* from the Greek word *lithos,* meaning "stone." The solid part, literally onesphere," extends 30 or 50 km beneath the surface. Beneath that, a zone called the *mantle* consists of heavy rock-like material that flows like sticky gum when under pressure, lie deeper the rock, the hotter it is. The core at the center is molten or, as believed by some geophysicists, so highly compressed that it consists of solid metal, mainly iron and nickel.

The outer portion of the earth is sometimes called the *crust* It consists of two layers, the *subcontinental* or inner layer and le *continental* or outer layer. The subcontircntal layer lies beneath the continents and makes up the ocean floor. It consists mainly of a type of rock called basalt. The continental layer forms all the continents. It is mostly a type of rock called granite.

About 70 per cent of the earth's surface is water and about 30 per cent is land. Slightly less than 10 percent of the land area is, in Antarctica, too cold to be populated by man. The soils of the crust were made from erosion of the rocks by the action of water, ice, and wind. Glaciers tear away pieces of rock, winds literally sand-blast the surfaces, and rivers wear down rock and soil to deposit millions of tons of rocks, mud, sand, and silt in the oceans and other bodies of water every day.

The earth is constantly changing, not only by the erosion of mountains but also by internal distortions that lift parts of the crust to higher levels. Thus, while the material of the mountains is slowly ground away and deposited in the oceans, some of the same material again becomes part of the dry land of mountains and plains.

The history of the earth is written in its crust. Twisting, tilting, upheaval, and distortion clearly show the release of tension* in the earth that changed the surface drastically over millions of years. Volcanic

[1] Department of Chemistry, Saifia Science College, (B.U.), Bhopal (India).
[2] Department of Chemistry, Gitanjali Girls College, (B.U.), Bhopal (India).

eruptions repeatedly laid new rock over the old. The rocks show the record of glaciers that advanced and receded several times and covered large parts of North America and Europe. There is strong evidence that entire continents, once part of the same land mass, drifted apart over a long period of time. This is called *continental drift.* The history of life on earth from the emergence of primitive plants and animals to the appearance of advanced forms contemporary with man is imbedded in the rock; that were formed by sedi-ments containing the remains of living forms.

The Biogeochemical Environment

The earth's crust contains all of the 88 natural elements in various combinations. The most abundant elements are oxygen and silicon. They comprise 40 per cent and 28 per cent, respec-tively, of the total composition of the crust. The other elements are present in smaller amounts (Table 2.1). Analysis of the average granite rock shows, that it consists of 67 per cent silicon dioxide (SIO_2), 15 per cent aluminum oxide, and about 14 per cent oxides of iron, calcium, sodium, and potassium. The remaining elements make up less than 4 per cent of the rock.

Table 2.1. Elements in the Earth's Crust

Element	Uses	Abundance in Earth's Crust in Earth's Crust parts per million (ppm)
Aluminium		81,300
Iron		50,000
Magnesium		20,900
Chromium	Stainless steel	200
Nickel	Stainless steel	80
Tungsten	Abrasion resistance	69
Copper	Electrical wire	70
Lead	Storage batteries	15
Zinc	Galvanized steel	220
Tin	Tinplate for cans	40
Molybdenum	Hardening steel	15
Mercury	Chlorine production	0.5
Silver	Photographic film	0.1

The soils of the earth.arc created by the decomposition of rocks. They furnish much of the nourishment for all life on land. Most of the green plants are dependent on the soil for the primary nutrients: nitiogen. Phosphorus, and potassium. "The abundance of these minerals in any particular place determines in large measure the productivity of the soil. A critical factor in the growth of plants is the availability of the so-called micionutrients, such as zinc, iron, and manganese. When even one of these essential elements is present in insufficient amounts or in a form that is not readily available, plants do not thrive. When deficiencies exist, all the living components of the ecosystem are affected.

Organisms are not only dependent on the geochemical environment; they modify it in many ways. For example, plants growing in a squarefoot plot of soil build it into a microcosm entirely different from its original state. Small plants grow in the shelter of the larger ones; fungi and insects proliferate in the discarded plant parts; worms find a congenial home in the organically impregnated soil; and bacteria and other micro-organisms multiply. The soil changes in structure and in nutrient content. The decomposed rock becomes a biological metropolis teeming with visible and invisible life. Similar changes take place in forests and valleys but on a grander scale. Conversely, the removal of plant growth by whatever means produces drastic changes. The soil becomes compacted and hard. Its capacity to absorb moisture is reduced, increasing run-off and erosion. Nutrients are lost, some of them irretriev-ably. Plants can be reestablished only with great difficulty. Having no plants, the land is shunned by animals and micro-organisms that depend on plant life for sustenance.

Trace Elements

Many of elements are present in rocks in small quantities and are, therefore, usually components of soils and water in only trace amounts. For this reason, they are often called *trace elements,* even though they may occasionally be found concentrated in large deposits. A number of the trace elements are essential to the growth, development, and health of plants; Some of them are needed in such minute amounts that they are called micronutrients.

Several of the trace elements are also essential for the growth and development of animals. Iron, manganese, and iodine are needed in small amounts. Others arc needed in even smaller quantities. These include copper, zinc, molybdenum and cobalt. laments such as vanadium, barium, strontium, silicon, chromium and nickel are required by some species of animals. Still other elements, although present in plant and animal tissues, do not have any known biological function. Such elements are lead, mercury, cadmium, aluminum, and tin.

Some of the elements that are essential for the growth and development of plants and animals can be highly injurious when they are present in the environment at excessively high concentrations, and there is often a narrow margin between the beneficial and injurious amounts.

The Carbon Cycle

The main sources of carbon dioxide to the atmosphere are .the decomposition of organic matter by micro-organisms, gas exchange in; the ocean, deforestation, respiration by animals, and the burning of coal and petroleum (Fig. 2.1). Photosynthesis converts CO_2 into the structural material for plants. When plants shed their leaves, the structural material is decomposed by microorganisms and is then either released as CO_2 or incorporated into, soil organic matter. The organic matter may remain in[1] the soil or it may be leached into rivers and eventually carried to the sea.

The annual release of carbon dioxide to the atmosphere from the burning of coal and petroleum is equal to about 10 percent of the carbon dioxide used during natural photosynthesis by terrestrial plants. This release has increased the amount of carbon dioxide in the global atmosphere from early nineteenth century values of approximately 265 30 parts per million ppm) by volume to: 352. ppm in 1989. Estimates of the future combustion of fossil fuels based on an energy growth of 2 and 3 percent each year imply that the atmospheric CO_2 concentration will be double sometime in the next century.

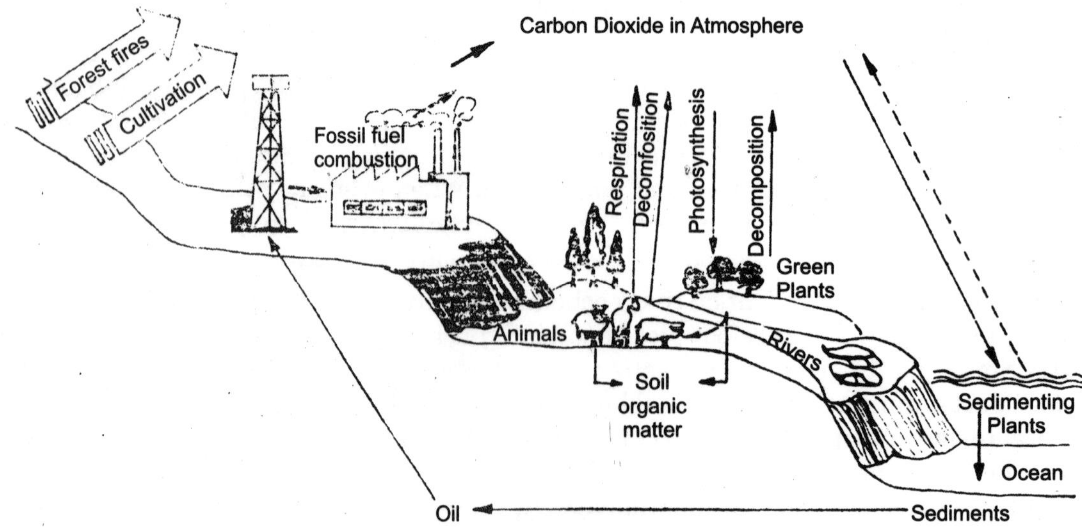

Fig. 2.1. The Carbon Cycle

Because carbon dioxide is one of the atmospheric gases important in determining the Earth's, climate, a doubling of CO_2, might cause a global surface air warming. These increases are predicted to vary considerably over latitude and over the seasons. The warming, of the polar regions could be two to three times as great as that in the equatorial regions, and the warming of solar winters will probably be greater than of polar summers.

In addition to CO_2, the atmospheric concentration of several other "greenhouse" gases produced by human activity also appears to be increasing. These gases are nitrous oxide (N_2O), methane, and ozone. If increases of all these gases are forecast, their synergistic effect will probably cause climate changes to occur significantly earlier than if CO_2, were acting alone. This has been discussed in detail in chapter 8.

The Sulphur Cycle

Atmospheric sulphur is derived from a number of sources, including volcanoes, the combustion of fossil fuels, and microorganism activity in. tidal floods and the water-logged soils of Swamps and bogs (Fig. 2.2). The sulphur cycle perhaps best exemplifies the difficulties that impede our understanding of how man's waste emissions disturb the natural biogeochemical 'cycles. Sulphur is emitted into the atmosphere from natural and anthropogenic (manmade) sources, but many of the effects of both types of emissions are identical.

Anthropogenic sulphur emissions primarily from the burning of coal and, petroleum probably exceed those from natural sources, especially in industrial regions. In the atmosphere, sulphur oxides are converted to sulphuric acid and conveyed to the Earth's surface by precipitation or dry deposition. The rapid removal of this element from the atmosphere restricts, the transport of anthropogenic emissions to short distances. Unlike fossil-fuel-produccd CO_2, which lies a global distribution, sulphur originating from' coal and petroleum only disperses to a maximum of a few thousand kilomenters.

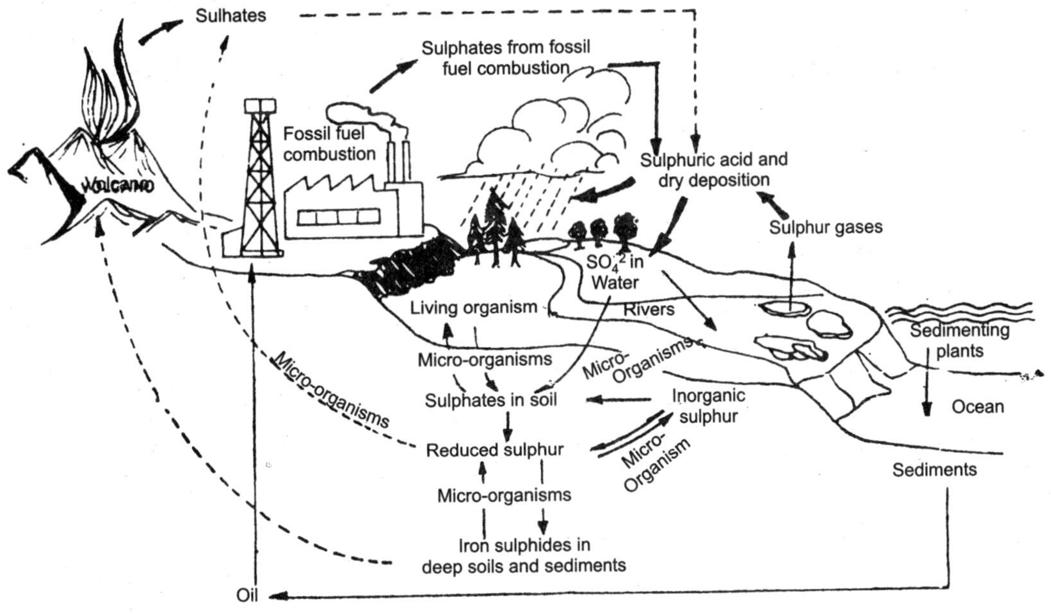

Fig. 2.2. The Sulphur Cycle.

Sulphur dioxide and other gases (e.g., ozone) have been shown to reduce plant growth near major sites of fossil fuel combustion.

The Nitrogen Cycle

Like carbon nitrogen is an essential constituent of living biomass, primarily as proteins and amino acids, and changes in nitrogen cycling. Nitrogen compounds in the atmosphere originate from natural biological activity (denitrification by bacteria, for example), volcanoes, lightning, combustion of fossil fuels, and waste products of domestic animals, as shown in Fig. 2.3.

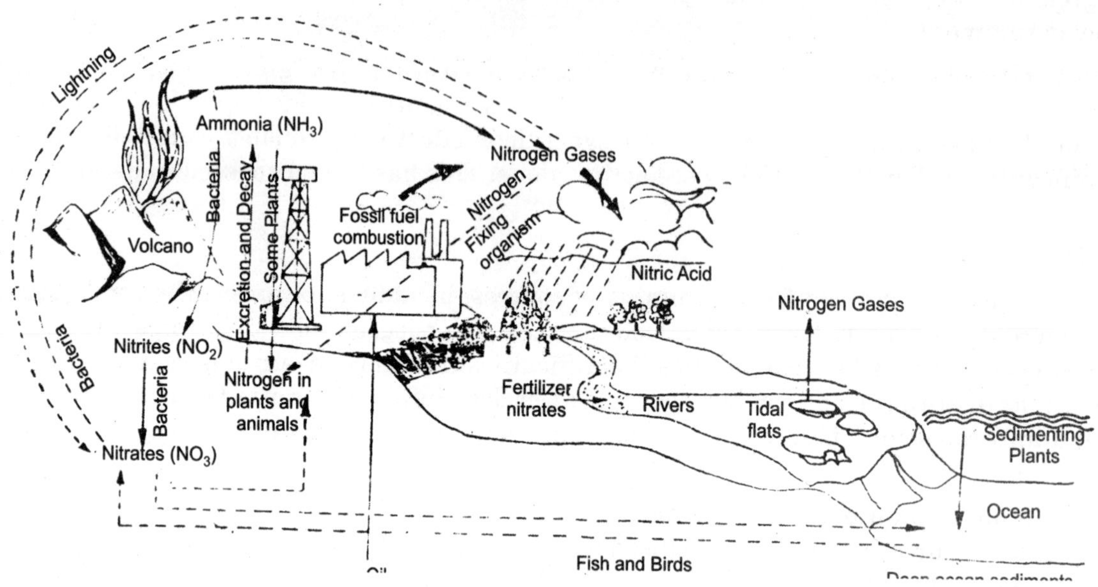

Fig. 2.3. The Nitrogen Cycle

The amounts of both nitrogen oxides and ammonia in the atmosphere are significantly influenced by combustion of fossil fules and by wastes from domestic animal feed lots, respectively. Like the sulphur oxides, nitrogen oxides react in the atmosphere to form acidic compounds that contribute to acidic deposition. Nitrogen is also removed relatively quickly from the atmosphere. This results in impacts in industrial areas only.

The amount of nitrogen (as nitrogen oxides and nitrate) formed by the burning of coal and petroleum and also by fertilizer manufacturing is equal to about one-half of that produced naturally by the biosphere.

The Phosphorus Cycle

Unlike carbon, nitrogen and sulphur, phosphorus does not have a significant gaseous phase in the atmosphere (Fig. 2.4). Atmospheric phosphorus is, however, largely in the form of an aerosol (fine suspended particle) that originates from crustal weathering, the oceans, and such human activities as detergent use and sewage processing.

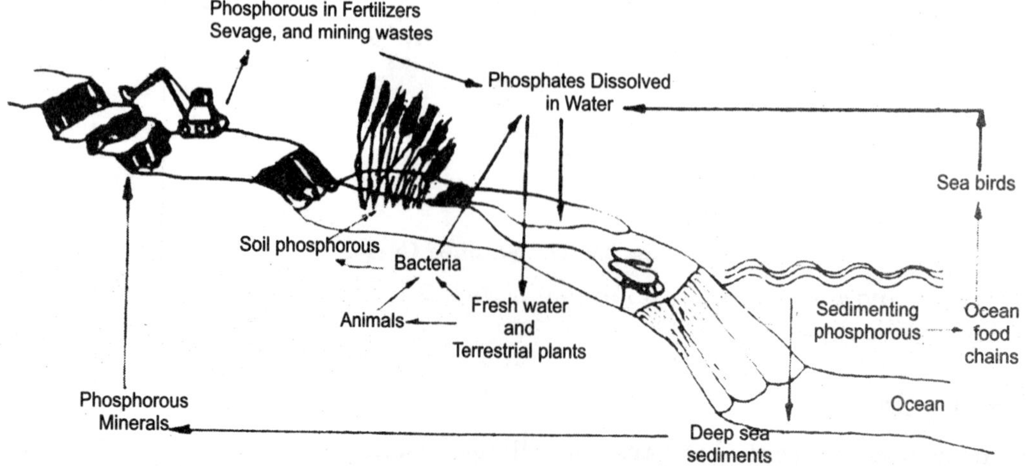

Fig. 2.4. The Phosphorus Cycle

The major impact of man on the phosphorus cycle is through the use of phosphorus fertilizers on agricultural land. Fertilizer use is necessary when the phosphorus in soil is naturally low or when it has been depleted due to intensive farming. Phosphorus is eventually removed via air or water transport to the ocean, where it is taken up by organisms and eventually incorporated into bottom sediments. After several million years, phosphorus in plant and fish remains is converted into phosphate minerals.

Nature, however, has the power to recover from her own ravages. Following the industrial revolution, humans have caused an upsetting of nature's air ecology. If our rush for increased comforts continues as in the past, we shall not only suffocate ourselves, but also destroy all life around us. Air pollution, basically the presence of foreign substances in the air, has become a serious problem because the concentration and qualities of these substances are injurious not only to property, but also to vegetation and animal life.

Motor vehicles are by far our worst polluters, followed by industry and power plants. Carbon monoxide constitutes the greatest mass of any air pollutant, followed by sulphur dioxide, hydrocarbons, nitrogen oxides, and particulates. Most air pollutants are toxic to plant and animal life. The effects are discussed separately in subsequent chapters.

The balance of pollutants in the atmosphere is illustrated in Fig. 2.5; the direction of pollution are marked by dark arrows; white arrows indicate-the process of purification.

Major Air Pollutants

Carbon Monoxide, CO

Carbon monoxide, at colourless, odourless poisonous gas, is the product of incomplete combustion of carbon and its compounds. Created largely by the automobile, CO constitutes more than half the-pollution caused by humans. Carbon monoxide is very dangerous to humans because it combines with haemoglobin in the blood to form carboxyhaemoglobin. If taken in sufficient quantities, CO prevents the haemoglobin from carrying the needed oxygen to the tissues.

On an annual basis about ten times as muc h carbon monoxide.entersr the atmosphere from natural sources as from transportation and other human activities. Most of this (77.6 percent) becomes from atmospheric oxidation of methane arising.

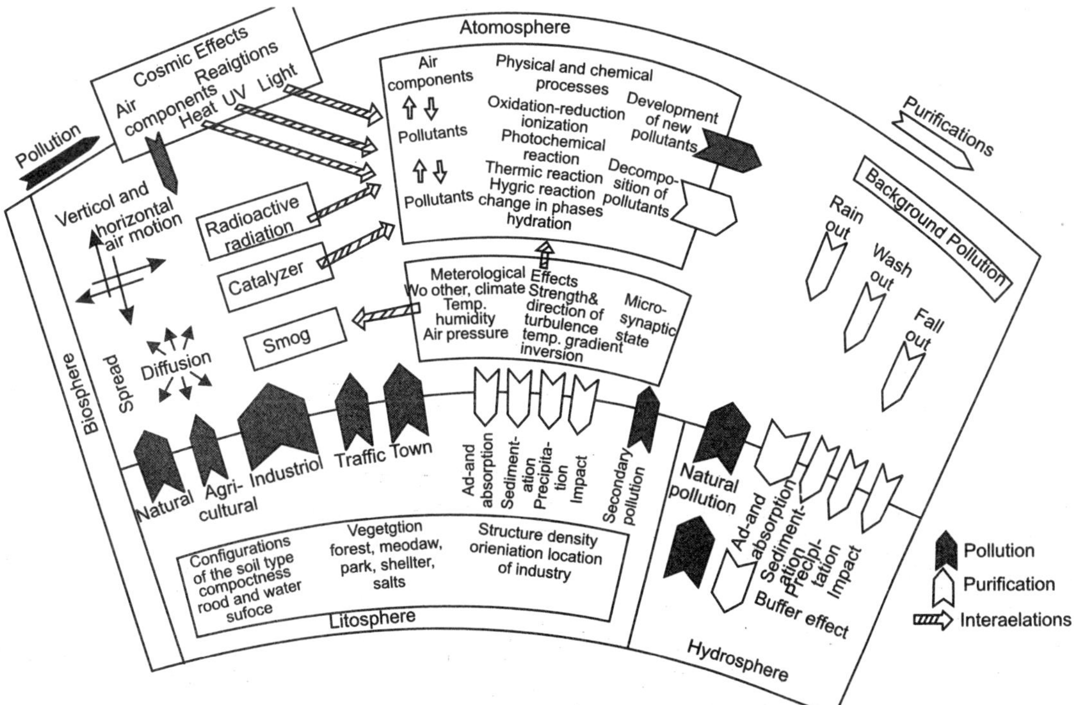

Fig. 2.5. The balance of pollutants in the atmosphere

AIR POLLUTION

The air is one of the most used and abused parts of the biosphere. Industrial, automotive, and domestic activities have resulted in increasingly outrageous insults to the atmosphere. Yet the air is finite; it cannot be manufactured and replenished as the need for it increases. It is at the same time our most precious and most fragile resource. Even animals and plants that live in the sea depend upon dissolved gases from the atmosphere. The enormous abuse of the atmosphere has become a health hazard and a threat to life, damaging both plants and animals in areas polluted with poisonous fumes, dust, and smoke.

Air pollution sometimes has natural causes, such as from forest, and grass fires caused by lightning. Volcanoes are a source of air pollution with ash, dust, sulphur compounds, and other gases that can be irritating or even fatal. The ash from smoking volcanoes often becomes distributed over wide areas and causes distress due to irritation of the respiratory tract and possibly aggravation of respiratory inflammations. The eruption in 1883 of the volcano on Krakatoa, an island between Sumatra and Java, blew most of the island away and started a tidal wave that killed 36,000 people. Volcanic dust from the explosion spread around the earth and was visible for Many months breakdown of organic residues, especially in tropical regions. Smaller contributions come from the oceans (3.9 per cent) and 'from the growth and Breakdown of chlorophyll in plants (2.6 per cent). This naturally generated carbon monoxide does not pose any serious problems because the rate of emission at any given place is relatively low. However, man-made carbon monoxide tends to be released rapidly in urban areas. This results in localized concentrations 50 to 100 times higher than the global average.

The concentration of carbon monoxide in the atmosphere at any time depends not only on the rate of production but also 'on the rate of removal. This occurs mainly in the soil, though a small amount is oxidized to carbon dioxide in the lower atmos-phere. In the soil, fourteen species of fungi have been identified as the active agents oxidizing carbon monoxide to carbon dio-xide. The activity of these fungi varies with the type of soil, being least active in desert soils and most active in the tropics.

Cultivated soils arc less-satisfactory to these fungi than arc those covered with natural vegetation. The total capacity of soil fungi for carbon monoxide oxidation seems to be more than adequate to deal with the carbon monoxide produced on a global basis. Unfortunately, the urban areas, have among the poorest soil reservoirs of these fungi which, therefore, cannot make an effective contribution to lowering the localized high concentrations in these areas. One study showed that in big citites, the air in heavy traffic contained more than 30 ppm carbon monoxide, a concentration that approaches the threshold for acute poisoning. Underground garages have been found to contain 100–200 ppm carbon monoxide in air.

Carbon Dioxide, CO_2

All organic substances used for fuel—petroleum, coal, natural gas, wood are the products directly or indirectly, of photosynthesis. Petroleum, natural gas and coal are the fossil remains of ancient living organisms, modified by the conditions to which they were subjected under the pressure of sediment and rock deposited by prehistoric seas. Petroleum, natural gas, coal and wood all contain an abundance of carbon that was dioxide, it was taken up initially by plants in the process of photosynthesis. Since carbon dioxide and water are the principal products of the complete combustion of fuel, the organic carbon is returned to the atmosphere in its ancient form (see carbon cycle. Fig. 1)

Carbon dioxide comprises a minute fraction of the gases in the atmosphere—about 3 parts in 10,000. Yet it has a profound effect on the temperature of the atmosphere, and hence, on climate. The gas is substantially soluble in water, so that the ocean contains about 1 part in 10,000. This is a fortunate circumstance for the photosysnthetic acitivity of marine and aquatic phytoplankton. Because of the small amount of carbon dioxide in the biospher,e and its relatively low toxicity, if would not ordinarily be thought of as a pollutant. However, carbon dioxide does not enter into the photochemical reactions in the atmosphere and its removal is dependent on biological and chemical deposition. The equilibrium brought about by the normal uptake and output from natural source during the present and recent past may be upset drastically by the enormous increase in the combustion of fuels during this century.

In the geologic past, there were large fluctations in atmospheric carbon dioxide. Volcanic eruptions injected enormous quantities of carbon dioxide into the atmosphere. The total amount during several billion years is estimated to have been at least 40,000 times the amount now in the air. Most of the carbon dioxide was precipitated from the ocean and other bodies of water as slats of calcium or magnesium in the form of limestone or dolomite (magnesium-calcium carbonate). About one-fourth of the total carbon dioxide was taken up by plants and was buried in the sediments. A small fraction of this was concentrated in deposits of coal, oil, sands, and gas pockets that became the fossil fuels of our industrial age.

As long as man continued to use mainly wood for fuel, the contribution of carbon dioxide to the atmosphere was negligible and burning affected the normal carbon dioxide cycle only slightly. Now, however, man is burning fossil fuels that were deposited by living organisms during the past several hundred million years. These geologic reserves of locked up carbon fuel are being oxidized for industrial purposes at a rate that is causing measurable and significant changes in the carbon dioxide content of the atmosphere.

Global Carbon Dioxide Concentrations

About 10 times as much carbon dioxide is injected into the atmosphere from coal fires and furnaces as from the process of breathing. These sources alone would double the carbon dioxide within 500 years if there were no natural process of removal. If the recoverable reserves of fossil fuels—coal, lignite, petroleum, natural gas—are 2,971 billion metric tons, the carbon dioxide, equivalent of the reserves is more than 300 per cent of all the carbon dioxide in the atmosphere present in 1950.

The production of carbon dioxide since 1950 has averaged 9 'billion tons per year, resulting in an increase in concentration of about 1; 6 ppm per year. At this rate, it is estimated that by the year 2000 there will be about 25 per cent more carbon dioxide in the atmosphere than at present. But the use of fossil fuels is increasing. Estimates of the rate of increase in combustion vary between 3.2 percent and 5 per cent per year. If the higher rate continues without further acceleration, the quantity of carbon dioxide that would enter the atmosphere during the next 30 years will be an astounding 60 per cent of all the carbon dioxide that was in the atmosphere in 1950. And if the current rate of combustion were continued, about 13,000 billion metric tons of fossil fuel would be used up during the next 1.000 years, and would add 17 times as much carbon dioxide as the present amount.

Physical Effects of Carbon Dioxide

Building, stones, especially limestone materials such as marble, are susceptible to acid-forming substances. The calcium carbonate of limestone is converted by carbon dioxide and water to bicarbonate, which is water soluble. "Stone cancer" is a term that has been used to describe the erosion of buildings and art objects in the polluted atmosphere of Venice, Italy. Much of the loss there, however, may also be the result of sulphur and sulphide combustion products in the industrial effluent.

Effects on Animals and Plants

Although carbon dioxide is essential to the physiology of animals, and more especially to that of plants, the concentra-tion is normally not critical. For example, uncontaminated air has a carbon dioxide content of about 300 ppm yet man can tolerate nearly 5,000 ppm. without drastically adverse effects on respiration. Moderately higher concentrations than normal bring about increases in the rate of photosynthesis in plants, but an increase of several fold is required before deleterious effects are noted. However, carbon dioxide has a physical role in the stability of the biosphere that may be as important, indirectly, as the direct effect on animal and plant physiology. Carbon dioxide forms a blanket of insulation around the earth. Changes in temperature caused by an increase in the carbon dioxide content *of* the atmosphere could alter world climate in ways that would drastically affect all living organisms. It is theorized that thele changes can be caused by a greenhouse-like effect. (See chapter 8 for more details)

Sulphur Oxides, SO_2

Sulphur pollutants are belched into the air in enormous quantities as part of the industrial effluent. Atmospheric sulphur comes mainly from the sulphur content of fuels. Nearly 80 per cent of the sulphur in sulphur dioxide is initially emitted as hydrogen sulphide and is converted to sulphur dioxide in the atmosphere. Sulphur dioxide and the products of its reactions ire highly damaging to structural and other materials, as well as required degree, for a conscious, objectively oriented, collective effort toward optimization of this activity.

The problem posed by man's chemical activities and the need for nature preservation is vast and many-sided. Among a great number of aspects and questions are the contamination of rivers by industrial wastes, and extensive and virtually uncon-trolled use of chemical fertilizers and herbicides in agriculture. There are scientific problems which are far beyond the present state of man's knowledge such as, the widespread use of anti-biotics; there are problems which are obvious insofar as objec-tives but obscure insofar as method of implementation *e.g.,* the creation of an adequately diversified and safe complex of contraceptive agents.

THE HYDROSPHERE

Water is the most useful natural resource on earth, economi-cally, culturally, and biologically. We drink it, bathe in it, relax in it, fish in it, cook in it, cool with it, irrigate plants with it, and use it for energy, power, transportation, and recreation. Though water is seemingly abundant, the uneven distribution of usable water creates a serious conservation problem in many places where it is vitally needed. In such areas the purity of the water becomes critical. Rivers and lakes in highly industrialized locations may carry an intolerable burden of chemical and human waste products to the point that aquatic life in its natural habitat is wiped out and human health is threatened. Even waters used for irrigation of crops may have excessively high concentrations of salts as the result of leaching or, along coastal areas, from underground seawater intrusion. Wells in such areas become nearly useless as sources of water for agri-cultural or domestic purposes.

About 300 B.C., Alexander the Great and his army of tough adventurers thrust deeply into Persia and the unknown land beyond. They came to a mighty river with more than twice the flow of the Nile, and on its banks was an ancient civilization. A thousand years before Alexander, the Aryans had invaded the continent and settled on the river. They did not give it a parti-cular name but called it the Indus, the Aryan word for "river".

The conquered land became India, the "land of the river." The Indus was destined to irrigate 2.1 million acres of land, the largest irrigated agricultural region on earth. Civilizations sprang up along the other waterways. In Egypt, the middle East, India, and China, as well as in South and Central America; ancient civilizations had their origins in the development of the water supply for irrigation and the production of a dependable food supply.

Modern civilization is dependent on water for irrigation,, industry, domestic needs, shipping, and, of increasing importance for sanitation and waste disposal. Most of the areas of the world that are without developed water remain in the hunting and gathering or grazing stage. Civilization's further advance, and possibly the survival of our cultures, will require intensive study of the development of water supplies and careful atten-tion to the protection of water quantity and quality on a worldwide scale.

The Wet Planet

A man in space can look down on the earth and see that the surface of the planet is mainly an ocean. It is continuous except for interuptions by numerous islands. If Mount Everest, stand-ing at 29,028 feet, were put into the deepest spot in the ocean, its peak would be more than a mile beneath the surface. The ocean occupies 70 per cent of the surface and contains 97 per cent of all the water on earth. Much of the remainder is frozen in the icecaps and glaciers. By comparison, the water in rivers and lakes is small. Less than 1 per cent is in the form of ice-free fresh water in rivers, lakes, and aquifers. Yet this relatively negligible portion of the planet's water is crucially important to all forms of terrestrial and aquatic life. There is also a large underground supply of water. Much of it remains locked deep underground for long periods of time. But the soils near the surface also serve as reservoirs for enormous quantities of water, eventually lost through evaporation or by seepage into underground storage.

The Vapour State of Water

Some of the earth's water is in the atmosphere. The warm of the sun and the air currents evaporate about 0.03 per of the water on the surface each year. The vapour condenses a returns as rain and snow and eventually rinds its way to the bodies of water to complete the cycle (Figs. 4.1 and 4.2).

The Solid State of Water

Much of the earth's water is in cold storage. Glaciers and the icecaps cover 11 per cent of the world's land area; icebergs and pack ice occupy 25 per cent of the ocean area. Permafrost permanently frozen ground-holds another 10 per cent of the land area in its grip, while 30 to :0 per cent of the land is covered with snow at any given time. Three-fourths of all fresh water is locked up as ice, mostly in Antarctica and Greenland. The cold regions of the earth contain vast resources in minerals, petroleum, timber, and water. For this reason the rapidly increasing demand will surely bring more intensive development and industrial activity in those areas. However, environmental problems of the cold regions differ greatly from those in the temperate and tropical parts of the world. Disturbances of the environment that may be mildly disruptive in tropical and temperate areas could be excessively damaging in cold regions.

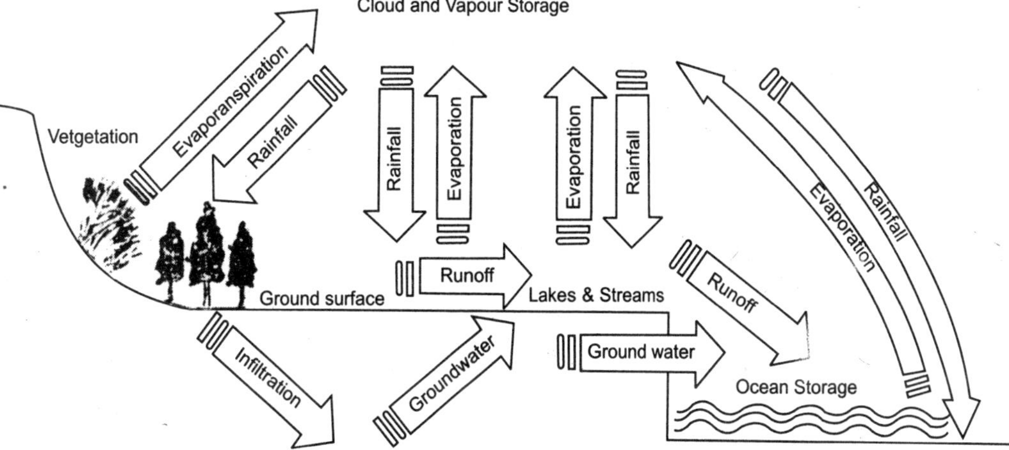

Fig. 2.6. The hydrogen cycle

Small changes in world climate drastically affect the earth's distribution of water supplies. Glaciers are highly sensitive to climatic trends. Observations indicate that many alpine and valley glaciers have been shrinking during the past 100 years, although some have been increasing. As yet, there is not enough data to prove whether the major ice sheets are shrink-ing, grjwing, or in a state of equilibrium. The information might provide a means for predicting future trends. Drilling ice cores in glaciers to

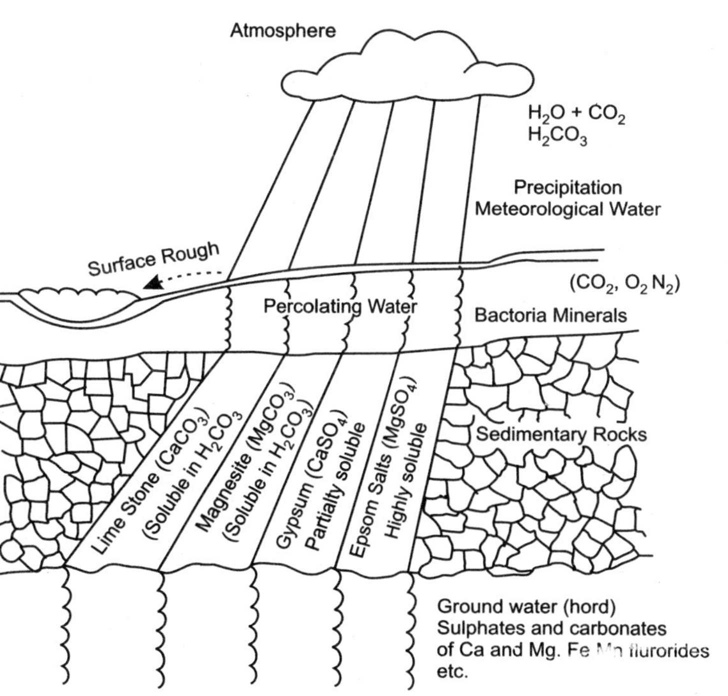

Fig. 2.7. Relationship between surface water and ground waters.

compare the water, dust, and isotope content of different layers makes it possible to determine pre-cipitation during past centuries. For further discussion of climate.

Liquid of Life

Water is the essence of the living process. It is the most abundant and the most versatile of the chemicals of life. It is not surprising that in its physical and chemical properties alone, water is the most uniquely remarkable substance known. Its importance to life demands an understanding of its role in the process of living organisms and in human life.

Water forms most of our flesh and blood. Protoplasm is mostly water. The content varies in different tissues of the organism and in different plants and animals. In jellyfish, the protoplasm contains 95 per cent water. Human blood is 90 per cent water. Muscles and nerves contain 80 percent or more water. The bones contain smaller amounts of water, and even the enamel of the teeth contains 2 per cent water. The total human body consists of 60 per cent to 90 per cent water. The variability being due to the fact that the tissues tend to become dehydrated with age.

With rare exceptions, animals and plants must have an abundant supply of moisture. Animals and plants that are fugitives from the sea and have managed to escape from the limitations of an aqueous existence cannot survive for long without access to moisture either in the air, the soil, or in food. Some organisms die within seconds if deprived of water. Others must have it within their systems but can survive for years in a dessicated environment. Some animals, for example certain rodents, can manufacture their own water. Humans can with stand a loss of no more than 20 per cent of the water in the body without suffering an agonizing death. A healthy person can normally withstand no more than 7 to 10 days without water, but survival time depends largely on the amount of excretion and loss of water from heat and exposure.

The "Universal" Solvent

It is commmon knowledge that water easily dissolves sugar and salt as well as many other substances. Lakes, streams, rivers, springs, and seas are not pure water but contain a great deal of dissolved substances in solution. The ocean contains an average of about 3.5 per cent dissolved solids derived from the land through billions of years of erosion and run-off. The material in solution is mostly common salt (sodium chloride), but there is also an abundance of other mineral constituents, such as calcium, magnesium, sulphates, carbonates, and many metallic elements. Their concentrations are crucially important to the forms of plant and animal life that cither select the water as their habitat or merely consume it for survival. The ability to take up and carry materials that are normally solid substances is perhaps the most important property of water in its role as the solvent. Water is the best solvent known. The solvent power of water enables it to carry the vital chemicals of life —minerals, salts, amino acids, and other organic substances— to the cells of the body. The return trip may be made with a load of waste products, and the water may be eliminated along with the refuse.

Terrestrial plants cannot take up mineral nutrients from the soil in the absence of water. Phytoplanktons cannot absorb their mineral requirements unless the minerals are dissolved in water of their environs. Food must be dissolved before it can enter the blood stream of animals. Even the oxygen and carbon dioxide needed by aquatic and marine organisms must be made available to most of them in solution. The products of metabolism are transported within plants in the aqueous solution called sap and within animals they are dissolved in the water of plasma. Many of the waste products of metabolism are carried away dissolved in water.

Transparency of Water

In the oceans, as on land, photosynthesis is the first stage in the nourishment of the food chain. Phytoplankton, like land vegetation, thrive on sunlight, carbon dioxide, and mineral nutrients. Fortunately, water is transparent to light. Yet photosynthesis by the producers in the ocean is limited by the relatively poor penetration of light through water compared to air. Some light penetrates clean ocean water to a depth of about 60 metres. However, most of the photosynthesis takes place in the upper few feet. In clear ocean water, the blue and green portions of sunlight penetrate the best, but these are the wavelengths of visible light that are the least efficient In photosynthesis. Red light is selectively absorbed and therefore is active only in a narrow zone very near the surface.

Clear water, such as that found in much of the open seas and in many tropical waters, is relatively free of plankton, but there is the compensating factor that light penetration is also Skater, resulting in photosynthesis taking place at greater depths. Tropical waters have a high rate of biological turnover because of the higher temperatures. But production of plankton is generally greater in cool waters than in tropical waters. A positive factor is the relatively high solubility of carbon dioxide and oxygen in water of low temperatures.

Ice

It is well known that ice floats on water. This is strange behaviour for a chemical substance but it is of great biological importance. Most materials that can be frozen and melted are heavier in their cold or solid state than in their warmer, liquid mate. For example, a chunk of iron will sink in a pot of the molten metal.

The ability of ice to float is explained by the molecular behaviour of water. The molecules of water are constantly vibrating. When the water is cooled, the vibrations gradually slow down. As cooling proceeds, the molecules crowd together and form an increasingly dense pack. When a temperature of 4°C is reached, water is in its most dense condition. Below this point a sudden change takes place. The depressed molecular vibration combined with the strengthened attraction of the *hydrogen bond* causes the molecules to shift their positions to a geometric formation that forms a light, expansive latticework structure. As the water cools further, it continues to expand. When it freezes into ice, it is less dense than water in its liquid form and readily floats instead of sinking to the bottom. This can be explained as follows. As water is cooled, it becomes denser because the movement of the water molecules slow down as temperature dorps. Water reaches its maximum density at 4°C. But as the temperature drops further from 4° to 0°C, the attraction between the electropositive hydrogen atoms and the electronegative oxygen atoms—called hydrogen bonding—increases, and the water molecules become bonded together in a crystalline lattice, forming ice. Because of the greater space between the molecules, ice is less dense than water and occupies about 9 per cent more volume.

The fact that frozen water floats instead of sinking to the bottom is a profoundly fovourable factor for life on the earth. If ice were heavier than liquid water, it would sink to the bottom of rivers, lakes, and forozen seas where it would not receive enouth heat from the sun to melt it. Much of the water on earth would be solid ice. Large quantities of the earth's water would be entrapped in an unusable form. Evaporation and precipitation would be greatly reduced. Without moisture in the air, there would be little moderating effect on the sun's radiation and there would be extreme fluctuations in temperature. The world's climate would be drastically altered. Life would be difficult if not impossible. The biosphre as we know would not exist.

Heating and Cooling

When we heat water, its temperature increases until it boils. Continued heating causes no further increase in temperature. The average velocity of the molecules remains the same no matter how hot the vessel is. All additional heat is abosorbed, and the energy is put to work in breaking up the hydrogen bonds between the molecules, which must take place before the molecules can evaporate.

When the vapour returns to the liquid state it must give up its heat of evaporation. If 1 gram of steam condenses at 100°C, it gives off 540 calories (2.2 kilo Joules) of heat, exactly what it absorbed when it evaporated. This principle accounts for the warmth imparted to objects when vapour condenses on them, and the cooling effect (withdrawal of heat) when moisture evaporates from a surface. This has an important moderating effect in biological systems such as the functioning of the sweat glands; when the temperature rises, the cooling effect of evaporation prevents an excessive increase in body temperature for the cooling effect in areas of vegetation. It has a cooling effect on moist soil and prevents the surface from becoming as hot as it other wise would from the direct rays of the sun. It has a similar homeostatic influence on the temperature of bodies of water and thus has an important moderating effect on land temperature and world climate.

Specific Heat

The ability of water to store heat is a characteristic that accounts for much of its biological importance. Water can absorb great amounts of heat while the temperature increase is very little. If we walk bare feet over sand, rock, or pavement on a hot day, the heat may soon become unbearable and we will be relieved to step into a pool of water. Although the water has received the same amount of the sun's radiation, it remains refreshingly cool. On the other hand, in the evening the sand and pavement lose their heat but the water stays about the same temperature that it was during the heat of the day. Upon cooling, the temperature of sand drops five times faster than that of water. This great capacity of water to absorb heat, the slowness of water to warm up and cool off, and its ability to give up great quantities of stored heat is summed up in a property called its *specific heat.* The specific heat of a substance is the number of calories required to raise the temperature of 1 gram of the substance through 1°C.

The water in a lake or in the sea gives up 5 times as much heat as the same amount of soil, sand, or rock. That is why a large body of water has a moderating effect on the temperature of the surrounding area. During hot weather when the water receives large amounts of the sun's radiation, the water abosorbs great quantities of heat. Equally large quantities of heat are given back to the air during cold weather. Much of the life on earth is dependent on the moderating effect produced by the three-fourths of the globe that is covered by water.

REFERENCES

Hutchinson, G. E. 1970. The Biosphere, *Scientific American,* 223(3), 45-53.

Jesson, N. M. 1970, *Biosphere, A Study of Life.* Prentice Hall, Englewood Cliffs, N. J.

Odum, E. P. 1971. *Foundations of Ecology. W. B. Saunders, Philadelphia.*

Southwick, C. H. 1976. *Ecology and the Quality of Our Environment* (2nd Ed.) Willam Grant Press, Boston.

Kvenen, P. H. 1963. *Realms of Water.* John Wiley & Sons. New Yourk.

Press. F. and R. Siever. 1974. *Earth.* W. H. Freeman and Company, San Francisco.

CHAPTER 3

Green House Effect

Dr. M.N. Khan and Ms. Meena Iqbal[1]

Much of the sun energy is emitted as heat rays, consisting of radiation in the infrared portion of the electro magnetic spectrum. The temperature at the surface of the earth is determined by the energy balance between the sun's rays that strike the planet and the heat that is radiated back in to space, The near-infrared rays of sun light penetrate the earth's atmosphere relatively unimpeded, and some of the heat is absorbed and retained by the earth or objects on the surface. The heated earth then radiates this absorbed energy as radiations of longer wave length, mainly in the middle-infrared portion of the spectrum. Much of this does not pass through the air envelop to outer space but is absorbed by the carbon dioxide and water vapour in the atmosphere and adds to the heat that is already present. Though carbon dioxide is almost completely transparent to visible light, and partially to near-infrared, it strongly absorbs and radiates the rays of longer wavelength, especially heat rays having wave lengths of 12×10^{-6} to 18×10^{-6} m. thus carbon dioxide acts like the glass of greenhouse, and on a global scale, trends to warm the air in the lower levels of the atmosphere. This is called the greenhouse effect (fig. 5.3). Water vapour and ozone are also absorbers of infrared radiation and help to keep the earth warm. Water vapour is and efficient absorber of infrared of around 15×10^{-6} mand of another band of wavelength around 6.3×10^{-6}m. Ozone, an absorber of infrared of around 9.36×10^{-6} m, is relatively minor in its effect.

(*a*) Heat form the sun readily penetrates the glass room of a green house in the form of a near-infrared rays; these are absorbed by soil, plants, and other objects and radiated as longer wavelength heat rays. However the glass is not transparent to these radiations of longer wavelength; thus much of the heat is held inside the greenhouse. (*b*) In a similar way, the atmospheric carbon dioxide, water, and ozone are transparent to the near-infrared rays. Of the sun but absorb the reflected longer wave length wave length heat rays, thus (*c*) holding the heat within the boundaries of the earth's atmosphere and having a warming effect.

Weather and Air Pollution

Even though it is not certain that pollution effects global climate, it does effect local climate. A city tends to be a warmer, cloudier, rainier, and has lower visibility than the surrounding countryside. The first day frost comes later in the city. Concrete walls and streets absorb heat during the day and then radiate it at night. Industries also help to increase a city's temperature. Although heat produced by combustion does have a local effect, especially in the heat inlands forming over cities, it appears to have no global significance. Fears have been expressed about the depletion of oxygen because its consumption in the combustion of

[1]Faculty of Science, Saifia College of Science & Education, Bhopal (India)

fuels is though to be greater than its supply from photosynthesis. Recent studies have found to evidence that the concentration of O_2 is changing. However, if the phytoplankton that supplies most of our O_2 is destroyed by either water pollution or an increase in ultraviolet radiation, we may have real reason for concern.

Weather determines that happens to pollutant after they reach the atmosphere. Surface winds help scatter pollutants; the greater the speed, the better the dispersal. Convective currents also help. During daylight hours, the earth's surface is warmed by the sun, causing currents of warm air to rise upward, carrying pollutants with them. As the warm air rises, it cools, because the pressure on it is decreased, allowing it to expand and cool. It continues upward until it becomes the same temperature as the surroundings air. The distance between the earth and this altitude is known as the mixing depth. It determines the depth to which pollutants will be mixed. Its height is continually changing, and it is higher in the summer than in the winter.

Temperature Inversion

In the troposphere, air normally decreased in temperature with an increase in altitude. However, during a temperature inversion, a layer o warm air lies over the cooler air and acts as a lid, keeping the smoke and other pollutants down near the earth's surface. Temperature inversions occur under several different meteorological conditions. When a warm front moves under several static cooler air, a temperature inversion is formed. On clear cool nights, the ground radiates its heat skyward and cools the air at low levels while warmer air prevails aloft. The disastrous smog episode in London in December 1952 was due to a third condition, which is caused by subsidence. This inversion results when upper layers of air descent or subside during a developing anticyclone or high-pressure area. The air warms as it descends, and the warming is greater at higher levels – than near the ground. This results in a temperature inversion. If the anticyclone remains stationary, the inversion may persist and the pollutants accumulate until severe conditions develop. Perhaps the most widely recognized temperature inversions are those that occur over Los Angels. They usually occur on at least 300 days during a year. These inversion are caused by local meteorological conditions augmented by the mountainous topography of the area. The cold Humboldt Ocean current flows along the California coastline west of Low Angeles. AS the prevailing westerlies with their warm moist air from the south Pacific pass over this current, the lower air layers are cooled. often forming fog, whereas the upper layers remain warm Thus, when the air moves eastward over Los Angeles, which is surrounded by mountains' the cooler lower air layers with fog and the city's air pollutants become trapped non only by the lid of warm air above it but also by the surrounding mountains.

The Climate Debate

In this final decade of the 20th century the greenhouse effect and the prospects of global warming is the subject of scientific and political controversy and a cause for widespread concern. A quantitative assessment of global warming trends can be made on the basis of historical temperature records are compiled through the World Weather Watch System, a global cooperative net work of national meteorological services. Observations in previous centuries were largely compiled by individual observers working without coordination. About 10m years ago, in the face of growing global climate concerns, Phillip D.jones and Tom M. L. Wigley and coworkers at the Climatic Research Unit of the University of East Anglia, Norwich, England, initiated a project to collect and analyze, once and for all, every available historical temperature record. This was sponsored by the U. S. Department of Energy. Raymond S. Bradly of the University of Massachusetts at Amherst and Henry F. Diaz of the Environmental Resource laboratory of the National Oceanic and Atmospheric Administration (NOAA); USA were their collaborators. Their assessment is that the earth has experienced an overall warming trend of half a degree Celsius since the late 19th century.

The relevance to climate of the condition of Earth's surface is obvious. The most important features are:

- Presence or absence of water;
- Reflectivity, or albedo;

- Ability to transfer water to the atmosphere;
- Capacity to store heat; and
- Topography and texture.

These interrelated features determine how much of the incident radiant energy is captured, how it is distributed between surface and atmosphere, how the surface interacts with the winds, and how the shoreline interacts with ocean currents. Civilization can influence the surface conditioned by changing or removing vegetation, by damming and diverting rivers to form shallow seas as bays, by covering land with pavement and buildings, by spilling oil on water, snow, and ice, and by releasing previously stored energy as heat.

The earliest significant human influence on climate was undoubtedly that of vegetation changes brought about the human use of fire, deforestation, cultivation, and grazing. Subsequently, changes brought about the irritation projects joined this category of climatic impacts associated with the provision of food and fiber.

As far as is known, the actual changes in climate associated with such surface alterations have been mainly regional, rather than global. One reason for this may be that some human activities, such as irritation of arid lands, actually decrease albedo which compensates to some degree for man-made increases in the albedo elsewhere.

Regional changes can themselves be devastating. There has been particular interesting possible climatologically explanations of the prolonged drought of the late 1960s and early 1970 in sub-Saharan Africa.

The activities of civilization certainly have modified Earth's surface significantly, and they appear to be modifying the stratosphere. What of the troposphere, where mort of the mass of the atmosphere reside and most of the motion that "produces" climate take place? Apart from the influences already mentioned, civilization's direct impact on the troposphere is mainly of three kind: an increased in the concentration of carbon dioxide, which is world wide and uniformly distributed; and increased in the concentration of the particles in the atmosphere over large regions; and the production of high-altitude clouds in initiated as get aircraft contrails at the tropopause. The principal impact of all these phenomena occurs through the interaction of the contaminants with incoming solar radiation and out going terrestrial radiation.

Carbon Dioxide

Beginning in 1957, the Mauna Loa Climate Observatory at Hawaii, USA colleted data which revealed a systematic increase in atmosphere carbon dioxide. To data the change from 290 parts per million (ppm) in 1880 to 352 parts per million in 1989 represents more than a 20% increase over the course of the part century. Efforts to unravel the climatic consequences of increasing carbon dioxide emissions have been going on since long. The world famous mathematician John Von Neumann at the institute for advanced study in Princeton, N. J. (USA) made the first attempt to represent the atmosphere mathematically on digital computers in 1950. In 1963, an unusual laboratory of the National Oceanic and Atmosphere Administration (NOAA) was established at Princeton University under the leader ship of Joseph Smagorinsky. The laboratory was devoted to the mathematical modeling of the atmosphere using the largest and the fastest digital computers available. Called the Geophysical Fluid Dynamics Laboratory, the centre kept researches from many nations interested in this new approach to the study of the atmosphere. Among them was a young Japanese scientist, Sykuro Manabe who developed the first climate model in collaborations with his colleague Richard T. Wetherald in the 1960's. In 1975 they calculated that a doubling of the carbon dioxide of then atmosphere world produced a global climate warming of about 3°C average over the surface of the earth. This calculation has been verified in many different laboratories and has changed substantially. Studies have since been under taken in many parts of the world, including Europe and Soviet Union. In the US the National Research Council contact studies in 1966, 1977, 1979, 1983 and 1987.

Although the mathematical models of the groups yielded simile results, the details of the geographic distribution of climate changes differed from one model to the other. All projected that an increased in carbon dioxide would bring about a gradual warming, but the timing of this warming would depend on the rate of global energy use. They all agreed that if reasonable assumptions were made about future global energy consumption, it would be around the middle of the nest century that the carbon dioxide content of the atmosphere might double.

Just now much this doubling of carbon dioxide would increase temperatures, however, varied greatly from model to model. Some as little as 1°C increase of as much as much 5°C.

These projected temperature changes may appear innocuous because variations of this magnitude are experienced in the normal course of daily and seasonal weather their full implications can be appreciated by noting that it took only 1°C averages decrease in temperatures in Europe to cause the run of several frigid centuries known as the Little Ice Age. Five degrees C believed to be the difference in temperature that separates the end of the last great ice age 12,000 year ago from the present. Further' the projections (fig. 3.4, 3.5, 3.6) indicate that the Northern Hemisphere would experience in just a half century an unprecedented temperature change 10 to 50 times faster than the change since the last the ice age.

Indicate that the Northern Hemisphere would experience in just a half century an unprecedented change, 10 to 50 times faster than the change since the last ice age.

Since the carbon dioxide content of the atmosphere has increased by more than 20% over the past century, we thought to be able to detect the climate warming in global temperature record during the same period. Researches have sought to do this, but is is a much more difficult task than it first appears. The problem is that climate is always in state of natural fluctuation. Separating out the changes that are caused by increasing carbon dioxide from the natural change is difficult. Moreover, the climatic temperature record is based on scattered in irregular observation not taken specifically for the purposes of determining climatic conditions.

Even so, careful analysis of these temperature records by scientists in the U. S. and in the U. K. sought to detect whether a climate warming is consistent with the prediction of the models. The prevailing view is that the climatic record over the past century for the entire globe reveals a net increase in temperature ranging from 0.3 to 0.8°C. But set against this conclusion is the disturbing result that similar increases in temperature cannot be detected over the past century in the U. S., where observations are numerous and accurate.

Even if the temperature rise is real, a puzzle remains that workers have not been able to solve: Is the rise in global temperatures a natural fluctuation or a result of the increase in green house gases? All the can be said is that the observed increase is consistent with the lower end of the temperature increases predicted by the computer models. Consequently, the temperature records, as well as the predictions of mathematical models provide support to both vies points: those who believe the evidence is still weak.

Particles

Particles in the atmosphere are of many sizes and compositions, depending their origins: smoke from forest fires, agricultural burning, and the combustion of fossil fuels; solid and liquid particles that began as gaseous emissions from combustion and natural processes; sea salt; dust from volcanic explosions and from agricultural activity.

Unlike carbon dioxide, particles generally do not become uniformly mixed in the atmosphere; their life expectancy before falling out is too short. There fore, the concentration of particles in the atmosphere tends to be high for some distance down wind of the major sources and else where. The smallest particles remain in the atmosphere longest and therefore have the biggest effects; this means that sources of small particles, whose contribution to emission may not be very large by weight, make a disproportionate contribution in terms of climatic impact. There is no doubt that major volcanic eruption put a considerable quantity of debris into the stratospheres, where it remains long enough to spread over an entire hemisphere. Climatic records indicate that uncommonly cold years have followed some of the major volcanic eruptions of the past 200 years, presumably because the particles have screened out some sunlight.

Particle play an important role in climatic processes by serving as condensation centers for the formation of water droplets or ice crystals when the relative humidity exceeds 100%. In general, there is virtually always a surplus of atmospheric particles that could serve as condensation nuclei if the relative humidity were very high. However the actual dynamics of cloud serve as condensation whey the relative humidity first barely exceeds 100% that class of particles called cloud nuclei, is defined as particles that are active condensation centers below 101% relative humidity. Particles that have the property of initiating freezing in a droplet that contains them are called ice nuclei. Under many circumstances, an increase in the atmospheric concentration of cloud nuclei can increase cloud cover but decrease rain fall because more but smaller droplets are formed, and small droplets do not fall as readily as large once. Man-made particles can also have the effects of increasing rain fall and decreasing cloud cover –this is what is done intentionally in cloud seeding programs to stimulate rain fall—but those phenomena seem to be rarer than those with the opposite effect.

A large amount of CO_2 get introduced into the atmosphere from fossil fuel burning. Furnaces and breathing of animals. From fossil fuel alone more than $2.5 \times 10_3$ tones of CO_2 is being emitted into the atmosphere each year.

CO_2 (present level 325 ppm), although a relatively insignificant non- pollutant species in atmosphere but when due to pollution its amount is increased it is of serious environmental concern.

Not all of the CO_2 injected into the atmosphere remains there, about half of life of it gets utilized by plant life or absorbed by water of the oceans. Part of the CO_2 dissolved in the ocean may get precipitated or incorporated in the marine organisms. In this respect aquatic plants in the ocean are playing an important role in maintaining CO_2 equilibrium between the atmosphere and surface layers of the ocean (up to 100m deep).

Part of CO_2 taken up by terrestrial plants gets deposited in dead vegetation and humus on the forest floor. Some of it, in the forms of organic plant parts, has been eaten by herbivorous animals and gets deposited on or in the soil. Thus the major sink in the ocean which contains the bulk of dissolved CO_2 as bicarbonates. Another sink is the biomass viz. green plants.

Among the constituents of the atmosphere only CO_2 and water vapour absorbs infra red radiations (14000 - 25000 nm) and effectively block a large fraction of earths emitted radiations. The radiations thus absorbed by CO_2 and H_2O vapours is partly reemitted to the earth's surface. The net result is that the earth's surface gets heated up by a phenomenon called the green house effect.

On a global time scale, the known amount of CO_2 in limestone and fossil sediments suggest that normal residence time of CO_2 in the atmosphere has been probably around 100,000 years.

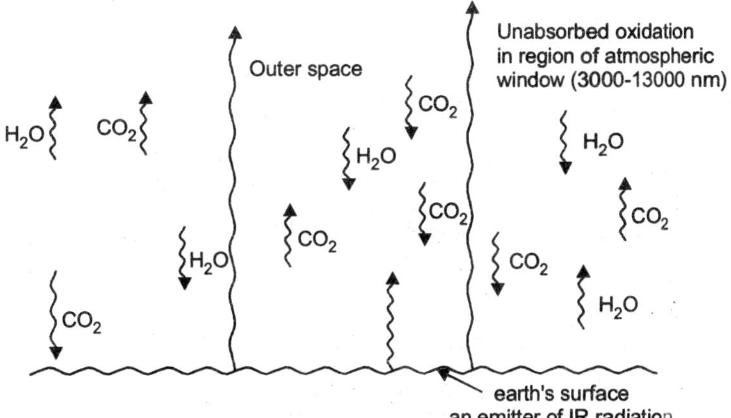

Fig. 3.1.

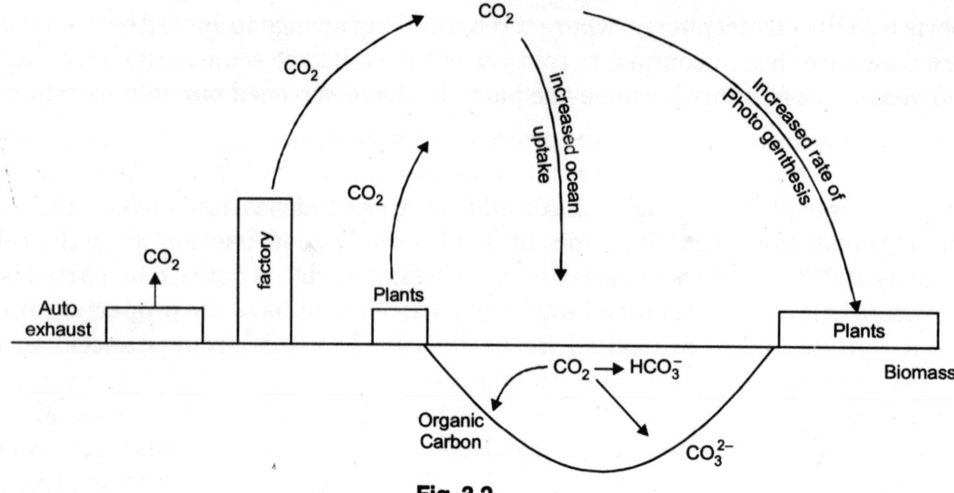

Fig. 3.2.

Sources and Sinks of CO_2

CO_2 gets confined exclusively to troposphere. In dense concentration it can act as serious pollutant. The temperature at the surface of the earth has been maintained by the energy balance of the sun's rays that strike the planet and the heat that get radiated back into the space. Some of the sun's rays that penetrate the thick layer of CO_2 are able to strike the earth and get converted into the heat. The heated earth is able to reradiate this absorbed energy as radiations of longer wavelengths. Much of this doesn't pass through CO_2 layer to outer space but gets absorbed by this CO_2 and water in the atmosphere and adds to heat that has been already present. Thus the earth's atmosphere gets heated up.

CO_2 thus acts like the glass of green house and on a global scale, tends to warm the air in the lower levels of the atmosphere.

An increased heating of earth would cause recede of glaciers, disappearance of ice caps, such as those found over Antartic and Greenland and rise in ocean level. It has been estimated that if all the ice on the earth should melt, 200 feet of water would be added to the surface of all oceans and low lying coastal cities such as Banglore and Venice would get immedated. Only a rise in sea level of 50 to 100 an caused by oceans because of increased CO_2, there been likely to be more hurricanes and cyclones and early snow melts in mountains causing more floods during monsoons. According to Mr. Stephen kecks, a Yugoslavian marine biologist, head of United Nations Environment Programmes centre on ocean and coastal areas, with in 30 years, rising seas will be able to wash away entire countries and flood cities from Boston to Bombay. He calculated that seas will rise by 1.5 to 3.5 m with in 3 decades.

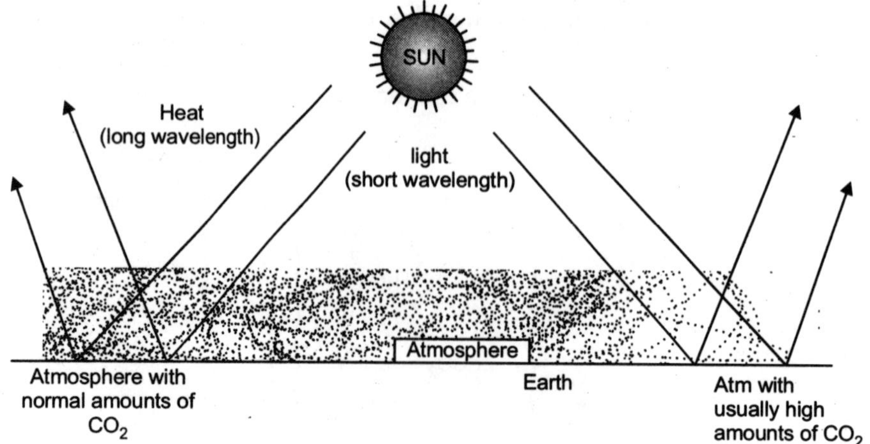

Fig. 3.3. Green House Effect of CO_2

It may be noted that a slight increase in the surface temperature, say 1°C can adversely affect the world food production. Thus the wheel growing zones in the northern latitude will be shifted from the USSR and Canada to the poles i.e., from fertile to poor soils (in the north pole).

The Consequences

If climate change is historical fact and a future inevitability, why worry? The argument is often put for word that people have survived climate change in the past and will continue to do so. Moreover, if civilization adds its own influences to natural climatic trends, how do we known the result won't an improvement?

Today, more over the absolute magnitude of the human misery that could be caused by climate change that is much larger than it ever was because the human populations in virtually all regions of the world are much larger than they were before. Far more people would be affected by any given regional climatic change, and there is not likely to be any room in adjacent regions into which they could flee.

The part of the ecosphere most sensitive to climatic change —that is the element most easily disrupted to the point directly influencing human well –being –is agriculture. Farmers know very well that a year's crop can be ruined by rain that comes too early or too late, or in too great quantity or too little, by too much hog weather or too much cold, by early frosts, and by other vagaries of weather, the consequences of changed climatic condition can be seen in historical records, one of the recent climatic events was the disastrous drought of the summer of the USA. During this drought of the worst on record, the water in the Mississippi River fell so low that navigation was impossible over long stretches urban water supplies were threatened and crops through out the grain belt were devastated. Both officials and public wandered weather this was the greenhouse effect manifest. Record show that in the U.S. five of the years of the 1980's were among the hottest on record, and the average temperature for the decade as the whole was the warmest since instrumental records have been kept. The temperature went up to 32°C. The worst affected was the Corn Belt. Leaves of corn plants curled up to conserve water. Pollens missed the flowers and consequently no corn was produced.

In fact we do not know enough o predict the severity of the consequence, because the warming would not be uniform over the face of the earth. Some parts of the earth would become warmer, some wetter and some drier. On the basis of the evidence at hand climatologists are not able to predict who would benefit and who would lose in such a global redistribution of so, called climatic resources.

Some aspects of global climate warming would be greatly beneficial according to agricultural researchers. Increased carbon dioxide will enhance more active photosynthesis and improve crop growth. Abut the lowered plant requirements for water in a CO_2 enhance atmosphere nothing can be said with certainty. Models do however; agree that Polar Regions of the world would under go greater increases in temperature than would be tropics. Some of the projections of temperature increases in polar areas are starting in their magnitude, predicting as much as 10°C on the average in the Northern Hemisphere and only slight increases in tropical regions. What are the general consequences of such a change in the temperature difference between equatorial and Polar Regions? If arctic regions were to under go significantly greater warming than equatorial regions and if precipitation belts were to move farther North, countries in the north temperature and polar were to move farther North, countries in the north temperature and polar zones would probably benefit greatly. Their growing season would lengthen, and air precipitation would increase. With suitable soils, agriculture might thrive. The polar ice caps would melt resulting in rise in sea level by 2 to 3 ft; this would mean loss of coastal land. Low-lying island nations such as the Maldives could be completely wiped out creating the words first "Greenhouse Refugees". Fires, floods, and prolonged droughts may be more common. These are however, speculations.

Changes in climatic patterns could lead great increase to human misery in altering the abundance and the geographical distribution of various disease-producing organisms. This is true of pathogens that attack crops and trees, as well as those that attack people.

The Alternatives

If the climate changes, the expectation is that it will do so gradually. We should be able to see the initial evidence of coastal inundation in an increasing frequency of high tides and in the undercutting of seacoasts. Climate warming itself should be evident in a rising frequency of heat waves or in other weather anomalies. The effects of a global climate warming are likely to take 30 to 50 years to become serious, and that is a long enough span in which actions to adapt to these changes should be possible.

Major investments in non fossil-energy sources are desirable. The circumstances favour significant new investment in possibly safe, publicity acceptable nuclear power. Further development of forms of solar energy—photovoltaic or biomass, for example-makes good sense. Reforestation and forest preservation yields many ecological and climatic benefits. Research aimed at producing stress and disease –resistant crops would also be recommended. It is likely that humanity will have to adapt to some climate changes. Modes of adaptation by society have not been well studied, Individuals, corporations and communities can adapt to climate change in numerous ways. Farmers can change crops water use can be regulated and management practices can be altered.

Other modes of adaptation would be needed if climate changes were severe. Sea-level rise, which is one of the predicted consequences of a climate warming, might inundate low-lying coastal areas and cause salt water to intrude into fresh water bodies. If this were to occur, in society would have to decide weather to invest in protective structures along coast or adapt by changing land use patterns. The North Sea dikes in the Netherlands are an out standing example of adaptation to relative rise in sea level.

The developing nations are not severely affected so far. They can escape the severity of such climatic changes by not adapting the same technology as the Western countries. The choice may be very expensive. However, a simple check could be to control the population growth.

Our global environment is under attack on many fronts. Climate warming is but one, perhaps the most complex, of these issues. If the change occurring in our atmosphere is likely to cause consequences, we must under stand the problems and promote sensible policies to remedy them.

REFERENCES

1. Keith Bucher, *Global Climate*, Wiley, New York (1976).
2. Murray J. McEwan and Leon F. Philips, *Chemistry of Atmosphere*, Halsted (Weley), New York (1975).
3. Joseph S. Weisberg, *Meteorology*, Houghton Mifflin, Boston (1976).
4. J. Heichlen, *Atmospheric Chemistry*, Academic Press, New York (1976).
5. B. J. Pitts and J. N. Pitts, *Atmospheric Chemistry*, Wiley, N. Y. (1986).
6. R. P. Wayne, *The Chemistry of the Atmosphere*, Oxford University Press, N. Y. (1991).
7. Inderjeet Sethi, M. S. Sethi and S. A. Iqbal, *Environmental Pollution Causes Effects and Control,* 1(1), Commonwealth Publishers, N. D. India (1991).
8. Y. Mido, M. Satake, M. S. Sethi and S. A. Iqbal, *Chemicals in the Environment,* 1, Discovery Publishing House, N.D., India (1995).
9. S. S. Dara, *Text Book of Environmental Chemistry and Pollution Control,* S. Chand and Company, N.D., India (2000).
10. Ajay Kumar Bhagi and G. R. Chatwal, *Environmental Chemistry*, Himalaya Publishing House, Mumbai, India (2003).
11. S. P. Aginhotri, *Environment Conservation Management and Planning*, Chugh Publication, Allahabad, India (1992).
12. Y. Mido, M. Satake, S. A. Iqbal and M. S. Sethi, *Chemistry for Agriculture and Ecology*, Discovery Publishing House, N.D. India (1995).
13. V. P. Kudesia, *Water Pollution*, Pragati Prakashan, Meerut, India (2002).
14. M. Satake, Y. Mido, M. S. Sethi and S. A. Iqbal, *Environmental Chemistry,* Discovery Publishing House, N.D., India (1995).
15. The Hindu, *Survey of the Environment*, India (2004).
16. Current *World Environment,* 4(2), Oriental Scientific Publishing Company, India (2009).
17. S. A. Iqbal and Y. Mido, *Chemistry of Air and Air Pollution*, Discovery Publishing House, India (2003).
18. M. Satake, Taguchi and S. A. Iqbal, *Environmental Toxicology*, Discovery Publishing House, India (2003).
19. M. S. Sethi and S. A. Iqbal, Encyclopedia of Nature and Environment, Discovery of Publishing House, India (2008).

CHAPTER 4

Ozone Layer Depletion

Ms. Aradhana Verma[1]

INTRODUCTION

Nowadays, modern man has initiated a number of abrupt changes beyond the capacity of the ecosystem to adopt and adjust. He has greatly modified or replaced natural processes that control stability and balance within the ecosystem with new processes and manufactured products, which the environment can not absorb. Human activities due to over population and increasing industrialization day-by-day are transforming and changing the global environment. These changes are of different types : climatic changes, deforestation, acid-rain, ozone depletion and increased atmospheric concentration of gases (Green House Effect) that absorb heat and may warm the global atmosphere. Ozone layer is the outer of the gas mixture that protects the surface of the earth from harmful radiation. Leak in the ozone layer at the. surface of the earth's gaseous envelope have been identified. The depletion of the ozone layer would lead to the entrance of the dangerous ultraviolet rays into the earth's atmosphere, which cause damage to the human body. This paper highlights about the ozone depletion a burning issue and the preventive measure to be followed for maintaining the green house' effect. Ozone occurs naturally in the atmosphere and shields the earth from almost all the sun's potentially harmful UV rays. Ozone concentrates between 20-25 Kms. The Ozone content rises in spring becomes lover in autumn season.

The compounds we use as refrigerants destroy the ozone layer and thus increase the intensity of the harmful radiation reaching the earth-surface the improvement of quality of life, refrigerants, cold storage and allied industries have a rapid expansion thus the release of the harmful refrigerants has also increased. Another ozone destructing agent is water vapour, which occurs giving a collection of hydrogen radicals HOx (Maitra S N 1998) Leak in the ozone layer at certain points of earth's gaseous envelope have been reported. Even in India, Variation have been observed with regards to the location and seasons.

OZONE DEPLETION

OZONE LAYER AND ITS IMPORTANCE

The ozone layer is a concentration of ozone orone molecules in the stratosphere. About 90 % of the planet's ozone is in the ozone layer. The layer of the Earth's atmosphere that surround us is called the troposphere. The stratosphere, the next higher layer, extends about 10-50 kilometer above the Earth's surface. Stratospheric ozone is a naturally-occurring gas that filters the sun's ultraviolet (UV) radiation. A diminished ozone layer allows more radiation to reach the Earth's surface. For people, over exposure to UV rays can lead to skin cancer, cataracts, and weakened immune systems. Increased UV can also lead to reduced crop yield and disruptions in the marine food chain. UV also has other harmful effects.

[1]Department of Chemistry Govt. Geetanjali Girls P.G. College, Bhopal (India)

Occurrence of Ozone Depletion

It is caused by the release of chlorofluorocarbon and other ozone-depleting substance, which were used widely as refrigerants, insulating foams, and solvent. The discussion below focuses on CFCs, but is relevant to all ODS. Although CFCs are heavier than air, they are eventually carried into the stratosphere in a process that can take as long as 2 to 5 years. Measurement of CFCs in the stratosphere are made from balloons, aircrafts, and satellite. When CFCs reach the stratosphere, the ultraviolet radiation from the sun causes them to break apart and release chlorine atoms which react with ozone, starting chemical cycle of ozone destruction that deplete the ozone layer. One chlorine atom can break apart more than 100,000 ozone molecules.

Other chemical that damage the ozone layer include methyl bromide (used as a pesticide), halons (used in fire extinguisher) and methyl chloroform (used as a solvent in industrial processes). As methyl bromide and halons are broken apart. they release bromine atoms, which are 40 times more destructive to ozone molecules than chlorine atoms.

EFFECTS OF OZONE-DEPLETION

It is universally accepted that all ozone layer in the stratosphere protects us form harmful UV radiations coming from sun. Ozone accounts for only **three parts in ten millions** of Earth's atmosphere, but plays several crucial roles in the radiation balance of the planet.

The depletion of the ozone layer by modern man's industrial activities have imparted serious implications by letting **"Ozone-eaters"**

Effect on Human Beings

1. With the depletion in atmospheric ozone there is danger of the increase in the flux of ultraviolet radiation over earth's biosphere. The range of wavelengths particularly affected by the changes in atmospheric ozone is from 2900 A to 3200 A. All the known effects of these radiation are harmful for man's life.
2. UV radiation -the narrow spectrum band which is thought to cause most biological damage-appears to trigger two quite distinct immunological effects. One is confined to patches of skin that are actually irradiated while the other develops in the immune system as a whole.
3. The three kinds of skin cancer-basal cell carcinoma, squamous cell carcinoma and melanoma are rapidly climbing the list of human disease caused by UV rays.
4. Langerhans cells in the epidermis of human skin are key player in immune surveillance. UV radiation get them first, breaking damaging UV radiations. But most fair complexioned skin people do not produce of enough melanin to protect them from excessive exposure to sunlight and are affected by several skin damages. Thus sun-bathing on the beach may not be pleasure anymore in England etc.
5. UV radiation cause blood vessels near the skin's surface to carry more blood, making the skin hot, swollen or red, causing **sun burns.**
6. Most epidemiologist concur on the causal relation between UV rays and non-melanoma skin cancers. Studies have shown correlation between these cancers and latitudes and by implication UV radiation levels. The closer a fair-skinned person lives to the equator, the more likely to get non-melanoma cancer by UV rays.
7. Curiously melanoma is caused by intermittently exposing of the body to relatively high doses of UV radiation and is often associated with burning sensation and skin aging.
8. Long exposure to UV radiation caused by *Oz* depletion creates handful of cancer that defy traditional links between poverty and disease.
9. UV radiation causes leukaemia and breast cancer, although the reasons are obscure. studies shown that a 10 % decrease in stratospheric ozone leads to 20-30 increase in cancer. Nearly 7000 people die of such diseases in USA each year. Such cases have increased by 10% in Australia and New Zealand.

10. Quantitatively, the effects of increased UV radiation on a biological specimen is described by a parameter, called the **erythemal does,** which is :

$$D = \int_{\lambda} E(\lambda).H\Delta\lambda(\lambda).F(\lambda)d\lambda$$

Where $E(\lambda)$ describes the relative response of a biological specimen to UV-radiation as a function of wavelength $H_{\Delta\lambda}$ (λ) is the solar UV flux, F (λ) is the factor increase in UV-flux intensity due to ozone reduction.

Calculations show that the erythemal does increases by about 23% for a 10% increase in ozone reduction content and related almost in linear manner to the increase in skin cancer.

11. UV radiation are also absorbed by cornea and lens in the eye leading to photo keratitis and cataracts. since the radiation is not sensed by the visual receptors o the eye, the damage is caused without the individual knowing about it.
12. Ozone at ground level enters the body through inhalation and exerts its toxic effects directly on the lungs. Severe lung injury is associated with edema and hemorrahage. Results of experiments indicates a marked genetic alteration on the lung upon exposure to 0.7 ppm. Ozone continuously for five days. (American Journal of Physiology, June 1991)
13. Ozone at low concentration is also known to cause accumulation of inflammatory cells at the site of lung injury causing severe damage in the lung.
14. The capacity of lung phagocytes which normally fight bacterial infection is also affected resulting in increasing incidence of respiratory infections.
15. **Emphysema** a destructive lung disease, and chronic obstructive lung disease such as chronic bronchitis and development of asthma might be the ultimate results of chronic ambient ozone exposure.
16. Exposure to ozone has been shown to be associated with **lung cancer, DNA** breakage, inhibition and alteration of its replication and formation of DNA adduct, which has been implicated in premature again and finally cell death.
17. Ozone exposure has also been implicated in dizziness and visual impairment - a sigh of central nervous system damage enlargement of spleen and thymus impairment of the immune system.
18. **'Photochemical Smog'** is the major cause of ozone -exposure causing urban air pollution posing a threat to human health.
19. Ozone, in highly populated area occurs with concentration ranging from 0.04 ppm. To 0.7 ppm. In airs, while in upper atmosphere its concentration is more tan 1.0 ppm. So the crew and passengers of flying commercial air crafts often suffer adverse reaction from ozone present in unflltered air cabin.
20. Any increased concentration of ozone brings about change in the nucleic acids, DNA and RNA, so increased UV absorption will have drastic result.

Table 4.1. Effects Of Ozone On Human Health

Concentration (ppm.)	Effects Observation
0.2	No severe effect;
0.3	Nose and throat irritation
0.8(long exposure)	Genetic alteration in the lung within five days
1.0 to 3.0	Extreme fatigue after two hours
9.0	Chronic Pulmonary edema

21. Ozone has been reported to be a strong irritant and is supposed to reach the lungs and respiratory tract much faster than the oxides of sulphur. Even its low concentration causes pulmonary edema. How increased ozone concentration causes ill effects on human health, is shown in the table 4.1.

Effects on Biotic Community

1. Many micro-phytoplanktons would die because of their exposure to UV solar radiation.
2. The marked reduction in the productivity of phytoplankotns would in turn adversely affect zoo planktons. The marine animal, fishes etc. will starve in the absence of sufficient supply of food.
3. The loss of fish population would directly affect the inhabitants of coastal areas.
4. Studies carried out on micro organisms indicate that both irreversible and photo reversible types of injury are caused.
5. Ozone is reported to be highly toxic to fish in the concentrations ranging from 0.1 to 1.0 ppm Anaerobic breakdown of organic phosphorus compounds results in the formation and phosphine and its? .6ppm concentration is highly lethal to fishes.
6. The increased UV radiation will increase the mortality rate of larvae of zooplanktons. Enhanced radiation also impairs fish productivity.

Effects on plants

1. Exposure to air containing ozone results in the lesions to plants, usually confined to the upper surface of leaves. These lesions are charaterised by the uniformly distributed white or brown flecks and stipples in irregularly distributed blotches.
2. Ozone flecking is observed with the plants of grapes, citrus and tobacco. At 0.02 ppm. It damages tomato, pea and other plants. In pine seedling it cause tip burn.
3. Plants proteins are also susceptible to UV injury, because they absorb strongly around 280 nm. 20-50 % chlorophyll reduction and harmful mutation have also been observed.
4. In USA and **California** fruits and vegetable yield have reduced due to ozone pollution. In USA, air pollution causes a crop loss worth two billion dollars. Grapes are no longer produced mainly because of ozone pollution.
5. Ozone alongwith, other pollutants like SO_2 and No_x affecting crop losses of over 50 % in **European countries. In Denmark**, Os afect spinach, Potato,clover and alfalfa etc.
6. In limited pockets Oz level can be potentially harmul. In **West Germany** 100 to 250 ug/m^3 ozone is not uncommon.
7. In **Netherlands** Oz level was high enough to reduce yield of beans, potato and poplars. In UK alone, Os-conentration exceeded 400 g/m^3 in 1976 due to industrial pollution.
8. In plants O3 enters through stomata. It causes visible damage to leaves, thereby reducing **their photosynthetic rate.**

ADVANCED RESEARCH TO PROTECT THE OZONE-UMBRELLA

In **USA, the space shuttle** has been planned which will be equipped with a space laboratory. It will have an atmospheric science facility for observating the latitudinal and distribution of minor constituent throughout the mesosphere and stratosphere (Scientific Uses of the space shuttle, Report of Summer study, National Academy of Sciences Washington DC)

In **Belgium** and **Canada,** the balloon born spectrometer measurements of NOx profile in the stratosphere have been made, which have indicated a significant stratospheric pollution depleting ozone concentration.

New CFCs to Lower the Degree of Pollution

Indian Institute of Technology, Madras is reported to have built a vapour absorption refrigeration system which is friendly to environment and facilitates energy conservation. Absorption system have specific advantages over compression system as they harness low potential and raised heat sources and use environment - friendly working fluids. The new system uses the refrigerants 'R22' which has a low ozone depletion potential. It works on solar energy, helping conservational source of energy.

In refrigeration technology, chloro-fluoro carbon (CFCs) used in compression system deplete the ozone level. In the wild search for alternatives, one of the refrigerants cleared or use till the year 2030 is 'R22', a hydrofluoro carbon. This has an ozone depletion potential od 0.05 compared to that of CFCs. 'R22' works well with organic solvent - absorbents like DMA (dimethyl acetamide). Another hydro - fluro carbon, 'R134A' is used as refrigerant in an absorption system. This "totally safe" refrigerant with an ozone depletion potential of zero had been introduced abroad only recently and that too only in compression system. It is made by a joint project with university of Karlsruhe Germany.

The ozone hole (over Antartica) is growing larger, Spreading northward, causing well-grounded apprehensions, writes YeBorisenkon, Director of the Main Geodelic Observatory.

Remedy and Policies to control ozone Depletion

Prudent policy action - prevention, adaptation and research - should be carried out. The Montreal PROTOCOL to limit emissions of CFCs, signed in the all of 1987 by 31 nations including the United states, is an international co-operation to prevent global environmental deteriorations.

The provisions of the Montreal PROTOCOL negotiated in sept. 1987- 50 % reduction of CFCs production from 1986 levels up to 1999- could be called on to urge deeper cuts in the production of CFCs, to acceleration the time table for their reduction, and to urge all countries to sign and enforce the protocol.

The Environment Protection Administration (EPA) is now estimating an additional 2 lac Americans will die of skin cancer in the next 50 years because twice as much of the ozone layer in the stratosphere is being destroyed as had been first estimated. Although the ozone layer has been depleted at a greater rate than previously believed over most of the globe, the greatest damage has been done over the band from 40 to 50° latitude north, an agency spokesman said.

The ozone layer begins about 9.6 Kms. above the earth's surface and extends as high as about 48 kms. It limits the amount of high energy UV radiation that reaches our earth. The increase in radiation could damage plants and aquatic life as well as human beings. According to an international treaty (1990), policy should be developed to check the use of CFCs, halons and CCLt by the year 2000. Methyl chloroform will be phased out by 2005 under the new pro vision of the treaty.

The ozone layer is essential to life because it shields us from UV radiations that can damage the human immune system and cause cataracts and skin cancer (about 304 lakhs new cases reported in the United State) Plats responses to UV radiation include reduce leaf size, poor seed quality, increased susceptibility to weeds, disease and pests. Significant amount of UV radiation can kill phythoplanktons andmicroscopic margine algae and can affect marine ecosystems.

The ozone hole created by pollution will be filled by special ozonator machines running by solar energy, which will be produced excess of ozone by Os filled balloons. the device was developed by Ozone Help Institution, London, convened by the scientist Zonat Clicherey. He said that by the end of the year 1993, 3 balloons filled with ozone, costing I lacpounds will be left in the stratosphere near the ozone 'hole' The machines will be conducted by 300 small panel and 100 ozonators or ozonizers.

REFERENCES

1. Bartell, L.S. and Ritz C.L. (1974) stratospheric Ozone destruction by man – made chlorofluiromethanes, ***Science 185***, 1163 – 1164.
2. Bowman, K.P. (1988) Global trends in total ozone, ***science 239***, 48-50.
3. Hammond, A.L. (1975) Ozone destruction Problem's scope grows, its urgency recedes, ***science, 187***, 1188-1183.

4. Jone, Robin Russel and Tom M.L. wigley (eds) (1989) Ozone depletion ***Health and Environment consequences John wiley & Sons, England.***
5. UK Review Group on stratospheric Ozone '***Stratospheric Ozone*** HMSO, London (1987)
6. U.K review Group on '***stratospheric Ozone'*** HMSO, London (1988).
7. UK Review Group on stratospheric Ozone ' ***Stratospheric Ozone'*** HMSO, London (1990)
8. Cicerone, R.J. S.Walters, and R.S. Strolarski. (1975) chlorine Compounds and stratospheric ozone, ***science 188*** 378-379
9. Murry J.McEwan and Leon F.Philips, ***chemistry of Atmosphere,*** Halsted (wiley) New York (1975)
10. Seinfeld, John H, Pandis, Spyros N., Atmospheric Chemistry and Physics : From Air Pollution to Climate Change. John Wiley and Sons, Inc ISBN 0-471-17816-0 (1998).
11. Concise Handbook of Sc. & Tech. in India, New Delhi 1999.
12. Economic Survey, 1998-1999 & 1999-2000, GOI.
13. Khitoliya (Dr.) R.K., *et.al.,* Ozone Depletion—A Burning Issue & Solution, 13th National Convention of Environmental Engineers, Bhubneshwar.
14. Maitra, S.N. *et.al.,* Ozone Depletion Oer Antaretica Journal of ASCE-IS, June-July (1998)
15. Various supplements on SC & Tech. of The Hindu.

CHAPTER 5

Radiations, Effect on Human Health and Management

Dr. Zia-Ul-Hasan[1]

The biosphere and ecosystem are self sustaining, nature balance the land, water, air all the livning organisms including man of the world. Any imbalance in biosphere is called environmental pollution, which causes the hazards to Human health and plants. Environmental protection Agency of USA has revealed that radiation in the range of radio and microwave frequency affect biological life. Now question has raised, weather people working on or living near to transmitting towers, microovens, working on X-rays units are safe. In civil departments for persons working with such sourses no standard or precautionary measures are generally adopted. Even standards laid down in military are very vague and are mainly range of UHF radiations. But in our country we do not have any standard measures to guide and protect public life against radiation hazards. So for we have no knowledge about the level of electromagnetic radiations.

Now at present it is necessitate the prohibition of using or installation of more than one radio frequency and microwave transmitters, which are responsible for contributing to an excess of radiations in the environment.

Wherever we discuss radiation exposure, the year 1945 comes in discussion, with Heroshima and Nagasaki when horrible nature of atom bomb shocked the world. This radiation exposure responsible for the discovery of X-rays by Roentgen in 1895. Soon after the discovery of X-rays their harmful effects eryphema, dermitilis and skin cancer were discover in 1920.

X-ray radiation, which consist of high energy photons which alter nuclic acid of living system of a cell both direct and indirect manner. In the direct effect there is a rupture of chemical bonds of the basis and deoxyribase as well as of the back bone of the DNA molecule.

In the indirect effect ionizing radiation produced free radicals either from water (H, OH) or from organic molecules. There free radicles attack the constituents of DNA. The destructive effects are greater for a given dose in the presence of oxygen, which seems to be due to the formation of an H_2O radiced.

In 1928, the International Commissions on Radiological protection and in following year National Committee on Radiation protection and measurements (NCRP) were set up to devise methods of protection from radioactive radiations NCRP Published a series on the effect of radiation on biological system and guidelines for protection.

[1]Department of Botany, Saifia Science College (B.U.), Bhopal (India).

Radiation from Space

Space radiation exposure from space missions and space labs. At present these missions are longer (like mass mission) and increased number of missions with more astronauts are involved. Exposures to radiation in space mission depend on many factors viz. space craft trajectory and inclination, altitude, mission duration nature and thickness of shielding of spacecraft (Bremsstrahlung radiations are produced by interaction between electrons and wall material of the craft). Expouser also depends on time of mission in the 11 years solar cycle movements of astronauts inside and outside craft, radiation energies and radio biological effects. Radiation exposures in space are three types :

(*i*) **Trapped particle radiations :** Consist of mainly of electrons and protons moving in closed orbits in the earths magnetic field.

(*ii*) **Galactic cosmic radiations :** This type of radiation composed mostly of protons with a small mixture of helium and even smaller component of heavier ions.

(*iii*) **Solar particle radiation :** This type of radiation consist of mostly protons with a small amount of helium ions and heavier ions. With the particle energy high and sporadic and galactic radiation energies are low and slow and varying.

It is not possible to predict which solare flora produce protons, which reach the orbit of the earth. It is also not known at which date solar event will reach the orbit of the earth. But it is known, whether a potential flora producing region exists on the sun. When a solar flora occures, the possibility of prediction of a subsequent (SPE) solar particle event improves. The characteristics of the flora, such as the energy which solar X-rays released or type of radio emissions, can be used to predict the SPE occurence.

Terristrial Radiation

Source of terristrial radiation are :

(*i*) Medical exposures.

(*ii*) Radio nuclides within the body and radon inhald into the lungs.

(*iii*) Small exposure from radiations operations,

(*iv*) Reactors (Leakage from radio active source)

Radiation exposure occurs from many of the sources mentioned previously. Directly ionizing radiation carries an electric change that interacts directly with atoms in the tissue or medium by electrostatic attraction or repulsion. Indirectly ionizing radiation is not electrically charged, but results in production of charged particles by which its energy is absorbed. A characteristic of charged particles produced directly or indirectly is linear energy transfer (LET), the energy loss per unit of distance traveled, expressed in kilo electron volts (keV) per micrometer (tim). The LET, depends on the velocity and charge of the particles produced and varies from about 0.2 to > 100 keV/um.

Many particles spend virtually all their energy at LETs of less than a few keV/um. The most significant of these particles are the principal components of primary cosmic radiation and high-energy electrons, such as those emitted by beta radiation. These high energy electrons as well as the indirectly ionizing radiation such as X-rays and gamma rays that produce them, are referred to as low- LET radiation. Low energy electrons, produced by both direct and indirect ionizing radiation are intermediate in LET.

High-LET radiation also contributes to the environmental radiation load. Alpha radiation emitted by internally deposited radionuclide is probably the most important directly ionizing high LET radiation. Neutron radiation is the principal indirectly ionizing high-LET radiation.

Irradiation induces formation of reactive chemical products when it enters a biological system. For example, super oxide radicals can be generated in the reaction of hydrated electrons or hydrogen electrons or hydrogen atoms with dissolved oxygen following gamma radiation of aqueous solutions. This radiolytically formed free *radical* is involved in oxidative chain reactions with the possibility of interconversion and postirradiation generation of other forms of activated oxygen, leading indirectly to further irradiation induced cellular damage.

Low-level ionizing radiation of living cells results in the formation of free radicals and hydroxyl radicals. Much of the cellular DNA damage inflicted by ionizing radiation is due to the formation of free radicals and hydroxyl radicals. Like-wise, the formation of hydroxyl radicals by ionizing radiation can initiate lipid peroxidation, leading to cell membrane damage and the formation of cytotoxic aldehydes.

Effect of Radiations on Human Health

The electrons, protons and heavy ions have characteristics radiation exposure effects. For protons and heavy ions, the available information of effects in human organs is not well known. Due to irradiations of matter with X-rays and gamma-rays electrons are set into motion with a good amount of energy to ionise atoms. Hence the effect of X-rays and gamma-rays and their electrons are similar on human and other biological systems, lose of energy from protons, due to interaction with atomic electrons as they pass through tissues.Secondary protons, neutrons, heavy particles and gamma rays are produced. As protons are slowed down, the rate of loss of energy increased. The penetration of depth dependent on the energy. The effect of protons of diverse energies have been studies.

The neutrons particles are uncharged, so highly penteraling, Neutrons intract with nuclie of atoms, whereas X-rays intract with orbital electrons. Atomic nuclie of neutrons are elastic *seatering* so not exit the thermal energy.

Hydrogen atoms, which are in abundance in tissues and cells, having large collission cross-sections, intract with neutrons, causing large transfer of neutron energy which produce recoil protons. The latter lose energy by ionisation and excitation as they traverse the cells. Neutrons and protons have the same mass. A proton can acquire all the energy of neutron in a single head on 7 elastic collision with hydrogen. Now interaction between neutrons and other elements except hydrogen also produce recoiling of heavy nuclei. The contribution of doses in tissues are smaller than for interactions with hydrogen.

The source of radiations in radio frequency range are many and varied. But effect of human health and plants with other biological systems are not known exactly. Experiments with human and plants have revealed that there are difinet biological effects. Some of there findings are :

1. Catrat production and loss of sight.
2. Chromosomal abnormalities with loss of a part of chromosome, Macromutations.
3. Micromutations, change in DNA, with various changes in the cell structure and density.
4. In human if 2.450 MHz radiation with absorb of 3 to 8 ca/gm. observed male thermal lethality called as Teratogenesis.
5. Exposer of 0.4 to 2.8 mW/sq.cm. al 1.3 to 1.5 GHz (pulsed) radiation induced the adverse motor coordination and balance and docility.
6. With increase of temperature and exposer of radiation of 20 to 60 mW/sq.cm. for 30 to 60 mintutes at 2.450 MHz induced neuroendocina and hormonal alteration.
7. Exposer of radiation at 10 mW/sq.cm. at 2,450 MHz for 5 hours daily for 15 to 17 days gestation period, adverse effect on parental baby and brain weight are observed, called parental impairement of body and brain weight.
8. When a living system expose to 10 mW/sq cm. at 2,450 MHz blood brain barrier alteration were observed.
9. Central nervous system effected when expose to 1 to 2 mW/sq.cm. at 147 mHz radiation dose with increase in calcium ion release.
10. With higher doses of radiations, mortility was observed.

Some of biological and clinical effects observed in human when exposed to microwave and radio frequency are:

(*i*) Catract production and loss of sight.

(*ii*) Epidemiological studies shows increase in lense apacities and retinal lesions.

(*iii*) Effect on auditory nerve, adververse effect of auditory nerve response when human exposed to micro/radiowave.

(*iv*) Many haemalogical charges in blood and vein instability have been observed after exposure to radio and microwave radiations.

It is now well known that, infants, children and pregnant women are more valnerable to the radiation in radio/microwave range. Further effect of these radiations are mutagenic, oncogenic and teratogenic, which are most harmful and require complete safety measures against them.

Effect of varying public health implication approximate threshold level dose between 1 to 10 mW/ sq.cm. Radiation non continuous dose exposures range 10 MHz to 100 GHz. There level of doses capable of producing a variety of biological abnormalities. On the basis of experiment conducted on human, it may concluded that the minimum permissible level to which human can be safely exposed to radiation energy is 0.5 mW/sq.cm. As the frequency range of 30 MHz to to 300 GHz is nonionising portion of electro magentic spectrum, its effects are less than X-rays and gamma-rays.

After the above studies in different countries scientist of the world thought about the real danger of radio frequency (RF) and Microwave (MW) radiation energy to public life. Now it is necessary to study the effect of RF and MW electromagnetic radiation range to collect data for formulation of International/ National standard. So that each country may be guided by them. The standards will enable us to evolve ways and means to exposer of the radiation at safe level. In the absence of such data, it would be safe for us to maintain the permissible level of RF to MW radiation at not exceeding 10% of indicated occupational permissible level i.e. 50 MW/Sq.cm.

Environmental radiations also cause mutations in human system, both by causing inherited defects resulting from germ cell mutation and by acting as carcinogens affecting the some cells a process which may be also involved in senescene. The environmental mutagens may be synthetic, Industrial or naturally occuring ones. The last group is present in plants and fungi or is produced by cooking or storage of food or formed endogenously. A synthesis required of the information obtained at the molecular level with practical epidemiologic aspects.

Effect on Plants

Plants and lower animals are much more resistant to the effects of radiation than humans (Eisenbud, 1987) Nevertheless, when a mixed forest on Long Island was subjected to chronic gamma irradiation during the growing season for 12 years a phenomenon known as retrogression occurred. The results of the irradiation of an oak-pine forest by a radiation source containing 9500 $Ci^{137}Cs$ have been documented. Within 6 months a vegetation gradient developed. What was originally on oak-pine forest become an oak forest, with the elimination of the pine. The oak forest then became a shurb zone, followed by a sedge zone. Near the radiation source, the sedge gave way to a central devastated area, where only mosses and lichens survived. The changes were similar to those seen along gradients of increasing severity of climatic changes, such as increasing altitude on a mountain. The chronic exposure to gamma radiation reversed the ecological succession.

After 12 years of chronic irradiation, only lichens and green algae survived up to a distance of 20m from the source, an area receiving as much as 3000oentgen (R)/d. Only sedges and grass lived from 20 to 75m from the source, where the dose was 20 to 160 R/d. From 75 to 85m, with a dose of 10 to 20 R/d, blueberry bushes died and no oaks survived. From 85 to 115m from the source, where 5 to 10 R/d were received, oaks survived but pines did not. Scraggly, stunted pines were seen at a distance of 125m, where 2 R/d of gamma radiation was received. Even as far away as 150m from the source, the pine needles were shorter than normal and the diameters of the trees were reduced.

The only animals left on the Long Island experimental plot were insects. Leaf-eating insects attacked the oak trees and shrubs. Bark insects and wood borers developed larger populations in response to an abundance of food and an absence of predators. All other animals died or left the denuded areas.

These observations agree with other studies (Odum, 1971) on the effects of gamma radiation on whole communities and ecosystems in a tropical rain forest of Puerto Rico and in the Nevada desert. The effects of mixed gamma/ neutron radiation have been studied on fields and forests in Georgia and at the Oak Ridge National Laboratory in Tennessee. Short-term effects of gamma radiation have been studied at the Savannah River Ecology Laboratory in South Carolina. The Oak Ridge National Laboratory has also conducted a low level chronic radiation study of a lake bed community.

In all of these studies, the differential sensitivity of a species in an ecosystem is of considerable interest (Odum, 1971). If an ecosystem receives a higher level of radiation than was confronted in its evolation, the elimination of sensitive strains or species may result.

In gamma source experiments, such as the ones mentioned above, weeds and grasses proved relatively tolerant of radiation and field crops were only slightly more sensitive. Exposure to sustained irradiation from a large gamma source is not an exact simulation of chronic exposure to short-lived radioisotopes, but it does provide an indication of the effects of an acute, high-level radioisotope exposure.

Radionuclides such as ^{210}Ra, ^{226}Ra, and ^{222}Rn that occur or are deposited in the soil are translocated by plants. In addition to root uptake radioiosotopes are absorbed following direct deposition on foliar surfaces. Although the mechanism of radionuclide adborption by plants is not well understood, it is known that individual radionuclides are translocated from the root or the leaves to the remainder of the plant Mean ^{232}Th concentrations of 0.018 ± 0.022 pCi/kg have been found in the edible portions of 25 vegetables, including beans, carrots, corn, potatoes, and squash. Flora near the summit of the Morro do Ferro, a hill in the state of Minas Gerais, Brazil have absorbed so much ^{228}Ra that they can be autoradiographed easily (Eisenbud, 1987).

Individual plant species vary widely in their susceptibility to the damaging effects of ionizing radiation. Exposure of the mixed vegetation of a Long Island forest to gamma rays demonstrated the vulnerability of conifers compared to the radiotolerance of grasses. Exposure in early stages of development or during the growing season, when cells are dividing, increases the radiosensitivity of the plant. During cell division, cell death or damage is induced at much lower irradiation levels than those required during the interphase stage of the cell cycle. Investigations of plant damage from gamma ray sources demonstrated that radiation causes a reduction in the number of cells per meristem (growing point). This reduction in cells varies directly with the number of cells displaying chromosome damage.

In higher plants, sensitivity to ionizing radiation is directly proportional to the chromosome volume of the cell nucleus (Odum, 1971). However, certain community attributes such as biomass and diversity also become determinants of species vulnerability. Indeed, the "unshielded" biomass above ground is a major determinant in that plants sprouting from seeds or from shielded under-ground parts have an increased chance of recovery.

Effects on Animals

In higher animals, unlike in plants, there is no simple, direct relationship between nuclear chromosome volume and sensitivity to ionizing radiation. Rather, the effects of irradiation on specific organ systems are more critical (Odum, 1971).

Developmental Stages

As in plants, proliferating cells are much more sensitive to irradiation than differentiated, non-dividing cells. An organism is more radiosensitive during its early stages of development, when most cells are dividing, than at any other stage of its life. The LD_{50} for fish embryos is 16 to 18 times smaller than required in later life. For insects, the LD_{50} in adults in about 1000 times larger than during the developmental stages.

Instead of lethality, morphological abnormalities are associated with irradiation during the middle stages of development. Irradiation-induced anomalies occurring during this time may result in death of the organism or abnormal development-of one or more organ systems. Exposure during this period may result in gross malformations, growth retardation at term or as an adult, and structural neuropathology. In the human, most major organogenesis occurs during the first trimester of pregnancy, with embryonic death and congenital abnormalities resulting from irradiation exposure during this period.

During late organogenesis and in the perinatal period, just before and just after birth, radiation damage tends to be functional rather than structural. Perinatal irradiation with X-rays and gamma rays (140 to 180 cGy) induces changes in tissue enzyme activity.

Reproduction

During a period of about 20 years in the early part of this century, radiation was used in an attempt to increase fertility. Exposure normally was to 1.5 to 2.25 Gy over a period of 3 weeks. These levels apparently had little effect on fertility or on any later conceived children.

In recent years, the safety of radiation levels is more often questioned since the threshold for the effects of ionizing radiation on male reproduction is difficult to predict. Although fully developed sperm cells and primary spermatocytes are relatively radio resistant, the proliferating spermatogonial cells of the testis are highly sensitive to ionizing radiation.

Occupational radiation exposure of the testis produces a significant decrease in serum gonadotropins and significant changes in semen production and morphology. However, the effects of ionizing radiation on spermatogenesis are normally reversible with the recovery of fertility predictable. Although an acute irradiation dose of 6 Gy to the testis is likely to produce permanent sterility, conception has occurred for males after years of either aspermic or hypospermic conditions following absorbed doses between 2.3 and 3.7 Gy (Robertson, 1989).

Genetic Effects

Mutations are structural changes occurring in genes as a result of exposure to a number of environmental agents including chemicals, heat and ionizing radiation (Eisenbud, 1987). Partially due to the efforts of the film industry, its post-World War II movies portray inaccurately the genetic effects of irradiation. In fact, the genetic effects of irradiation dominated the philosophy of radiation risk and radiation protection. Only during the past two decades has the scientific community increased its appreciation of the somatic effects of ionizing radiation. Data evaluation during this time also suggested that the previous estimates of genetic risk were too high.

Damage to DNA along a low LET radiation tract is the same kind that occurs spontaneously. It will likely consist of single strand breaks in the double helix and may be repaired by cellular enzymes. The ultimate effects depend more on the effectiveness of the repair than upon the initial break. Chromosomal breaks from low-LET radiations have a small probability of resulting in a translocation during repair of a lesion. In contrast, high-LET radiation lesions have a high probability of interaction and result in a greater number of unrepaired or misrepaired lesions. These lesions may be amplified many times during transcription and translation and are the major contributors to the genetic damage resulting from irradiation.

Current studies indicate that there may have been one possible mutation in the 78,000 children of people who survived the atomic bombings at Hiroshima and Nagasaki. When the mutation rate of the group of exposed parentsis estimated, this single mutation has a high probability of being unrelated to radiation exposure.

Available information suggests that the radiation lose required to double the human mutation rate varies between 0.8 and 2.4 Gy (UNSCEAR, 1988).

Radiation Carcinogenesis

Radiation carcinogenesis is considered by most radiobiologists to the most important effect of exposure to levels of ionizing radiation below 1 Gy. Ionizing radiation increases the incidence of virtually every type of neoplasm, whether benign or malignant, and the carcinogenic effects of radiation have been observed in practically all species.

The development of cancer appears to be a multistage process involving an initiator and at least one promotor. Promotors may elicit the production of activated forms of oxygen, including superoxide radicals and peroxides, and hydroxyl radicals, which either directly or indirectly affect DNA. Certainly, radiation causes strand breaks and modification of DNA bases. It is also well known that radiation of living cells elicits the formation of free radicals, and hydroxyl radicals, and it is well established that hydroxyl radicals are responsible for the radiation-induced DNA damage.

At times it becomes difficult to characterize ionizing radiation as either initiator or promotor. For example, cigarette smoking uranium miners exposed to radon have radiation-induced lung cancer at five times the rate of non-smoking miners. It is difficult to assign the role of initiator or promotor to either radon or cigarette smoke since both have been implicated in the etiology of lung cancer.

Acute Radiation Syndrom

There have been fewer than 25 documented fatalities world wide between 1946 and 1985 that can be attributed to radiation accidents. Although exposure of the whole body to lethal amounts of ionizing radiation is very rare, any discussion of the biological effects of ionizing radiation would be incomplete without mentioning acute radiation sickness and acute radiation syndrome (ARS).

Acute radiation sickness is manifest in characteristic clinical sequelae known as ARS, a combination of syndromes determined primarily by the total radiation dose received, the rate of exposure, and the distribution of the radiation in the body. Signs and symptoms of ARS result from injury to bone marrow, gastrointestinal system, cardiovascular system, CNS, gonads, and skin. The variation in radiation sensitivity of these tissues causes the signs and symptoms of ARS to occur in three successive phases: an initial prodromal phase a subsequent latent period and the manifest illness phase. The length of each phase may vary directly with the radiation dose, and the time between each phase may vary indirectly with the dose, so at an extremely high dose of radiation, the phases will blend with the latent period disappearing completely.

The initial prodromal phase is characterized by a combination of gastrointestinal and neuromuscular symptoms such as anorexia, nausea, vomiting, diarrhea, apathy, tachycardia, fever, headaches, insomnia, dizziness, and vertigo. The pathogenesis of the prodromal phase is not known, but several casual factors have been suggested, including direct radiation effects on the central and autonomic nervous systems, disturbance of endocrine balance, and the production and release of various chemical mediators.

The manifest illness phase of the ARS is classically divided into three major syndromes traditionally known as the hemopoietic syndrome, gastrointestinal (GI) syndrome, and central nerous system (CNS) syndrome. However the current view replaces the CNS syndrome with the neurovascular syndrome.

The hemopoietic syndrome occurs following exposure to 200 to 700 cGy. Radiation doses of 100 cGY or more can significantly damage the blood-forming capability of the body. Approximately 50% of individuals exposed to 300 cGY will die within 2 months. The signs and symptoms result from radiation damage to the bone marrow, lymphatic organs, and immune system. The pathophysiological effects of damage to bone marrow include increased susceptibility to infection, bleeding, anemia, and lowered immunity. Death usually results from hemoorrhage and infection.

At radiation doses of 7 to 50 Gy, injury to the GI tract inhibits the renewal of the cell lining. The intestinal epithelial stem cell is the target of radiation damage, and the resulting decrease in mitotic activity leads to denudation of the intestinal mucosa, fluid and electrolyte imbalance, and bacteremia. The symptoms of the GI syndrome include lethargy, diarrhea, dehydration, and sepsis. At doses of 3 to 8 Gy, temporary injury to the tight junctions between epithelial cells of the mucosal lining permits the escape of bacterial endotoxins into the blood stream. As the dose increases, the epithelial lining is more extensively depleted. With doses of 10 to 15 Gy, denudation of the mucosa exacerbates the loss of fluid and electrolytes. At doeses of 12.5 Gy and above, early mortality occurs due to dehydration, and electrolyte imbalance, with death occurring 4 to 5 d after exposure.

A *tolerance (or threshold) level* is a level below which no damage occurs.

The effects of a high dose of radiation to the whole body are a matter of scientific agreement : death. The United Nations Scientific Committee on the Effects of Atomic Radiation has determined the effects of different whole-body doses, which are listed below :

- Doses of 100 grays (*100 sieverts*) :
  death through central nervous system damage in hours or days.
- 10 to 50 grays (*10 to 50 sieverts*) :
  death though gastrointestinal damage in about one to two weeks.
- 3 to 5 grays (*3 to 5 sieverts*) :
  death for half of the people exposed in one to two months (bone marrow damage).

Effects of lower doses are :

- 150 to 250 rems (*1.5 to 2.5 sieverts*) :
  nausa, vomiting, probable skin burns, foetal or embryonic death if pregnant; long term, people in ill-halth may not survive, people in good health will apparently recover, though with possible permanent health damage.
- 50 to 150 rems (*0.5 to 1.5 sieverts*) :
  less severe radiation sickness and burns, spontancous abortion or stillbirth if pregnant; long term, possible benign or malignant tumours, premature ageing, shortened lifespan, genetic damage to offspring.
- 10 to 50 rems (*0.1 to 0.5 sieverts*) :
  most people experience little immediate reaction; sensitive people may have radiation sickness; long term, possible premature ageing, genetic effects and some risk of tumours.
- under 10 terms (*under 0.1 sieverts*) :
  no immediate effects; long term, premature ageing, some off-spring mutation, risk of tumours.

Different Tolerances of Radiation

10 sievert	(= 10,000 millisieverts)	death
below 1 sievert	(= 1,000 millisieverts)	no 'earth' death
below 0.1 sievert	(= 100 millisieverts)	no radiation sickness in sensitive individuals
0.03 sieverts	(= 30 millisieverts)	one X-ray
0.01 sievert	(= 1 millisievert)	one year's does from background radiation
0.0001 sievert	(= 0.01 millisievert)	one modern X-ray

All the figures, excepting the figure for background radiation refer to whole body doses over a short period of time. If the dose is spread over a longer period, the body is better able to cope. Because of the natural background radiation that we are all exposed to in our everyday lives, our bodies have developed repair mechanisms for radiation damage.

According to the UK Atomic Energy Authority, a single dose of 4 sieverts would result in a one in two chance of death : the same dose delivered gradually over a year would probably be tolerated because of the body's natural repair processes, although with possible long-term consequences.

There is a variation in radiation sensitivity between the different organs of the body. Adult tissues are relatively robust : the kidneys, bladder and cartilage can take relaively high levels of radiation, though red bone marrow, eyes, testes and ovearies are more sensitive.

Sensitivity differs between individuals. Some adults, and all children, are particularly at risk. Quite small doses of radiation to a child's cartilage can slow or halt the growth of bones and lead to deformity. The younger the child, the more severe the stunting. Irradiation of children's brains during radiotherapy can cause changes of character, loss of memory and, in very young children, dementia and idiocy.

Unborn children are particularly prone to brain damage, if their mothers are irradiated between the eightth and fifteenth week of pegnancy. This danger extends to X-ray, which can cause severe mental retardation to an unbron child.

Large doses of radiation are given only to patients who are already so ill that the risk of the treatment is justified.

Risks and Benefits of Radiation Exposure

Exposure to radiation should be treated as a risk, to be quantified against the benefits to be gained by such exposure. Any unnecessary exposure to radiation should be avoided.

This statement is accepted by the international bdoes who set radiation protection limits.

- Radiotherapy, in which large doses of radiation are given to treat serious illnesses such as cancer, is accepted as a reasonable exppsure to radiation. The risk is large, but in many cases the risk of not receiving the treatment is larger.
- The use of X-ray for medical diagnosis is also accepted as valid, provided that they are used for specific reasons and not as, say, a screening system for new employees in a firm.
- Natural background radiation is a variable about which little can be done, although a debate is opening as to how 'natural' should be defined.

Safety Standards

Safety standards for exposure to radiation are set by the International Commission on Radiological Protection (ICRP). In 1952, the ICRP issued its recommended exposure limits to radiation, which was the same standard agreed by nuclear physicists after the Second World War 5 rem (50 millisieverts) to the whole body.

The revised limits in 1959 recommended that workers 5 rem (50 millisieverts) whole-body exposure should be calculated by adding external and internal exposure together (internal exposure occurs by swallowing or breathing in radioactive materials). The limits have stayed the same, except for an upward revision in 1977 of exposure to certain parts of the body, to the present day.

Exposure limit for radiation workers are based on the assumption that they know the risks of what they are doing, and are healthy, fit, and capable of withstanding a relatively high exposure. The ICRP recommended that the public, for whom greater protection is necessary, should receive no more than one-tenth of the occupational exposure, *i.e.* a whole-body dose of 0.5 rem (5 millisieverts) over a yar.

The ICRP accept that there is no threshold limit for damage from radiation exposure, and that all doses can cause some damage to the body. The 5 millisieverts exposure limit for the general public was considered the absolute maximum, while a limit of 1 millisievert was considered a better value to adopt. West Germany decided on a 0.35 millisievert limit, the USA took 0.25 to be its limit—but the UK adopted a limit of 5.0 millisieverts.

REFERENCES

1. Allaby, M. (ed). (1988) *Dictionary of the Environment, MacMillan,* Basingstoke. Burn, D. (1978) *Nuclear Power and the Energy Crisis,* MacMillan, Basingstoke. Eisenbud, M. (1987) *Environmental Radioactivity from Natural, Industrial and Military Sources* (3rd edition), Academic Press, Orlando.
2. Lilienthal, D. (1980) *Atomic Energy : A New Start,* Harper and Row.
3. Marshall, W. (ed.) (1983) *Nuclear Power Technology,* Vol. 3, *Nuclear Radiation,* Clarendon Press, Oxford.
4. Nuclear Installations Inspectorate (1992) *Safety Assessment Principles for Nuclear Plants,* NII, London.
5. UK Royal Commission (1976) *6th Report. Nuclear Power and the Environment,* HMSO, London.
6. *Yearbook of Science and Technology* (1992) McGraw-Hill, p. 297.

CHAPTER 6

Chemical Toxicology & Environment

Dr. S.A. Iqbal[1] and Dr. Mamta Bhattacharya[2]

CHEMICAL TOXICOLOGY

Introduction

Industrial development and agricultural professionalisation that took many decade in the West, is occurring at much faster speed in Third world countries like India (due to demographic pressure) thus making problem certainly very complex. Human eagerness to perform better and better with respect to production of food, energy and convenience products in order to ameliorate the way of living is the causes of chemical pollution. This eagerness led to tremendous growth in production of chemicals.

For example in the last two or three decades the production of pesticides only in India increased more than 40 fold, whereas the production of dyes 30 fold, drugs 5-10 fold, petrochemicals 40 fold, fertilizers 30 fold and metal 3 fold increased as well. It is evident such increased population has also lead to the release of toxic and hazardous wastes in to the environment in solid, liquid or gaseous formThis cosmopolitan problem of chemical pollution of our environment has now been recognized as a real problem at global level.

'All substances are poisons the difference is in the dose'. The above aphorism is attributed to Paracelsus. It illustrates that the potential for harm is widespread and all chemicals could be toxic but the degree of harm that a chemical can inflict on a human or any other living being depends on the dose or the degree of exposure as well as on other factors. In other words the risk (i.e. that product of the likelihood and the severity of harm) from a toxic hazard depends on the exposure. This account is intended for those with little or no background in toxicology. Toxicology is a complex and difficult science. In an attempt to make it more understandable, many broad generalizations are made, without detailing the mechanisms or addressing the exceptions.

Toxicology is the study of the nature and action of poisons. *Toxicity* is the ability of a chemical molecule or compound to produce injury once it reaches a susceptible site in or on the body. Toxicity hazard is the probability that injury will occur considering the manner in which the substance is used.

Chemicals play a major role in our every day lives. They are part of what we eat, where we work, and how we live. Despite their prevalence in our lives, many chemicals are hazardous, or toxic. Toxic chemicals can be found in our soil, water, air, and bodies. This contamination has seriously effected the health of humans and wildlife everywhere.

[1]Department of Chemistry, Saifia Science College, (B.U.), Bhopal (India).
[2]Department of Chemistry, Sadhu Vaswani College, (B.U.), Bhopal (India)

Children are often more vulnerable than adults to the harmful effects of chemical pollutants because they are growing and developing rapidly. In addition, children's behavior, including increased hand to mouth activity, a tendency to crawl and play in spaces that could be contaminated, and a lack of awareness about proper safety and sanitary habits, all put children at a higher risk. In some cases, childhood exposure to toxins can cause serious health damage to an individual later on in life.

Definitions

The following terms describe the states of matter in which chemical contaminants may occur in the environment:

Gas : A formless fluid that completely occupies the space of an. enclosure at 25°C.

Vapour : The gaseous phase of a material that is liquid or solid at 25'C.

Aerosol: A dispersion of particles of microscopic size in a gaseous medium, these may be solid particles (dust, fume, smoke) or liquid particles (mist, fog).

Dust: Airborne solid particles (an aerosol) that range in size from 0.1 to 50 μ and larger in diameter. A person with normal eyesight can detect dust particles as small as 50 μ i diameter. Smaller airborne particles cannot be detected by unaided eyes unless strong light is reflected from the particles. Dust of respirablc size (below 10 μ) cannot be seen without the aid of a microscope.

Fume: An aerosol of solid particles generated by condensation from the gaseous state, generally after volatilization from molten metals. The solid particles that make up a fume are extremely fine, usually less than 1.0 u, in diameter. In most cases, the volatilized solid reacts with oxygen in the air to form an oxide. A common example is cadmium oxide lumc.

Smoke : An aerosol of carbon or soot particles less than 0.1 μ in diameter that results from the incomplete combustion of carbonaceous materials such as coal or oil. Smoke generally contains droplets as well as dry particles.

Mist: An aerosol of suspended liquid droplets generated by conden-sation from the gaseous to the liquid state or by the breaking up of a liquid into a dispersed state, such as by splashing, foaming, or atomizing Mist is formed when a finely divided liquic is suspended in the atmosphere. Examples are the oil mist during cutt ng and grinding operations, acid mists from electroplating, acid or alkali mists from pickling operations, and paint spray mist from spraying procedures.

Fog : A visible aerosol of a liquid, formed by condensation. The following terms of measurerient arc commonly used :

ppm : Parts of vapor or gas per million parts of contaminated air by volume.

mg/m^3 : Milligrams of a substance per cubic meter of air. mppcf: Millions of particles of a paniculate per cubic fool of air (mainly historical use).

Toxicology is the study of the effects of poisonous substances on living organisms. Above a certain concentration, toxicants have detrimen-tal effects on some biological function. The concentration at which a significant detrimental effect occurs is determined by the *dose response.* The critical (or threshold) dose at wh:ch toxicity occurs differs between species, sexes and individuals within a species due to genetic and other factors such as the composition of the diet and some illnesses.

The dose, or degree of expose of an organism to a toxicant, can be expressed as:

(a) the amount of toxicant present in the organism (units of mass of toxicant per unit weight of body mass of the organism);

(b) the amount of the toxicant entering the organism (in the diet, drinking water, or inhaled air in animals and absorbed through the roots or through the leaf cuticle in the case of plants);

(c) the concentration in the environment of the organism (duration of exposure is important).

The effect of the toxicant dose is called the *response.* This can vary from no effect to death. Toxicity is commonly categorized on the basic ol the duration of exposure, that is acute, chronic and subchronic.

Acute exposure involves a single dose whereas chronic exposure refers to exposure over a long time period (often almost a lifetime-two years in the case of lest rodents). Sub-chronic exposure is dosing over a shorter time period-fraction of a lifetime, such as one eight of an experimental rodent's lifetime.

Acute toxicity is caused by fast poisons which include both synthetic and naturally occurring compounds, such as the botulinum toxins produced by the soil bacterium Clostridium botulinum, the venom of certain snakes (e.g. rattle snake and cobra) or species of spider (e.g. black widow spider) and plant-derived toxins such as strychnine and nicotine and some synthetic chemicals, including organo-phosphorous compounds

Phosphine (PH_3), phosgene ($COCl_2$) and sodium fluoracetate. These sub-stances are classed as supertoxins because they cause lethal effects in humans at doses of less than f mg/kg body weight.

The subchronic effects of chemicals are determined by investigating the biochemical and other changes which take place over a period of months. Investigations on chronic effects will examine effects on the lifespan of the organism, cancer induction, changes in geriatric conditions and effects on the offspring caused by exposure of the parent to toxic chemicals. However, it is important to note that the acute and sub-acute effects of toxins determined in laboratory experiments cannot always be relied on to predict responses to the same chemicals in the environment. This is due to instruction between pollutants (antagonistic and synergistic effects) and reactions of the toxicants with the components of the environ-ment (such as adsorption, photodecomposition, acidification and dissolu-tion). In the case of non-carcingens, it is possible that there may safe or threshold dose levels of toxins (NOAEL = no observed adverse effect level). However, carcinogens arc not considered to have safe or threshold concentrations because a single genetic change may lead to an uncon-trolled reaction.

Toxicants in the Environment

Our built environments, including schools, residential areas, and places of work, can all be sites of hazardous chemical contamination. In fact, many toxic chemicals can be found right in your home.

- Household products such as detergent, floor and furniture polish, paints, and various cleaning products for glass, wood, metal, ovens, toilets, and drains may contain hazardous chemicals such as ammonia, sulfuric and phosphoric acids, lye, chlorine, formaldehyde, and phenol. Air fresheners can also contain chemicals that are harmful to health. Art supplies, such as markers, paint, and glue, may also contain toxic materials. When not properly handled, these products can make the home environment a dangerous place, especially for kids.
- Home furnishings, such as carpets, curtains, wall decorations, and some furniture, may be treated with chemicals and are potentially dangerous. It is important to note that a few days after installation, new carpets emit volatile organic compounds, which are chemicals associated with carpet manufacturing that can be harmful to humans and the environment. Gas and wood stoves and kerosene heaters may also release dangerous chemicals.
- Building materials such as particleboard, insulation, asbestos, and treated wood (used for decks and outdoor furniture), can also pose health threats. Some play sets and toys, as well as outdoor swing sets and play grounds, may also be treated with toxic chemicals, made from toxic plastics, or include hazardous materials. The more time that children spend playing in such an environment, the higher their exposure to toxic chemicals, and the greater a risk to their health.

Toxicants in Air : There are a large number of toxicants exist in our environment. Some of them are not identified yet. Following is the list of 24 extremely hazardous chemicals present in the environment-Arsenic, Asbestos, Acrylonitrile, Berilium, Benzen Chlorinated solvents, Chromates,,Cadmium, CFC's, Coke oven emissions, Diethylstilbestrol, Dibromochloropene,Ethelene dibromide, Ethelene oxide, Lead, Mercury, Ozone, Nitrosoamines, PCB's, PBB's, Sulfurdioxide, Vinyl chloride, Radiations and Toxic waste disposal emissions and leachets.

Toxicants in Water : can be grouped into two categories—

(*i*) Toxic trace elements in to natural water and wastewater- some of these are essential at low level, act as a nutrient for animals and phyto life.

For eg. As, Be, B, Cu,Cd, Cr, F, Pb,Hg,Mn,Mo,Ni,Se, Zn.

(*ii*) Pesticides in water-these may be - Chlorinated hydrocarbons and organo phosphates.

Classification of Toxicants : Toxicants can be classified both Chemically and Biologically (Physiologically)

Chemical Classifications of Toxic Materials : Chemically Toxic Materials can be classified on the basis of their type and chemical structure—

Table 6.1.

Category	Examples
Metals, and metalloids	Arsenic, cadmium, lead, mercury, nickel, tin, etc.
Inorganics (other)	Asbestos, carbon monoxide, hydrogen sulphide
Hydrocarbons - aliphatic	Propane, butane, pentane, hexane
Aliphatic alcohols, ketones, ethers, aldehydes and acids	Ethyl alcohol (ethanol), acetone, diethyl ether, formaldehyde, acetic acid
Hydrocarbons - aromatic	Benzene, toluene, xylene, naphthalene
Phenols	Phenol, pentachlorophenol
Chlorinated volatile organic compounds	Perchlorethylene (tetrachloroethene), trichloroethylene (trichloroethene), vinyl chloride
Chlorinated non-volatile organic compounds	Chlorinated dioxins and dibenzofurans, polychlorinated biphenyls, pesticides such as chlordane and DDT
Miscellaneous organic compounds compounds	Acrylonitrile, benzidine, aniline, di-isocyanates, organophosphates

Physiological Classifications of Toxic Materials

(*i*) *Irritants* are materials that cause inflammation of mucous membranes with which they come in contact. Inflammation of tissue results from concentration far below those needed to cause corrosion. Examples include:

ammonia, alkaline dusts and mists, hydrogen chloride, halogens, ozone, hydrogen fluoride, phosgene, diethyl/dimethyl sulfate, nitrogen dioxide, phosphorus chlorides.

Irritants can also cause changes in the mechanics of respiration and lung function.

Examples include: Formaldehyde, acetic acid, sulfur dioxide, formic acid, sulfuric acid, acrolein, iodine.

Long term exposure to irritants can result in increased mucous secretions and chronic bronchitis.

A primary irritant exerts no systemic toxic action because the products formed on the tissue of the respiratory tract are non-toxic or because the irritant action is far in excess of any systemic toxic action. Example: hydrogen chloride.

A secondary irritant's effect on mucous membranes is over-shadowed by a systemic effect resulting from absorption. Examples include: hydrogen sulfide,aromatic hydrocarbons

Exposure to a secondary irritant can result in pulmonary edema, hemorrhage, and tissue necrosis.

(*ii*) *Asphyxiants* have the ability to deprive tissue of oxygen. Simple asphyxiants are inert gases that displace oxygen. Examples include: nitrogen,nitrous oxide,carbon dioxide,hydrogen.

Chemical asphyxiants render the body incapable of utilizing an adequate oxygen supply. They are toxic at very low concentrations (few ppm). Examples include: carbon monoxide, cyanides, hydrogen sulfide.

(iii) *Corrosives* are chemicals, which may cause visible destruction of, or irreversible alterations in living tissue by chemical action at the site of contact. Examples include: sulfuric acid, potassium hydroxide,chromic acid,sodium hydroxide.

(iv) *Primary anesthetics* have a depressant effect upon the central nervous system, Particularly the brain. Examples include: halogenated hydrocarbons, alcohols etc.

(v) *Hepatotoxic agents* cause damage to the liver. Examples include: carbon tetrachloride, nitrosamines, tetrachloroethane.

(vi) *Nephrotoxic agents* cause damage to the kidneys. Examples include: halogenated hydrocarbons, uranium compounds.

(vii) *Neurotoxic agents* damage the nervous system. The nervous system is especially sensitive to organometallic compounds and certain sulfide compounds. Examples include: trialkyl tin compounds, methyl mercury organic phosphorus insecticides, manganese tetraethyl lead, carbon disulfide, thallium

(viii) *Hematopoietic (blood) system agents* either directly affect blood cells or bone marrow. Examples include: nitrites, aniline, toluidine, nitrobenzene benzene.

(ix) *Pulmonary tissue (lungs) agents* can be toxic, through other mean than by immediate irritant action. Fibrotic changes can be caused by free crystalline silica and asbestos. Other dusts can cause a restrictive disease called pneumoconiosis. Examples include:

silica, asbestos,coal dust, cotton dust, wood dust.

(x) *A teratogen* (embryo toxic or fetotoxic agent) is an agent which interferes with normal embryonic development without damage to the mother or lethal effect on the fetus. Effects are not hereditary. Examples include: lead, dibromo dichloropropane

(xi) *A mutagen* is a chemical agent which may able to react with nucleophilic structures such as DNA. Mutations can occur on the gene level (gene mutations) when, for example, one nucleotide base-pair is change to another. Mutations can also occur on the chromosomal level (chromosomal mutations) when the number of chromosomal units or their morphological structure is altered. Examples of mutagens include: most radioisotopes, barium, permanganate, methyl isocyanate.

(xii) *A sensitizer* causes a substantial proportion of exposed people to develop an allergic reaction in normal tissue after repeated exposure to the chemical. The reaction may be as mild as a rash (contact dermatitis) or as serious as anaphylactic shock. Examples include: formaldehyde, epoxides, nickel compounds, chromium compounds, chlorinated hydrocarbons, amines toluene diisocyanate

Factors Affecting the Environmental Concentration of Toxicants

In the environment the concentration, transport, transformation and disposition of a chemical is controlled by following factors :

1. **The physical and chemical properties of a compound :** It include molecular structure, solubility in water and vapour pressure, rate constant for hydrolysis, photolysis, bio-degradation, evaporation, sorption and depuration by organisms. Partition coefficient also provides significant informations.
2. **The physical, chemical and biological properties of ecosystem :** The properties of environment may affect the fate of a chemical. Surface area to volume relationship, temperature, salinity, pH, depth, flow, amount of suspended material, sediments, particle size and carbon in sediments.
3. **The source and rate of the input of chemical in to environment :** Knowledge of average rates of input and occurrence of single large "sluges" of chemicals entering in environment is important for predicting environmental concentrations. Above data is helpful in determining
   - Mobility of chemical and parts of the environment in which it would more likely be distributed.
   - The kinds of chemical and biological reactions take place during transport and after deposition.
   - The eventual chemical form and its persistence.

Affects of Physical Class on Toxicity : When considering the toxicity of gases and vapors, the solubility of the substance is a key factor.

1. Highly soluble materials like ammonia irritate the upper respiratory tract.
2. On the other hand, relatively insoluble materials like nitrogen dioxide penetrate deep into the lung.
3. Fat-soluble materials, like pesticides, tend to have longer residence times in the body.
4. An aerosol is composed of solid or liquid particles of microscopic size dispersed in a gaseous medium. The toxic potential of an aerosol is only partially described by its concentration in milligrams per cubic meter (mg/m3). For a proper assessment of the toxic hazard, the size of the aerosol's particles is important. Particles above 1 micrometer tend to deposit in the upper respiratory tract. Particles less than 1 micrometers in diameter enter the lung. Very small particles (< 0.2 ?m) are generally not deposited.

Affects of Speciation : The exact chemical identity (the 'species') of a substance can make a very big difference as regards its toxicity. This concept is called *'speciation'*.

For example asbestos is a complex chemical compound containing atoms such as potassium, sodium, magnesium, aluminium, silicon and others, which in a different context would have a much lower toxicity than they exhibit in the compound of asbestos.

Chromium in the hexavalent state (Cr VI) is a human carcinogen (as in the orange coloured potassium dichromate or bichromate) while trivalent chromium (Cr III) (as in the green coloured chromium III chloride) appears not to be. If the chromium is in the Cr III form and oxidation to Cr VI is prevented, exposure should present no cancer risks.

Nickel tetracarbonyl ($Ni(CO)_4$) is a highly toxic gas inflicting severe damage to the lungs and heart, while nickel carbonate ($NiCO_3$) is a solid which is much less hazardous. Metallic nickel probably poses no cancer risk at all while nickel subsulphide is almost certainly a very highly carcinogenic and dangerous compound which has been responsible for many sad deaths.

Other Factors Affecting Toxicity

Rate of entry and route of exposure; that is, how fast is the toxic dose delivered and by what means. Age can affect the capacity to repair tissue damage. Previous exposures can lead to tolerance, increased sensitivity or make no difference. State of health, physical condition, and life style, can affect the toxic response. Pre-existing disease can result in increased sensitivity. Environmental factors such as temperature and pressure may also affect the exposed individual as well as host factors including genetic predisposition and the sex of the exposed individual.

Dose-response Relationships

The potential toxicity (harmful action) inherent in a substance is manifest only when that substance comes in contact with a living biological system. A chemical normally thought of as "harmless" will evoke a toxic response if added to a biological system in sufficient amount. The toxic potency of a chemical is defined by the relationship between the dose (the amount) of the chemical and the response that is produced in a biological system.

Routes of Entry into the Body

There are four main routes by which hazardous chemicals enter the body:

- Inhalation through the respiratory tract. This is the most important in terms of severity. Absorption through the respiratory tract through inhalation;
- Skin absorption or absorption through the mucous membrane;

- Ingestion through the digestive tract. This can occur through eating or smoking with contaminated hands or in contaminated work areas;
- Injection of a toxin into the bloodstream by accidental needle stick or puncture of the skin with a sharp object.

Most exposure standards, Threshold Limit Values (TLVs) and Permissible Exposure Limits (PELs), are based on the inhalation route of exposure. They are normally expressed in terms of either parts per million (ppm) or milligrams per cubic meter (mg/m^3) concentration in air. If a significant route of exposure for a substance is through skin contact, the MSDS will have a "skin" notation. Examples include: pesticides, carbon disulfide, phenol, carbon tetrachloride, dioxane, mercury, thallium compounds, xylene, and hydrogen cyanide.

Toxicokinetics or How the Body Handles Poisons

(*a*) Absorption into the body : As a general rule, fat soluble liquids are readily absorbed through the skin and fat soluble vapours are readily absorbed through the lungs. Notably these routes apply to organic solvents such as hexane, toluene, trichlorethylene and many others.

(*b*) Distribution within the body : Many factors affect the distribution of a toxic substance but water or fat solubility is very important. Thus for example water soluble compounds of lead are found (amongst other places) in the red blood cells, while fat soluble ones concentrate in the central nervous system (CNS).

The distribution of a toxic substance determines its concentration at a particular tissue and therefore the number and type of cells exposed to high concentrations of it.

(*c*) Metaholism/ biotransformation of toxic substances : Toxic substances may be converted into other substances (metabolites) by organs such as the liver and kidneys.

Thus non-polar and therefore not water-soluble organic compounds tend to be oxidised within the liver e.g.: trichloroethane oxidised to trichloroethanol trichloroacetaldehyde and trichloroacetic acid dichloromethane (methylene chloride CH_2C_{12}) oxidised to carbon monoxide (CO) Water-soluble metabolites are then more easily excreted by the kidney.

Metabolism or biotransformation does not necessarily result in less toxic compounds. For example benzene may be oxidized to an expoxide which then inflicts damage on the DNA in genes, i.e. it is genotoxic and thence carcinogenic.

(*d*) Routes of elimination of toxic substances/or their metabolites : Kidneys - especially water soluble substances.

Lungs - especially fat soluble vapours e.g., alcohols, or gases such as carbon monoxide.

Toxicodynamics or what Poisons may do to the Body

Acute effects refer to the short term consequences of *exposure.Chronic* effects relate to a much longer time scale, while sub-acute are in between acute and chronic).

Some effects may be dose related-the higher the exposure the worse it gets e.g. irritant effects on the skin, asthma, asbestosis etc. Other effects are 'all or none' and for a given exposure there is an element of chance (stochastic) as to whether or not the disease develops e.g. the development of cancer (carcinogenesis) or some forms of developmental damage to the foetus (teratogenesis).

Irritant effects: Detergents may remove fat from the skin and cause dermatitis.

Cement dust being alkaline may irritate the skin, or cause more severe damage (chromates within cement may also cause sensitisation and allergic dermatitis).

Respiratory irritation may be caused by low concentrations of formaldehyde vapour.

More serious inflammation: more toxic agents and/or higher exposures may be associated with damage resulting in inflammation for example of terminal bronchioles and alveoli leading to a chemical pneumonitis and pulmonary oedema (e.g. from nitrogen dioxide NO_2).

Corrosive effects : severe local effects by contact e.g. caustics such as sodium hydroxide, or acids such as sulphuric, nitric or hydrochloric acid.

Narcotic and anesthetic effects : fat soluble solvents will behave as anaesthetics and cause drowsiness, nausea, headache, unconsciousness and death e.g. vapours from organic solvents such as ether or trichlorethylene.

Asphyxiation : Various gases can cause asphyxia by interfering with oxygen transport. Examples: Carbon monoxide, Hydrogen cyanide, Hydrogen sulphide. Carbon monoxide is present wherever there is incomplete combustion of carbon compounds. It is odourless, and will react with haemoglobin (Hb) to form COHb which cannot carry oxygen. Hydrogen sulphide might initially be detected by its smell at low concentrations but it paralyses the sense of smell and can effectively become odourless. Hydrogen cyanide: in the form of its salts sodium and potassium cyanide is used in many industries and the organic cyaniden acrylonitrile (vinyl cyanide) is used in the rubber industry. Absorption can also occur through the skin. At low concentrations these gases poison cytochromes and cause the rapid onset of headache, dizziness, vomiting and confusion. At high concentrations they are very rapidly lethal.

Types of Poisoning

- Acute poisoning is characterized by rapid absorption of the substance and the exposure is sudden and severe. Normally, a single large exposure is involved. Examples : carbon monoxide or cyanide poisoning.
- Chronic poisoning is characterized by prolonged or repeated exposures of a duration measured in days, months or years. Symptoms may not be immediately apparent. Examples : lead or mercury poisoning and pesticide exposure.
- Local poisoning refers to the site of action of an agent and means the action takes place at the point or area of contact. The site may be skin, mucous membranes, the respiratory tract, gastrointestinal system, eyes, etc. Absorption does not necessarily occur. Examples : strong acids or alkalis.
- Systemic poisoning refers to a site of action other than the point of contact and presupposes absorption has taken place, for example, an inhaled material may act on the liver, examples: arsenic affects the blood, nervous system, liver, kidneys and skin; benzene affects the bone marrow.
- Cumulative poisoning are characterized by materials that tend to build up in the body as a result of numerous chronic exposures. The effects are not seen until a critical body burden is reached. Examples: heavy metals.
- Synergistic poisoning responses: When two or more hazardous material exposures occur the resulting effect can be greater than the effect of the individual exposures. This is called a synergistic or potentiating effect. Example: exposure to both alcohol and chlorinated solvents, vapours e.g. - alcohols, or gases such as carbon monoxide.

Effects of Pollutants on Human and Other Mammals

Biochemical effects

(*a*) impairment of enzyme function by the binding of the toxicant to enzymes, coenzymes, metal activators, or enzyme substrates

Exposure limits o Hazardous Chemicals

Chemical	NIOSH REL's Recommendations	NIOSH Health Effect	NIOSH Remarks
1	2	3	4
Aldrin/dielrin	Ca Lowest reliably detectable level	Potential cancer in humans; produced tumors o the lungs liver, thyroid, and adrenal grand in animals	Aldrin/diedrin no longer produced in U.S.: prevent skin contact
Antimony	0.5 mg Sb/m3 TWA	Irritation; cardiovascular and lung effects	Chest X-ray, Pulmonary function testing, and electrocardiogram required .
Arsenic, inorganic	Ca 2µg As/m^3 (15 min)	Lung and lymphatic cancer; dermatitis	Chest x-ray required
Arsine	Ca 2µg As/m^3 (0.002 mg As/m^3	Sudden extensive hemolysis	Workers to be warned of working with arsenic compounds in presence freshly formed hydrogen
Asbestors	Ca 100.000 fibers/m^3 over 5 µm in length 8-hr TWA in a 400 litter air sample	Lung cancer; mesothelioma asbestosis	Chest x-ray and pulmonary function testing required
Benzene	Ca 0.1 ppm (0.32 mg/m^3) 8-hrTWA 1 ppm (3.2 mg//m^3)	Cancer (leukemia)	Prevent skin contact
Cadmium	Ca Reduce exposure to lowest feasible level	Lung cancer, prostatic cancer renal system effects	None
Chlorine	0.5 ppm (1.45 mg/m^3) ceiling in (15 min)	Eye and respiratory irritation	Chest X-ray required
Chromium (VI)	Ca Carcinogenic Cr (VI) 1µg/m^3 TWA other Cr VI) 25 µg/m^3 TWA; 50 µg/m^3 (15 min)	Lung cancer, skin ulcers, and lung irritation	Employer must demonstrate absence of carcinogenic Cr (VI); x
Coal tar Products	Ca 0.2 mg/m^3-TWA (cyclohexane-extractable function	Lung and skin cancer	Includes coal tar pitch; pulmonary function testing and chest X –rays required
DDT	Ca Lowest reliably detectable level; 0.5 mg/m^3 TWA by NIOSH- validated method	Potential for cancer in human produced tumors of the liver ... lung, and lymphatic system in animal	Prevent skin contact

1	2	3	4
Hydrogen cyanide and cyanide salts	4.7 ppm (5mg CN/m^3) ceiling (10 min)	Thyroid blood respiratory system effect	Concurrent measurement required for HCN when measuring for cyanide salt trained first aid personnel and first aid kits to be available during ase; prevent skin and eye contact.
Kepone	Ca 1µ/g^{3}TWA	Liver cancer; nervous system effects	Liver function testing required
Lead, inorganic	< 100 µg pb/m^{3}TWA; air level to be maintained so that worker blood lead Extensive work-practice and personal protection recommendations	Kidney, blood and nervous system effect primarily trauma and falls	Blood monitoring required Tetanus toxoid inoculation and first aid program to be instituted
Malathion	15 mg/m^{3}TWA	Nervous effect	Prevent skin contact blood monitoring requireds
Mercury, inorganic	0.05 mg Hg/m^3. 8 hr TWA	Central nervous system and effects	Work practices, sanitation monitoring and medical surveillance emphasized
Nickel, inorganic compounds	Ca 15 µg Ni m^3 TWA	Lung and nasal cancer; skin effect	Chest x-ray and pulmonary function testing required
Nitric acid	2 ppm (5mg/m^3) TWA	Dental erosion : nasal / lung irritation	Prevent skin and eye contact chest x-ray required
Organotin	0.1 mg/m^3 TWA	Eye skin; liver. nervous system and cardiovascular effect	Chest x-ray blood and urine monitoring eye test heart examination and nervous system testing required . prevent skin and eye contact
Parathion	0.05 mg/m^3 TWA	Nervous system effects	Prevent skin contact blood monitoring required
Pesticides		Wide range of toxicities considered; cancer nervous and reproductive system effect	Blood monitoring required for some grops; works to be warned of reproductive effect for some compounds; prevent skin contact prevent skin and eye contact
Phenol	5.2 ppm (20 mg/m^3) TWA 15.6 ppm (60 mmg/m^3) ceiling (15 min)	Skin eye central nervous system liver and kidney effects	Prevent skin and eye contact
Sulfuric acid	1 mg/m^{3}TWA	Pulmonary irritation	Prevent skin and eye contact
Vanadium	Vanadium compound 0.05 mg V/m^3 ceiling(15 min) metallic vanadium and vanadium carbide 1 mg V/m^3 TWA	Eye, skin and lung effect	Pulmonary function testing and chest X-ray required
Zinc Oxide	5 mg Zn0/m^3 TWA 15 mg Zno/m^3 ceilling (15 min)	Metal fume fever	

(*b*) alternation of cell membrane or carriers in cell membranes.
(*c*) interference with lipid metabolism, resulting in excess lipid accumulation
(*d*) interference with respiration
(*e*) interference with carbohydrate metabolism
(*f*) stopping or interfering with protein biosynthesis through toxic effects on DNA
(*g*) interference with regulatory processes mediated by hormones or enzymes

Clinical response to toxicants :

(*a*) alterations in the vital signs of temperature, pulse rate, respiratory rate and blood pressure.
(*b*) abnormal skin colour
(*c*) effects on the eye, which include:
miosis (excessive contraction of pupil)
mydriasis (excessive pupil dilation)
conjunctivitis (inflamation of the membrane covering the front of the eyeball)
nystagmus (involuntary movement of the eyeballs)
(*d*) gastrointestinal effects: pain, vomiting, paralytic ileus (stoppage of normal peristalsis)
(*e*) central nervous system effects: convulsions, paralysis, hal-lucinations, ataxia and coma

Sub-clinical effects of toxicants

(*a*) damage to the immune system
(*b*) chromosomal abnormalities
(*c*) modification of the functions of lwer enzymes
(*d*) slowing of the conduction of nervous impulses

Teratogenesis, Mutagenesis, Carcinogenesis, and Immune System Defects

Teratogenesis is the creation of birth defect arising from damage to embryonic or foetal cells or from nutations in egg or sperm ells. The biochemical mechanisms of tcratogenesis include: enzyme inhibition by xenobiotics. deprivation of essential nutrients and alternation of the placental membrane.

Mutagenesis is the creation of mutations by chemicals or ionizing radiation which bring about alterations to DNA to produce inheritable traits. Although mutations can occur naturally in the absence of xenobiotic* substances, most mutations are harmful. Mechanisms of mulagcnicity are similar to those of Carcinogenesis and teratogenesis.

Chemical Carcinogenesis (or cancer) occurs when xenobiotic sub-stances cause uncontrolled cell replication (i.e. cancer). From the layman's view points, carcino-genesis is the most commonly associated toxie effect of hazardous substances.

Assessment of Short-term Lethality and Acute Toxicity

The most commonly used assessment of toxicity is the measurement of short term lethality. For a given substance, this involves determining the median concentration which is lethal to a fixed proportion, usually 50%, of a test population of organisms after continuous exposure for a fixed time, usually 48 or 96 hrs. This is interpolated from the sigmoidal dose response curve which is obtained by plotting percentage mortality at each dose against dose.

The lethal concentration (LC) for 50% of a population after continuous exposure for 48 h is referred to as the 48hLC_{50} or LD_{50} (*LI*) = lethal dose). This abbreviation may be altered appropriately to correspond to different exposure times and proportions of the population. Usually the concentration referred to is that in aqueous solution but it may also be a concentration in air. If the test compound is insoluble or sparingly soluble in water, it must be uniformly dispersed for a rcpeatable LC_{50} to be obtained. If an emulsifier or other solubilizing agent is used for this purpose, it should be chosen carefully to ensure a minimal contribution to the toxicily of the system.

Difficulties arise where organisms arc exposed to harmful substances in association with particulate matter, or in their prey or other food. What then is the effective concentration to which they are exposed'? The nature of the particulate matter will determine whether the substances are ingested or not. It will also determine the potential for dissociation of toxic-substances following ingestion. Normally, only dissociated material can exert toxic effects. In a food organism, toxic substances may be converted to derivatives of greater or less toxicity and may be localized in specific parts of the organism, some of which may be selectively eaten or discarded by its predators. Therefore, in these cases it is almost impossible to know the effective concentration to which the predator is exposed. The best one can do is to copy the natural situation as closely as possible in the laboratory and express the LC_{50} as a notional concentration, taken as the total available toxic material divided by the weight of particulate matter or by the weight of the food organism. If one knows that a certain derivative of the toxic substance is much more lethal than any of the others, and suitable methods of analysis are available, the derivative concentration in the food organism should be determined since this will be the most significant factor. In medical usage the LC_{50} frequently refers to an injected concentration, and this must be borne in mind when attempting to extrapolate from such studies.

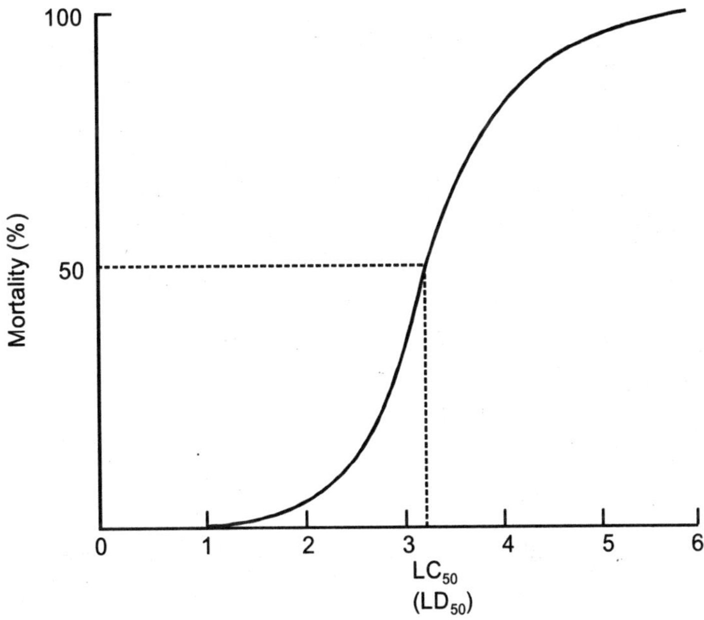

Fig. 6.1. Graph showing typical relationship between mortality and toxicant concentra-tion, or dose, following continuous exposure of organisms for a short fixed time, usually 48 h or 96 h.

Lethal concentrations have been expressed in a variety of units, meet frequently in milligrams per litre or per kilogram body weight. Where possible, molarity (moles per litre) or molality (moles per kilogram) should be used as this will give a uniform chemical basis for comparing toxicity. The most information the LC_{50} gives is an idea of the order of magnitude of the lethal dose under specified conditions. However, it is relatively quick and cheap to determine and provides a basis, however rough, for initial assessment of the likely hazard from a toxicant and of the effects of various parameters on its toxicity.

Table 6.4. Summary of the analytical methods used for soil contaminants

Contaminant	Method
Heavy metals (in acid digests or partial extracts)	Flame atomic absorption spectrophotometry (FAAS) or inductively coupled plasma-atomic emission spectrometry (ICP-AES)
As, Bi, Hg, Se, Sn and Te (in acid digests or partial extracts)	Hydride generation atomic absorption spectroscopy (HGAAS) using sodium borobydride in NaOH.
Borate (water soluble)	ICP-AES (using quartz or plastic apparatus instead of glass)
Organic pollutants (in organic solvents)	Gas chromatography (GC) or combined with Mass Spectrometry (GC-MS)
Cyanides	Colorimetrically-pyridine pyrazalone (blue) reaction
Sulphates (Water soluble)	Elution though cation exchange resin followed by titration of eluate with standard NaOH.
Sulphate (total)	Dissolution in HCl precipitation of Al and Fe followed by gravimetric determination using $BaCl_2$
Chlorides (in HNO_3)	Volhard's Method, back titration with ammonium thiocyanate after initial precipitation with $AgNO_3$.
PH (in distilled water or $CaCl_2$ or KCl)	Electrometrically using glass electrode

THE EFFECTS OF SOIL CONTAMINATION &

Soil contamination can restrict the options available for the use of land because of the potential hazards posted by the contaminants. Some examples of the hazards are:

Table 6.5.

Hazard	Examples of Contaminants
1. Direct ingestion of contaminated soil (mainly children or animals)	As, Cd, Pb, CN^-, cost tars phenols
2. Inhalation of dust and vaporus from contaminated soil	Toluene, benzene, xylene, various organic solvents Rn, Hg, metal-rich particles
3. Uptake by plants of contaminants hazardous to animals and people through food chain	As, Cd, Pb, Ti, PAHs
4. Phytotoxicity	SO_4^{2-}, Cu, Ni, Zn, CH_4
5. Deterioration of building materials and services	SO_4^{2-}, SO_3^{2-}, Cl^-, coal tar, phenol, mineral oils, organic solvents
6. Fires and explosions	CH_4,S, coal dust, oils, petroleum, tar, rubber, high calorific organic wastes (old landfills)
7. Contact of people with contaminants during dersolitions	Coal tar, phenols, asbestos, radionuclides, PAHs TCDDS, etc.
8. Contamination of water	CN^-, SO_4^{2-}, soluble metals, solvents, pesticides.

Contaminants in soil can affect humans by absorption into the body through oral, inhalation, or cutaneous pathways and their effects are usually related to the amount al sorbed into the body. Inhalation is an important route of intake for volatile compounds such as organic solvents and mercurial compounds. Dust particles < 7 μm can penetrate alveoli in the lungs and are a hazard to workers involved in moving contaminated soil and possibly to people using playing fields constructed on contaminated land. Oral intake is a particularly serious problem with young children playing on contaminated land when they lick their fingers. Oral intake in adults is mainly through the consumption of soil on vegetables (and also the contaminants accumulated in the plants). Cutaneous absorption is only a problem with lipophyllic organic solvent.

CRITICAL CONCENTRATIONS FOR CONTAMINANTS IN SOILS

Having established that an area of soil is contaminated, it is necessary to make a decision about the action that needs to be taken in order to avoid unnecessary risk of health effects or of damage to structures. For this purpose, various sets of critical concentrations are in use around the world. The ranges of critical concentrations used in two European countries (UK and The Netherlands) for the interpretation of contaminated land analytical data are given as examples in Tables 8.2.

In the Netherlands, a system has been used which involves three indicative values: A—the 'normal' reference value, B-the test value to determine the need for further investigations, and C-the value above which the soil definitely needs cleaning up. Examples of these indicative values for three typical contaminants are (in $\mu g\ g^{-1}$): for Cd, A = 1, B = 5, and C = 20; for total complex cyanides, A = 5, B = 50, and C = 500; and for total PCBs, A = 0.05, B = 1.0, and C = 10.

Target Organ Categorization of Effects

The following is a target organ categorization of effects, which may occur from exposure to hazardous chemicals, including examples of signs and symptoms and chemicals which have been found to cause such effects. The list of chemicals is not all enclusive.

Prevention and Treatment

Preventing exposure: Through the practice of good occupational hygiene.

Treatment: Decontamination e.g. eye washes, showers, etc.

Antidotes e.g. methylene blue for treating methaemoglobinaemisa caused by aniline Other treatment e.g. oxygen for asphyxia.

Toxic use reduction is an important aspect of Toxicant reduction in the environment.

Table 6.6.

Target Organ	Signs/Symptoms	Chemical of Concern
Hepatotoxins (liver)	Jaundice liver enlargement	Carbon tetrachloride, toluene chloroform, cresol, dirnethylsulfate; nirtosamines, perchloroethylene
Nephrotoxins (kidney)	Edema proteinuria	Chloroform,dimethyl sulfate, halogenated hydrocarbons, mercury, uranium
Neurotoxins (nervous system)	Narcosis behavior change decreased muscle coordination	Benzene, carbon disulfide, carbon tetrachloride, lead, mercury
Hematopoietic (blood)	Cyanosis loss of consciousness	Aniline, arsenic, benzene carbon monoxide cyanides, toluene
Pulmonary (lung)	Cough shortness of breath tightness of chest	Alcohol, asbestos, chromium hydrogen sulfide, nickel, nitrogen dioxide, ozone, silica
Reproductive (mutations/ teratogenesis	Birth defects sterility	Dibromo-dichloropropane lead
Dermal	Defeating rashes irritation	Alcohols, chlorinated compounds, ketones, nickel, phenol trichloroethylene
Vision	Corneal damage conjunctivitis	Acids, bases, organic solvents, quinone

Source reduction : Source reduction involves effort to reduce hazardous waste and other materials by modifying the industrial production. It involves change in manufacturing technology, raw material inputs and product formulations. At times the term "pollution prevention" refers to "Source reduction".

Another method of source reduction is to increase incentives for recycling. It will be helpful in reducing the size of municipal waste. Source reduction is measured by efficiencies and cutbacks in waste products.

Educating People to Minimize Use of Toxicants

We can help protect children from the hazards of toxic chemicals at home and in school.

Make an effort to use nontoxic products. Use natural products for cleaning agents such as baking soda, soda ash, vinegar, and cream of tarter. A variety of nontoxic products can be found in health food stores and some supermarkets. Make sure to check labels carefully and keep all hazardous products well-marked and away from children. Check to see if various school and household appliances, toys, and learning materials are hazardous. Keep living, playing, and learning areas, especially the kitchen and bathroom, clean and safe.

Educate yourself! Learn more about toxic chemicals by visiting informational websites. Become active with groups working to promote policies designed to protect children from toxic chemicals. and will react with haemoglobin (Hb) to form COHb which cannot carry oxygen.. Hydrogen sulphide might initially be detected by its smell at low concentrations but it paralyses the sense of smell and can effectively become odourless. Hydrogen cyanide: in the form of its salts sodium and potassium cyanide is used in many industries and the organic cyaniden acrylonitrile (vinyl cyanide) is used in the rubber industry. Absorption can also occur through the skin. At low concentrations these gases poison cytochromes and cause the rapid onset of headache, dizziness, vomiting and confusion. At high concentrations they are very rapidly lethal.

Table 6.7. UK Department of the Environment Trigger Concentrations for Environmental Contaminantsl7,19 (total concentrations except where indicated)

Conteninant	Proposed uses	Threshold (Triger Concentrations $\mu g\, g^{-1}$)	Action
5 Contaminants which may pase hazards to health			+
As	Garens, allotments* parks, playing fields, open space	1040	
Cd	Gardns, allotmentsparks, playing fields, open space	315	
Cr (hexavalent)	Gardens, allotmentsParks, playing fields, open space	251000	
Cr	Gardens, allotmentsParks, playing fields, open space	6001000	
Pb	Gardens, allotments	500	
	Parks, palying fields; open space	2000	
Hg	Gardens, allotmentsParks, playing fields, open space	120	
Se	Gardnes, allotmentsParks, playing fields, open space	36	
(*hexavalent Cr extracted by 0.1 M HCl adjusted to pH 1 at 37.5°C)			
Contaminants which are phytotoxic but not normally hazardous to health			
B (water soluble)	Any uses where plants grown	3	
Cu (total) (extractable)	Any use where plants grown	13050	
Ni (total) (extractable)	Any use where plants grown	7020	
Zn (total) (extractable)	Any use where plants grown	300130	

** Extracted in 0.05 m EDTA
+ Action concentration yet to be specified.

REFERENCES

1. B.J. Alloway, 'heavy Metals in soils', ed., B.J. Alloway, Blackie, Glasgow, 1990.
2. G.W. Brummer, 'The Imfiortance of Chemical Speciation in Environmental Processes', Springer Verlag, Berlin, 1986.
3. I. Kogel-Knabncr, P. Knabaer, and H. Deschauer, 'Contaminated Soil '90', ed.
4. F. Arendt, M. Ilinsenvcld, and W.J. van den Brink, Kluwcr Academic, Dordrecht, 1990, p. 323.
5. H. Kishi and Y. Hasimoto, 'Contaminated Soil '90', ed. F. Arendt, M. Hinsenveld, and W.J. van den Brink, Kluwcr Academic, Dordrecht, 1990, p.331.
6. MJ. Beckett and D.L. Sims, 'Contaminated Land', ed. J.W. Assink and W. J. van den Brink, Martinus Nijhoff, Dordrecht, 19S6, p.285.
7. Interdepartmental Committee on the Redevelopment of Contaminated Land, 'Guidance on the Assessment and Redevelopment of Contaminated Land', Guidance Note 59/83, Department of the Environment, London, 1987.
8. W.L. Lindsay, 'Chemical Equilibria in soils', John Wiley, Chichester, 1979.
9. D. Sauerbeck, 'Scientific Basis for Soil Protection in Europe' ed. H. Barth and P. L' Hermite, Elsevier, Amsterdam, 1987, p. 181.
10. S. Ross, 'Soil Process'. Routledge, London, 1989.
11. R.f. White, 'Introduction lo the Principles and Practice of Soil Science', 2nd Edn. Blackwells, Oxford, 1987.
12. N. Brady, 'The nature and properties of Soils', 10th Edn., Macmillan, New York, 1990.
13. MJ. Singer and D.N. Munns, 'Soils: An Introduction', Macmillan, New York, 1987.
14. F.A.M. de Haan., 'Scientific Basis for Soil Protection in Europe', cd. H. Barth and P.L'Hermite, Elsevicr. Amsterdam, 1987, p. 211.
15. E.M. Bridges, 'World Soils', 2nd Edn., Cambridge University Press, Cambridge, 1978.

CHAPTER 7

Fly Ash-Generation, Characterization and Recycling Potential

Mohini Saxena[1] and P. Asokan[2]

Summary

It is envisaged that the total annual fly ash generation in India from thermal power stations is to the tune of 160 million tonnes. Conventionally, aggregates, cement, steel, brick, block, tiles, paint, timber etc., are being used as major building components in construction sector. All these materials have been produced from the existing natural resources. To save the energy, contribute to the economy and reduce the environmental pollution, the huge quantity of already accumulated and ever increasing quantity fly ash can be recycled and used effectively as a major raw materials, filler and binder in developing alternative building materials. This Article addresses the quantity, characteristics, recycling potential and other environmental importance of the safe disposal of Fly Ash and effective management with specific reference to the processes and technologies developed in India. As a result of R&D efforts from various institutes like CSIR, Academic institute and other Govt. Organizations like Fly ash Mission, TIFAC, DST and non Govt. Organizations, fly ash utilization has been increased from about 5% in the nineties to about 40% at present in various applications such as building materials, road & embankments, mine filling, agriculture, human settlement on abandoned ponds and other value added applications. Indeed fly ash is considered to be a resource material rather than as a waste which may contribute to save the natural resources leading to reduction in carbon dioxide sequestration and global warming resulting in sustaining the clean and green environment.

Introduction

India's 82 utility and more than 25 captive thermal power plants contribute more than 70% to the country's total electric power installed capacity (approx 100,000 MW). Due to vast coal reserves (about 211 billion tonnes), coal is being used as the largest source of energy. In fact about, 240 million tonnes of coal is being used every year to generate electricity. Indian coals though low in sulphur, radio active elements and heavy metals content, yet are rich in incombustible siliceous material and other inorganic matter which comes out as ash on combustion.

What is Fly Ash

Fly ash is a collective term referring to the residues produced during the combustion of coal for the generation of electricity. The coal combustion residues (CSIR) include fly ash, bottom ash, boiler slag, and other solid fine particles. These particles are so intimately mixed with the coal and have the specific

[1]Scientists, Advanced Materials and Processes Research Institute (CSIR), Hoshangabad Road, Bhopal (M.P.)

gravity in the range close to that of coal. India is currently producing about 160 million tones of ash per year. The figure is likely to go up in view of developing nature of Indian economy, which involves large number of energy intensive infrastructure projects. It is estimated that fly ash generation would increase to around 170 million tones by 2012 (Vimal Kumar & Mukesh Mathur 2005).

There has been a considerable increase in power generation in India in the last five decades, from 1350 MW in 1947 to about 140,000 MW in 2008, of which about 90% is from the coal based thermal power stations. As a consequence of increase in power generation to meet the demand in India, leading to increase in fly ash generation from 15 million tonnes in 1992 to about 160 million tones in 2009. As per the ASTM standards, Indian bituminous and sub bituminous coal produces class 'F'ash and lignite coal produces class 'C' ash having high degree of self-hardening capacity. Physical, chemical and mineralogical, morphological and radioactive properties of CCRs in general varies as they are influenced by coal source/ quality, combustion process, degree of weathering, particle size and age of the ash.

Environmental Implication

Environmental pollution by the coal based thermal power plants all over the world are cited to be one of the major source of pollution affecting the general aesthetics of environment in terms of land use, health hazards, air, soil and water in particular and thus leads to environmental dangers. Hence, safe disposal and gainful utilisation is of prime concern.

Fly ash, like soil, contains trace concentrations of many heavy metals that are known to be detrimental to health if excess in their concentration. These include nickel, vanadium, arsenic, beryllium, cadmium, barium, chromium, copper, molybdenum, zinc, lead, selenium, uranium, thorium, and radium. Though these elements are found in extremely low concentrations in fly ash, their mere presence has prompted some to sound alarm.

The United State Environmental Protection Agency (US EPA) has said in the past that coal fly ash does not need to be regulated as a hazardous waste. Studies by the US. Geological Survey and others conclude that fly ash compares with common soils or rocks and should not be the source of alarm (U.S. Geological Survey). Crystalline silica and lime are the major components of exposure concern. However, fly ash is neither toxic / poisonous, nor it is considered hazardous. However, the fine crystalline silica present in fly ash may affect the lungs. Also the fly ash component like calcium oxide is of some concern as it may reacts with water to form calcium hydroxide [Ca(OH)2], increasing the fly ash pH between 10 - 12. This may also cause lung damage if present in excess.

These hazards, due to mismanagement of fly ash collection, handling and disposal, may be minimised by effective emissions control devices and handling equipment. Air pollution is due to high particulate matter emission labels due to burning of inferior grade coal which leads to generation of large quantity of fly ash. Environmental issues in coal based power generation deals with the emission of SO_2, NOx and green house gas, (CO_2) are also matter of concern (Table 7.1). Water pollution is mainly caused by the effluent discharge from ash pond, condenser cooling/ cooling tower, De-mineralized plant and boiler blow down. High noise pollution is faced due to release of high pressure steam and running of fans and motors. More than 160 tones of fly ash is generated presently by use of coal fired energy. The disposal of such large quantity of fly ash has occupied more than 65 thousand hectares of land which includes agricultural and forest land too and more than 630 million cubic metre water is required for disposal of coal ash in slurry form per annum (Vimal Kumar & Mukesh Mathur 2005; Pandey et.al. 2006).

Table 7.1. Pollution load from coal based thermal power plant

Pollutants	Emissions (in tones/day)
CO_2	424650
Particulate Matter	4374
SO_2	3311
NO_x	4966

Percentage share of sulphur dioxide emission as compared to different industries is given in graph-1.

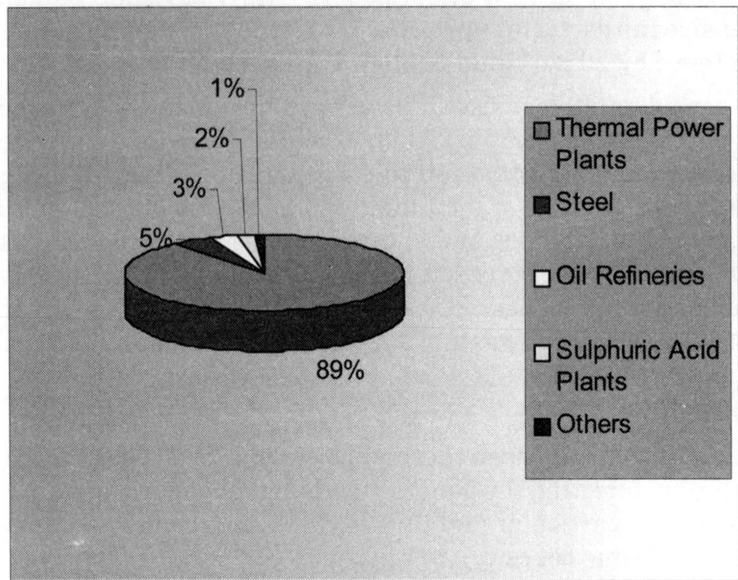

Fig. 7.1

Reasons for non-compliance of environmental standards in thermal power plants are due to inconsistent supply of coal, high resistivity of coal, inefficient operation of ESPs, delay in supply of ESPs, low specific collection area (SCA) of ESPs, inefficient management of ash ponds, large quantities of ash generation etc. To meet increasing demand of power with minimal environmental impact for sustainable development, adoption of clean coal technologies with enhanced power plant efficiency, fuel switching, use of washed coal, efficient pollution control systems and proper by-product and waste handling & utilization, is necessary.

Coal Combustion Residues (CCRs) includes fly ash, bottom ash, boiler slag, and fluidised bed combustion ash and other solid fine particles (Asokan, 2001; Keefer, 1993). As per the ASTM standards India bituminous and sub bituminous coal results class 'F'ash and lignite coal produces class 'C' ash having high degree of self-hardening capacity. Physical, chemical and mineralogical, morphological and radioactive properties of CCRs in general varies as they are influenced by coal source/ quality, combustion process, degree of weathering, particle size and age of the ash (Asokan, 2000).

Presently, from all these thermal power plants dry fly ash have been collected through Electro Static Precipitator (ESP) in dry state condition as well as pond ash in semi-wet condition. In India, most of the thermal power plants do not have the facility for automatic dry ash collection system. Commonly both fly ash and bottom ash and discharged together as slurry to the ash pond/ lagoon.

The world scenario of CCRs generation from thermal power plants are in exponential stage and is expected to reach 520 MMT by the year 2010. Developed countries have well defined quality of segregated CCRs at different stages and utilise more than 33% of the CCRs generated in their country. Literature review indicates that CCRs has many potential uses in back fill, cement, concrete, adhesives, wall board, and soil amelioration (Asokan et al., 2005;Keefer, 1993).

Use of CCRs increase various environmental benefits besides the economy. It is reported that during the production of each ton of cement, ~1 ton of CO_2 is released along with NO_x and CH_4, and CCRs incorporation in cement manufacturing contributes for such emission reduction. The other studies indicate that use of 1 ton of fly ash in concrete will avoid 2 tons of CO_2 emission from cement production and reduces green house effect and global warming (Krishnamoorthy; 2000).

CHARACTERISTICS OF FLY ASH

Physical Properties

The fly ash particles are generally grey in colour and some of the pond ashes are blackish grey. But when fly ash is mixed with bottom ash, overall carbon contents in the fly ash conglomerates and varies between ranges 3-5%. Other important physical features of fly ash include size distribution and surface area. Fly ash constitute an assemblage of particles of wide variety of shape and size, ranging from coarse sand to clays.

The texture of the Fly ash varies distinctly based upon the source, disposal site and location from where the ash is collected. As per the United States Department of Agriculture standards, ~55 % of Indian dry fly ash collected from ESP falls within the silt and clay sized particles and the rest is sand sized particles. The physical properties of fly ash is shown below.

Table 7.2

Sl. No.	Parameters	Range
1.	Colour	Greyish
2.	Bulk Density (kg/m^3)	960-1500
3.	Porosity (%)	30-55
4.	Water Holding Capacity (%)	35-55
5.	Sand (%)	60- 80
6.	Silt (%)	10-35
7.	Clay (%)	0.5-15
8.	pH	3.5-12.5
9.	Electrical Conductivity (dS/m)	0.075-1.0

Chemical Properties

The chemical characteristics of fly ash vary from power plant to power plants depending on the quality of coal used, conditions during combustion and age of the fly ash. The major constituents of fly ash are silica, alumina and iron oxides (~87%). One of the major concerns with Fly ash disposal is the leaching of heavy metals to surface and underground water source, which may contaminate the ground water quality near by the ash disposal area. The trace elements in fly ash, such as Zn, Cd, Pb, Mo, Ni, As, Se and B are important due to their environmental significance. But the ultimate impact of each trace element will depend upon their state in fly ash and toxicity, mobility and availability of this particular form in the ecosystem. The chemical characterization of the fly ash is shown below.

Engineering and Geo-technical Properties

The specific gravity of most of the fly ash are considerably less than that of soil due to variation in particle size, shape, chemical composition and mineralogy etc. Specific gravity of Fly ash play an important role in geo technical application and it varies from 1.66 to 2.86 with high shear strength. The co-efficient of permeability of Fly ash varies from 10^{-4} mm/ sec to 10^{-3} mm /sec. Lime reactivity of fly ash is much greater than that of bottom ash and pond ash. The lime reactivity depends on the proportion of silica content. High percentage of free lime content in fly ash contributes in increasing its compressive strength with the curing time due to the pozzolanic reactivity.

The California Bearing Ratio (CBR) value of soaked fly ash varied from 6.8-13.5% and unsoaked fly ash varied from 10.8-15.4%. But the CBR value of bottom ash (15.3- 36.5%) found higher than that of fly ash. But the pond ash has 20% CBR value. Compacted fly ash has the requisite properties for use in load bearing fills or highway sub-bases. Low unit weight, of Fly ash is quite suitable for structural fills over soils with low pre consolidation pressures. Though pond ash has very lower compacted dry density it exhibits higher shear strength. When mixed with mooram, it develops a fairly high CBR value.

Table 7.3. Chemical characterization of typical Indian fly ash

Sl. No.	Parameters	Fly Ssh
Range (%)		
1.	Phosphorous (P)	0.06 - 0.3
2.	Potassium (K)	0 - 1.8
3.	Sulphur (S)	0.03 - 0.055
4.	Manganese (Mn)	0.002 - 0.84
5.	Calcium (Ca)	0.37 - 0.76
6.	Magnesium (Mg)	0.02 - 0.9
7.	Sodium (Na)	0.07 - 0.71
Range (ppm)		
8.	Barium (Ba)	26 - 1275
9.	Copper (Cu)	39 - 1000
10.	Zinc (Zn)	10 - 250
11.	Chromium(Cr)	10 - 353
12.	Cobalt (Co)	7 - 128
13.	Cadmium (Cd)	1 - 26
14.	Lead (Pb)	10 - 144
15.	Nickel (Ni)	29 - 265
16.	Boron (B)	100 - 1000
17.	Molybdenum (Mo)	8 - 100
18.	Mercury (Hg)	0 - 0.005
19.	Arsenic (As)	5 - 68
20.	Selenium (Se)	1 - 10
21.	Scandium(Sc)	0.5 - 106
22.	Vanadium(V)	40 - 190

Morphology

Several morphological classes of fly ash particles observed by scanning electron microscopy showed that most of the fly ash particles are regular in shape and size. Some of them are spherical, hollow shaped and cenospheric in nature. Also, some of the smaller size fly ash particles are adhered / inhered in to bigger size particles.

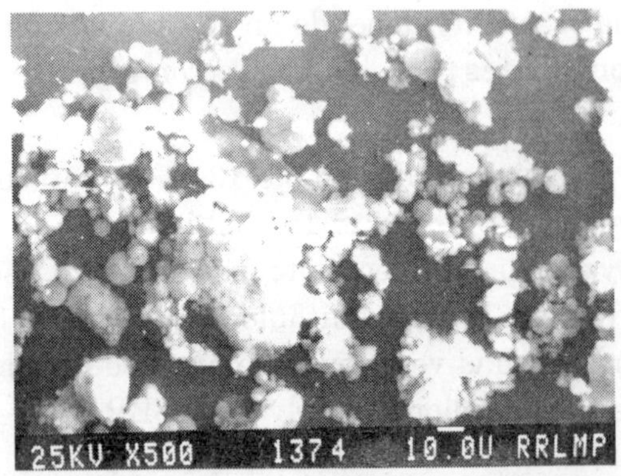

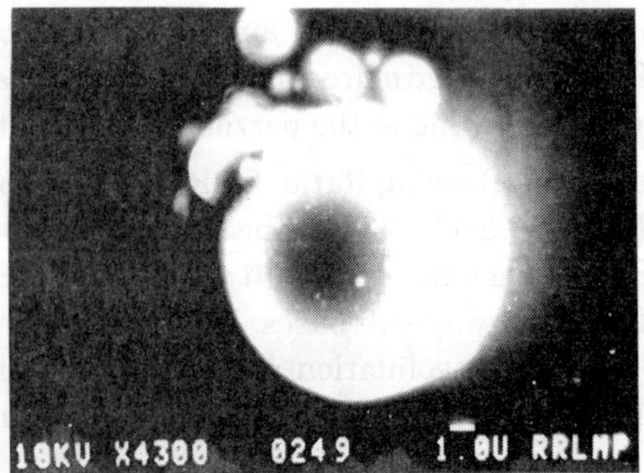

Fig. 7.2. The smooth and spherical particles of fly ash are shown in SEM micrographs

Mineralogy

Mineralogical analysis revealed that CCRs can be separated into three major matrices: glass mullite, quartz and magnetic spinal. In Indian CCRs, quartz, hematite found predominate phase constituents, which influences the concentration of aluminium, silica and iron oxide. The other mineral phases present in fly ash are Albite ($KalSi_3O_8$), Mullite ($Al_6Si_2O_{13}$), Esperite (CaPb)$ZnSiO_4$, Nepoutite (NiMg) $3Si_2O_{15}$ $(OH)_4$ and Tenorite(CuO) (Saxena et al., 1998)

Radioactivity

The studies carried out by Vijayan, (1999) indicate that the radio activity level of Indian CCRs and pond ash are almost similar to that of normal soil and the radio active level in lignite CCRs found less than that of bituminous and sub bituminous coal. The measured level of the radionuclides in Indian CCRs is below the limits specified from environmental point of view. The upper limit for naturally occurring radionuclides such as ^{232}Th (parent radionuclide of ^{228}Ac), ^{226}Ra and ^{40}K are 259 bq/kg, 370 bq/kg and 925 bq/kg respectively (Moghisi et al., 1978). Also the radio activity concentration of Indian CCRs are similar to the concentration reported for the CCRs of United States and England and less than the corresponding level in the CCRs of Poland, Denmark and Australia (UNSCEAR, 1982).

Fly ash Utilisation International Scenario

Countries like USA, UK, Germany, France, Poland and China utilise about 50% of fly ash generated in their respective country. *In Japan more than 9 MMT of fly ash is discharged from power plant every year. Out of which about 50 % of ash being utilised in cement industry, concrete additives and road construction. But Sea disposal has increased in Japan due to insufficient disposal land.* The major utilisation of ash is in construction materials and land fills. In landfills and structural fills, fly ash application lowered the compressibility, higher ground density and strength. Fly ash also can be used for the stability of soil against liquefaction during earthquake.

Work carried out in USA showes that value added products like magnetite, pozzolanic materials, cenospheres, and unburnt carbon can be recovered after processing. Removing unburnt carbon in fly ash also enhances the quality of the remaining fly ash especially for use in blending cement. Literature available also indicates that attempts are being made for the improvement in the pozzolanic property of fly ash through grinding and classification. The grinding technique increases the surface area of particles for more reactivity in cement concrete mixes. There are more than 18 patents been filed in USA on beneficiation of fly ash for improving its quality for utilization in various applications. It is also reported that processed and segregated fly ash would improve the quality of the fly ash based building products. Still various R&D activities are under ways for effective management of fly ash.

Fly ash utilization Scenario in Indian Context

Looking to the physico-chemical, engineering, mineralogical and morphological properties of ash, the Bureau of Indian standard has released IS10153 indicating various applications of CCRs. Presently in India, CCRs being used as raw materials in cement, cellular concrete, fly ash lime bricks, fly ash lime gypsum block, building tiles; as admixture in cement concrete, timber substitute products; as aggregate in concrete, road and building block; as pozzolana in lime pozzolana mortars/ plasters, portland pozzolana cement; as stabiliser in soil stabilisation, road construction; as filler in consolidation of ground, land and mine filling. The other applications of CCRs are metals extraction, cenosphere, soil amendment/ soil modifier, fertiliser and waste water treatment (Asokan, 1999; Chandrasekar, 1997; Iyer and Scott, 2000).

In India, several laboratories of Council of Scientific and Industrial Research, Agricultural Universities, Indian Institutes of Technology, Tata Energy Research Institute, National Thermal Power Corporation, various Governmental and Non Governmental organisation are actively involved in conducting various in-depth experiments and demonstration trails in recycling and use of CCRs effectively. As a result, in

India the CCRs utilisation rate has increased considerably and how ~15 % have been used in various applications (Asokan, 2001; TIFAC, Home Page, 2002). The details of fly ash utilization R&D in India, especially in Council of Scientific & Industrial Research are as follows.

UTILISATION OF CCRS IN BUILDING CONSTRUCTION MATERIALS

Cement and Asbestos Industries

Experimental investigations indicate that in cement Industry upto 25% of dry fly ash is being used as major raw materials for the production of CCRs blended cement. Studies revealed that fly ash based blended cement is much superior to ordinary portland cement on account of its higher resistance to lime leaching, alkali aggregate reactions, higher resistance to carbonation, smoother surface, lower water permeability and penetration by chloride and sulphate ions (Kamal, 2000). Silica, Alumina and iron oxide are the major chemical constituents contribute in achieving superior quality of blended cement as per the IS 3812- 1981. The presence of SiO_2 + Al_2O_3 (70%), SiO_2 (35%), MgO (5%), SiO_3 (2.75%), and Na_2O (1.5%) in CCRs are the requisite properties for making pozzolona cement (Roongta, 2000). Production of CCRs based cement also increases over all availability of cement production and which is cost effective. In Cement Manufacture / Substitution, about 49% of the ash generated is being used every year in India (Vimal Kumar & Mukesh Mathur 2005).

CCRs Based Bricks

The CCRs bricks can be broadly categorised in to two types namely clay CCRs brick (Sintered bricks) and CCRs sand lime bricks (Calcium silicate bricks). The CCRs sintered brick contributes in replacing the topsoil and thus silica and oxides of iron and aluminium play an important role. The presence of unburned carbon in the CCRs becomes an advantage as it saves the fuel consumption. On the other hand in case of air and water cured calcium silicate bricks, CCRs play an important role as a pozzolonic materials and presence of CaO, soluble silica, Al_2O_3 and higher surface area, helps in improving the bricks/ block quality (Kumar et al., 1999). But the presence of organic carbon hampers the quality of bricks. The compressive strength of clay CCRs brick is as high as 120kg/cm^2, water absorption is less than 18% and shrinkage is less than 10% (Karade et al., 1995). 16 prototype houses were constructed in the premises of Advanced Materials and Processes Research Institute, Bhopal using clay fly ash bricks (J. Prabakar et.al. 1995).

Fig. 7.3. Demonstration house constructed at AMPRI, Bhopal using fly ash bricks manufactured

Another fly ash based pozzolona product is FaL-G brick, in which 60-75% of fly ash is being used. The compressive strength of FaL-G brick is varying from 80-160 kg/cm^2 (Bhanumathidas, 1997). There are various methods of manufacturing ash bricks with application of ash content varying from 30% for ash clay bricks to 80% for fly ash-lime-gypsum bricks without compromising the quality. The states of Tamil Nadu, Maharastra, Punjab, Uttar Pradesh, Andhra Pradesh in India use significant amount of CCRs in manufacturing different types of bricks. It is reported that fly ash-lime bricks have better crushing strength than clay bricks.

Fly ash Blocks

Cement concrete building blocks are appropriate materials for construction of walls (load bearing and non-load bearing). The commonly used sizes of these building blocks are 40x20x20 cm, which is equal to nearly 8 burnt clay bricks. Fly Ash can be used in hollow blocks as a replacement of cement.

The advantageous characteristics of these blocks are lighter in weight and are easier in handling, Being hollow from inside, it creates air cavities in masonry, which is a bad conductor of heat and insulates the rooms better, Having smooth finish from outside, it does not require further plastering, Considering total cost of construction, hollow blocks masonry is cheaper than bricks, Walls of lesser width (in comparison with bricks) are constructed and increases effective area. This should be considered while assessing monetary advantages, More crushing strength than bricks, All engineering properties of pozzolana are achieved in construction (Bhatnagar et. al., 2005).

Concrete blocks were developed from stone dust waste with 50% CCRs. The compressive strength of block is 80-130 Kg/cm^2; water absorption is 5-10% (Karade et al., 1995). The combined effect of particle shape, grading and particle density of CCRs causes a substantial reduction in the water demand of concrete mix. Such concrete gives much higher long-term strength, lower permeability and increased resistance to chemical attack. Ash being used as a major filler and binder materials due to the pozzolanic properties and meet the IS 456 specification. But studies carried out by National Council for Cement & Building Material, India showed that use of CCRs from 15% to 25% in normal concrete and even more for mass concrete works resulted in impairing the total compressive strength.

CCRs Based Binder in Concrete

It is reported that conditioned fly ash can be used as binder and it will find large-scale utilisation for replacement of portland cement and admixture (McCarthy and Dhir, 1999). In construction work, where cementatious binders are required, CCRs can be used and will give equivalent or improved properties as compared to portland cement binder. Literature illustrates that fly ash-lime-phosphogypsum based binder decreased the strength and loss in weight of binder with increase in temperature 27-60^{0}C (Mridul et al., 1996). Presently good quality of fly ash conforming to IS 3812 is available in most of the modern thermal plants in India. Fly ash may be used in concrete as a raw material for cement production, as an ingredient in blended cement, and as a partial replacement for cement in concrete. Sometimes fly ash is also used as a partial replacement of fine aggregate as well as in the product ion of light weight aggregate for concrete. The present state of the art includes its usage with respect to the use of fly ash as a cementitious component or mineral admixture in concrete. Fly ash in concrete has been tired with encouraging results on concrete using rice husk ash, silica fume which enhanced the durability of concrete as well as utilizes the industrial by products (Shiban Raina 2004)

Wood Substitute Products from Fly Ash Polymer Composites

To avoid deforestation and increasing environmental hazards, the use of timber is restricted in India. Timber substitute products such as doors shutters, windows frame, false ceilings and partition walls etc., have been developed using organic fibre as a reinforcement and fly ash as filler in polymer matrix composites. This composite is the result of an extensive R&D carried out by Advanced Materials and Processes Research Institute (CSIR), Bhopal in association with Building Materials and Technology

Promotion Council (BMTPC), New Delhi and National Aluminium Company Ltd. This is environmental friendly technology, quality is better than timber, and it is 100% timber free product in which upto 50% CCRs was utilised (Saxena and Prabakar, 2000). The wood substitute products are developed using fly ash natural fibres and polymer based composite. The Salient features of this composites are stronger than wood, weather resistant and durable, corrosion resistant, termite fungus, rot and rodent resistant, fire retardant, self-extinguishing nature and cost effective. The fly ash based polymer composite timber substitute products are comparable to natural wood and thus could be used as a wood substitute for doors, windows, ceilings, flooring, partitions and furniture (Mohini Saxena et.al, 2008). The products are cost effective and no further maintenance is required. This is an environment friendly product with fruitful utilisation of fly ash.

Fig. 7.4. Wood Substitute Products

The Technology Enabling Centre (TEC) is being set-up at AMPRI Bhopal with support of MOEF, BMTPC, CSIR for developing customized products and also would serve as an enabling centre for budding entrepreneurs for manufacturing industrial waste based wood substitute products including fly ash. Photographic view of TEC equipment and machineries is shown below.

Fig. 7.5. Photographic view of Technology Enabling Centre for making fly ash polymer composites

CCRs as Base Course in Embankments and Roads Construction

Pond ash and bottom ash which range in particle size from fine to coarse sand have been used as a granular sub base material for construction of embankment and road (Sikdar, 2000). A mixture of local soil and CCRs stabilised with 3-5% lime provides good sub base course. Utilisation in structural fill, back fills for reclamation of undulated land and abundant mine found the most effective for bulk utilisation (Saxena et al., 2000). Under water placement of the CCRs slurry has been found suitable for stable ground construction such as harbour and airport construction (Sumio et al., 2000). Compacted pond ash and bottom ash possess good bearing strength and also meet gradation requirements for use as a sub base material (Martin, 1990).

CCRs added to cement concrete mix permits easier placement and finishing in which upto 50% of sand can be replaced with CCRs in road construction. In lime CCRs bound macadam, lime CCRs mix was used as filler in the Water Bound Macadam (WBM) construction to provide additional stability. The pond ash has been found as a very useful material for the replacement of soil for making of embankments and for rising of outer bunds of ash dump areas. During 1999-2000, about 10 lakh tonnes of CCRs was used for raising of the ash pond at NTPC, Rihand, Korba and Badarpur Captive Power Plant (CPP), India. CPP, National Aluminium Company Limited Orissa utilised about 2 lacs tonnes of CCRs both in ash pond raising and land reclamation. For widening the embankments of Nizzamuddin Bridge, New Delhi about 1.5 lakh tonnes of ash was used along with soil cover of about one metre thickness to protect the ash filling. In addition CCRs have also been used for embankments for various fly over bridges in Delhi (Sikdar, 2000).

The other study showed that bottom ash was used successfully as sub-base course for roads at NTPC's Dadri in association with Central Roads Research Institute (CRRI), New Delhi. About 20,000 tonne of ash was used by NTPC for road works at Talcher-Kaniha, Orissa during 1999-2000 (Mathur, 2000). Apart more than 7 lacs tonnes of CCRs being utilised for construction of roads and ash pond by most of the NTPC located at different part of India.

More recently, fly ash has been used as a component in geopolymers, where the reactivity of the fly ash glasses is used to generate a binder comparable to a hydrated Portland cement in appearance and properties, but with dramatically reduced CO2 emissions.

Another new application is using fly ash in roller compacted concrete dams. This has been demonstrated in the Ghatghar Dam Project in India.

CCRs in Land Reclamation and Agriculture

The advantages and disadvantages of CCRs applications to improve the soil fertility have been published by large number of researchers in India and abroad (Menon et al., 1993; Schwab, 1993; TIFAC Home Page, 2001). Long term studies, on the effect of fly ash on soil fertility and crop yield, carried out by Saxena and Asokan, (2000) revealed that CCRs can be used (up to 1170 t/ ha.) as a enriched medium in improving the productivity of wasteland soil and increase the yield of most of the crops, vegetables and cereals without affecting the food quality and soil fertility (Fig.6). These effects are usually observed when CCRs overcome nutrients deficiency in the soil to which it has been introduced. CCRs are also known to improve crop growth by neutralising soil acidity (Keefer, 1993). On the contrary, few investigations involving use of CCRs in agriculture show that CCRs produced undesirable effects on crop yield and on development of plants. The most frequently cited cause of these effects are heavy metals and boron toxicity (Ferraiolo et al., 1990). In some cases, CCRs is shown to induce P deficiency, salt injury, pozzolonic effects and heavy metal toxicity to crops (Shukla and Mishra, 1986). But our experience as well as review indicates that the effect of CCRs in agriculture may depends on soil and CCRs texture, structure, pH, moisture content, reactivity of ash and soil, ion exchange capacity, method of application and percentage addition of ash.

One of the studies carried out by Saxena and Asokan (2000) showed that for reclamation of land ~ 40,000 t/ha. of ash was used at NTPC, Rihand Nagar, Northern India (TIFAC Home Page, 2001). It is also reported that about 3.0 lakh tonnes of ash was used as filler materials for reclamation of low-lying area at NTPC; Badarpur Delhi (Mathur, 2000) during 1995- 2000. NTPC, located at different part of the country, used more than 15 lakh tonnes of ash for various land filling and vegetative activities.

Multiplier Effect

The technology dissemination activities like Kisan Melas, Farmers Awareness campaigns etc. were organised at projects sites in each of the cropping seasons as part of the project activity during different stages of crop growth. The project demonstrations could generate confidence among the local farming community and the rural masses on the beneficial nature of Pond ash for wasteland reclamation and agriculture. The farmers at various Kisan Melas have requested the concerned power plant officials for the supply of Pond ash to their fields. This has subsequently led to the transfer of Technology to the farmers' fields in Jhansi and kanpur (U.P) and Angul (orissa). The feedback from the farmers has beenquite encouraging. The farming community has been reaping the benefits of Pond ash application in their fields every season.

CCRs for Reclamation Abandoned Mine

Bulk quantities of CCRs have been used to replace the conventionally used sand for reclaiming underground mines. During 1999-2000, NTPC used about 60,000 tonnes of ash for backfilling underground mines of Singrouli Collory Company Limited, Southern India in collaboration with Central Mining Research Institute (CSIR), India, (Mathur, 2000). The potential application of CCRs in reclaiming abundant coal mine is of great practical significance in India. Research and Development are still in progress for commercial use of such huge quantum of CCRs as mine-fill material. Since about 80 % coal is produced from opencast mines, Coal India Ltd. is in crucial stage of handling excess overburden and planning for CCRs back filling in the abandoned mine for eco- engineering development with viable plant life. On the contrary, coal India itself is facing problems for disposal of the abundant overburden wastes (~6000 million cubic meter) as against their in-situ volume of the available open cast mine pits (4000 million meter cube) and regaining the configuration of the landscape (Pan, 2000). Since, there are various limitations and threats towards environmental degradation, effective scientific inventions are to be made before a firm decision is taken for bulk use of CCRs in reclaiming abandoned mines.

CCRs in Paint and Environmental Application

It is reported that paints can be developed using CCRs for protection of metallic and non-metallic structures and for the protection of corrosive environment. CCRs were used (30%) as extender due to its chemical inertness, abrasion resistance, less oil absorption and low specific gravity. The CCRs paints showed resistant to water, acid, alkalis, organic solvents and has improved abrasion resistance (Tiwari and Saxena, 1999).

Earlier studies indicate that a series of aromatic substrates were adsorbed on the surface of ash and found to be a source for the extraction of chlorinated compounds (Robert and Kenneth, 1990). Addition of ash in soils contributes to detoxification of lindane residues (Albanis, et al., 1988). Application of CCRs is also there in environmental field for reducing the polluting content of waste waters, for de watering biological sludge and for cleaning oil polluted seawaters (Ferraiolo, et al., 1990).

In chemical and paper industries the COD and absorbency reduction induced by different addition of CCRs. Fine particles of CCRs were used for removal of dye under acidic condition at low temperature. The presence of unburned carbon in the CCRs could be activated to further improve the adsorption capacity. Even CCRs were used in liner materials for waste disposal to augment the contamination of soil (Iyer and Scott, 2001). Apart from the above, literature review indicates that CCRs are being used in synthesis of mullite, zeolite, granite, extraction of alumina and germanium, making ceramics products and solidification and stabilisation of hazardous wastes. (Chandrasekar, 1997; Mondragon et al., 1990; Rajasekhar et al., 1995; Roongta 2000; Shrivastava, 2000).

Constraints and Future of CCRs Application

Indian CCRs has lower lime reactivity (pozzolonicity), higher unburned carbon and crystalline phase in comparison to European and North American CCRs. The characteristics of CCRs are not uniform in nature and quality control parameters such as quality of coal, combustion process, ash handing system, design of ash pond etc., is to be maintained efficiently in each TPS.

In India ~19 % of the total cement production is FAPPC. Inspite of the PPC demand in southern India, the availability is very less. On the contrary in the northern part of the country, there is no demand for FAPPC and the manufacturer are finding difficulties in marketing FAPPC. Such gap should be fulfilled to increase the CCRs utilisation. If about 50% of the total cement production in India is FAPPC with 35% ash substitution, about 10 million tonnes of CCRs can be utilised annually (Krishnamoorthy, 2000). Indian standard (IS 1480-1996) for making PPC and reinforcement concrete (IS456) are more specification oriented (physico chemical properties) rather than performance and ignore the long-term durability and hence, customary standards are required to meet the compelling need of the large scale utilisation of CCRs.

Setting up of CCRs based building products unit are capital intensive. For setting up of brick making plant, capacity of 5.4 million brick per day, to utilise 10,000 tons of ash an investment of Rs.600 crore is required. Also for utilising same quantity of CCRs for manufacturing sintered lightweight aggregates or aerated cellular concrete block making plant an investment of Rs.350 crore and Rs.2000 crore respectively is required (Krishnamoorthy, 2000). Entrepreneur/ manufacturer are not showing much interest to invest such a large amount. Cost benefit analysis of CCRs Vs conventional building materials is required to be significantly evaluated for a concrete recommendation. The existing CCRs disposal system in India should be modified and suitable technique to be made for size wise segregation of ash and make available finer dry ash for cement production and coarse ash for building, road and mine reclamation.

CCRs transportation cost for reclamation of abundant mine and road construction is a major constrain on the whole. Restriction of excavation of earth for filling low-lying areas and construction of embankment with in 100 km radius is essential. There should be made mandatory in the policy by legislation to use CCRs in place of soil for such applications.

Lack of awareness on the advantages of CCRs based products among end user is limiting new initiatives and market potential. There should be an integrated approach by the coordination of technologist, architects and manufacturers for the production of superior quality of CCRs based products to meet the consumer acceptability and increase marketability. In addition, in association with scientist, policy maker and CCRs generators, adverse publicity and awareness of the quality parameters and beneficial effects of CCRs based building materials should be made clear to the utility perception for mass consumption and effective utilisation of CCRs.

Keeping in view of the complexity of CCRs disposal, Fly Ash Mission, Government of India has taken several effort in "Confidence building" in CCRs based technology and to provide a solution to this mammoth task of safe disposal and fruitful utilisation of CCRs (TIFAC Home Page, 2001). Under Fly Ash Mission, 55 technology demonstration projects have been implemented at 21 locations in India so far for CCRs effective utilisation and safe management in building materials components, underground mine fills, road embankments, reclamation of ash pond, and dam construction etc., Ministry of Environment and Forest, Building Materials Technology & Promotion Council and Housing and Urban Development Corporation, Government of India has taken various initiatives and contributed not only for the financial and technical support to carryout R& D work but also in promoting of CCRs based Building Materials for large-scale utilisation.

As a result, the Bureau of Indian standards, 3812 and 456, has been revised to use CCRs as pozzolana and admixture. By substituting ash, the quality parameters have been further tightened. Apart from fiscal incentives to encourage the entrepreneur, Government of India has made legislation for restricting the use of topsoil for construction activities and to use 25% of ash in bricks, blocks, tiles within a radius of 50km from thermal power plants without excise duty. All these efforts have significantly contributed in increasing the CCRs utilisation in India. Now the liability of CCRs safe disposal involves the CCRs generator but also the scientist and technocrat as well as the people of the country as a whole.

Presently, thermal power plants release ~160 million tonnes of solid wastes as CCRs/ by products. This huge quantity of wastes, indeed resources, which have to be recycled and used in a effective manner with Life Cycle Assessment Studies. Presently 33% of the CCRs produced worldwide find market applications. In India, attempts have been made to evaluate its characteristics variations and understand its environmental significance through several lab and pilot scale experiments. As a results, today CCRs are being used up to 40% of total generation in India in building materials, road and embankment, land development and agriculture, extraction of metal and cenospheric ash, paints and waste treatment. Attempts are also been made to recycle and use huge quantity of CCRs for reclamation of abundant coal mine for socio-techno-economic development. For long terms prospective of CCRs management it is quite imperative that the CCRs generators, scientists, technocrats, entrepreneurs, consumers along with decision/ policy maker should jointly put an effort for the design of paradigm shift in effective management of Coal Combustion Residues to achieve the target of 100% fly ash utilization in coming years.

The technologies developed so far showed major utilisation potentials which can significantly contribute towards increasing the employment opportunity and economy of the urban / rural people. However, yet there are several impediments for effective implementation of such technologies for optimum benefit. Especially the lack of awareness, interaction and confidence among the general community, heavy initial investment, intellectual perception and participation in effective implementation are the major hindrances. At this juncture, technologists along with the policy makers, bureaucrats, entrepreneur and the rural mass should make a venture jointly to implement a coordinated programme with the support of major funding agencies to surmount the existing situation and make a flip to fulfill the requirement and meet the demand of the society. Such effort will certainly contribute towards employment generation for the sustainable development in an environmentally sound manner.

Ministry of Environment and Forests has issued guideless under Environment (Protection) Act, 1986 and Gazette Notification 1999 which are to be followed for utilisation of fly ash by thermal power plants. Accordingly, new power plants should utilize 30% fly ash within next 3 years of commissoing and 100% fly ash by the 9th years. However, existing power plants should utilize 20% fly ash within 3 year and 100% fly ash within 15 years.

Fly ash has generally been considered as a waste material. Nevertheless, some people categorized it as hazardous industrial waste, though which it is not hazardous waste. The Government of India recognising the importance/ potential of fly ash, established "Fly Ash Mission" in the year 1994 to undertake concerted efforts in selective Thrust Areas of fly ash utilisation. The initiatives undertaken along with major stake holder agencies through technology demonstration projects have turned around the perception of fly ash from a "waste material" to that of "resource material".

The efforts over the last decade have made significant impact. A meagre 3% utilisation of 40 million tonnes of Fly ash generation in 1994 has risen to 38% of 112 million tonne generated in 2004-05. The generation of fly ash is expected to increase to 170 million tonne by the end of XI five year plan.

This thrust imparted by the Fly Ash Mission is now being continued by the fly ash Utilization Programme (FAUP), under Technology Information, Forecasting and Assessment Council (TIFAC), Department of Science and Technology (DST), GOI along with other stake holders, Thermal Power Plants, Council of Scientific & Industrial Research, Research & Academic Institutes and the Industry, other agencies including Ministry of Power, Ministry of Environment & Forests, Ministry of Water Resources, Ministry of Road Transport & Highways, Ministry of Urban Development, Ministry of Agriculture, Ministry of Coal, Ministry of Mines.

Conclusions

Keeping in view of depletion of available natural resources like forests, fine & coarse aggregates and environmental restriction, it is necessary to use the available natural resources suitably for various engineering applications. To meet the shelter component of the people, timber, bricks / blocks, cement, steel are being extensively used as a major construction materials in the building industries. Increase in population and industrialization leads to major shortage of such building materials. The processes and products developed using fly ash as alternative materials have massive potentials for the sustainable livelihood development of urban /rural masses. Development of innovative, durable, cost effective and environmental friendly alternative building construction materials from fly ash and scale up of laboratory processes for commercial applications has a long way to go to meet the country's demand. Fly ash characteristics such as pozzolanic activity, fineness and various other useful properties has many applications in different areas of building industry and make it imperative to tap the potential of the accumulated and undoubtedly available fly ash that would cater to the demand of building material in the future. The technology demonstrations on use of fly ash in agriculture, road and embankment, mine filing, etc., around India have generated confidence on bulk utilization of fly ash in diversified fields. There is further scope for maximizing the use of fly ash in multidisciplinary areas which will contribute for protecting environmental, natural resources and increase/ improve the employment potentials/economy of rural mass.

Acknowledgements

The authors are thankful to Dr. Anil K. Gupta, Director, AMPRI, CSIR, Bhopal for the permission and support to submit this article. Authors are also thankful to Mr. Pavan K. Srivastava and Mr. Dharam Raj Yadav and Building Materials Development Group of AMPRI Bhopal for the assistance at different stages in carrying out the experiments and preparing this article. Most of the research work, at AMPRI Bhopal on fly ash utilization have been carried out with the support of CSIR, BMTPC, MOEF, NALCO, NBO, NTPC, TIFAC and BHEL and is gratefully acknowledged.

REFERENCES

1. Albanis, T. A., Pomonis, P. J., and Sdoukos, A. T., 1988. The influence of fly ash on pesticide fate in the environmental hydrolysis, degradation and adsorption of lindane in aqueous mixtures of soil with fly ash. Toxicol. and Environ. Chem., 19: 161-169.
2. Asokan. P., 2001. Environmental implications of coal combustion residues disposal. Ph.D. course work seminar report, Indian Institute of Technology, Bombay, India.
3. Asokan, P., 2000. Evaluation of coal combustion residues disposal site and toxicity leachate characteristic studies. M.Tech Thesis, Maulana Azad College of Technology, Bhopal, India.
4. Asokan, P., Saxena, M., and Asolekar, S. R., Coal Combustion Residues- Environmental Implications and Recycling Potentials. Resources, Conservation & Recycling, 43 (4), 239-252, 2005.
5. Asokan, P., Saxena, M., and Aparna, C., 1999. Contribution of coal ash in the enhancement of vegetation. In: American Coal Ash Association. Use and Management of Coal Combustion Products (CCPs). Proc.13th Int. Symposium on Use and Management of CCPs, 11-15 January 1999, Orlando, USA., 1(13): 1-13.
6. Bhanumathidas, N. and Kalidas, N., 1997. New Trend in Bricks and Blocks: The role of FaL-G. In: Karnataka Power Corporation Ltd and Associated Cement Companies Ltd. Fly ash utilization. Proc. Symposium on fly ash utilization, 10 January 1997, Bangalore, India, 82-90.
7. Chandrasekar, B. K., 1997. Innovation in fly ash utilization. In: Karnataka Power Corporation limited and Associated Cement Companies Ltd. Fly ash utilization. Proc. Symposium on fly ash utilization, 10 January 1997, Bangalore, India, 82-90.
8. Duxson, P.; Provis, J.L. & Lukey, G.C. et al. (2007), "The role of inorganic polymer technology in the development of 'Green concrete'", Cement and Concrete Research 37 (12): 1590-1597.
9. Ferraiolo, G., Zilli, M., and Converti, A., 1990. Flyash disposal and utilisation. J. Chemical Biotech, 47: 281-305.
10. Garg, M., Singh. M., and Kumar, R., 1996. Some aspects of the durability of a phosphogypsum-lime-fly ash binder. Construction and Building Materials, 10(4)): 273-279.
11. G. K. Pandey, S. K. Tyagi and B. Sengupta. Management of thermal power plants in India, BAQ 2006, Yogykarta, Indonesia.
12. Iyer, R. S. and Scott, J. A., 2001. Power station fly ash – A review of value- added utilisation outside of the construction industry. Resources, Conservation and Recycling, 31: 217-228.
13. J.M. Bhatnagar, R.K. Goel, L.P. Singh and Jaswinder Singh (2005). Use of Fly Ash in Clay Brick Manufacturing. Proc. National Seminar cum Business Meet on Use of Fly Ash in BUILDING COMPONENTS. New Delhi.
14. Prabhakar, J., R.S. Ahirwar, M. Saxena and S.K. Bose, "Application of Fly Ash in Building Construction", Proc. Workshop on Fly ash Utilisation organised by CBIP at RRL, Bhopal, pp. 107, 1995.
15. Kamal K., 2000. Fly ash utilisation in cement industries. In:. Proc. Workshop on fly ash utilisation: Issues and strategies, organised by Institution of Engineers (India), Bhopal Chapter, 15 September 2000, Bhopal, India, 38-40.
16. Karade, S. R., Morchhale, R. K., Saxena, M., and Khazanchi, A. C., 1995. Fly ash utilisation with Indian soils for making bricks. In: Regional Research Laboratory, Bhopal. Fly ash utilisation. Proc. Workshop on Fly ash utilisation, 16 September 1995, Bhopal; India, 103-104.
17. Keefer, R. F., 1993. Coal ashes-industrial wastes or beneficial by-products. In: Trace element in coal and coal combustion residues, Keefer, R. F. and Sajwan, K. (Ed.), Advances in Trace Substances Research, Lewis Publishers. CRC Press, Florida, 3-9.
18. Keefer, R. F., 1993. Coal ashes-industrial wastes or beneficial by-products. In: Trace element in coal and coal combustion residues, Keefer, R. F. and Sajwan, K. (Ed.), Advances in Trace Substances Research, Lewis Publishers. CRC Press, Florida, 3-9.
19. Krishnamoorthy, R., 2000. Ash utilisation in India-Prospect and problems. In: Barrier and utilisation option for large volume application of fly Ash in India, Hajela, V. (Ed.). Proc. Workshop on USAID/ India Greenhouse Gas Pollution Prevention Project, 5 February 2000, New Delhi, 63-67.

20. Kumar, V., Mathur, M., Jha, C. N., and Goswami, G., 1999. Characterisation of Fly Ash-A Multifacet Resource Materials. In: Proc. National seminar on fly ash characterization and its geotechnical applications, organised by Indian Institute of Science, Bangalore, 30 August 1999, Bangalore, India, 45-50.
21. Martin, J. P., Collins, R. A., Browning, J. S., and Biehl, F. J., 1990. Properties and use of fly ashes for embankments. J. Energy Engg., 116(2): 71-86.
22. Mathur A. K., 2000. Ash utilisation in NTPC. In: Proc. Workshop on fly ash utilisation: Issues and strategies, organised by Institution of Engineers (India), Bhopal Chapter, 15 September 2000, Bhopal, India, 41-45.
23. McCarty, M. J and Dhir, R. K., 1999. Towards maximizing the use of fly ash as a binder. fuel, 78: 121-132.
24. Menon, M. P., Sajwan, K. S., Ghuman, G. S., and Chandra, K., 1993. Element in coal and coal ash residues and their potential for agricultural crops. In: Keefer, R. F. and Sajwan, K. (Ed.), Trace element in coal and coal combustion residues. Advances in Trace Substances Research, Lewis Publishers. CRC Press, Florida, 259-283.
25. Moghissis, A. A., Paras, P., Carter, M. W., and Baker, R. F., 1978. Radioactivity in consumer products. U.S. Nuclear Regulatory commission, Washington.
26. Mohini Saxena, R. K. Morchhale, P. Asokan and B. K. Prasad (2008). Plant fibre - Industrial wastes reinforced polymer composites as a potential wood substitute materials, Journal of Composite Materials. 42 (4): 367-384.
27. Mondragon, F., Rincon, F., Sierra, L., Escober, J., Ramirez, J., and Farnandez, J., 1990. New perspective of fly ash utilization: Synthesis of zeolite material. Fuel, 69: 4781-4786.
28. Pan, P. N., 2000. Prospect of utilisation of fly ash in coal mines- A Prospective. In: Proc. Workshop on fly ash utilisation: Issues and strategies, organised by Institution of Engineers (India), Bhopal Chapter, 15 September 2000, Bhopal, India, 57-62.
29. Rajasekhar, C., 1995. Retention and permeability characteristics of clays and clay-fly ash systems subjected to flow of contaminants. Ph.D. Thesis, Indian Institute of Science, Bangalore, India.
30. Robert, L. J. and Kenneth, R. O., 1990. Fly ash use as a sedimentation studies. J. Soil Sci., 54: 855-859.
31. Roongta, 2000. Fly ash utilisation in cement grinding. In: Proc. Workshop on fly ash utilisation: Issues and strategies, organised by Institution of Engineers (India), Bhopal Chapter, 15 September 2000, Bhopal, India, 46-56.
32. Saxena, M., Asokan, P., Mandal, S., and Chauhan, A., 1998. Impact of fly ash phase constituents on wasteland soils. J. Environ.& Energy Conservation, 4(4): 229-234.
33. Saxena, M. and Prabakar, J., 2000. Emerging Technologies for Third Millianium on wood substitute and paint from coal ash. In: Verma C. V. J., Rao, S. V., Kumar, V., and Krishnamoorthy, R. (Editors), Fly Ash Disposal and Utilisation. Proc. 2nd Int. Conf. on Fly Ash Disposal and Utilisation, 2-4 February 2000, New Delhi, India 1(II-7): 26-31.
34. Schwab, A. P., 1993. Extractable and plant concentrations of metals in amended coal ash. In: Keefer, R. F. and Sajwan, K. (Ed.), Trace element in coal and coal combustion residues. Advances in Trace Substances Research, Lewis Publishers. CRC Press, Florida, 185-211.
35. Shiban Raina (2004). Use of fly ash in building materials- recent trends Proc. Seminar on Recent Trends in Building Materials, pp.275-276.
36. Shrivastava, A. P., 2000. Generation and possible utilisation of fly ash from thermal power plants. In: Proc. Workshop on fly ash utilisation: Issues and strategies, organised by Institution of Engineers (India), Bhopal Chapter, 15 September 2000, Bhopal, India, 22-26.
37. Shukla, K. N. and Mishra, L. C., 1986. Effect of fly ash extract on growth and development of corn and soyabean seedling. Water, Air, Soil Pollution, 27: 155-167.
38. Sikdar, P. K., Guru Vittal, U. K., and Kumar, S., 2000. Use of fly ash in road embankment. In: Proc. 2nd Int. Conf. on Fly Ash Disposal and Utilisation, Verma, C. V. J., Rao, S. V., Kumar, V., and Krishnamoorthy, R. (Ed.), 2-4 Feb.,2000, New Delhi, India, 1(VIII-7): 45-56.
39. Sumio, H., Kawaguchi, M., and Yasuhara, K., 2000. Effective use of fly ash slurry as fill materials. J. Hazardous Materials, 76: 301-337.
40. TIFAC Home Page, October 2001. http://www.tifac.org.in/news/fly ash mgn.htm
41. TIFAC Home Page, October 2002. http://www.tifac.org.in/news/fly ash mgn.htm.
42. Tiwari, S. and Saxena, M., 1999. Use of fly ash in high performance industrial coatings. British Corrosion J., 3(34): 184-191.
43. UNSCEAR, 1982. Ionising Radiation: Sources in Biological effect. United Nation, New York.
44. Vijayan, V. and Behera, S. N., 1999. Characterisation of Natural Radioactivity in Coal Ash. In: Proc. National Seminar on fly ash characterization and its geotechnical applications, organised by Indian Institute of Science, Bangalore, 30 August 1999, Bangalore, India, 139-144.
45. Vimal Kumar & Mukesh Mathur 2005. Fly ash : Building Material for Sustainable Development. Proc. National Seminar cum Business Meet on Use of Fly Ash in building components. New Delhi.

CHAPTER 8

Eco-friendly Shelter Design

Ar. S.M. Husain[1]

Abstract

Building construction industry is one of the largest energy consuming sectors in India. Efficient use of energy is important since global energy resources are limited. In Modern buildings significant amount of energy is consumed to keep the building environment comfortable. Consumption of energy in construction activity is likely to rise even further due to the increase in urbanization, Industrialization and living standards. This is an initiative with the use of appropriate materials and techniques for producing energy efficient and environment friendly houses.

Introduction

The purpose of constructing a building is to provide shelter—an artificial environment, which is more comfortable for human occupancy than the natural environment. Energy consumption for production of materials and construction of buildings is increasing in India. Since, the required energy is mainly derived from fossil fuels; the building sector has become a major factor to the environment. The designing of buildings with energy efficient considerations is necessary right from the planning stage. To achieve the collective objective of reduction in the use of energy and environmental protection, eco-sensitive buildings should be planned and designed.

The environmental impact resulting from the energy consumption in building industry has become a global problem. In this scenario Energy Efficient Building design concept has become essential for achieving energy saving and environmental protection.

Role of Architecture in Heating and Cooling of Buildings

Ancient architecture, all over the world, had many characteristics, which led to thermal comfort in building. The buildings were shaped and different parts of the building *e.g.* indoor spaces, door, windows etc. were located and oriented in such a way that they take maximum advantage of the climate and the role of the trees, vegetation and the water around the building in determining the thermal comfort was well taken into consideration. The massive wall and clustered residences, to reduce the surface to volume ratio, for reducing the effect of temperature swings, were also commonly employed. Due to energy crisis originated in 1974, there has been renewed interest in those aspects of Architecture which lead to minimum energy consumption from conventional sources to keep the space inside the building at desired level of comfort. Recent efforts of integrating the ancient concept with the vast knowledge, the present day science offers along with the availability of newer materials, have lead to the recognition of a new discipline as solar passive Architecture.

[1]'Shelter' Architects, Bhopal-462 001 (India)

Energy Conscious Architecture in Ancient Building

In ancient times most of the designers constructed building which provided more or less comfortable living conditions without the use of sophisticated technical devices. They were familiar with the techniques by which they could get maximum benefit from climatic features, for comfortable design, merely by means of building shapes, location and orientation. Additional elements were also integrated into the building design, such as vegetation and water to improve "micro-climatic-conditions" for the dwelling of inhabitation. In India, ancient architecture shows that buildings exhibit ingenuity in their design, by the use of locally available materials and techniques, to erect buildings, which are well adapted to the local climate. Many such proven methods have been ignored in the modern buildings which need special means for heating and cooling of building incurring high cost of equipment and energy.

For example in Rajasthan, a hot and dry region in the north-west of India, the climate is characterized by high daytime temperatures and uncomfortable low night temperatures. The solution best suited to such wide temperatures fluctuations is to delay entry of heat into the building, such that it reaches the interior of the building when it is least bothersome. The inhabitants of this area achieved this desired thermal performance by using thick wall and materials having high thermal capacity, such as mud and stone. Further, the houses were closely spaced in order to achieve fewer surfaces to volume radio. By these means, a considerable reduction in solar heat absorption is achieved and it takes longer time for walls to heat up completely, there by keeping the interiors of the building cool during the day, and after sun set most of the heat stored in the walls in these buildings is radiated out to the sky, while a small amount warms the indoor space during night.

In warm and humid climates, the diurnal temperature variation is small, so the materials of low heat storage capacity are appropriate. Therefore in ancient times people constructed their houses with lighter materials such as palm leaves and twigs, allowing air to circulate, avoiding stagnant humid air and providing comfort by evaporative cooling.

In cold climates such as in Shimla, people used to construct buildings with their walls made from stone and timber to compensate for wide temperature fluctuations and orienting large face of buildings towards south for maximum heat gain.

The great pyramid at Gizeh was built on efficient solar passive technique. The great mass of this naturally air-conditioned structure maintains the king's and queen's chambers at 23ºC through out year. Ventilation air also enters the chambers at same temperature.

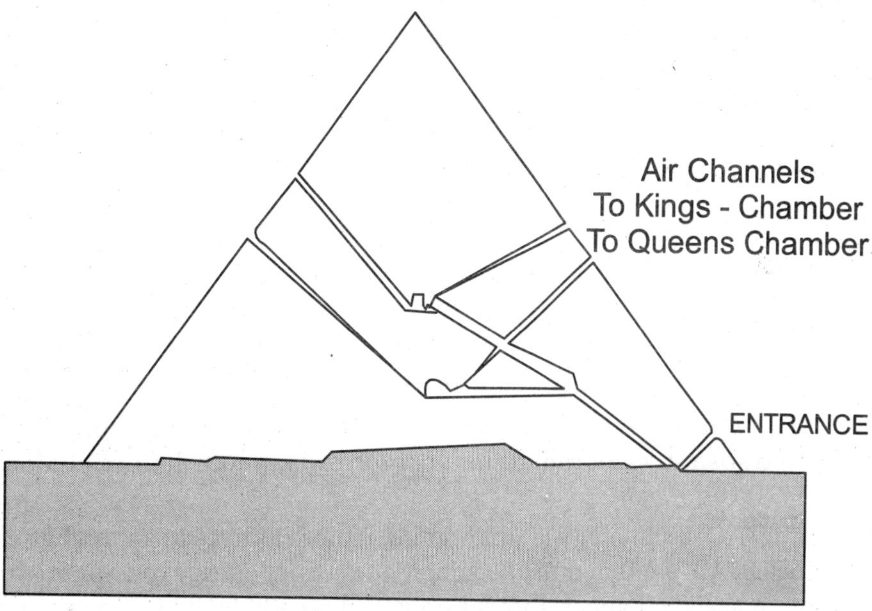

GREAT PYRAMID AT GIZEH

Fig. 8.1

MATERIALS AND METHODS

Energy Efficient Design Concept

The design of energy efficient building is an integrated Approach. Providing thermal comfort is a basic requirement in the design of buildings for efficiency and health of its occupants. For creating comfortable environment in the interior of buildings, the methods are:

1. Natural systems based on utilization of solar and wind energy.
2. Use of electrically operated mechanical devices, such as, air conditioners heaters, blowers etc.

Since, building sector is a major consumer of electricity, it is essential to use building designs that would utilize solar and wind energy for balancing thermal environment inside buildings. The second one more dependable, the use of electricity in heating, ventilating and air conditioning, lighting and water heating is about 30% of the total electricity consumption in the country. In the present energy scenario in India the demand of electrical energy is continuously increasing, the awareness must be created about efficient use of energy. The objective involves three steps:

1. Identification of the climate at the building site;
2. Determination of the comfort requirements of the relevant climate; and
3. Selection of appropriate materials and design features including space planning, orientation, location and size of fenestration, shading devices, treatment of building envelops etc.

Design Concepts for Energy Efficiency in Buildings

Energy efficiency in buildings can be achieved in three ways:

(*i*) Avoiding wastage of energy due to undesirable electrical appliances
(*ii*) Development of energy efficient appliances and
(*iii*) Optimum use of non-conventional sources of energy by proper planning & designing of buildings.

Various design options for achieving the above objectives are:

1. The passive solar techniques will reduce the heat penetration through building envelope and provision of fenestration for inducing desired natural ventilation.
2. Use of renewal energy technology
3. Use of environment-friendly low energy materials
4. Construction techniques for energy efficiency
5. Design of energy lighting and HVAC systems

1. Passive Solar Techniques in Building Design

(*a*) Suitable orientation of building with respect to the sun
(*b*) Vegetation overhangs and sunscreen: minimum disturbance to landscape & site conditions.
(*c*) Maximize ventilation to reject heat in hot climate.
(*d*) Use of approximate energy efficient windows and skylights to provide daylight.
(*e*) Thermal insulation of external walls, roofs, floors and double pane windows can reduce heating
(*f*) For cold climate "TIM" (transparent insulation material) is a new development. TIMs are characterized by their ability to transmit solar radiation when providing sufficient insulation against heat to losses.
(*g*) For air conditioning (AC) technology is undergoing major changes to get higher efficiency & better environment. Efficient A.C system must be adopted to reduce energy consumption of A.C. buildings. Solar cooling system based on vapor absorption is technically feasible but uneconomical.
(*h*) Energy efficient light/lamp consume considerably less electricity than an incandesant lamp. Its initial cost is high but in long run it costs less. Energy efficient glazing : (insulating glasses, solar

controlled glazing, low energy glazing etc.) can contribute to efficient utilization of energy by three ways.

(*i*) Reducing capacity of installed A.C. system

(*ii*) Lowering electricity consumption for its operation and

(*iii*) Reducing electricity consumption by using natural daylight in buildings

2. Use of Renewal Energy Technology

Alternate source of energy, solar photovoltaic, solar cookers & water heaters: wind mills, biomass gasification, biogas plants etc.

(*a*) **Solar water heaters :** can be easily installed and generate hot water at 60-80°C.The technology is economical and can save electrical energy.

(*b*) **Solar air heaters :** can supplement requirement of heating of building during day time, in cold and sunny climate. This can reduce demand of electricity and fire-load for space heating.

(*c*) **Solar electricity :** can be generated by solar cells in buildings, on roofs or by installing grid connected photovoltaic (PV) power plants. It is expensive where grid connected electricity is available.

3. Use of Environment-friendly Low Energy Materials

Reduce energy consumption by using natural materials such as earth, stones & wood. Construction consumes a variety of building materials. The most common building materials used in construction activity today are cement, steel, bricks, stones, glass, aluminum, timber etc. An analysis of the "embodied energy" indifferent r-materials, construction methodologies and building types suggest some rules for decision making for selecting materials for design in energy efficient way, are

(*a*) Choose less toxic materials if two materials perform same function for the same price.

(*b*) As for as possible choose material with their natural state.

(*c*) Use locally available materials whenever possible

(*d*) Design for minimum energy consumption

4. Construction Techniques for Energy Efficiency

(*i*) Building envelope : The main source of heat on building envelop is solar radiation. Heat reduction of exterior surface of building is necessary for keeping the indoor surface temperature at a low value. Transparent window facing sun also permits direct entry of sun. This also contributes to the rise in the temperature of indoor surfaces. Hence, control of direct entry of sun through windows is an essential requirement for preventing the rise in interior surface temperature. Based on these criteria, various methods for minimizing heat flow through building envelope can be adopted.

(*ii*) Orientation : The amount daily solar radiation incident per unit area on 'east and west facing wall is much more as compared to that on the walls facing other directions. So, for minimum solar heat gain by the building envelope, it is suggested that the shorter axis of building should lie along north-south direction. Also, the effect of orientation of a building depends on the aspect ratio, i.e. length / breadth of the building. For a building with rectangular plan building with aspect ratio 2:1, the fabric load is reduced by @ 30%.

(*iii*) Shading of widows : The direct entry of sun into the room especially during summer is controlled by providing louvers, overhangs on windows. Windows of the same dimensions but oriented differently should have different dimensions of louvers. It is studied that overhang with suitable dimensions can produce cooling load reduction of 12.7% in summer.

(*iv*) Exterior surface color : Surface color of the external wall affects the percentage of solar radiation absorbed by the external surface. So the heat transmitted into the building is considerably reduced when external surface is painted with a light light color. It will result in saving of electrical cal energy by about 40% to 50%.

***(v)* Insulation of wall and roof** : The thermal resistance and conduction of heat flow through the building envelope is increased by insulation on wall and roofs. Introduction of air cavity in a wall also increases its thermal resistance Studies on estimation of thermal properties of such a wall revealed that the overall heat transmission co-efficient u value of a 27.5 cm brick cavity wall (11.25 cm brick + 5.0 cm air gap +11.25 cm brick) is 1.63 W/m2 k while that of a 22.5 cm solid brick wall with 1.25 cm cement plaster on both the side U value is 2.26 W/m2 k. here it is specified that the thermal performance of performance of the above cavity wall is slightly better than that of a 35 cm solid brick wall.

***(vi)* Energy efficient windows** : The window design to conserve energy is done by deciding the window size and location. Windows on East and West faces should be minimized as these are not the suitable orientations from the heat gain point of view, in air conditioned buildings, windows are considerably less insulating than other parts of the structure, it is observed that for a single glazed window system the U value is 5.22 W/m2 K which is less than the desired value. The U value is considerably less (3 W/m2 K) for a window system consisting of a double glazing with an air gap of 12mm - 18 mm. providing such a system reduces heat gain by at least about 10%.

***(vii)* Indoor air** :

1. For getting maximum benefit from natural wind, buildings may be oriented at any convenient angle between 0° and 30°. If the prevailing wind is from East or West, buildings can be oriented at 45° to the incident wind for minimizing the solar heat gain.
2. At least one window should be provided on windward wall and the other on leeward wall.
3. In rooms of normal size having identical windows on opposite walls, the average indoor air speed increase rapidly by increasing the width of window up to about 2/3 of the wall width; beyond that the increase is in much smaller proportion than the increase of the window width.
4. For a total fenestration area (inlet plus outlet) of 20% 30°% of floor area, average indoor wind velocity around 27% of outdoor velocity. Further increase in window size increases the velocity but not in the same proportion. in fact, even under ideal conditions the maximum average indoor wind velocity does not exceed 40°% of the outdoor velocity.
5. In regions having fairly constant wind direction, the size of the inlet should be kept within 30% to 50% of the total area of fenestration and building should be oriented perpendicular to the incident wind, in case of room with only one wall exposed to outside, provision of two windows is preferred to that of a single window.

Examples from the World

The following are the upcoming green, eco-friendly projects around the world:

Pyramid Farm

Fig. 8.2

It uses a heating and pressurization system to separate sewage into water and carbon which will fuel both the machinery and lighting. It would use only 10-percent of the water and 5-percent of the land needed by conventional farms.

Sustainable zoo- South Korean

Fig. 8.3

It is a sustainable zoo with renewable energy harvesting capabilities that could house animals and be treated as nature reserves. Rainwater will be collected and all waste would either be reused or composted for use as biofuel and fertilizer.

The Dragonfly

Fig. 8.4

It is a 128-floor vertical farm concept which will support housing, offices, laboratories and twenty-eight different agricultural fields. It completely sustains itself using solar-power, wind-power, and captured rain water.

Seawater Vertical Farm

Fig. 8.5

To deal with the lack of fresh water in Dubai and other Arabian nations the ***Seawater Vertical Farm*** was created which cool and humidify greenhouses. This innovative concept produces adequate humidity to convert seawater into fresh water, necessary for irrigation.

Anti-Smog Building, Paris:

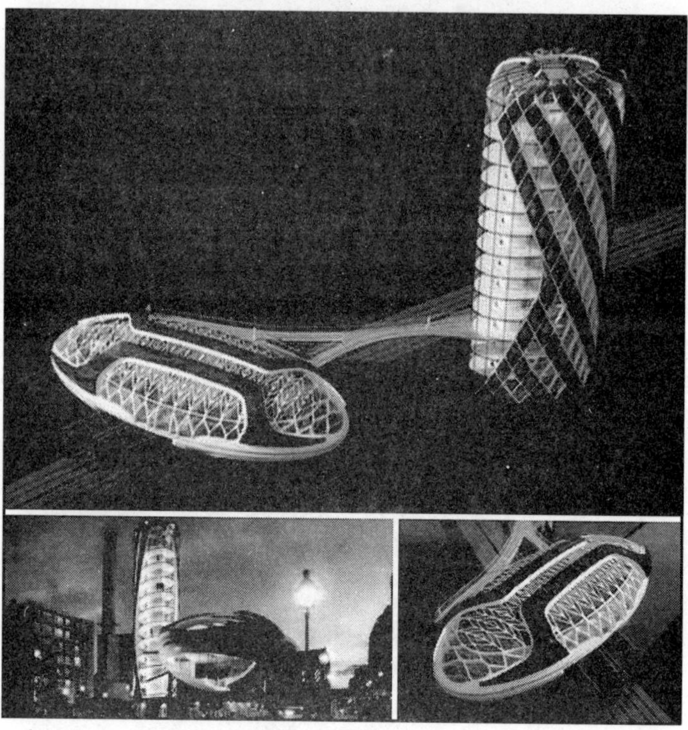

Fig. 8.6

It's a mixed-use building, designed using green technologies that actually suck the smog from the streets and turn it into useful recycled energy resources. A natural lagoon and rooftop view of Paris make people want to spend more time in this eco-friendly building.

The COR Building-MIAMI

Fig. 8.7

The building will incorporate mixed-use residential and commercial space, integrating green technologies including wind turbines, photovoltaic panels, and solar hot water generation. The building's exoskeleton is a hyper-efficient structure that provides thermal mass for insulation, shade for residents, and architectural elements such as terraces and armatures that support turbines.

Webstar

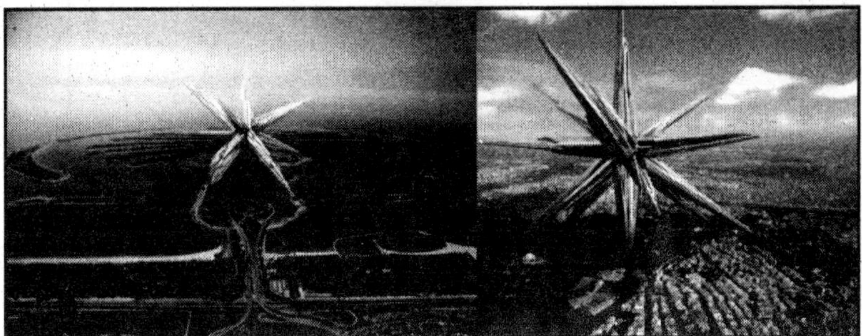

Fig. 8.8

It is giant floating star, which can camp above any city in the world and can house a perfect green city of 15,000 people. It has an ability to produce power, utilize resources by recycling them.

Sky Village

Fig. 8.9

This new and unique structure is a model that can be considered as a blueprint for many eco-structures in the future. It takes the shape of many blocks called 'pixels' being put together like green hubs and the building can expand strategically without occupying too much space at the base. Green parks put up in the sky, a concept that could do great for every edifice in future.

Precinct 4

Fig. 8.10

It is a city of delightful eco-friendly structures that stand like beacons of green architecture and amazing design. The design is special with green features like terraces, sunshades, natural ventilation and integrated green space and amazing architectural splendor inspired by marine life.

Fog Tower

Fig. 8.11

There is no green structure that in sheer concepts stands singularly alone as this Fog Tower does on the edge of the Atacama Desert. While just a concept for now, this gigantic tower in pristine while aims to harvest land that is barren in the desert by tapping into the fog on the land. The giant screw design progressively collects and condenses fog into water and supplies it down for irrigation.

Fusionopolis

Fig. 8.12

Fusionopolis will be Singapore's most eco-friendly building, what makes it so green is a vertical spine of planting that rises up through the 15-storey building.

Sinosteel International Plaza

Fig. 8.13

The design is modern, eccentric and still green with ample ventilation and sustainable features that are integrated fashionably into the design and form. Construction is underway in Tinajin, China and its impeccable design makes sure that there is no extra energy wasted on insulation.

Carbon Neutral Pyramid of Dubai

Fig. 8.14

A merger in architecture of Ancient Egyptian pyramids and Middle Eastern ziggurats along with modern green features makes it one of the best green structures in the world. The pyramid shaped structures run on solar power and are completely carbon neutral. They can sustain an entire community of a million people in them and all with green energy.

Rotating Wind Towers of Dubai:

Fig. 8.15

It is the spectacular wind-powered rotating skyscraper which twirls around and uses wind and solar energy to produce power that is sufficient to power another 5 skyscrapers along with power to supports its own residents' energy needs. This is almost like a giant swiveling structure on a beam at center.

Lilly Pad

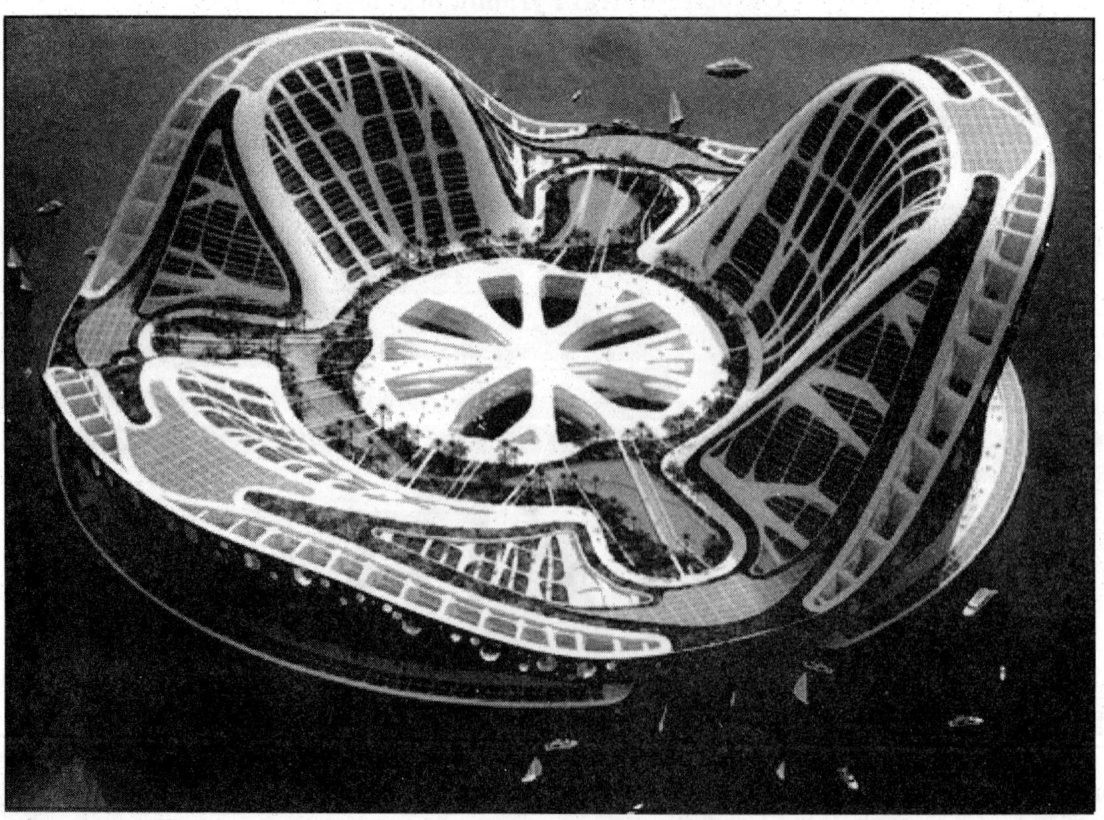

Fig. 8.16

It is a floating green "Ecopolis" that can accommodate up to 50,000 climate refugees.

The Sky Farm

Fig. 8.17

The Sky Farm proposed in Toronto can produce as much as a thousand acre farm, feeding 35 thousand people per year. The main concept behind this to produce crops in the center of the city with the minimum use of the ground surface and maximize the use of renewable sources of energy.

Helix Hotel, Dubai

Fig. 8.18

The entire structure garners green energy from both the sun and the wind and the specially designed GROW panels made from 100% recyclable polyethylene on the exterior will harness both solar and wind energy for the power needs inside.

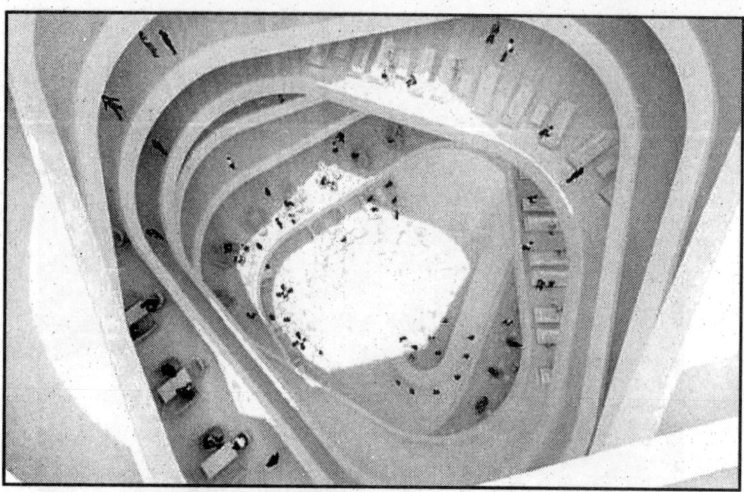

Fig. 8.19. STEP SKYSCRAPER

Fig. 8.20

The concept places urban farms on the outer fringes of residential apartments. Some floors are enclosed for year-round production of greenhouse crops, while others include terraces for seasonal items such as orchards. The ground floor would contain a farmers' market where residents could sell to one another and the general public.

Conclusions

(*i*) By the use of solar passive design concepts it is possible to save a substantial amount of energy for providing comfortable living conditions in the buildings.

(*ii*) It is possible during summers to maintain a temperature of upto 32ÚC inside the building, when ambient temperature is 40.5ÚC by using solar passive concepts. And upto this temperature comfort can be provided only by using fans and A.C 's are not required at all.

(*iii*) For summer comfort it is more important to insulate roof by providing roof top treatments such as earthen pots roof treatment, glazed surface roof treatment etc. as most of the heating effect of solar radiation is through roof during summers. In winters heat gain is achieved by vertical surfaces, hence roof top treatments have no effect on winter comfort conditions.

(*iv*) Orienting longer face of the building towards south is best orientation as it provides winter heat gain and blocks summer sun as a result of solar geometry, which is achieved by providing sun control devices for required shadow throws.

(*v*) Prevailing wind direction during summers and winters is different which helps in creating comfort conditions by allowing cool breezes in summer to penetrate inside the building and blocking cold winds of winter by providing openings in suitable direction and by tight fitting doors and windows.

(*vi*) East and West face of the building should be protected from summer heat gain by providing deciduous trees so that they may receive heat gain through bare branches of trees during winters.

Recommendations

Keeping in view the energy crises and environmental problems, which are caused during production of conventional energy the following recommendations are made.

1. the Government Departments which deals with the construction and design of the buildings such as PWD, Housing Board, Architecture Department and Urban Development Authorities, should be made aware of the concepts of Solar passive building design so that all Government buildings designed by them may be based on the solar passive design concepts and should use minimum energy for their comfortable use, by providing passive cooling, passive cooling, passive heating, ventilation and day lighting.
2. The Building contractors, Architects and masses in general should be made aware of Solar passive design concepts and their uses. And they should be made aware of how much energy and money they can save by using these concepts without any extra cost.
3. Department of non conventional energy sources should promote the use of non conventional energy devices such as solar water heating systems, solar cookers, solar electric generators, photo-voltaic, bio-gas, wind mill etc. so that the conventional source of energy can be saved.
4. Various state government agencies responsible for building construction should undertake Model House Construction on Pilot based for field testing and demonstration and motivating public for adopting these concepts.
5. Since these concepts are comparatively new to most of the designers, builders, and public, so wide publicity through fairs, internet, TV, news paper and periodicals is required.

CHAPTER 9

Carbon Trading (The Antipollution Business)

Dr. S.A. Iqbal[1], Syed Azam Husain[2]

Global Warming—The Issue

The Earth has an atmosphere of the proper depth and chemical composition. About 30% of incoming energy from the sun is reflected back to space while the rest reaches the earth, resulting in warming the air, oceans, and land, and maintaining an average surface temperature of about15°C.

The chemical composition of the atmosphere is also responsible for nurturing life on our planet. Most of it is nitrogen (78%); about 21% is oxygen, which all animals need to survive; and only a small percentage (0.036%) is made up of carbon dioxide which plants require for photosynthesis.

The atmosphere carries out the critical function of maintaining life-sustaining conditions on Earth, in the following way: each day, energy from the sun is absorbed by the land, seas, mountains, etc. If all this energy were to be absorbed completely, the earth would gradually become hotter and hotter. But actually, the earth both absorbs and, simultaneously releases it in the form of infra red waves . All this rising heat is not lost to space, but is partly absorbed by some gases present in very-small (or trace) quantities in the atmosphere, called greenhouse gases (GHGs).

Greenhouse gases (for example, carbon dioxide (CO_2), methane (CH_4), nitrous oxide (N_2O), water vapour), re-emit some of this heat to the earth's surface. If they did not perform this useful function, most of the heat energy would escape, leaving the earth cold (about –18 °C) and unfit to support life.

However, ever since the industrial revolution began about 150 years ago, man-made activities have added significant quantities of GHGs to the atmosphere. The atmospheric concentrations of carbon dioxide (CO_2), methane (CH_4), and nitrous oxide (N_2O) have grown by about 31%, 151% and 17%, respectively, between 1750 and 2000 (Intergovernmental Panel on Climate Change, IPCC 2001).

As the GHGs are transparent to incoming solar radiation, but opaque to outgoing longwave radiation, an increase in the levels of GHGs could lead to greater warming, which, in turn, could have an impact on the world's climate, leading to the phenomenon known as climate change. Indeed, scientists have observed that over the 20th century, the mean global surface temperature increased by 0.6°C (IPCC 2001). They also observed that since 1860 (the year temperature began to be recorded systematically using a thermometer), the 1990's have been the warmest decade.

Important greenhouse gases are : carbon dioxide (CO_2), methane (CH_4), nitrous oxide (N_2O), hydrofluorocarbons (HFC), perfluorocarbons (PFC), and sulfur hexafluoride (SF_6). Water vapor is also

[1]Department of Chemistry, Saifia Science College (B.U.), Bhopal (India).
[2]Asian Biotech Research Centre, Idzak Hills, Bhopal (India)

an important greenhouse gas, but since human beings do not generally have a direct effect on water vapour concentration in the atmosphere, it is not included in this paper. Because each greenhouse gas traps different amounts of heat and stays in the atmosphere for different lengths of time, studies use measures of global warming potential (GWP) to compare between gases. Carbon dioxide is used as the benchmark, so all other gases are measured in carbon dioxide equivalence (CO_2).

Table 9.1. The global warming potential of six major greenhouse gases (This measure takes into account the heat trapping abilities and the time the gas stays in the atmosphere (IPCC 2001 a, 2001 b))

Gas	Global Warming Potential	Atmospheric Life (years)
CO_2	1	5 to 200
CH_4	21	12
N_2O	310	114
HFC	140 to 11,700	1.4 to 260
RFC or PFC	6,500 to 9,200	10,000 to 50,000+
SF_6	23,900	3200

Natural and Anthropogenic Causes of Global Warming

Another IPCC publication states that there is a "very high confidence" that human activities have caused a net warming of the planet (IPCC 2007a).

Kyoto Protocol

Presently, a variety of approaches are being implemented to reduce carbon emissions. These range from efforts by individuals and firms to reduce their climate footprints to initiatives at city, state, regional and global levels. Among these are the commitments of governments to reduce emissions through the 1992 United Nations Framework Convention on Climate Change (UNFCCC) and its 1997 Kyoto Protocol.

In 1992 famous Rio earth summit, United Nation Framework Convention on Climate Change (UNFCCC) was adopted with an objective to stabilize atmospheric concentration of GHG at levels that would prevent dangerous human interference with climate system. The UNFCCC came into effect on 21st March, 1994 according to which industrialized countries shall have the main responsibility to mitigate climate change. Such countries are listed as Annex-I countries. Under UNFCCC all the member countries were to report on their national GHG emissions inventories and propose climate change mitigation strategies. After two and half years of intense negotiation between Annex-l countries, an agreement was struck at the now famous Kyoto protocol on 11 December 1997 in Kyoto, Japan. Born in the 1997 World Earth Summit held at Kyoto, Japan, this Protocol is making miracles in society today. The convention, participated by 160 countries of the world, was to negotiate binding limitations on greenhouse gases for the developed nations pursuant to the objective of the Framework Convention on Climate Change of 1992.

Under the Kyoto Protocol, emission caps were set for each Annex-1 countries, amounting in total to an average reduction of 5.2% below the aggregate emission level in 1990. Each country has a predetermined target of emission reduction as compared to 1990 level. No emission cap is imposed on Non-Annex I countries. However, to encourage the participation of Non-Annex I in emission reduction process a mechanism known as Clean Development Mechanism (CDM) has been provided.

The carbon markets are a prominent part of the response to climate change and have an opportunity to demonstrate that they can be a credible and central tool for future climate mitigation.

The outcome was the Kyoto Protocol, in which the developed nations agreed to limit their greenhouse gas emissions, relative to the levels emitted in 1990 or pay a price to those that do. At this point comes the carbon trading.

Carbon Credits

Carbon trading (or Emissions trading) is an administrative appraoch used to control pollution by providing economic incentives for achieving reductions in the emissions of pollutants. The international agreement (Kyoto Protocol) linked to the United Nations Framework Convention on the Climate Change (UNFCCC), sets the binding target for reducing green house gas (GHG) emissions, a commodity was created in the form of emission reductions or removals. Since carbon di-oxide is the principal green house gas, the term has become simply of trading in carbon. Carbon is now tracked and traded like any other commodity, is known as "Carbon Trading", is one of the Kyoto mechanism of "Carbon Market".

The Mission

Under Kyoto protocol all developed countries will have to cut down their emissions by 5.2% below emissions in 1990 else they have to pay heavy fines. Now, one way of measuring how much they are polluting the air, is by calculating the GHG emissions due to their production. They have various ways to aggregate these units called "CER" or "Certified Emission Reduction" or 'one tonne of CO_2 equivalent'.

Working

A central authority sets a limit on the amount of a pollutant that can be emitted. Companies or other groups are issued emission permits and are required to hold an equivalent number of allowances (or credits) which represent the right to emit a specific amount. The total amount of allowances and credits cannot exceed the cap, limiting total emissions to that level. Companies that need to increase their emission allowance must buy credits from those who pollute less. The transfer of allowances is referred to as a trade. In effect, the buyer is paying a charge for polluting; while the seller is being rewarded for having reduced emissions by more than was needed. Thus, in theory, those that can easily reduce emissions most cheaply will do so, achieving the pollution reduction at the lowest possible cost to society.

Carbon Trading V/S Carbon Tax

Carbon trading is sometimes seen as a better approach than a direct carbon tax or direct regulation. By solely aiming at the cap it avoids the consequences and compromises that often accompany other methods. It can be cheaper, and politically preferable. In addition, most of the money in the system is spent on environmental activities, and the investment directed at sustainable projects that earn credits in the developing world can contribute to the Millennium Development Goals. Critics of emissions trading point to problems of complexity, monitoring, enforcement and sometimes dispute the initial allocation methods and cap.

Marginal Abatement Cost

It is possible for a country to reduce emissions using a Command-Control approach, such as regulation, direct and indirect taxes. But the approach is more costly for some countries than for others. That is because the Marginal Abatement Cost (MAC) that is the cost of eliminating an additional unit of pollution differs from country to country. It might cost China $2 to eliminate a ton of CO_2, but it would probably cost Sweden or the U.S. much more. International emission-trading markets were created precisely to exploit differing MACs. Its advantages are :

- Provide information about system-wide of abatement.
- Can help minimize overall cost of abatement.
- In absense of a marginal damage function, policy makers, can relate the cost of desirable abatement level.
- Can aid energy and environmental modelling.

The World Bank has built itself a role in this market as a referee, broker and macro-manager of international fund flows. The scheme has been entitled Clean Development Mechanism, or more commonly, Carbon Trading.

CDM Project Types

Carbon Credits are sold to entities in Annex-I countries, like power utilities, who have emission reduction targets to achieve & find it cheaper to buy 'offsetting' certificate rather than do a clean-up in their backyard.

Type of projects, which are being applied for COM and which can be of valuable potential, are:

Energy Efficiency Projects

- Increasing building efficiency (Concept of Green Building/LEED Rating), *e.g.* Technopolis Building Kolkata
- Increasing commercial/industrial energy efficiency (Renovation & Modernization of old power plants)
- Fuel switching from more carbon intensive fuels to less carbon intensive fuels; and
- Also includes re-powering, upgrading instrumentation, controls, and/or equipment

Transport

- Improvements in vehicle fuel efficiency by the introduction of new technologies
- Changes in vehicles and/or fuel type, for example, switch to electric cars or fuel cell vehicles (CNG/ Bio fuels)
- Switch of transport mode, *e.g.* changing to less carbon intensive means of transport like trains (Metro in Delhi); and
- Reducing the frequency of the transport activity

Methane Recovery

- Animal waste methane recovery & utilization
  * Installing an anaerobic digester & utilizing methane to produce energy
- Coal mine methane recovery
  * Collection & utilization of fugitive methane from coal mining;
- Capture of biogas
  * Landfill methane recovery and utilization
- Capture & utilization of fugitive gas from gas pipelines
- Methane collection and utilization from sewage/industrial waste treatment facilities

Industrial Process Changes

- Any industrial process change resulting in the reduction of any category greenhouse gas emissions

Cogeneration

- Use of waste heat from electric generation, such as exhaust from gas turbines, for industrial purposes or heating (*e.g.* Distillery-Molasses/bagasse)

Agricultural Sector

- Energy efficiency improvements or switching to less carbon intensive energy sources for water pumps (irrigation)
- Methane reductions in rice cultivation
- Reducing animal waste or using produced animal waste for energy generation (see also under methane recovery) and
- Any other changes in an agricultural practices resulting in reduction of any category of greenhouse gas emissions

Indian Scenario—Favouring Points

(*a*) India-high potential of carbon credits

(*b*) India can capture 10% of Global COM market

(*c*) Annual revenue estimated range from US$1 0 million to 330 million

(*d*) Wide spectrum of projects with different sizes

(*e*) Vast technical human resource

(*f*) Strong industrial base

(*g*) Dynamic, transparent & speedy processing by Indian DNA (NCDMA) for host country approval

(*h*) MoU Signed between MoP and GTZ (Oct 2006)—Indo German Energy program (IGEN)

- Baseline CO_2 Emissions from Power Sector already in place-first COM country
- Improvement in EE
- COM in Power Sector

Policies and Way Ahead

Greenhouse gas abatement policy design is exceedingly difficult because GHG emissions result from nearly all modern human activities. It involves every sector of the economy as well as habits and choices of individuals. Economics is more than just a study of business, it is the science which studies human behavior as a relationship between aspirations and the scarce means to reach those goals. Individuals make decisions every day that influence the amount of greenhouse gases that enter the atmosphere. If a stable climate is one objective among the many to which society aspires, then economics is a tool well-suited to understand how those decisions are made and how efficient and effective outcomes can be reached.

Indian Forum

India is a Party to the United Nations Framework Convention on Climate Change (UNFCCC) and the objective of the Convention is to achieve stabilization of greenhouse gas concentrations in the atmosphere at a level that would prevent dangerous anthropogenic interference with the climate system.

To strengthen the developed country commitments under the Convention, the Parties adopted Kyoto Protocol in 1997, which commits developed country Parties to return their emissions of greenhouse gases to an average of approximately 5.2% below 1990 levels over the period 2008-12.

The Seventh Conference of Parties (COP-7) to the UNFCCC decided that Parties participating in CDM should designate a National Authority for the CDM and as per the CDM project cycle, a project proposal should include written approval of voluntary participation from the Designated National Authority of each country and confirmation that the project activity assists the host country in achieving sustainable development.

Accordingly, the Central Government constituted the National Clean Development Mechanism (CDM) Authority for the purpose of protecting and improving the quality of environment in terms of the Kyoto Protocol.

The CDM Authority has the powers:

(*a*) to invite officials and experts from government, financial institutions, consultancy organizations, non-governmental organizations, civil society, legal profession, industry and commerce, as it may deem necessary for technical and professional inputs and may co-opt other members depending upon need.

(*b*) to interact with concerned authorities, institutions, individual stockholders for matters relating to CDM.

(*c*) to take up any environmental issues pertaining to CDM or Sustainable Development projects as may be referred to it by the Central Government, and

(*d*) to recommend guidelines to the Central Government for consideration of projects and principles to be followed for according host country approval.

As discussed above, India has a vast opportunity to explore in terms of CDM and carbon-credits. Through its giant ongoing Infrastructure projects and projects on non-conventional energy sources, a new phase of development is still to be observed, moderate start of which has already begun.

The present scenario of CO_2 Emission

The emissions of CO_2 go unabated in countries like U.S.A., Australia, Japan, Canada, France, Germany, U.K., Russia, Saudi Arabia, Italy, China and even in India. USA tops the test with 5902 million metric tonnes of annual emissions with per capacity emission of 19.78 tonnes, followed by Australia with 417 million metric tonnes and 20.58 tonnes total emission of CO_2 as per capita emission respectively.

The table 9.2 clearly exhibits of CO_2 emission in top 12 countries of the world.

Total 9.2. Annual CO_2 emissions and per capita emission of top 12 countries of the world

S.No.	Country	Total Emissions (million metric tonnes)	Per capita emission (tonnes)	Rank per capita emission
1.	U.S.A.	5902	19.78	II
2.	Australia	417	20.58	I
3.	Japan	1246	9.73	VII
4.	Canada	614	18.81	III
5.	France	417	6.60	X
6.	Germany	857	10.40	VI
7.	U.K.	585	9.66	VIII
8.	Russia	1704	12.0	V
9.	Saudi Arabia	424	15.70	IV
10.	Italy	468	8.05	IX
11.	China	6017	4.58	XI
12.	India	1293	1.16	XII

From the point of per capita emission of CO_2 Australia tops the test with 20.58 tonnes, followed by U.S.A. (19.78), Canda (18.81), Saudi Arabia (15.70), Russia (12.0), where as China (4.58) and India (1.16) tonnes per capita emission stand at the XI & XII positions respectively.

Responses of the above Countries at Climate Change Copenhagen Meet (2009)

U.S.A. : Refused to rectify Kyoto protocol, has blocked all efforts at Copenhagen, now offers meager cuts.

Australia : Rectified the prototol only in 2007, since then has worked to undermine it and wants a new treaty.

Japan : Failed to meet 1990 commitments and was foot trigger till Hatoyama said 25% cut in emissions.

Canada : Instead of cutting emission, they increased by 30% since 1990, now wants to scuttle Kyoto.

France : Pushing for trade sanctions against developing countries who do not fall in line ine emission cuts.

Germany : Failed to pull its weight and convince U.S. not to be deal breaker, emissions have started rising.

Russia : Got away lightly on the protocol because of forests still clean up on carbon emissions front.

Saudi Arabia : Have played double game by ensuring oil prices stay low enough to thwart alternate fuel technologies.

Italy : Its emission rose sharply and it is proving to be the bad boy of Europe, unwilling to commit anything.

China : Taking shelter behind developing countries China's emission have exploded, offers cut only now.

India : Per capita emissions are eighteen times lower than the U.S.A., has offered cut but wants finance from west.

REFERENCES

1. Clean Development Project Opportunities in India, TERI New Delhi, January 2001.
2. Kyoto Protocol to the United Nations Framework Convention on Climate Change, United Nations, 1998.

CHAPTER 10

Chemistry of MIC with Reference to Bhopal Gas Disaster

Dr. S.A. Iqbal[1]

The world had never heard of Bhopal, a town with a population of just under one million people, located in central India and the capital city of the State of Madhya Pradesh, until the morning of the 3rd of December 1984. From that day onwards for at least a week, Bhopal dominated the front pages of the press world-wide and also received extensive television and radio coverage. Ironically, the Bhopal disaster occurred towards the end of the Golden Jubilee celebrations of Union Carbide India Limited; Today, almost everyone everywhere knows what Bhopal is, where it is, and what happed there.

The Bhopal nightmare happened on the night of December 3, 1984, but it was no nightmare. It was real, methyl isocyanate (MIC) gas leaked from three storage tanks of the Union Carbide Factory exposing thousands of inhabitants. It was the worst ever industrial disaster. The number of dead (estimates range from 2,000 to over 5,000 but the precise number is not known) exceeds the death reported for ALL the industrial accidents in the past forty years. The nightmare was, mercifully, a brief one for the dead. But for the survivors who were exposed to the deadly MIC gas, the nightmare may be lifelong and even beyond because of the serious genetic effects.

It is really ironic to think that the solution was so simple. A balloon to show the wind direction and the simple action of breathing through a wet towel could have saved hundreds of lives.

Methyl isocyanate is an intermediate used in the manufacture of carbamate type of pesticides. Carbon monoxide, CO, obtained by partial oxidation of coal is combined with chlorine gas in presence of activated carbon to form phosgene, $COCl_2$. Phosgene is a poisonous gas and was used as a chemical weapon in World War I. Phosgene and methyl amine combine to form methyl isocyanate, the product (MIC) is stored in tanks for further use to produce the carbamate insecticide.

$$\underset{\text{Phosgene}}{COCl_2} + \underset{\text{methyl amine}}{CH_3NH_2} - \underset{\text{methyl isocyanate}}{CH_3NCO} + \underset{\text{hydrochloric acid}}{2HCl}$$

Inhalation of MIC causes burning sensation in the eyes, removes oxygen from the lungs and death due to choking. The threshold limiting value (TLV) for MIC for man is 0.02 ppm and the LD_{50} (LD = lethal dose) is 2 ppm in air.

[1]Department of Chemistry, Saifia Science College, (Barkatullah University), Bhopal (India)

Table 10.1. Survey of Selected Air Pollution Episodes

Date	Location	Pollutants	Effects
December 1-5, 1930	Meuse Valley, Belgium	Particulates, SO_2 (9-38 ppm)	63 morethan deaths, cough, chest pain, eye and nasal irritation-all age groups.
1 December 5-9, 1952	London, UK	Particulates (4000 fig m^{-2}) SO_2(1.3ppm)	4000 morethan deaths
December 5- 10, 1961	London, UK	SO_2 (2.0 ppm)	700 morethan deaths
January 29-February 12, 1963	New York, USA	Particulates, SO_2 (0.5 ppm)	200-400 morethan deaths
November 24-30, 1966	New York, USA	Particulates SO_2(1.02 ppm)	1 68 morethan deaths
December 3rd, 1984	Bhopal, India	Methyl Isocyanate (MIC)	10000 morethan deaths, Blindness all age groups.

The Bhopal catastrophe is certainly not merely a case of the proverbial 'Once-in-a-100 years' event. It was not a freak accident. The Bhopal plant was unprofitable, mainly due to poor capacity utilization. In fact discussions had taken place before December 3rd 1984 on dismantling the plant (Economic Times, 1985). But this never happened.

PHYSICO-CHEMICAL PROPERTIES OF MIC

Physical Data

1. Molecular weight: 57.1
2. Boiling point (760 mm Hg): 39°C (102° F)
3. Specific gravity (water = 1): 0.96.
4. Vapour density (air = 1 at boiling point of methyl isocyanate): 2.0.
5. Melting point: -80°C (-112°F)
6. Vapour pressure at 20°C (68° F) : 348 mm Hg.
7. Solubility in water, g/100g water at 20°C (68° F): 6.7 (reacts slowly).
8. Evaporation rate (butyl acetate = 1): 26.8.

Reactivity

1. ***Conditions contributing to instability:*** Elevated temperatures may cause methyl isocyanate to polymerize and burst container.
2. ***Incompatibilities:*** Contact with water causes for mation of carbon dioxide and memylamine gases. The reaction is much more rapid in the presence of acids, alkalies, and amines. Contact with iron, tin, copper (or salts of these elements.) and with certain other catalysts (such as triphenylarsenic oxide, triethylphosphine, and tributyltin oxide) may cause violet polymerization.
3. ***Hazardous decomposition products:*** Toxic gases and vapors (such as hydrogen cyanide, oxides of nitrogen, and carbon monoxide) may be released in a fire involving methyl isocyanate.
4. ***Special precautions:*** Methyl isocyanate will attack some forms of plastics, rubber, and coatings.

Flammability

1. *Flash point:* Less than - 18°C (0°F) (Open cup).
2. *Auto ignition temperature:* 535°C (995°F).
3. *Flammable limits in air, % by volume:* Lower: 5.3; Upper: 26.
4. *Extinguishant:* Carbon dioxide, dry chemical, foam.

Warning Properties

Odor Threshold : The Documentation of TLV's states that human beings exposed to 0.4 ppm methyl isocyanate for periods up to 5 minutes could not detect the odour. Even at a concentration of 2 ppm, the odour was not perceived.

Irritation Levels : At a concentration of 0.4 ppm, eye, nose, and throat irritation was not experienced by human subjects who were exposed for period up to 5 minutes. The Documentation of TLV's states that at 0.2 ppm no odour was detected, but the subjects experienced irritation and lacrimation. At 4 ppm the symptoms of irritation were more marked. Exposure was unbreakable at 2.1 ppm."

Evaluation of Warning Properties : Since the gas and irritation thresholds of methyl isocyanate are within 3 times the permissible exposure limit, for the purpose of this guideline, methyl isocyanate is treated as a material with poor warning properties.

EMERGENCY FIRST AND PROCEDURES

In the event of an emergency, institute firs aid procedures and send for first aid or medical assistance.

- **Eye Exposure :** If liquid methyl isocyanate gets into the eyes, wash eyes immediately with large amounts of water, lifting the lower and upper lids occasionally. Get medical attention immediately. Contact lenses should not be worn when working with this chemical.
- **Skin Exposure :** If liquid methyl isocyanate gets on the skin, immediately flush the contaminated skin with large amounts of water. If liquid methyl isocyanate soaks through the clothing, remove the clothing immediately and flush the skin with large amounts of water. If irritation persists after washing, get medical attention immediately.
- **Breathing :** If a person breathes in large amounts of methyl isocyanate, move the exposed person to fresh air at once. If breathing has stopped, perform artificial respiration. Keep the affected person warm and at rest. Get medical attention as soon as possible.
- **Swallowing :** When liquid methyl isocyanate has been swallowed and the person is conscious, give the person large quantities of water immediately. After the water has been swallowed, try to get the person to vomit by having him touch the back of his throat with his finger. Do not make an unconscious person vomit: Get medical attention immediately.
- **Rescue :** Move the affected person from the hazardous exposure. If the exposed person has been overcome, notify someone else and put into effect the established emergency rescue procedures. Do not become a casualty. Understand the facility's emergency rescue procedures and know the locations of rescue equipment before the need arises.

SPILL, LEAK AND DISPOSAL PROCEDURES

- Persons not wearing protective equipment and clothing should be restricted from areas of spills or leaks until cleanup has been completed.
- If methyl isocyanate is spilled or leaked, the following steps should be taken:
  1. Remove all ignition sources.
  2. Ventilate area of spill or leak.
  3. For small quantities, absorb on paper towels. Evaporate in a safe place (such as a fume hood). Allow sufficient time for evaporating vapors to completely clear the hood ductwork. Burn the paper in a suitable location away from combustible materials. Large quantities can be reclaimed or collected and atomized in a suitable combustion chamber equipped with an appropriate effluent gas cleaning device. Methyl isocyanate should not be allowed to enter a confined space, such as a sewer, because of the possibility of an explosion. Sewers designed to preclude the formation of explosive concentration of methyl isocyanate vapors are permitted.

Methyl Isocyanate Chemistry

The electrophilic nature of isocyanates as indicated by their reactivity with a variety of nucleophilic reagents is well-recognized. The most characteristic reactions of isocyanates, and methyl isocyanate (MIC) in particular, are those involving compounds with an active hydrogen. These include hydroxylated compounds such as water, alcohols, phenols, and oximes, and also amino compounds, among others.

Water reacts exothermically with MIC to form 1,3 di-methylurea (I) and 1, 3, 5-trimethylbiuret (II) with evolution of carbon dioxide. Excessive water leads to prominently I and limited amounts of water (excess MIC) to

Excess water

$$\underset{\text{MIC}}{CH_3N{=}C{=}O} + H_2O \rightarrow [CH_3NHCOOH] \rightarrow \underset{\text{Methyl amine}}{CH_3NH_2} + CO_2$$

$$CH_3NH_2 + CH_3N{=}C{=}O \rightarrow \underset{\text{I}}{CH_3NHCONHCH_3}$$

Excess MIC

$$\underset{\text{I}}{CH_3NHCONHCH_3} + CH_3\,N{=}C{=}O \rightarrow \underset{\text{II}}{CH_3NHCONCH_3CONHCH_3}$$

Methyl isocyanate is also known to react with itself under a variety of conditions to form cyclic dimer and trimer, as well as, Linear polymers, with cyclotrimetrization being the most common. Although linear polymers of isocyanates are formed at very low temperatures, methyl isocyanate does not normally polymerize under these conditions. Hexamethylenetetramine is claimed to be a unique catalyst to effect the polymerization of MIC, even at ambient temperatures. Highly purified MIC, even at ambient temperatures. Highly purified MIC, however, has been observed to polymerize . Iron has been found to inhibit polymerization.

REPORT ON THE DISCUSSION ON RESIDUES OF SEVIN TAR AND NAPHTHOL TAR FOR SAFE DISPOSAL

The residues in the premises of M/s. Union Carbide Factory Bhopal, namely Sevin tar & Naphthol tar whose incineration composition and the evolved gases are reported to have carbon dioxide, Oxygen, Carbob-monoxide, Hydrogen, Methane, Nitrogen. The analysis dta submitted by Indian Institute of Chemical Technology (IICT). Hydrabad further reveals that the composition of these gases at 100°C, were found as—

Table 10.2

	Naphthol tar	Sevin tar
Carbon dioxide	5.6%	3.4%
Oxygen	18.4%	21.6%
Carbon monoxide	Negligible	0.2%
Hydrogen	0.6%	5.6%
Methane	3.4%	0.4%
Nitrogen (by difference)	72.0%	68.54%
& Solids	90%	—

1. The presence of the above gases which are formed at 100°C are highly inflammable particularly carbon monoxide and methane. The maximum temperature which reaches in summer is about 50°C. Thus the residues are safe & may not catch fire at atmospheric temperature.

2. If the residues incidently catch fire the above mentioned hazardous gases may get produced which will be harmful to human beings living around the factory.
3. There is a possibility that the residues if come in contact with water may under-go hydrolysis & methylamine, carbondioxide & some polymers may be formed.
4. As far as the report of IICT which was made available, indicate that the atmospheric chlorine may react with the residues at elevated temperature & may from dioxane ($C_4H_8O_2$) which is also very toxic & hazardous to human health.
5. (*i*) The materials should not be kept in open to avoid coming in contact with atmospheric gases, water & weather effects.

   (*ii*) Trial experiments should be carried out to see the probability of formation of dioxane in presence of chlorine in different concentrations.

   (*iii*) However the committee* feels it very essential that the residual material should be sent to three reputed chemical laboratories such as Indian Institute of Chemical Technology (IICT), Hydrabad, National Chemical Laboratory (NCL), Pune and National Environment Engineering Research Institute (NEERI), Nagpur, so that an exact composition is known in the present state. This will help in drawing further conclusions.

STUDIES ON SAFE DISPOSAL OF NAPHTHOL & SEVIN TAR SAMPLES

Following studies have to be carried out with a view to suggest safe method of disposal of hazardous waste material available in M/s Union Carbide Plant premises at Bhopal.

1. Proximate analysis
2. Ultimate analysis
3. Calorific value
4. Fabrication of an equipment for carrying out combustion studies.
5. Gas analysis by Orsat & GLC.
6. Studies are envisaged to be done in an open Bhatta for safe disposal of these materials.
7. Discussion with outside agencies in this regard.

The following methods are generally employed for the disposal of waste materials :

1. Incineration
2. Bio-degradation
3. Landfill
4. Dumping in sea

Among the above mentioned methods, Incineration is considered to be the most popular one. However, Philip W Powers has reported in Environmental Technology Handbook No. 4, page 387-407 that pesticide containing Nitrogen when incinerated will invariably produce oxides of nitrogen and along with the photochemical smog. Both are quite dangerous and are generally recommended to be avoided in a work place. Further, Sevin is not destroyed effectively even on incineration. The reported data is given as under :

Temperature : 600°C, 700°C, 800°C, 900°C and 1000°C.

Decomposed : 88.7%, 88.8%, 88.8%, 89.1% and 89.5%.

From the above, it could be seen that depending on the temperature partial decomposition of Sevin is possible via incineration.

*Meeting of M.P.P.C.B. Bhopal regarding UCIL—residues.

Further, there is a possibility that the dioxame may also be formed as traces of chlorine will also be available in atmosphere which may react with the aromatic compound and may yield dioxane which is many more times toxic than sevin-considering the above facts, it is not advisable to have the incineration of sevin lying at Union Carbide, Bhopal premises.

Generally, it is degraded by Microbial degradation when brought in alkaline solutions. Some biodegradation work has to be done in this regard, and then the detailed methods will be suggested.

The Problem

MIC is an aliphatic isocyanate with excellent warning properties as it is highly irritating on mucous membranes of the eyes and respiratory system. It is a hazardous material by all means of contact. Adverse effects are evident almost immediately but edema can be delayed upto a couple of hours.

Relative Health Hazard Rating

	Swallowing	Skin Penetration	Skin Irritation	Breathing	Eye
Methyl Isocyanate	4	3	4	5	5

3. Minor residual injury may result in spite of prompt treatment.
4. Minor residual injury may result in spite of prompt treatment.
5. Major residual injury is likely in spite of prompt treatment.

Symptoms

- Tearing (Lacrymation) of eyes.
- Easily recognized nasal irritation and burning of threat at low levels.
- Intensely irritating to breathe, causes chest pain, coughing, choking.
- Severe bronchospasm and asthma like breathing (wheezing). Chemical pneumonitis/pulmonary oedema.
- Oral and contact poison.
- Skin contact causes severe burns.
- Serious injury to eyes even when diluted to 10% concentration.
- Idiocyneratic reactions.
- Sensitization may develop.

First Aid

Any area of human contact/skin/eyes should be flushed with plenty of water for at least 15 mins. while removing contaminated clothing and shoes.

If inhaled move to fresh air. Give oxygen if breathing is difficult. If not breathing give artificial respiration. Call a physician at once.

Treatment

Adverse effects are evident almost immediately but edema can be delayed upto a couple of hours. The victim should be observed for a variable period of time depending on severity of exposure and degree of symptoms.

Lungs

Treatment should be symptomatic. Also it will be helped by IPPB with Bronkosol, Bronchial dilatation and oxygen treatment if he/she has bronchial spasm. Chemical contamination on eye, skin, etc. should be washed out immediately. Being a high irritant and so unbearable, that duration of exposure is usually short.

Treatment of MIC—Pulmonary Complication

1. I. V. Hydrocortisone 1 gm. of equivalent (Prednisolone) stat and after 24 hours.
2. I.P.P.B. (Oxygen inhalation) intermittently.
3. If cyanide poisoning is suspected use Amyl Nitrite.
   If no effect - Sod. Nitrite - 0.3 gms. and Sod. Thiosulphate 12.5 gms.
   I.V. in 2-4 minutes.
   Can be repeated (Half dose) as a prophylactic measure.
4. Observation period 24-48 hours.
5. Meth haemoglobin should be kept below 40%.

OCCUPATIONAL HEALTH GUIDELINE FOR METHYL ISOCYANATE

This guideline is intended as a source of information for employees, employers, physicians, industrial hygienists, and other occupational health professionals who may have a need for such information. It does not attempt to present all data; rather, it presents pertinent information and data in summary form.

Substance Identification

- Formula: CH_3NCO
- Synonyms: None
- Appearance and odour: Colorless liquid with a sharp color that causes tears.

Permissible Exposure Limit (PEL)

The current OSHA standard for methyl isocyanate is CO_2 part of methyl isocyanate per million parts of air (ppm) averaged over an eight-hour work shift. This may also be expressed as 0.05 milligram of methyl oxyanate per cubic meter of air (mg/m^3).

Health Hazard Information

- Routes of exposure
- Methyl isocyanate can affect the body if it is inhaled or it comes in contact with the eyes or skin. It can also affect the body if it is swallowed.
- Effects of overexposure.

Inhalation of methyl isocyanate vapors may cause irritation of the yes, nose, throat and lungs. Cough, shortness of breath, increases phlegm and chest pain may be present. The liquid splashed in the eyes may cause permanent damage. The liquid splashed on the skin may cause irritation. Exposure to methyl isocyanate may cause a person to become allergic to it so that extremely low levels of exposure may cause an asthmatic attack.

Reporting Signs and Symptoms

A physician should be contacted if anyone develops any signs or symptoms and suspects that they are caused by exposure to methyl isocyanate.

Recommended Medical Surveillance

The following medical procedures should be made available to each employee who is exposed to methyl isocyanate at potentially hazardous levels:

A. Initial Medical Examination:

A complete history and physical examination: The purpose is to detect pre-existing conditions that might place the exposed employee at increased risk, and to establish a baseline for future health monitoring. Persons with a history of asthma, allergies, or known sensitization to methyl isocyanate would be expected to be at increased risk from exposure. Examination of the eyes and respiratory tract should be stressed. The skin should be examined for evidence of chronic disorders.

14″ × 17″ chest roentgenogram: Methyl isocyanate causes lung damage in animals. Surveillance of the lungs is indicated.

FVC and FEV (I Sec): Methyl isocyanate is a respiratory irritant. Persons with impaired pulmonary function may be at increased risk from exposure. Periodic surveillance is indicated.

B. Periodic Medical Examination:

The aforementioned medical examinations should be repeated on an annual basis, except that an x-ray is necessary only when indicated by the results of pulmonary function testing or by signs and symptoms of respiratory disease.

Summary of Toxicology

Methyl isocyanate vapor is an intense lacrimator and irritates the eyes, mucous membranes and skin. It can cause pulmonary irritation and sensitization. In rats exposed for 4 hours, the LC50 was 5 ppm; effects were injury to the lungs and subsequent pulmonary edema. Exposure of humans to high concentrations can cause cough, dyspnea, increased secretions, and chest pain. Isocyanates cause pulmonary sensitization in susceptible individuals; should this occur, further exposure should be avoided, since extremely low levels of exposure may trigger an asthmatic episode; cross sensitization to unrelated materials probably does not occur. Experimental exposure of four human subjects for 1 to 5 minutes caused the following effects: 0.4 ppm, no effects; 2 ppm, lacrimation, irritation of the nose and throat; 4 ppm, symptoms of irritation more marked: 21 ppm, unbearable irritation of eyes, nose, and throat. A cotton plug saturated with the liquid was applied to the ear of a rabbit for 30 minutes and caused erythema, edema, necrosis, and perforation: a few drops of the liquid on the car of a rabbit caused destruction of tissue. The liquid in contact with the eye may cause permanent damage.

REFERENCES

1. International Organization of Consumers Unions, 1985. The Lessons of Bhopal—A Community actions resources manual on hazardous technologies. Penag, Malaysia, August 1985, pp. 151.
2. Kharbanda, O.P. 1985. The Bhopal Nightmare. *Chem. Econ. and Eng. Rev.* 17. 40-41.
3. Kharbanda, O.P., 1985, Another Bhopal? Never Again. *The Chem. Engr.* (UK) July/August 1985, p. 40.
4. McCarrol, J. and W. Bradley. 1966. Excess Mortality as an Indicator of Helath Effects of Air Pollution. *Am. J. Public Health.* 56, pp. 1933-1942.
5. *McGraw-Hill Encyclopaedia of Environmental Sciences,* 1974, McGraw-Hill Book Company, New York.
6. Sheldon, J.M., R.C. Lovell, and K.P. Mathews. 1967. *A Manual of Clinical Allergy.* 2nd ed. W.B. Saunders Company, Philadelphia.
7. *The Chemical Industry After Bhopal.* An International Symposium held in London on 7/8th Nov. 1985, IBC Technical Services Ltd., London.
8. Wagner, R.H., 1978. *Environment and Man.* 3rd, ed., W.W. Norton & Company, Inc., New York.
9. Wise, W., 1971. *Killer Smog.* Audubon-Ballantine Books, New York.

CHAPTER **11**

Acid Rain

Ms. Asma Fareed[1] and Ms. Meena Iqbal[2]

Introduction

Acid rain *is* rain or any other form of precipitation that *is* unusually acidic. It has harmful effects on plants, aquatic animals, and infrastructure. Acid rain is mostly caused by human emissions of sulfur and nitrogen compounds which react in the atmosphere to produce acids. In recent years, many governments have introduced laws to reduce these emissions.

Definition

"Acid rain" is a popular term referring to the deposition of wet (rain, snow, sleet, fog and cloud water, dew) and dry (acidifying particles and gases) acidic components. A more accurate term is "acid deposition". Distilled water, which contains no carbon dioxide, has a neutral pH of 7. Liquids with a pH less than 7 are acidic, and those with a pH greater than 7 are basic. "Clean" or unpolluted rain has a slightly acidic pH of about 5.2, because carbon dioxide and water in the air react together to form carbonic acid, a weak acid (pH 5.6 in distilled water), but unpolluted rain also contains other chemicals.

$$H_2O\ (1) + CO_2\ (g) \rightarrow H_2CO_3\ (aq)$$

Carbonic acid then can ionize in water forming low concentrations of hydronium ions:

$$2H_2O\ (1) + H_2CO_3\ (aq) \ \square\ CO_3(aq) + 2H_3O^+(aq)$$

The extra acidity in rain comes from the reaction of primary air pollutants, primarily sulfur oxides and nitrogen oxides, with water in the air to form strong acids (like sulfuric and nitric acid) . The main sources of these pollutants are industrial power-generating plants and vehicles.

History

Since the Industrial Revolution, emissions of sulphur dioxide and nitrogen oxides to the atmosphere have increased. In 1852, Robert Angus Smith was the first to show the relationship between acid rain and atmospheric pollution in Manchester, England. Though acidic rain was discovered in 1852, it wasn't until the late 1960s that scientists began widely observing and studying the phenomenon. The term "acid rain" was generated in 1972. Canadian Harold Harvey was among the first to research a "dead" lake. Public awareness of acid rain in the U.S increased in the 1970s after the New York Times promulgated

[1]Asian Biotech. Research Centre, Bhopal-462 001 (India)

reports from the Hubbard Brook Experimental Forest in New Hampshire of the myriad deleterious environmental effects demonstrated to result from it.

Occasional pH readings in rain and fog water of well below 2.4 (the acidity of vinegar) have been reported in industrialized areas. Industrial acid rain is a substantial problem in Europe, China, Russia and areas down-wind from them. These areas all burn sulfur-containing coal to generate heat and electricity. The problem of acid rain not only has increased with population and industrial growth, but has become more widespread. The use of tall smokestacks to reduce local pollution has contributed to the spread of acid rain by releasing gases into regional atmospheric circulation. Often deposition occurs a considerable distance downwind of the emissions, with mountainous regions tending to receive the greatest deposition (simply because of their higher rainfall). An example of this effect is the low pH of rain (compared to the local emissions) which falls in Scandinavia.

What is Acid Rain?

Normal, unpolluted rain would contain almost pure water (H_2O)in which some carbon dioxide (CO_2)some ammonia(NH_3)originating from organic matter and existing in water as NH_4 and varying but small amounts of cations (Ca^{++}, Mg^{++}, K^+, and Na^+) and anions (Cl^-,SO_4^-,) would be dissolved.

Although the pH of pure water is neutral 7.0 the pH of normal "unpolluted" rain is usually 5.6. In other words it is already acidic. Such rain however, is considered "normal", and only when the pH of rain or snow is below 5.6 it is considered acidic or acid rain.

Acid rain is the result of human activities, primarily the combustion of fossil fuels and the melting of the sulphide ores. These activities release in the atmosphere large quantities of the sulphur and nitrogen oxides which when in contact with atmosphere moisture are converted in to two of the strongest acids known and fall to the ground in rain and snow.

Acid rain exerts a variety of influences by greatly increasing the solubility of all kinds of molecules and by directly or indirectly affecting many forms of life. The adverse effects of acid rain on the micro organisms, plants and fishes of rivers and lake have been well documented.

Emissions of Chemicals Leading to Acidification

The most important gas which leads to acidification is sulfur dioxide. Emissions of nitrogen oxides which are oxidized to form nitric acid are of increasing importance due to stricter controls on emissions of sulfur containing compounds. 70 Tg(S) per year in the form of SO_2 comes from fossil fuel combustion and industry, 2.8 Tgs(S) from wildfires and 7-8 Tg(S) per year from volcanoes.

Natural Phenomena

The principal natural phenomena that contribute acid-producing gases to the atmosphere are emissions from volcanoes and those from biological processes that occur on the land, in wetlands, and in the oceans. The major biological source of sulfur containing compounds is dimethyl sulfide.

Acidic deposits have been detected in glacial ice thousands of years old in remote parts of the globe.

HUMAN ACTIVITY

The Coal-fired Gavin Power Plant in Cheshire, Ohio

The principal cause of acid rain is sulfur and nitrogen compounds from human sources, such as electricity generation, factories, and motor vehicles. Coal power plants are one of the most polluting. The gases can be carried hundreds of kilometres in the atmosphere before they are converted to acids and deposited. In the past, factories had short funnels to let out smoke, but this caused many problems locally; thus, factories now have taller smoke funnels. However, dispersal from these taller stacks causes pollutants to be carried farther, causing widespread ecological damage.

Air pollutants and acids generated by industrial activities are now entering forests at an unprecedented scale and rate, greatly adding to these stresses carried over from the past. Many forests in Europe and North America now receive as much as 30 times more acidity than they would if rain and snow were falling through a 'pure' atmosphere. Ozone levels in many rural areas of Europe and North America are now in the range known to damage trees. Despite air quality improvements made during the seventies, the average concentration of sulphur dioxide in many areas is high enough to reduce the growth of the trees.

Forests and Other Vegetation

Needle sand leaves yellow and drop prematurely from branches, and tree crowns progressively become thin, and timately, trees die. Even trees that show no visible sign of damage may be declining in growth and productivity. More over, the tendency of the acid rain to leach nutrients from sensitive soils may under mine the health and productivity of forest long into the future. Taken together, these direct and indirect effects threaten not only future wood supplies but the integrity of whole ecosystems on which society depends.

The main pollutants that cause acid rain, industries eject Sulphur Dioxide and Nitrogen Oxide in to the atmosphere which becomes part of the clouds and forms acid rain.

Acid Rain and Forest

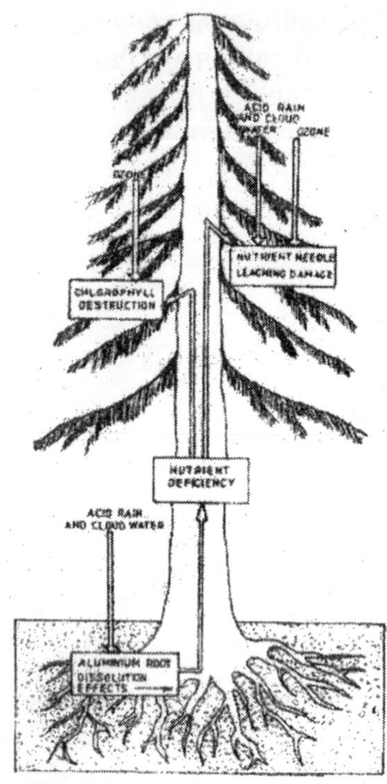

The role of acid rain and other forms of air pollution is under intensive investigation. Inspite of sedimentations of the forest damage, however a firm link has not been established. One can get some idea of the difficulties by contrasting by air pollutants. Smelters and chemical plants that emit sulphur dioxide, oxides of nitrogen or fluoride compounds are often girdled by dead timber. In such cases there is a clear correlation between tree damage, a specific pollution source and a threshold concentration of the pollutant. The forest that are dying, incondrast, are far from any source and are exposed to pollutants in concentrations well below the levels previously reported to injure trees. If air pollution and specifically acid rain, place a part in forest decline, it probably does so less as a lethal agent than as a stress.

The Chemistry of Acid Rain

Acid rain is a direct consequence of the atmosphere's self-cleansing nature. The tiny droplets particles and soluble trace gases. When precipitation coalesces from cloud water, it washes the impurities out of the atmosphere. Sulphur dioxide (SO_2) and oxides of nitrogen emitted in to the atmosphere are chemically converted in to forms that are readily incorporated into cloud droplets sulphuric and nitric acids.

The process that convert the gases in to acids and wash them from the atmosphere began operating long before human being started to burn large quantities of fossil fuels; sulphur and nitrogen compounds are also released by natural processes such as volcanism and the activity of soil bacteria, but human economic activity has made the reactions vastly more important .they are triggered by sun light and depend on the atmosphere's abundant supply of oxygen and water.

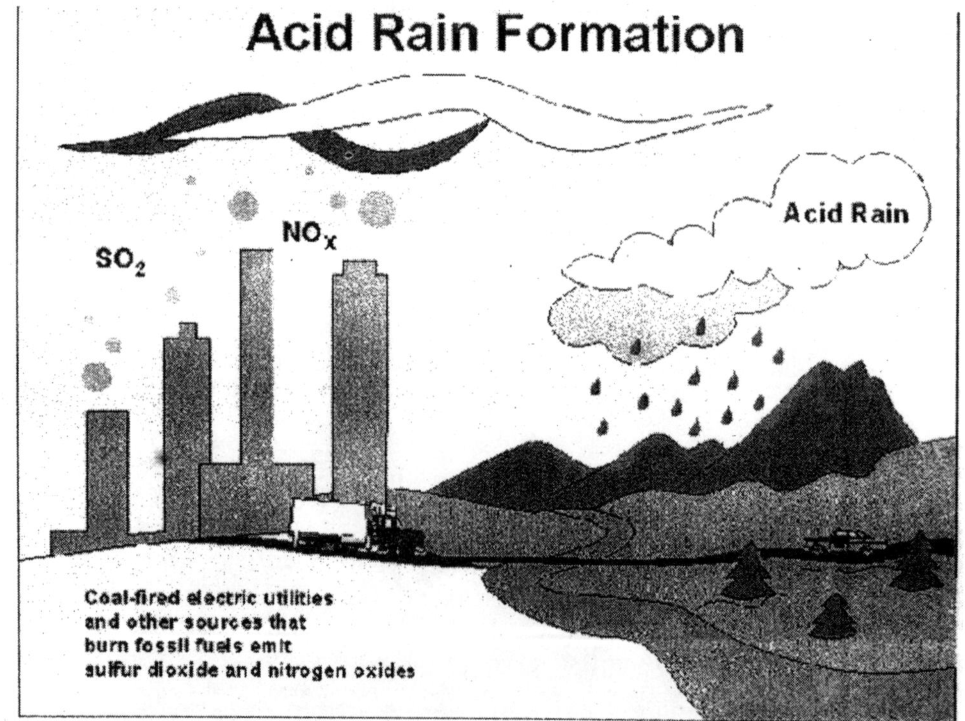

Fig. 11.1. Illustrates the formation of acid rain from sulphur dioxide and Oxides of nitrogen given off by industry and vehicle.

In exchange the ions of the calcium, magnesium and other metals found in many soils take the place of the acid's hydrogen ions.

Chemistry in Cloud Droplets

When clouds are present the loss rate of SO_2 is faster than can be explained by gas phase chemistry alone. This is due to reactions in the water droplets.

Hydrolysis

Sulfur dioxide dissolves in water and then, like carbon dioxide, hydrolyses in a series of equilibrium reactions:

$$SO_2\,(g) + H_2O \rightleftharpoons SO_2.H_2O$$
$$SO_2.H_2O \rightleftharpoons H^+ + HSO_3^-$$
$$HSO_3^- \rightleftharpoons H^+ + SO_3^{2-}$$

Oxidation

There are a large number of aqueous reactions that oxidize sulfur from S(IV) to S(VI), leading to the formation of sulfuric acid. The most important oxidation reactions are with ozone, hydrogen peroxide and oxygen (reactions with oxygen are catalyzed by iron and manganese in the cloud droplets).

Acid Deposition

Processes involved in acid deposition (note that only SO_2 and NO_X play a significant role in acid rain).

Wet Deposition

Wet deposition of acids occurs when any form of precipitation (rain, snow, etc.) removes acids from the atmosphere and delivers it to the Earth's surface. This can result from the deposition of acids produced in the raindrops (see aqueous phase chemistry above) or by the precipitation removing the acids either in clouds or below clouds. Wet removal of both gases and aerosols are both of importance for wet deposition.

Dry Deposition

Acid deposition also occurs via dry deposition in the absence of precipitation. This can be responsible for as much as 20 to 60% of total acid deposition.

This occurs when particles and gases stick to the ground, plants or other surfaces.

Adverse Effects

This chart shows that not all fish, or the insects that they can tolerate the same amount of acid; for example, frogs can tolerate water that is more acidic (i.e., has a lower pH) than trout.

Acid rain has been shown to have adverse impacts on forests, freshwaters and soils, killing insect and aquatic life forms as well as causing damage to buildings and having impacts on human health.

Surface Waters and Aquatic Animals

Both the lower pH and higher aluminum concentrations in surface water that occur as a result of acid rain can cause damage to fish and other aquatic animals. At pHs lower than 5 most fish eggs will not hatch and lower pHs can kill adult fish. As lakes and rivers become more acidic biodiversity is reduced. Acid rain has eliminated insect life and some fish species, including the brook trout in some lakes, streams, and creeks in geographically sensitive areas, such as the Adirondack Mountains of the United States. However, the extent to which acid rain contributes directly or indirectly via runoff from the catchments to lake and river acidity (i.e., depending on characteristics of the surrounding watershed) is variable. The United States Environmental Protection Agency's (EPA) website states: "Of the lakes and streams surveyed, acid rain caused acidity in 75 percent of the acidic lakes and about 50 percent of the acidic streams".

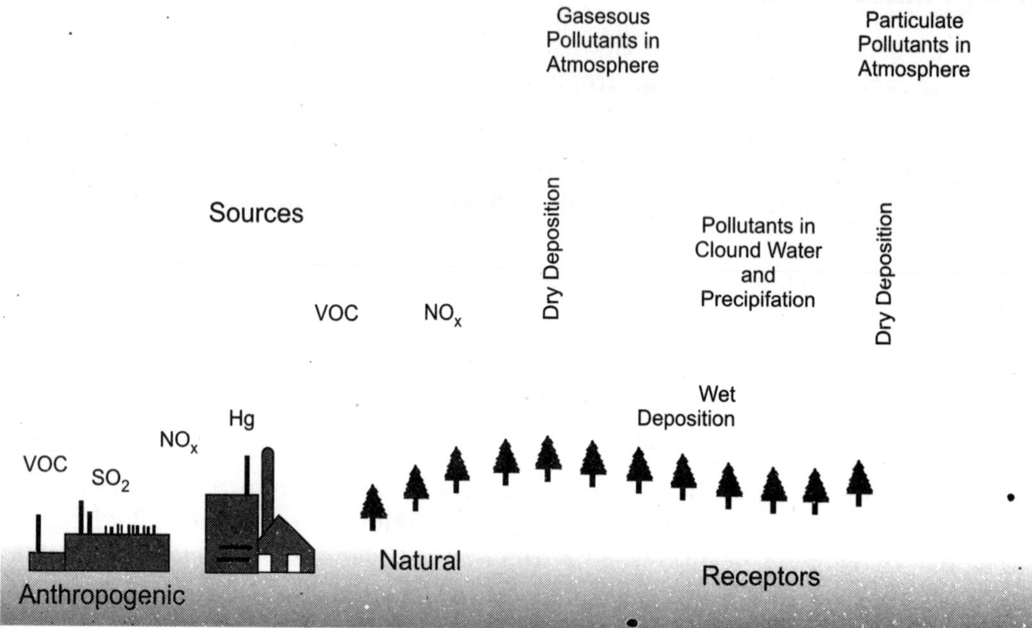

Fig. 11.2. Graphical representation of the adverse effect.

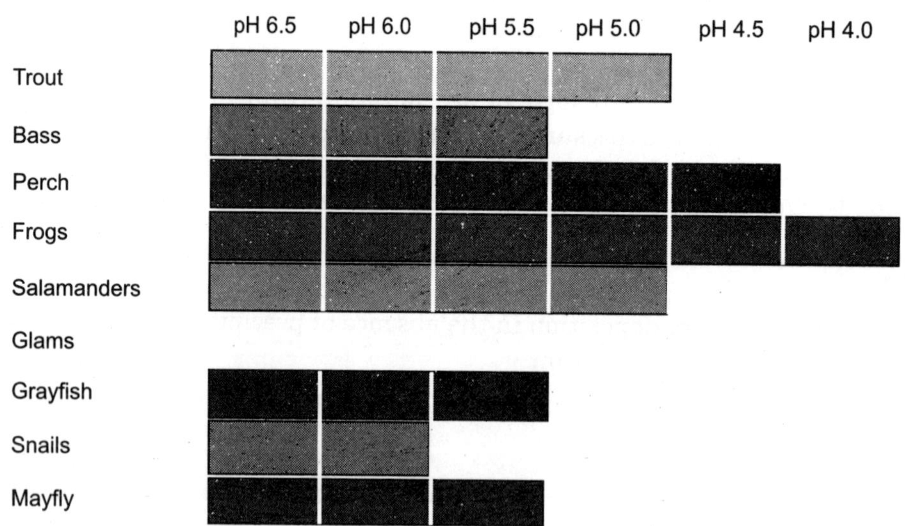

Fig. 11.3. Graphical representation of the adverse effect.

Soils

Soil biology and chemistry can be seriously damaged by acid rain. Some microbes are unable to tolerate changes to low pHs and are killed. The enzymes of these microbes are denatured (changed in shape so they no longer function) by the acid. The hydronium ions of acid rain also mobilize toxins, e.g. aluminum, and leach away essential nutrients and minerals.

$$2H^+ \text{ (aq)} + Mg^{2+} \text{ (clay)} \rightleftharpoons 2H^+ \text{ (clay)} + Mg^{2+}\text{(aq)}$$

Soil chemistry can be dramatically changed when base cations, such as calcium and magnesium, are leached by acid rain thereby affecting sensitive species, such as sugar maple (Acer saccharum).

The reaction cycle occurs in the troposphere the lowest 10 or 12 km's of the atmosphere. It begins at a photon of sun light strikes a molecule of ozone which may have mixed downward from the ozone layer in the stratosphere or may have been formed in the troposphere by the action of nitrogen, and carbon contains pollutants. The result is formation molecule of oxygen and highly reactive oxygen atom, which combines with a water molecule to form two hydroxyl radicals .This scarce but active species transform nitrogen dioxide into nitric acid and initiate the reactions that transform sulphur dioxide into sulphuric acid.

The sulphuric acid and nitric acids formed from gaseous pollutants can make their way into clouds acid is also formed directly in cloud droplets from dissolves in existing cloud droplet.

The sulphuric and nitric acids in cloud droplets can give them an extremely low pH. The sulphuric nitric acids formed from gaseous pollutant can easily make there way into clouds.

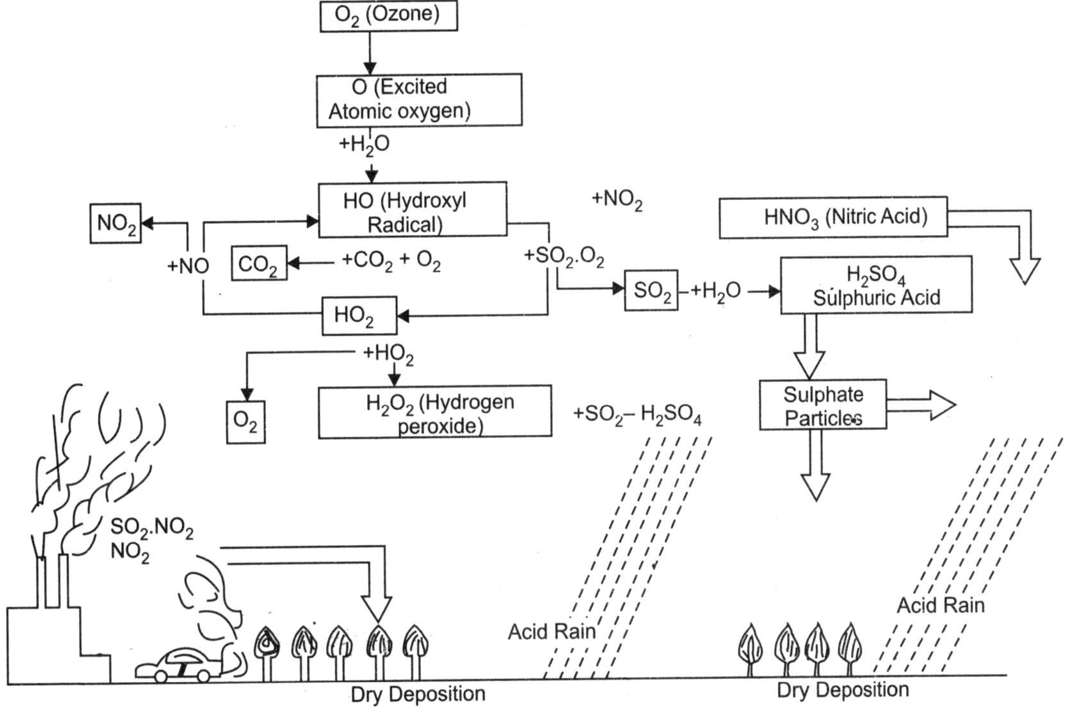

Fig. 11.4.

Nitric acid gas readily dissolves in existing cloud droplets. Sulphuric acid formed through gas-phase reactions condenses to form microscopic droplets, from roughly 1 to 2 micrometers diameter. Some these sulphate particles settle to the ground in a process known as dry deposition. Most of them however, are incorporated in clouds. Moister readily condenses on an existing surface—a condensation nucleus—and sulphate particles are ideal condensation nuclei. They grow into cloud droplets containing water collected near the base of clouds in the eastern U.S during the summer typically has a pH of about 3.6, but values as low as2.6 have been recorded. In some parts of the world the pH of fog has fallen a low as 2 which is the same 1 as that of the acid—lemon juice.

The acid rain may fall hundreds of miles from the pollution source. Wherever it lands, it under goes a new round of physical and chemical changes, which reduce the acidity and alter the chemical characteristics of the water that eventually, reaches lakes and streams. Alkalin soils, such as soils rich in lime-stone, can neutralize the acid directly. In the slightly acidic soils typical of the evergreen forest exposed to acid rain in the U.S., Canada and Europe two other processes can counter the effects of acid deposition. The acid can be immobilized as the soil or vegetation retains sulphate and nitrate ions. It can also be buffered through a process that is known as cation exchange.

Other Adverse Effects

Acid rain can also cause damage to certain building materials and historical monuments. This results when the sulfuric acid in the rain chemically reacts with the calcium compounds in the stones (limestone, sandstone, marble and granite) to create gypsum, which then flakes off.

$$CaCO_3 (s) + H_2SO_4 (aq) \rightleftharpoons CaSO_4 (aq) + CO_2 (g) + H_2O (1)$$

This result is also commonly seen on old gravestones where the acid rain can cause the inscription to become completely illegible. Acid rain also causes an increased rate of oxidation for iron. Visibility is also reduced by sulfate and nitrate aerosols and particles in the atmosphere.

Affected Areas

Particularly badly affected places around the globe include most of Europe (particularly Scandinavia with many lakes with acidic water containing no life and many trees dead) many parts of the United States (states like New York are very badly affected) and South Western Canada. Other affected areas include the South Eastern coast of China and Taiwan.

Potential Problem Areas in the Future

Places like much of South Asia (Indonesia, Malaysia and Thailand), Western South Africa (the country), Southern India and Sri Lanka and even West Africa (countries like Ghana, Togo and Nigeria) could all be prone to acidic rainfall in the future.

PREVENTION METHODS

Technical Solutions

In the United States, many coal-burning power plants use fuel gas desulfurization (FGD) to remove sulfur-containing gases from their stack gases. An example of FGD is the wet scrubber which is commonly used in the U.S. and many other countries. A wet scrubber is basically a reaction tower equipped with a fan that extracts hot smoke stack gases from a power plant into the tower. Lime or limestone in slurry form is also injected into the tower to mix with the stack gases and combine with the sulfur dioxide present. The calcium carbonate of the limestone produces pH-neutral calcium sulfate that is physically removed from the scrubber. That is, the scrubber turns sulfur pollution into industrial sulfates.

In some areas the sulfates are sold to chemical companies as gypsum when the purity of calcium sulfate is high. In others, they are placed in landfill. However, the effects of acid rain can last for generations, as the effects of pH level change can stimulate the continued leaching of undesirable chemicals into otherwise pristine water sources, killing off vulnerable insect and fish species and blocking efforts to restore native life.

Automobile emissions control reduces emissions of nitrogen oxides from motor vehicles.

REFERENCES

1. Abrahasmen G., Effect of acidic deposition on forest soil and vegetation, 305, 1984.
2. Balzhiser R.E. and Kurt E., Yeager; Coal-fired Power Plants for the future, 258, 92-96, 1958.
3. Blank, L.W., A New Type of Forest Decline in Germany, Nature, 314, 311-314, 1985.
4. Buttler, J.D., Air Pollution Chemistry, Academic Press, New York (1979).
5. Park, Chris, C. Acid Rain—Rheloric and Reality, Meuthen & Co. New York, (1987).
6. Woodman, M. James, E.B. Cowling, Airborne Chemicals and Forest Health Environmental Science and Technology 21, 121-126.

CHAPTER 12

Role of Biomarkers in Assessment of Environmental Pollution

Dr. Ayesha S. Ali, Julia Mitra[1] and Dr. Sharique A. Ali[2]

Introduction

Environment pollution is introduction of contaminants in an environment that causes instability, disorder, harm or discomfort to the ecosystem *i.e.,* physical system with living organism or organisms. It is also defined as the contamination of the physical and biological components of the earth/atmosphere system to such an extent that normal environmental process are adversely affected. Pollutants are the elements of pollution, can be foreign substances or agents of natural occurrence;where they are considered as contaminants when they exceed natural levels. Humankind has had some effect upon the environment since the Paleolithic era during which the ability to generate fire was acquired. In the Iron Age, the use of tools led to the practice of metal grinding on a small scale and resulted in minor accumulations of discarded materials which got probably easily dispersed without too much impact. Human waste in this category would have polluted rivers or water sources to same degree. The first advanced civilization of Egypt, India and China increased the use of industrial goods marking the advent of commercialization and industrialization. Core samples of glaciers in Greenland, indicate increase in air pollution associated with anthropogenic metal production. Towards the end of the middle ages populations grew and concentrated more within cities, creating pockets of readily evident contamination. In certain places air pollution levels were recognizable as health issues and water pollution in population centers was a serious medium for disease transmission from untreated human wastes.

The earliest known writings concerned with pollution were Arabic medial treatises written between 9th and 13th centuries by physicians, such as Al-Kindi(Alkindus) Muhamnad ibn Zakariya Razi(Rhazes), Issac Irraeli Ben Soloman and others whose works covered a number of subjects related to pollution such as air contamination, water contamination, soil contamination and environment assessment of certain localities. Later on *Chicago* and *Cincinnati* were the first two American cities to enact laws ensuring cleaner air in 1881. Other cities followed around the country until early in the 20th century, when the short lived Office of Air Pollution was created under the Department of the Interior. Extreme smog events were experienced by the cities of *Los Angeles* and *Donora, Pennsylvania* in the late 1940s, serving as another public reminder. Air pollution continued to be a problem in England especially later during the industrial revolution, the famous London smog in 1952 is a serious reminder to us. This same city also

[1]Corresponding Author. Email:drshariqali@yahoo.com
[2]Department of Biotechnology Saifia Science College, Bhopal-462001 (India).

recorded one of the realer extreme cases of water quality problems with great sink on the river Thames in 1858, it was this industrial revolution that gave birth to environmental pollution as we know it today. (Crossley 1970; Hogan *et al.*, 1973, Gupta *et al.*, 1982 ; Springer *et al.*, 2001;)

Environmental Pollution and Its Kinds

Pollution comes from both natural and manmade sources, principal stationary pollution sources include chemical plants, petrochemical plants, nuclear waste disposal activity, incinerators large live stocks farms (dairy, cow, pig, poultry etc.), PVC factories, metal production factories, plastic factories and other heavy industries. Agricultural air pollution comes from the contemporary practices which include clear felling and burning of natural vegetation as well as spraying of pesticides and herbicides (Spring *et al.*, 1962). Some of the more common soil contaminants are chlorinated hydrocarbons, heavy metals(such as chromium cadmium etc.) found in rechargeable batteries and lead found in fuels and other usables. There have also been some unusual releases of polychlorinated dibenzo dioxides, commmanly called dioxins for simplicity, such as TCDD[14] (Beychok *et al., 1987*). Though globally manmade pollutants from combustion construction, mining, agriculture and warfare are increasingly significant in the pollution equation. Motor vehicle emission is one of the leading causes of air pollution and apparently China, United States, Russia, Mexico, & Japan are the world leaders in air pollution emission. *(Environmental Performance Report 2001; Environmental Performance Report 2001;* Pollution and Society Marisa)

There are various kinds of pollutants however heavy metals and pesticides are the principal agents of environmental pollution. A heavy metal is a member of an ill-defined subset of elements that exhibit metallic properties, which would mainly include the transition metal, some metalloids, lanthanides and actinides The term heavy metals has been called “meaningless and misleading” in an IUPAC technical report due to the contradictory definitions and its lack of a “coherent scientific basis.” There is an alternative term toxic metal for which no consensus of exact definitions exist either various heavy metals occur naturally in the ecosystem with large variations in their concentrations, such as copper, iron, lead, mercury, cadmium, cobalt, cromium, nickel, zinc etc. (Duffus, 2002), consequently their effects on different organisms have been investigated.

To understand the relationship of living organisms with heavy metals it is necessary to evaluate the reasons of living organism requiring varying amounts of metals such as iron copper, manganese, molybdenum and zinc including the basic various biological activities in nature which have symbiotic relationships with metals, however excessive levels can be damaging to the organisms. For example heavy metals such as mercury, cadmium and lead are toxic which have no known vital or beneficial effects in organisms and their accumulation over time in the bodies of animals can cause serious illness However certain elements that are normally toxic to certain organisms or under certain conditions can be beneficial to some like-vanadium, tungsten, and even cadmium. Some of the heavy metals cited above can be very dangerous to health or to the environment as they cause corrosion, pollute catalysts, or can be extremely carcinogenic or toxic affecting, among others mainly the central nervous system, blood tissues, skin, bones or teeth (Zevenhoven *et al.,* 2001).

Heavy metal pollution can arise from many sources but most commonly it arises from the purification of metals through smelting processing or preparation of nuclear fuels. Unlike organic pollutants heavy metals do not decay and thus pose a different kind of challenge for remediation. Microorganisms as well as plants have been tentatively used to remove several heavy metals such as mercury, chromium nickel and other especially plants which exhibits hyper accumulation can be used to remove heavy metals even from soils by concentrating them; for instance incase of mining and tilling where the vegetation can be consequently incinerated to recover the heavy metals.

Metals like zinc, molybdenum, copper and others are very useful for us, as several mines and smelters in Indiahave been found to be active since last 100 years (Craddock *et al.,* 1983), which are an essential minerals of exceptional biological & public health importance, whose deficiency affects about millions of people in the developing world these are associated with many diseases. Consumption of excess of zinc can cause ataxia, lethargy and even copper deficiency. Using millions of tonnes of metallic zinc and zinc oxide from the 12th to 16th centuries. The use of zinc oxide for skin treatment dates back to Avicenna's zinc which was preferred treatment for a variety of skin conditions. The use of zinc in creams, calamine cream, anti-dandruff shampoo antiseptic ointments become a very normal deed. But although zinc is an essential requirement for good health, excess use of zinc can be harmful. Excessive absorption of zinc suppress copper and the U. S. Food & Drug administration (FDA) has stated that heavy metals especially zinc damages nerve receptors in the nose, which even can cause insomnia. Reports of insomnia were also observed as early in 1930s, recently the FDA has stated that consumers should stop using zinc based intranasal cold products, as the he loss of smell can be life threatening because people with impaired smell cannot detect making gas or smoke and cannot tell if food is spoiled before they eat it. *(Emsley 2001;* Prasad 2003; Boudreaux *et al.,* 2008;)

The other most threatening pollutant is the pesticide which is a substance or mixture of substances used to kill a pest. It is any chemical substance or mixture of substances intended for preventing, destroying, repelling or mitigating any pest. Pesticides may be chemical substances, biological agents (virus or bacteria) antibacterial or disinfectants used against pests, which include insects and mammals. Although there are several benefits from the uses of pesticides, there are also drawbacks, such as potential toxicity to humans and other animals as these substances may be administered to animals for the control of pests.

Historically since thousands of years humans have utilized pesticides to protect their crops, the first known pesticide was elemental sulphur dusting used in summer about 4,500 yrs ago by the 15th century, toxic chemicals such as arsenic, mercury and lead were being applied to crops to kill pests. In 17th century nicotine sulfate was extracted from tobacco leaves for use as an insecticide. The 19th century saw the introduction of two more natural pesticides, pyrethrum is derived from chrysanthmus, and rotenone from the roots of tropical vegetables(Miller *et al.,* 2002). Until the 1950s arsenic based pesticides were dominant. Paul Muller discovered that organochlorines like DDT were very effective insecticides, became globally dominant but were replaced in the US by oraanophosphates and carbamates in 1975. (Ritter 2009), thereafter herbicides became more common in the 1980s.

In the 1950s manufacturers began to produce large amounts of synthetic pesticides and their use widespread became very common. Some sources consider the 1940s & 1950s to have been the start of "pesticides". 75% of all pesticides in the world are used in developing countries and their use is still increasing. In the 1960s, it was discovered that DDT was preventing many fish eating birds from reproducing. Rachel Carson wrote the best selling book Silent Spring about biological magnification of pesticides. The agricultural use of DDT is now banned under the Stockholm convention as a persistent organism pollutant, but it is still used in some developing nations to prevent malaria and tropical diseases by spraying on interior walls to kill or repel mosquitoes. Many pesticides can be grouped into chemical families. Prominant insecticides families include organochlorine or organophosphates and carbamates. Their toxicities vary greatly but they have been phased out because of their persistence and potential to bioaccumulate particularly the organophosphates and carbamates have largely replaced organochlorines. Both operate through inhibiting the enzyme *acetylene cholinesterase* allowing acetylcholine to transfer nerve impulses indefinitely and causing a veriety of symptoms such as weakness or paralysis ;being quite toxic to all invertebrates and vertebrates (Kamrin *et al.*, 1997; Daly *et al.*, 1998; Miller 2004; Murphy *et. al.,* 2005; Lobe 2006).

Pesticides are used to control organisms which are considered harmful e.g., they are used to kill mosquitoes that can transmit potentially deadly diseases like West Nile viral fever, yellow fever and malaria. They can also kill bees, wasps or ants that can cause allergic reactions. Herbicides can be used to clear roadside weeds, trees and bushes. Sometimes uncontrolled pests which transmit and mould can damage structures such as houses. Pesticides are used in grocery stores and food storage facilities to manage rodents and insects that infect food such as cereals and grains. Each use of pesticides carries some associated risk, which can be assessed to a level deemed acceptable by pesticide regulatory agencies such as the United States Environmental Protection Agency (USEPA) & the Pest management Regulatory Agency (PMRA) Canada or the NEERI (India).

A recent (October 2007) survey has linked breast cancer with exposure to DDT prior to puberty showing that pesticides and other chlorinated hydrocarbons may enter the human food chain affecting animal tissues. The use of pesticides raises a number of environmental concerns. Over 95% of herbicides reach a destination other than their target species, including non-target species, air, water and soil For example pesticides use causes pollution of water and land, where some byproducts in the form of organic pollutants persist and contribute to water as well as soil contamination affecting the bio-diversity coupled with lower soil quality reduced nitrogen fixation, a significant pollinator decline can reduce habitat especially for birds and thus threatening nature. (Walter *et al.*, 2000; Eric *et al.*, 2001; Miller 2004; Rusty, Palmer & Hackenberg 2007).

By seeing consistently deteriorating effects of pesticides towards our health, the American Medical Association recommended limited exposure to pesticides and has suggested using safer alternatives. The WHO & UN environment programmes have estimated that more than, 3 million workers in agriculture in the developing world experience severe poisoning from pesticides, about 18,000 of whom die annualy (Miller 2004;Council on Scientific Affairs, American Medical Association. 2007). Organophospahate pesticides have increased in use because they are less damaging to the environment and less persistent than the organoclorine pesticides (Jaga *et al.,* 2003). However these are also associated with acute health problems such as abdominal pain, headache, dizziness, nausea, vomiting as well as skin and eye problems (Ecobichon 1996). Additionally many studies have indicated that with long term exposures health aspects such as respiratory problems, memory disorders, dermatologic conditions (Arcury 2003; Malley1997). cancer (Daniels 1997), depression(Beseler *et al.,* 2008), neurological effects(Kamel *et al.*, 2003; Firestone 2005), miscarriages and birth defects occur in vertebrates including human beings. (Schwartz *et al.*, 1986; Moses 1989; Eskenazi., *et al., 1999;* Das *et al., 2001;* Stallones *et al.,* 2002*;* García 2003; Van Maele-Fabry *et al.*, 2003; trong *et al.,* 2004).

According to researchers from the National Institute of health (NIH), licensed pesticides applicators who used chlorinated pesticides on more than 100 days in their life time were at greater risk of diabetes. An interesting study by Khan and Ali (1993) showed that a private pesticide formulating factory in India got several awards for following rules and regulations of environmental safety for their workers, however on actual research, the workers were found to suffer from serious abnormalities of haematological biochemical and neurological nature (Khan and Ali, 1993). Thus the paradox of making environmental laws and their implementation is a common practice in developing countries (Ali *et al.,* 1995; Ali & Ali 2004*)*.

Alternatives to pesticides are available which include new modified methods of cultivation, use of biological controls, pheromones and microbial and bio-pesticides, genetic engineering and modified methods of interfering with insect breeding and others such as application of composed yard wastes which has also been found to be control pests for example sterile male mosquitoes and their biotechnological use has been found quite effective in controlling malaria;thus reducing pesticides use. These innovative methods are becoming increasingly popular and often are safer than traditional chemical pesticides, release of other organisms that fight the pests is another example of an alternative to pesticide use. These organisms can include natural predators or parasites of the pests. Similarly the thermal treating of soil through steam can kill pests and increase soil health. (Montgomerry, 2008)

Up to here we came to know that some of the main etiological agents of environmental pollution are the heavy metals and pesticides, but there are some other factors which cause environmental pollution. And their introduction of contaminant into the environments that makes unhealthy habitation and on the extreme can even harm the ecosystem permanently. It can not only cause the disabilities but also psychological and behavioral disorders in people. Some of the other causes of pollution in environment are discussed below.

Air pollution, is the release of polluting chemicals and particulates into the atmosphere, which makes it unhealthy to breathe in such air. Common air pollutants are sulfur dioxide, chlorofluorocarbons (CFC), carbon monoxide, and nitrogen oxides produced by industry and motor vehicles. When hydrocarbons and nitrogen oxides react to sunlight, smog and photochemical ozones are formed when nitrogen oxides and hydrocarbons react to sunlight, causing severe problems of environmental pollution.

Water pollution, when waste products and potential contaminants such as metals pesticides and chemicals are released into rivers, ponds, lakes drainage systems and other water bodies they make the water unfit from quality point of view of the water for use by man or in habitation of water fauna and flora, we have problems of water pollution, as its consequences lasting for generations. *Soil Contamination,* occurs when chemicals are released by spill or underground leakage. Among the most significant soil contaminants are hydrocarbons, heavy metals, herbicides, pesticides and chlorinated hydrocarbons, organochemicals, plastics and several nondegenerable products. *Radioactive contamination,* resuls from 20th century activities in atomic physics, nuclear plants, carboratories natural source minings and others such as nuclear power generation and nuclear weapons, research, manufacture and deployment. *Noise pollution,* which encompasses roadway noise, aircraft noise, industrial noise as well as high-intensity sonar, noise, even daily chores including use of gadgets results in considerable noise pollution, and is the another source of environmental pollution. *Light pollution,* includes light trespass, over-illumination and astronomical interferences excessive use of office lights, leading to disturbed photoperiods for living organisms. *Visual pollution*, which can refer to the presence of overhead power lines, motorway billboards, scarred landforms, open storage of trash or municipal solid waste a new menace of over population in mega industrial cities. *Thermal pollution,* is a temperature change in natural water bodies caused by human influence, such as use of water as coolant in a power plant leading to geochemical disturbances all over the globe(Environmental Protection Agency).

In modern industrialized societies, fossil fuels (oil, gas, coal) have transcended virtually all imaginable barriers and firmly established themselves in our everyday lives, not only do we use fossil fuels for our obvious everyday needs (such as filling car) as well in the power generating industry they (specifically oil) they are present in a variety of products as well sorts of plastics, solvents detergent, lubricating oils and a wide range of chemicals for industrial use etc Fossil fuels also contribute to soil contamination and water pollution, for example, when oil is transported from the point of its production to further destination by pipelines, an oil leak from the pipeline may occur and pollute soil, and subsequently r when oil is transported by tanks via oceans oil spills may occur and pollute oceans making the fossil fuels are among the most serious sources of environmental poluution. Power generating plants and transport are also one of the biggest sources of fossil fuel pollution. Agriculture is worth mentioning as the largest generator of ammonia and methane emission resulting pollution of our environmental. Residential sector is another significant source of pollution generating solid municipal wastes that may end up in land fills or incinerators leading to soil contamination and air pollution. Environmental pollution is causing a lot of distress not only to humans but also animals, during which many animals species are becoming endangered & even becoming extinct. What is happening to several species of birds, amphibians and insects is well known.

Biomarkers for the Assessment of Environmental Pollution

The trans boundary nature of environmental pollution makes it even more difficult to manage our living: we cannot build stone walls along the borders of our country or put custom cabins at every point of entry to regulate the pollution and its flow into our country. Everything on our planet is interconnected, and while the nature supplies us with valuable environmental services without which we cannot exist, we all depend on each other's actions and the way we treat natural resources. It's widely recognized that we are hugely overspending our current budget of natural resources–at the existing rates of its exploitation, there is no way for the environment to recover in good time and continue "performing" well in the future. Longback, the environmentalists have warmed that the burdern and load of our activities will drown us, its only the help from the Almighty, that we are still surviving!

Perhaps we should adopt a holistic view of nature–it is *not* an entity that exists separately from us; the nature *is* us, we are an inalienable part of it, and we should care for it in the most appropriate manner. Only then can we possibly solve the problem of environmental pollution. The care of our own environment comes from not only table speaks, but actual research work done for months and years, to assess and evaluate our delicate factories of nature which are under constant threats of pollution from our own products of leisure, the heavy metals, pesticides, agriculture and industrial contaminants and alike. Scientific research which has always benefited mankind, has also been outstanding towards safety of our planet earth and its inhabitants, the concept of analysis has lead to the discovery of toxicological monitors particularly the *Biomarkers.*

In recent years we have observed a vast expansion in scientific literature in which the term 'biomarker' has been used which has grown in popularity day by day. With this growth has come a great increase in applications and a correspondingly increased diffuseness in the meaning of the term, to the extent that one of the major difficulties in the area of biomarker research is reconciling differing views on what constitutes an acceptable definition. The term biomarker has a significant and lengthy history. The meaning for which it is used clearly depends upon the context and this is reflected most clearly in the parameters of the database which is used as the basis for any search. Since the use of biomarkers as predictive tools in toxicity testing, and other medical testing- the focus was on biological, medical and toxicological uses of the term. Therefore, an appropriate database set was likely to be found in Medline and Toxline, so many databases were searched for the term "biomarker", with different results, we can observe the progress of a biomarker by the following flow diagram:

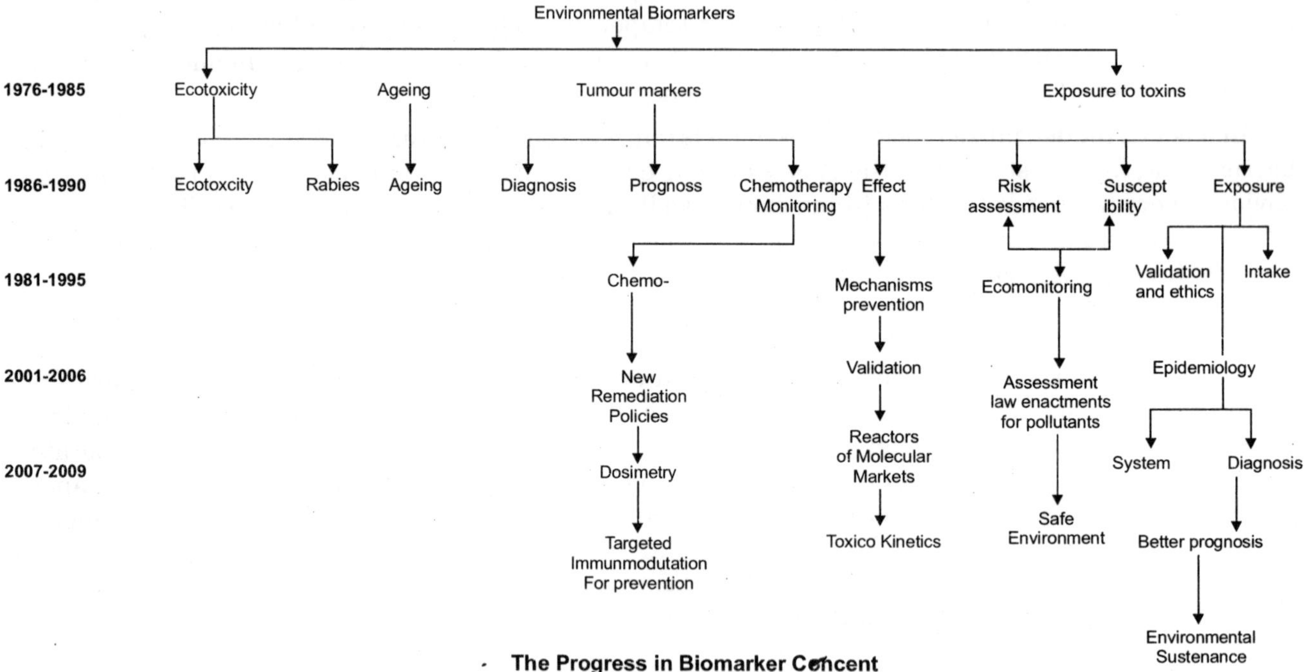

The Progress in Biomarker Concent

Biomarker', which is generally used in a broad sense to include almost any measurement reflecting an interaction between a biological system and a potential hazard, may be chemical, physical or biological (WHO, 1993). A biomarker is defined as a change in a biological response (ranging from molecular through cellular and physiological responses to behavioral changes) which can be related to exposure to or toxic effects of environmental chemicals (Peakall, 1994). Workers have redefined the terms 'biomarker', 'bioindicator' and 'ecological indicator', linking them to different levels of biological organization. They considered a biomarker as any biological response to an environmental chemical at the subindividual level, measured inside an organism or in its products (urine, faeces, hair, feathers, etc.), indicating a deviation from the normal status that cannot be detected in the intact organism. A bioindicator is defined as an organism giving information on the environmental conditions of its habitat by its presence. (van der Oost *et al.,* 2003)

Biomarkers applied in both the laboratory and the field, can provide an important linkage between laboratory toxicity and field assessment for the hazardous effects of environmental pollution, biomarker data may provide an important index of the total external load that is biologically available in the 'real world' exposure. In order to evaluate the strength and weaknesof biomarker, six criteria were proposed comprising the most important information that should be available or has to be established for each candidate biomarker (based upon the criteria formulated by (Stegeman *et al.*, 1992):

- The assay to quantify the biomarker should be reliable (with quality assurance (QA)), relatively cheap and easy to perform;
- The biomarker response should be sensitive to pollutant exposure and/or effects in order to serve as an early warning parameter;
- Baseline data of the biomarker should be well defined in order to distinguish between natural variability (noise) and contaminant-induced stress (signal);
- The impacts of confounding factors to the biomarker response should be well established;
- The underlying mechanism of the relationships between iomarker response and pollutant exposure (dosage and time) should be established;
- The toxicological significance of the biomarker, e. g. the relationships between its response and the (longterm) impact to the organism, should be established.

There are many studies in relation to ecological pollution assessment for example using earthworms and fishes as bioindicators from seasonally different areas and this has been recently shown by Ali *et al.*, (2007a, 2007b)which had from varying levels different status and types of contamination. The measured parameters were body length, body weight, Fulton condition index, hepatosomatic index and the enzymes In addition, several physical-chemical parameters of the soil and water were measured. A considerable seasonal variability of the biomarkers used was found. However and despite the variability found, the approach used was able to correctly rank the soil according to existing knowledge regarding their levels and main types of pollutants present. (Quintaneiro *et al.*, 2006; Monteiro *et al.*, 2007).

With exception of ALAD activity that according to our present knowledge only responds to lead, the most part of the biomarkers are responsive to several environmental contaminants. Even AChE activity that was considered as specific biomarker for organophosphate and carbamate pesticides, has been found to be inhibited for other classes of pollutants, such as metals, detergents and complex mixtures of pollutants in real scenarios. So, what can biomarkers tell about the specific chemicals responsible for the observed effects in the most part of situations? This is an important question in several practical works, for example, in ecosystems contaminated with a high number of different chemicals making highly cost or impossible to quantify all the chemicals present. (Guilhermino, 2007).

In recent years, biomarkers have proved to be good tools in detecting early biological changes caused by environmental pollution. Another biological early warning sign is the developmental instability during the ontogeny of an individual that is shown by many animals under various environmental stress situations (Leary & Allendorf 1989; Parsons & Pankakoski *et al* 1990 Biomarkers of human exposure, dose, and genetic effects in the general population have recently been studied with respect to exposures from environmental pollution (Reddy *et al.,* 1990; Perera *et al.*, 1992; Grzybowska *et al.,* 1993; Gallagher& Hemminki *et al.*, 1994; Autrup, Betti & Binkovi., *et al.,* 1995; Grzybowska *et al.,* 1997.*)*. Perera *et al.*, 1992 found that the exposure to environmental pollution in the Silesian region of Poland was associated with a significant increase of DNA adducts, SCE, and CA, as well as with a frequency of ras oncogene overexpression.

Biomarkers could also be related to an exposure that has already been identified or represents an alteration caused by the exposure that results in the disease. The most precarious situation is one in which the biomarker is related to some unknown factor that is also related to the exposure. This type of confounder, if unidentified, can decrease the validity of the association between the biomarker and the disease. The preceding account has briefly shown that there are various causes of environment pollution, but to detect this we can use different kinds of biomarkers which can be any kind of molecule indicating the living organisms, particularly in the field of geology and astrotology biomarkers are also known as biosignatures.

Types of Biomarkers

According to the NRC (1987), and WHO (1993), biomarkers can be subdivided into three classes:

Biomarkers of exposure : Covering the detection and measurement of an exogenous substance or its metabolite or the product of an interaction between a xenobiotic agent and some target molecule or cell that is measured in a compartment within an organism.

Biomarkers of effect : Including measurable biochemical, physiological or other alterations within tissues or body fluids of an organism that can be recognized as associated with an established or possible health impairment or disease.

Biomarkers of susceptibility : Indicating the inherent or acquired ability of an organism to respond to the challenge of exposure to a specific xenobiotic substance, including genetic factors and changes in receptors which alter the susceptibility of an organism to that exposure.

However the subdivision of biomarkers in the literature is rather confusing since biomarkers of exposure and those of effect are distinguished by the way they are used, not by an inherent dichotomy (Suter, 1993) The responses of biomarkers can be regarded as biological or biochemical effects after a certain toxicant exposure, which makes them theoretically useful as indicators of both exposure and effects. The bioaccumulation of certain persistent environmental contaminants in animal tissues may be considered to be a biomarker of exposure to these chemicals (NRC, 1987; WHO, 1993).

According to the definitions given by Van Gastel and Van Brummelen (1994), however, body burdens are not considered to be biomarkers or bioindicators since they do not provide information on deviations related to 'health'. Good biomarkers are sensitive indices of both environment pollutant bioavailability and early biological responses. Biomarkers may be used after exposure to dietary, environmental or occupational sources, to elucidate cause/effect and dose/effect relationships in health risk assessment, in clinical diagnoses and for monitoring purposes. Generally, biomarker responses are considered to be intermediates between pollutant sources and higher-level effects (Suter, 1990).

Biomarkers have been introduced in chronic disease epidemiology under the assumption that they could improve the investigation of health effects of air pollution and other exposures, by (i) improving exposure assessment, (ii) increasing the understanding of mechanisms, e. g. by measuring intermediate biomarkers, and (iii) allowing the investigation of individual susceptibility. Regarding the role of biomarkers, although there are examples of effective contribution of some of them to the understanding of the health effects of air pollution, there are still many aspects that need clarification, in particular reliability of markers. For example, 'bulky' DNA adducts have some degree of batch variation and inter-laboratory variationCurrently, we do not have direct studies on the effects of outdoor air pollution on biomarkers such as tumour mutations or promoter methylation in humans. Although studies on smokers or subjects exposed to indoor emissions Today better biomarkers are urgently needed to improve diagnosis, guide molecularly targeted therapy and monitor activity and therapeutic response across a wide spectrum of disease. Proteomics methods based on mass spectrometry hold special promise for the discovery of novel biomarkers that might form the foundation for new clinical blood tests, but to date their contribution to the diagnostic armamentarium has been disappointing. This is due in part to the lack of a coherent pipeline connecting marker discovery with well-established methods for validation. (Nader *et al.*, 2006).

There are clinical Biomarker –the biomarker which enable the characterization of patients population and quantification of the extent to which new drug reach the intented targets, alter proposed pathophysiological mechanism and achieve clinical outcomes. In genomics the biomarker challenge is to identify unique molecular signature in complex biological mixtures that can be unambiguously co-relate to biological events in order to validate novel drug responces. Biomarker can stratify patients populations or quantify drug benefit in primary prevention or disease modification studies in poorly served areas such as neurodegeneration and cancer. Clinically useful biomarkers are required to inform regulatory and therapeutic decision making regarding candidate drugs and their indications in order to help to bring medicines to the right patients faster than they are today.

Biomaker approach has been proven to be of high relevance in several types of studies. Despite of its high value, it has been criticized mainly due to high variability of results in long-term studies, low information regarding environmental contaminants and low ecological relevance. The results of several authors have been showing that it is possible to overtaken all these criticisms but more research is still necessary to find relationships among parameters at different levels of biological organization. Also, it is important to make clear to decision-makers the advantages of using the biomarker approach in addition to other types of studies. Another important aspect related with biomarkers use in coastal and brackish-water areas is to investigate why populations from close and similar ecosystems seem to respond differently to pollution and which mechanisms and life strategies have been developed to face and survive in polluted environments by gases, metals, chemicals and many others. Here, the so called "omics" approach can give a very important contribution (Ali S. A. 1984; Guilhermino, 2007).

Advances in methods and technology now enable construction of a comprehensive biomarker pipeline from six essential process components: candidate discovery, qualification, verification, research assay optimization, biomarker validation and commercialization. (Nader *et al.*, 2006). A biomarker can also be used to indicate exposes to various environmental substances in epidemiology and toxicology. Biomarkers have the ability to identify if an exposure has occurred, when dealing with exposure assessment, there is a special type of biomarker as biomarkers of effect are the quantifiable changes than are individual endures, which indicates an exposures to a compound and may indicate a resulting health effect for eg:- after exposure to DDT an organophosphate pesticides known to cause problem in the reproductive system an women may experience miscarriages, which can be linked to her previous exposure From breath to hair to saliva, almost every tissue in the body has been tested as a biomarker of exposure & almost every major environment pollutant can be identified by biomarker, including volatile organic chemicals, (VOCs) and metals like arsenic or lead. It all depend on the compound with the make up of its storage space. (*Biomarkers and surrogate endpoints: preferred definitions and conceptual framework* 2001).

APPLICATIONS OF SOME BIOMARKER STUDIES: MUTAGENICITY AND CYTOGENETIC EFFECTS IN HUMANS

Urinary mutagenicity was elevated in the *Salmonella* assay among non-smoking bus drivers exposed to polluted urban air, mainly traffic exhausts, as compared with mail carriers working in the same city. In addition, several but not all studies investigating cytogenetic effects in groups of healthy individuals from a wide variety of geographical locations with variable air pollution have reported positive findings, especially among traffic policemen (Chandrasekaran *et al.*, 1996; Zhao *et al.*, 1998; Knudsen *et al.*, 1999; Burgaz and Carere *et al.*, 2002; Leopardi *et al.*, 2003; Hrelia *et al.*, 2004.).

Biomarkers can be also useful to understand the mechanisms and to characterize high-risk groups. Although biomarkers are potentially useful, they have a number of important limitations over the years, many biomarkers have been developed that are claimed to be efficient at providing an early warning of deleterious effects on biological systems and for estimating biological effects due to contaminants Professional drivers are exposed to a higher level of ambient air pollution in an urban environment, as measured by total dust and PAH. Groups with extensive occupational exposure to ambient air pollution, especially that generated from motor vehicles, including diesel fumes, have an increased risk of lung, abdominal and laryngeal cancers. The increased cancer risk may be due to interaction of reactive chemical metabolites with DNA. It is possible to measure exposure to these genotoxic compounds present in ambient air by sensitive assays for DNA adducts, e.g. 32P-postlabelling. This analysis is the most sensitive for detection of bulky PAHDNA adducts and one of the most widely used biomarker assays for genotoxic exposure. Postlabelling has been used in a large number of studies on occupational exposures to PAH, for a review see and in a more limited number of studies on environmental exposure (Gupta *et al.*, 1982; Balrajan *et al.*, 1988; Hayes *et al.*, 1989; Steenlant *et al.*, 1990; Guberan *et al.*, 1992; Guillemin, Beach & Perara *et al.*, 1992; Nielsan *et al.*, 1996.).

Biomarker studies investigating *HPRT* gene mutations in healthy adults in association with air pollution did not find increased frequencies in peripheral blood lymphocytes. However, somatic *HPRT* mutation frequencies and aromatic DNA adducts were found to be correlated in cord blood samples from newborns of mothers living in polluted area in Poland, thus suggesting DNA damage in utero *HPRT* mutations and DNA adducts were not correlated in peripheral lymphocytes of the mothers. Air pollution contributes to the overall cancer risk, especially lung cancer, in urban areas. Though the proportion of lung cancers attributable to the air pollution is unknown, the bulk of studies point towards a smoking-adjusted risk in urban areas compared with rural areas of ~1.5. Molecular biomarkers of exposure have been developed during the last decades to measure the biological effective dose and to better describe the mechanisms of genotoxicants. (Farmer *et al.*, 1996; Kyrtopoulos *et al.*, 2001; Perera *et al.*, 2002).

Regarding the role of biomarkers, although there are examples of effective contribution of some of them to the understanding of the health effects of environmental pollution, there are still many aspects that need clarification, in particular reliability of markers. For example, 'bulky' DNA adducts have some degree of batch variation and inter-laboratory variation. Cancer develops due to smoke and environmental pollution occurs through a series of progressive pathological changes occurring in the respiratory epithelium. Molecular alterations, such as loss of heterozygosity, gene mutations and aberrant gene promoter methylation, have emerged as molecular biomarkers of lung carcinogenesis available for studies on groups or individuals at increased risk of cancer, smokers and involuntary smokers in particular. Indoor exposure to combustion emissions from smoky coal rich in PAHs is associated with lung cancer that carries *TP53* and *KRAS* mutations, with both genes exhibiting a mutation spectrum typical of PAHs. Gene promoter methylation is common in lung tumours and bronchial epithelium from lung cancer patients who are smokers, and also detectable in variable frequencies in bronchial epithelial cells from cancer-free smokers. Experimentally, particulate carcinogens such as diesel exhaust, carbon black and cigarette smoke have been observed in rodents to induce lung tumours exhibiting frequent aberrant methylation. (Hecht., 2003; DeMarini, Husgafvel, & Belinsky *et al.*, 2004; Peluso *et al.*, 2005)

Currently, we do not have direct studies on the effects of outdoor air pollution on biomarkers such as tumour mutations or promoter methylation in humans. Although studies on smokers or subjects exposed to indoor emissions rich in PAHs (polycyclic aromatic hydrocarbons) are valuable for understanding common mechanisms of lung carcinogenesis, not necessarily are all these biomarkers optimal for studying effects of outdoor air pollution in humans. In fact, it may be that downstream markers are not sensitive and specific enough for low-dose exposure to carcinogens, such as outdoor air pollution. Therefore, such biomarker studies contribute to carcinogenicity of outdoor air pollution mainly indirectly, via low-dose extrapolation from circumstances of higher exposure.

Histopathology has been recommended as a physiological approach to pollution investigation (Sinderrnann 1983). Hinton (1993) recommended this approach since it enables the researcher to examine multiple potential sites of injury rapidly and is applicable to field collection followed by examination. Histopathologic biomarkers have proven to be a valuable tool in laboratory (Meyers and Wdricks 1985, Wester and Canton 1991, Schwaiger *et al.* 1996) and field studies(Pierce *et al.* 1978, Malins, *et al.* 1998;Adams *et al.* 1989, Myers *et al.* 1994, Teh *et al.* 1997). Histopathological biomarkers at the tissue level of biological organization are intermediate benveen molecular and individual levels (Hinton *et al.*, 1992, Hinton 1997). They follow biochemical changes and resultant lesions can be observed histologically (Teh *et al.*, 1997). Histo-pathological biomarkers should be applied as one of a suite of biomarkers in field contaminant studies. The suite of biomarkers relate changes at one level to effects seen at other levels of biological organization (Depledge *et al.*, 1995, Luoma, 1996) and multiple biomarker parameters are required to provide information on the links between levels of biological organization

The environment is continuously loaded with foreign organic chemicals (xenobiotics) released by urban communities and industries. In the 20th century, many thousands of organic trace pollutants, such as polychlorinated biphenyls (PCBs), organochlorine pesticides (OCPs), polycyclic aromatic hydrocarbons (PAHs), polychlorinated dibenzofurans (PCDFs) and dibenzop-dioxins (PCDDs) have been produced and, in part, released into the environment. Since the early sixties mankind has become aware of the potential long-term adverse effects of these chemicals in general and their potential risks for aquatic and terrestrial ecosystems in particular. The ultimate sink for many of these contaminants is the aquatic environment, either due to direct discharges or to hydrologic and atmospheric processes (Stegeman and Hahn, 1994).

The presence of a xenobiotic compound in a segment of an aquatic ecosystem does not, by itself, indicate injurious effects. Connections must be established between external levels of exposure, internal levels of tissue contamination and early adverse effects. Many of the hydrophobic organic compounds and their metabolites, which contaminate aquatic ecosystems, have yet to be identified and their impact on aquatic life has yet to be determined. Therefore, the exposure, fate and effects of chemical contaminants or pollutants in the aquatic ecosystem have been extensively studied by environmental toxicologists. Ecological or environmental risk assessment (ERA) is defined as the procedure by which the likely or actual adverse effects of pollutants and other anthropogenic activities on ecosystems and their components are estimated with a known degree of certainty using scientific methodologies (Depledge and Fossi, 1994).

The National Research Council (NRC) has proposed criteria for a valid marker of (Environmental Tobacco Smoke) ETS in the air as follows. The markers a) should be unique or nearly unique for ETS so that other sources are minor in comparison, b) should be easily detectable, c) should be emitted at similar rates for a variety of tobacco products, and a should have a fairly constant ratio to other ETS components of interest under a range of environmental conditions encountered. Furthermore, the validity of a biomarker depends on the accuracy of the biologic fluid measurement in quantitating the intake of the marker chemical, which in turn may be influenced by individual differences in rates or patterns of metabolism or excretion, the presence of other sources (such as diet) of the chemical, and sensitivity and specificity of the analytic methods used to measure the chemical. Other issues of interest in assessing the risks of exposure to ETS are how well the biomarker indicates long-term exposure to ETS as well as whether a biomarker predicts the likelihood of ETS-related disease (NRC, ETC., 1986).

Conclusion

In the last decades, biomarkers have been widely used to diagnose environmental contamination and to assess their effects on living organisms including human beings. The so called "biomarker" approach has been proven to be very efficient in detecting early effects of pollutants that may have reflexes later in time in higher levels of biological organization levels. However, since these are parameters measured at a sub-individual level, their use in ecological risk assessment and other types of environmental studies have been questioned because they were considered as having low ecological relevance. The main criticisms that have been presented against the biomarker approach are the high seasonal variability that is frequently found in field studies based on biomarkers, the low information that this approach can provide regarding the specific chemicals responsible for the observed effects and the lack of ecological relevance of the biomarker approach. In principle, biomarkers could be valuable tools in toxicity testing protocols. Biomarkers that develop early in the course of repeat exposure, but that can be reliably correlated to the subsequent development of a lesion or an effect, would allow animal testing protocols to be refined and animal testing to be reduced. However, the mechanistic link to the lesion needs to be understood in order to define the uses and limitations of a particular biomarker. At the current time, there is insufficient understanding of mechanisms of toxicity to permit the identification of meaningful biomarkers for inclusion in routine toxicity testing protocols. For a very small number of compounds, some associations are known, such as liver glutathione content as an indicator of paracetamol hepatotoxicity, but these cannot be used in a general sense. Some of the "biomarkers" now used in clinical bio-chemistry, such as serum levels of liver transaminases, are more appropriately termed "organ function tests", and arguably should be defined as diagnostic markers rather than as biomarkers. Nevertheless, they are valuable because they demonstrate an important link between *in vivo* and *in vitro* testing, i. e. that the same endpoints can be used. (Mitchell *et al.*, 1973; Quintaneiro, *et al.*, 2006; Lehtonen *et al.*, 2006; Monteiro *et al.*, 2007).

Biomarkers have the potential either to be used as indicators of events at various stages in the progression of a toxic lesion, or to provide tools to test the relevance of proposed toxic mechanisms. The extent to which a biomarker is a predictive tool depends on its connection to the causal pathway leading from exposure to effect, and on the quantitative assessment of the consequences of changes in the biomarker in terms of toxicological endpoints. Our current understanding of mechanisms of toxicity is generally not sufficient to support development of biomarkers that could be early predictors of toxic effect. However, as our understanding increases, and with the enormous growth in the use of new techniques such as genomics and proteomics, which seek to determine the link between gene expression, gene function, protein synthesis and protein function and exposure to environmental and chemical agents, there is the potential to develop such predictive biomarkers in the future. The success of a genomics/proteomics based approach to biomarkers relies on the use of intelligent data manipulation, and on correlation between measurable changes in the functional properties of biological macromolecules and the properties of the cell, organ or organism. (Diana *et. al.*, 2000, Ali *et. al.,* 2007a, 2007b)

Although biomarkers currently cannot replace or reduce the use of animals in toxicity testing, future technological developments show great promise for making this possible. *In vitro* techniques may play an important role in the development of biomarkers of effect, as the greater control which is possible in *in vitro* studies facilitates investigation of mechanisms of toxicity, and therefore the identification of key events. A mechanistic understanding of the progression of a toxic effect is a necessary prerequisite for the use of biomarker approaches for the prediction of toxicity. It is possible to envisage a natural progression, in which a potential biomarker is identified during *in vitro* mechanistic studies, and its value tested in animal models, is then incorporated in *in vivo* testing protocols (refinement of animal testing), and is subsequently validated for use in *in vitro* toxicity tests (leading to reduction and replacement of animal testing). Once an acceptable level of use has been established for a new chemical, there is a potential for biomarkers to provide the continuity between initial toxicity testing and post-marketing surveillance, to ensure that the acceptable levels are adequate to protect the human population against adverse effects.

REFERENCES

1. Ali S. A., Khan S. A. (1993). Assessment of certain hematological factors in pesticides exposed factory workers, Bull. Environ. Contam. Toxicol, USA, Vol. No. 5, pp. 747-750.
2. Ali S. A. & Ali A. S. Environmental laws:- (2004)Have They Lost Their Teeth To Make A Bite ABST. National Seminar on Environmental Pollution and sustainable development 28-29.
3. Ali S. A. Khan I, Ali AS (2006). Friendly Earthworms Sciences Reporter, Vol. 43, No. 1, pp. 28-30.
4. Ali AS, Khan, SA Ali SA(-2007a) Toxicological Monitoring using Earthworms 64. In Int. Review of Toxicollogy, Pointer Publication Jaipur.
5. Ali A.S. Khan I and Ali S.A. (2007, b) Iranian J. of Toxicology Vol. II, No. 1 37-43.
6. Ali S. A. Khan SA and Ali AS(1995). Enforcement of Environmental Laws and Regulation Env. Cons. (Elsevier Science) Vol. 22 No (1) 77-78.
7. Ali S. A. (1984), Effect of Methyl Isocynate gas on the sleeping population of Bhopal Polish Journal of Environmental Science Vol. 23-17-21
8. *Arcury TA, Quandt SA, Mellen BG (2003). "An exploratory analysis of occupational skin disease among Latino migrant and seasonal farmworkers in North Carolina". J Agric Saf Health 9 (3): 221–32.*
9. Autrup H, Vestergaard AB, Okkels H. Transplacental transfer of environmental genotoxins: polycyclic aromatic hydrocarbons- albumin in non-smoking women, and the effect of maternal GSTMI genotype. Carcinogenesis 16:1305-1309 (1995).
10. Belinsky, S. A. (2004) Gene–promoter hypermethylation as a biomarker in lung cancer. *Nat. Rev. Cancer*, 4, 707–717.
11. Beychok, Milton R. (1987). "A data base for dioxin and furan emissions from refuse incinerators". Atmospheric Environment 21 (1): 29–36.
12. *Beseler CL, Stallones L, Hoppin JA,. (2008).* "Depression and pesticide exposures among private pesticide applicators enrolled in the Agricultural Health Study". *Environ. Health Perspect. 116 (12): 1713–19.*
13. Betti C, Davini T, Giannessi L, Loprieno N, Barale R. (1995) Comparative studies by comet test and SCE analysis in human lymphocytes from 200 healthy subjects. Mutat Res 345:201-207.
14. Binkovi B, Lewtas J, Mifkovi I, Lenfcek J, Grim R. DNA adducts and personal air monitoring of carcinogenic polycyclic aromatic hy[rocarbons in an environmentally exposed population. Carcinogenesis 16:1037-1046 (1995).
15. *Boudreaux, Kevin A. "Zinc, Sulfur". Angelo State University.* http://www. angelo. edu/faculty/kboudrea/demos/zinc_sulfur/ zinc_sulfur. htm. Retrieved 2008-10-08.

16. Chandrasekaran, R., Samy, P. L. and Murthy, P. B. (1996) Increased sister chromatid exchange (SCE) frequencies in lymphocytes from traffic policemen exposed to automobile exhaust pollution. *Hum. Exp. Toxicol.,* 15, 301–304.

17. Council on Scientific Affairs, American Medical Association. (1997). Educational and Informational Strategies to Reduce Pesticide Risks. Preventive Medicine, Volume 26, Number 2.

18. *Craddock, P. T. ; Gurjar L. K. ; Hegde K. T. M. (1983).* "Zinc production in medieval India". *World Archaeology 15 (2): 211.* http://www. jstor. org/pss/124653.

19. D. A. Crossley, *Roles of Microflora and fauna in soil systems*, International Symposium on Pesticides in Soils, Feb. 25, 1970, University of Michigan.

20. Diane J. Benford, A. Bryan Hanley, Krys Bottrill, Sarah Oehlschlager and, Michael Balls, (2000) Biomarkers as Predictive Tools in Toxicity Testing. ALTA28, pp 119-131.

21. Daly H, Doyen JT, and Purcell AH III (1998), *Introduction to insect biology and diversity*, 2nd edition. Oxford University Press. New York, New York. Chapter 14, Pages 279-300.

22. *Das R, Steege A, Baron S, Beckman J, Harrison R (2001). "Pesticide-related illness among migrant farm workers in the United States". Int J Occup Environ Health 7 (4): 303–12.*

23. *Daniels JL, Olshan AF, Savitz DA (October 1997).* "Pesticides and childhood cancers". *Environ. Health Perspect. 105 (10): 1068–77.*

24. DeMarini, D. M. (2004) Genotoxicity of tobacco smoke and tobacco smoke condensate: a review. Mutat Res., 567, 447–474.

25. *Eskenazi B, Bradman A, Castorina R (June 1999).* "Exposures of children to organophosphate pesticides and their potential adverse health effects". *Environ. Health Perspect. 107 Suppl 3: 409–19.*

26. *Engel LS, O'Meara ES, Schwartz SM (June 2000). "Maternal occupation in agriculture and risk of limb defects in Washington State, 1980-1993". Scand J Work Environ Health (193–8.)*

27. FDA says Zicam nasal products harm sense of smell, Los Angeles Times, June 17, 2009.

28. *Firestone JA, Smith-Weller T, Franklin G, Swanson P, Longstreth WT, Checkoway H (January 2005). "Pesticides and risk of Parkinson disease: a population-based case-control study". Arch. Neurol. 62 (1): 91–5.* doi:10. 1001/archneur. 62. 1. 91.

29. Food and Agriculture Organization of the United Nations (2002), International Code of Conduct on the Distribution and Use of Pesticides. Retrieved on 2007-10-25.

30. Gallagher JE, Everson RB, Lewtas J, George M, Lucier GW. Comparison of DNA adduct levels in human placenta from polychlorinated biphenyl exposed women and smokers in which CYP JAI levels are similarly elevated. Teratogenesis Carcinog Mutagen 14:183-192 (1994).

31. Grzybowska E, Hemminki K, Chorazy M. Seasonal variations in levels of DNA adducts and X-spots in human populations living in different parts of Poland. Environ Heafth Perspect 99:77-81 (1993).

32. Grzybowska E, Hemminki K, Szeliga J, Chorazy M. Seasonal variation of aromatic adducts in human lymphocytes and granulocytes. Carcinogenesis 14:2523-2526 (1993).

33. *García AM (December 2003). "Pesticide exposure and women's health". Am. J. Ind. Med. 584–94.*

34. Graeme Murphy (December 1 2005), Resistance Management - Pesticide Rotation. Ontario Ministry of Agriculture, Food and Rural Affairs. Retrieved on September 15, 2007.

http://www. beekeeping. com/articles/us/german_bee_monitoring. htm. Retrieved 2007-10-10.

35. *Haefeker, Walter (2000-08-12).* "Betrayed and sold out-German bee monitoring"*(inEnglish)* http://www. beekeeping. com/articles/us/german_bee_monitoring. htm. Retrieved 2007-10-10.

36. Hansen, A. M., Wallin, H., Binderup, M. L., Dybdahl, M., Autrup, H., Loft, S. and Knudsen, L. E. (2004) Urinary 1-hydroxypyrene and mutagenicity in bus drivers and mail carriers exposed to urban air pollution in Denmark. *Mutat. Res.,* 557.

37. *Hackenberg D (2007-03-14).* "Letter from David Hackenberg to American growers from March 14, 2007" *(in English). Plattform Imkerinnen —Austria.*

38. Hayes, R. B., Thomas. T., Silverman. D. T. *et al.* (1989) Lung cancer in motorexhaust-related occupations. [Published erratum appears in *Am. J. lnd Med.* 1991 19(1), 135]. *Am. J. lnd. Med,* 16, 685-695.

39. Hecht, S. S. (2003) Tobacco carcinogens, their biomarkers and tobacco-induced cancer. *Nat. Rev. Cancer*, 3, 733–744. Received July 6, 2005; revised August 12, 2005; accepted August 16, 2005.

40. Hemminki K, Zhang LF, Kruger J, Autrup H, Tornquist, Norbeck HE. Exposure of bus and taxi drivers to urban air pollutants as measured by DNA and protein adducts. Toxicol Lett 72:171-174 (1994).

41. Hrelia, P., Maffei, F., Angelini, S. and Forti, G. C. (2004) A molecular epidemiological approach to health risk assessment of urban air pollution. *Toxicol. Lett.,* 149, 261–267.

http://www. imkerinnen. at/Hauptseite/Menues/News Brief%20David%20Hackenberg%20307% 20engl. doc. Retrieved 2007-03-27.

42. Husgafvel-Pursiainen, K. (2004) Genotoxicity of environmental tobacco smoke: a review. *Mutat. Res.,* 567, 427–445.

43. Husgafvel-Pursiainen, K. (2004) Genotoxicity of environmental tobacco smoke: a review. *Mutat. Res.,* 567, 427–445.

44. John H. Duffus ""Heavy metals" a meaningless term? (IUPAC Technical Report)" Pure and Applied Chemistry, 2002, Vol. 74, pp. 793-807.

45. John H. Duffus ""Heavy metals" a meaningless term? (IUPAC Technical Report)" Pure and Applied Chemistry, 2002, Vol. 74, pp. 793-807.

46. *Johnston, AE (1986). "Soil organic-matter, effects on soils and crops". Soil Use Management 2: 97–105.*

47. *Jaga K, Dharmani C (2003).* "Sources of exposure to and public health implications of organophosphate pesticides". *Rev. Panam. Salud Publica 14 (3): 171–85.*

48. K., Gryzbowska, E. *et al.* (1992) Molecular and*Kamel F, et al. (2003).* "Neurobehavioral performance and work experience in Florida farmworkers". *Environmental Health Perspectives 111: 1765–1772.*

49. *Kamel F, et al. (2003).* "Neurobehavioral performance and work experience in Florida farmworkers". *Environmental Health Perspectives 111: 1765–1772.*

50. Kamrin MA. (1997). Pesticide Profiles: toxicity, environmental impact, and fate. CRC Press.

51. Knudsen, L. E., Norppa, H., Gamborg, M. O., Nielsen, P. S., Okkels, H., Soll-Johanning, H., Raffn, E., Jarventaus, H. and Autrup, H. (1999) Chromosomal aberrations in humans induced by urban air pollution: influence of DNA repair and polymorphisms of glutathione S-transferase M1 and N-acetyltransferase 2. *Cancer Epidemiol. Biomarkers Prev.,* 8, 303–310.

52. Kyrtopoulos, S. A., Georgiadis, P., Autrup, H. *et al.* (2001) Biomarkers of genotoxicity of urban air pollution. Overview and descriptive data from a molecular epidemiology study on populations exposed to moderate-to-low levels of polycyclic aromatic hydrocarbons: the AULIS project. *Mutat Res.,* 496, 207–228.

53. Leopardi, P., Zijno, A., Marcon, F., Conti, L., Carere, A., Verdina, A., Galati, R., Tomei, F., Baccolo, T. P. and Crebelli, R. (2003) Analysis of micronuclei in peripheral blood lymphocytes of traffic wardens: effects of exposure, metabolic genotypes, and inhibition of excision repair *in vitro* by ARA-C. *Environ. Mol. Mutagen.,* 41, 126–130.

54. Lobe, J (Sept 16, 2006), "WHO urges DDT for malaria control Strategies, " Inter Press Service, cited from Commondreams. org. Retrieved on September 15, 2007.

55. Lewtas J, Binkova B, Mftkovi I, Rossner P, Cerni M, Meyer S, Mumford J, Watts R, Lenicek J, Subrt P, grim RJ. Biomarkers and personal exposure to carcinogenic polycyclic aromatic hydrocarbons from air pollution. Cancer Epidemiol Biomarkers Prev (in press).

56. Michael Hogan, Leda Patmore, Gary Latshaw and Harry Seidman *Computer modelng* of pesticide transport in soil for five instrumented watersheds, prepared for the U. S. Environmental Protection Agency Southeast Water laboratory, Athens, Ga. by ESL Inc., Sunnyvale, California (1973).

57. Miller GT (2004), *Sustaining the Earth,* 6th edition. Thompson Learning, Inc. Pacific Grove, California. Chapter 9, Pages 211-216.

58. Miller, GT (2002). Living in the Environment (12th Ed.). Belmont: Wadsworth/Thomson Learning. ISBN 0-534-37697-5.

59. *Montgomery MP, Kamel F, Saldana TM, Alavanja MC, Sandler DP (May 2008). "Incident diabetes and pesticide exposure among licensed pesticide applicators: Agricultural Health Study, 1993-2003. ". Am J Epidemiol. 167 (10): 235–46...*

60. Mitchell, J. R., Jollow, D. J., Potter, W. Z., Davis, D. C., Gillette, J. R. &rodie, B. B. (1973). Aceta - minophen-induced hepatic necrosis. IVProtective role of glutathione. *Journal of Pharmacology and Experimental Therapeutics* 187, 185–194.

61. *Moses M (1989). "Pesticide-related health problems and farmworkers". Aaohn J (3): 115–30.*

62. Nader Rifai, Michael A Gillette & Steven A Carr(2006). Protein Biomarker discovery and validation the long and uncertain path to clinical utility, In Journal Nature Biotechnology Vol. 24, No. 8, pp-971-983.

63. Nature biotechnology volume 24 number 8 august 2006 971.

64. National Research Council. Environmental Tobacco Smoke. Measuring Exposures and Assessing Health Effects. Washington:National Academy Press, 1986.

65. Newswise: Long-term Pesticide Exposure May Increase Risk of Diabetes Retrieved on June 4, 2008.

66. *O'Malley MA (1997). "Skin reactions to pesticides". Occup Med 12 (2): 327–45.*

67. Palmer, WE, Bromley, PT, and Brandenburg, RL. Wildlife & pesticides - Peanuts. North Carolina Cooperative Extension Service. Retrieved on 2007-10-11.

68. Peluso, M., Hainaut, P., Airoldi, L. *et al.* (2005) Methodology of laboratory measurements in prospective studies on gene–environment interactions: the experience of Genair. *Mutat. Res.,* 574, 92–104.

69. Perera, F., Hemminki, K., Jedrychowski, W., Whyatt, R., Campbell, U., Hsu, Y., Santella, R., Albertini, R. and O'Neill, J. P. (2002) *In utero* DNA damage from environmental pollution is associated with somatic gene mutation in newborns. *Cancer Epidemiol. Biomarkers Prev.,* 11, 1134–1137.

70. Pollution and Society Marisa Buchanan and Carl Horwitz, University of Michigan.

71. *Prasad, A. S. (2003).* "Zinc deficiency". British Medical Journal *326: 409.*

72. Protein biomarker discovery and validation:the long and uncertain path to clinical utility.

73. Perera FP, Hemminki K, Grzybowska E, Motykiewicz G, Michalska J, Santella RM, Young TL, Dickey C, Brandt-Rau P, DeVivo I, Blaner W, Tsai WY, Chorazy M. Molecular and genetic damage in humans from environmental pollution in Poland. Nature 360:256-258 (1992).

74. R. McSorley and R. N. Gallaher, "Effect of Yard Waste Compost on Nematode Densities and Maize Yield", *J Nematology*, Vol. 2, No. 4S, pp. 655–660, Dec. 1996.

75. Rockets, Rusty (2007), Down On The Farm? Yields, Nutrients And Soil Quality. Scienceagogo. com. Retrieved on September 15, 2007.

76. Rainer Stegmann, *Treatment of Contaminated Soil: Fundamentals, Analysis, Applications*, Springer Verlag, Berlin 2001.

77. Reddy MV, Kenny PC, Randerath K. 32P-assay of DNA adducts in white blood cells and placentas of pregnant women: lack of residential wood combustion-related adducts but presence of tissue-specific endogenous adducts. Teratogenesis Carcinog Mutagen 10:373-384 (1990).

78. Ritter SR. (2009). Pinpointing Trends In Pesticide Use In 1939. *C&E News*.

79. Ron Zevenhoven, Pia Kilpinen: *Control of Pollutants in Flue Gases and Fuel Gases*. TKK, Espoo 2001.

80. S. K. Gupta, C. T. Kincaid, P. R. Mayer, C. A. Newbill and C. R. Cole, "A multidimensional finite element code for the analysis of coupled fluid, energy and solute transport", Battelle Pacific Northwest Laboratory PNL-2939, EPA contract 68-03-3116 (1982).

81. State of the Environment, Issue: Air Quality (Australian Government website page).

82. SteenlantLN. K., Silverman. D. T. and Hornung, R. W. (1990) Case-controlstudy of lung cancer and truck driving in the Teamsters Union. *Am. J. Public Health,* 80, 670-674.

83. *Stallones L, Beseler C (October 2002). "Pesticide illness, farm practices, and neurological symptoms among farm residents in Colorado". Environ. Res. 90 (2): 89–97.*

84. *Schwartz DA, Newsum LA, Heifetz RM (February 1986). "Parental occupation and birth outcome in an agricultural community". Scand J Work Environ Health 12 (1): 51–4..*

85. *Tong LL, Thompson B, Coronado GD, Griffith WC, Vigoren EM, Islas I (December 2004). "Health symptoms and exposure to organophosphate pesticides in farmworkers". Am. J. Ind. Med. 46 (6): 599–606.*

86. US Environmental Protection Agency (July 24, 2007), What is a pesticide? epa. gov. Retrieved on September 15, 2007.

87. *Van Maele-Fabry G, Willems JL (September 2003). "Occupation related pesticide exposure and cancer of the prostate: a meta-analysis". Occup Environ Med 60 (9): 634–42.*

88. *Wells, M (March 11, 2007).* "Vanishing bees threaten US crops" *(in English). www. bbc. co. uk (*BBC News*).* http:// news. bbc. co. uk/2/hi/americas/6438373. stm. Retrieved 2007-09-19.

89. World Health Organization (September 15, 2006), WHO gives indoor use of DDT a clean bill of health for controlling malaria. Retrieved on September 13, 2007.

90. *Zeissloff, Eric (2001).* "Schadet imidacloprid den bienen" *(in German).* http://www. beekeeping. com/artikel/ imidacloprid_1. htm. Retrieved 2007-10-10.

91. Zhao, X., Niu, J., Wang, Y., Yan, C., Wang, X. and Wang, J. (1998) Genotoxicity and chronic health effects of automobile exhaust: a study on the traffic policemen in the city of Lanzhou. *Mutat. Res.,* 415, 185–190.

CHAPTER 13

Corrosion and Environment

Dr. Ishaq Zaafrany[1]

Introduction

Corrosion involves the interaction (reaction) between a metal or alloy and its environment. Corrosion is affected by the properties of both the metal or alloy and the environment. In this discussion, only the environmental variables will be addressed the more important of which include.

- pH (acidity)
- Oxidizing power (potential)
- Temperature (heat transfer)
- Velocity (fluid flow)
- Concentration (solution constituents)

The concept of pH is complex. It is related to, but not synonymous with, hydrogen concentration or amount of acid.

While corrosion obeys well known lows of electrochemistry and thermodynamics many variables that influence the behavior of a metal in its environment can result in accelerated corrosion or failure in one case and complete protection in another, similar case.

Avoiding detrimental corrosion requires the interdisciplinary approach of the designer the metallurgist, and the chemist. Sooner or later, nearly everyone in these fields will be faced with major corrosion issues. It is necessary to learn to recognize the forms of corrosion and the parameter that must be controlled to avoid or mitigate corrosion.

The theory of corrosion from the thermodynamic and kinetic points of view covers the principles of electrochemistry, diffusion, and dissolution as they apply to aqueous corrosion and high-temperature corrosion in salts, liquid metals, and gases.

We can face the various forms of corrosion, and we must know how to recognize them, as well as the driving conditions or parameters that influence each form of the corrosion for it is the control of these parameters which can minimize or estimate corrosion.

All corrosion processes show some common features. Thermodynamic principles can be applied to determine which processes can occur and how strong the tendency is for the changes to take place. Kinetic laws then describe the rates of the reactions. There are however, substntial difference in the fundamentals of corrosion in such environments as aqueous solutions, non-aqueous liquids and gases.

[1]Department of Chemistry, Umm-Al-Qura University, Makkah-Al-Mukaramah (K.S.A.)

Corrosion and Environment

Corrosion in aqueous solutions. Although atmospheric air is the most common environment, aqueous solutions, including natural waters, atmospheric moisture and rain, as well as man-mae solutions, are the environments most frequently associated with corrosion problems.

Because of the ionic conductivity of the environment, corrosion is due to electrochemical reactions and is strongly effected by such factors as the electrode potential and acidity of the solution.

Corrosion of metals in aqueous environments. This type of corrosion is almost always electrochemical in nature. It occurs when two or more electrochemical reactions take place on a metal surface. As a result, some of the elements of the metal or alloy change from a metallic state into a non-metallic state.

The products of corrosion may be dissolved species or solid corrosion products. In either case, the energy of the system is lowered as the metal converts to a lower-energy form. Rusting of steel is the best-known example of a conversion of the metal (iron) into a nonmetallic corrosion product (rust). The change in the energy of the system is the driving force for the corrosion process and is a subject of thermodynamics.

Thermodynamics examines and quantifies the tendency for corrosion and its partial processes to occur. It does not predict if the changes actually will occur and at what rate. Thermodynamics can predict, however, under what conditions the metal is stable and corrosion cannot occur.

Corrosion in Molten Salts and Liquid Metals : These are more specific but important areas of corrosion in liquid environments. Both have been strongly associated with the nuclear industry, for which much of the research has been performed, but there are numerous non-nuclear applications as well.

Corrorrion in Gases : In gaseous corrosion the environment is nonconductive, and the ionic processes are restricted to the surface of the metal and the corrosion product layers.

Because the reaction rates of industrial metals with common gases are low at room temperature, gaseous corrosion, generically called oxidation is usually an industrial problem only at high temperatures when diffusion processes are dominant.

Forms of Corrosion

Over the years, corrosion scientists and engineers have recognized that corrosion manifests itself in forms that have certain similarities and therefore can be categorized into specific groups. However, many of these forms are not unique but involve mechanisms that have over lapping characteristics that may influence or control initiation or propagation of a specific type of corrosion.

The most familiar and often used categorization of corrosion is uniform attack, crevice corrosion, pitting, intergranular corrosion, selective teaching, erosion corrosion, stress corrosion, and hydrogen damage. This classification of corrosion was based on visual characteristics of the morphology of attack.

Other prominent corrosion authors have avoided a classification format and have simply discussed the classical types of corrosion (for example pitting and crevice corrosion) as they relate to specific metals and alloys.

Forms of corrosion are :

1. General corrosion

Atmospheric corrosion

Galvanic corrosion

Stray-current corrosion

General biological corrosion

Molten salf corrosion

Corrosion in liquid metals

2. High-temperature corrossion

Oxidation

Sulfidation

Carburization

Other forms(a)

3. Localized corrosion

Filiform corrosion

Crevice corrosion

Pitting corrosion

Localized bilogical corrosion

4. Metallurgically influenced corrosion

Intergranular corrosion

Dealloying corrosion

Mechanically assisted degradation

Erosion corrosion

Fretting corrosion

Cavitation and water drop impingement

Corrosion fatigue

5. Environmentally induced cracking

Stress-corrosion cracking

Hydrogen damage

Liquid metal embrittlement

Solid metal induced embrittlement

General Corrosion : General corrosion is defined as corrosive attack dominated by uniform thining. Although high-temperature attack in gaseous environments, liquid metals, and molten salts may menifest itself as various forms of corrosion, such as stress-corrosion cracking and de-alloying, high-temperature attack has been incorported under the term "General Corrosion" because it is often dominated by uniform thining.

Localized Corrosion : The forms of corrosion under this category need no explanation even though other forms could be placed in this category a should be noted, however, that localized biological corrosion often causes or accelerates pitting or crevice corrosion.

Metallurgically influenced corrosion was so classifed as a result of the significant role that metallurgy plays in these forms of attack. It is well understood that metallurgy is important in all forms of corrosion, but this classification is meant to emphasize its role in these specific forms of attack.

Mechanically assisted degradation groups those forms of corrosion that contain a mechanical component, such as velocity abrasion and hydrodynamics that has a significant effect on the corrosion behavior. Corrosion fatigue was include in this category because of the dynamic stress state, however, it could easily be categorized as a form of environmentally induced cracking.

The environmentally induced cracking follows the current trend in the literature of combining forms of cracking that are produced by corrosion in the presence of stress.

(*i*) they can be adjacent or widely separated so that, for example, if two metals are in contact one metal can be the anode and other the cathode, leading to galvanic corrosion of the more anodic metal.

(*ii*) these can exist variations over the surface of oxygen concentration in the environment that result in the establishment of an anode at those sites exposed to environment contriving the lower oxygen content differential aeration corrosion

(*iii*) or similarly variations in the concentration of metal ions or other species in the environment arising because of the spatial orientation of the corroding metal and opacity, or finally

(*iv*) variations in the homogeneity of the metal surface, due the pressure of inclusions, different phases, grain boundaries, disturbed metal and other causes, can lead to the establishment of anodic and cathodic sites.

The process occurring at the anodic sites is the dissolution of metal as metallic ions in the electrolyte or the conversion of these ions to insoluble corrosion produces such as rust. This is the destructive process called corrosion. The flow of electrons between the corroding anode and the non-corroding cathodes forms the corrosion current the values of which is determined by the rate of production of electrons by the anodic reaction and their consumption by the cathodic reaction. The rates of electron production and consumption of course must be equal a buildup of charge would occur.

Driving Force, Corrosion Tendencies

A driving force is necessary for electrons to flow between the anodes and the cathodes. This driving force is the difference in potential between the anodic and cathodic sites. This difference exists because each oxidation or reduction reaction has associated with a potential determined by the tendency of the reaction to take place spontaneously. The potential is a measure of this tendency.

Types of Corrosion

The breakdown and repair of passivity process just described is involved in many but not all of the various types of corrosion.

Pitting Corrosion

The initiation of a pit occurs when electrochemical or chemical breakdown exposes a small local site on a metal surface to damaging species such as chloride ions. The sites where pits initiate are not completely understood but some possibilities are at scratches, surface compositional heterogeneities (inclusions), or places. Where environmental variation exist.

Crevice Corrosion

Crevice corrosion results when a portion of a metal surface is shielded in such a way that the shielded portion has limited access to the surrounding environment. Such surrounding environments contain damaging corrosion species, usually chloride ions. A typical example of crevice corrosion is the crevice formed at the area between two metal surfaces in close contact with gasket or another metal surface. The environment that eventually forms in the crevice is similar to that formed under the precipitated corrosion product that covers a pit. Similarly an electrochemical corrosion cell is formed from the couple between the unshielded surface and the crevice interior exposed to an environment with a lower oxygen concentration compared with the surrounding medium. This combination of being the anode of a corrosion cell and existing in an acidic high chloride environment where repassivation is difficult makes the crevice interior subject to corrosive attack.

An idealized picture of the environment that develops in a crevice by the corrosion cell produced on iron by an anode in a crevice and a cathode outside of the crevice.

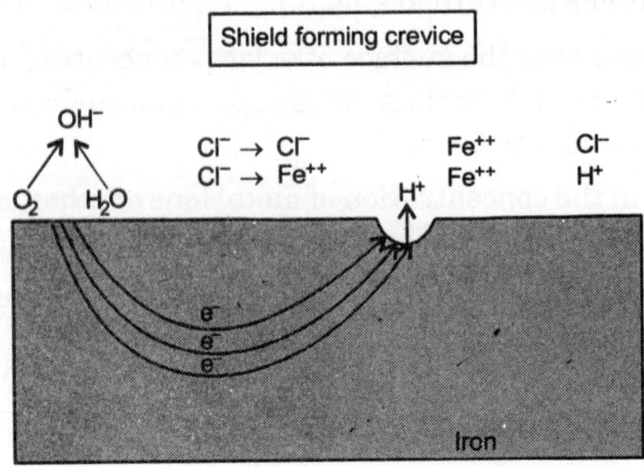

Fig. 13.1.

Stress Corrosion Cracking, Hydrogen Damage, and Corrosion Fatigue

Stress corrosion cracking (SCC) is a form of localized corrosion, which produces crack in metals by the simultaneous action of a corrodent and tensile stress. The electrochemical cell between the exterior and interior environments of a crack is similar to that described above for a crevice because of the necessary for the application of stress, the breakdown of the passive layer on a metal surface in SCC it generally ascribed to mechanical causes but many suggests that electrochemistry is a significant factor because it controls the rate of repair of the passive layer ruptured by mechanical stess.

Intergranular Corrosion

With the exception of metallic glasses the metal used in practical device are made up of small crystals (grains) whose surfaces join the surfaces of other grains form grain boundaries. Such boundaries or the small regions adjacent to these boundaries can under certain conditions be considerably more reactive (by being more anodic) then the interior of the grains. The resulting corrosion is called intergranular corrosion. It can result in a loss of strength of metal part of production of debits.

Corrosion is the destructive attack of a material by reaction with its environment. The serious concequences of the corrosion process have become a problem of worldwide significance. In addition to our everyday encounters with this form of degradation, corrosion causes plant shutdowns, waste of valuable resources, loss or contamination of product, reduction in efficiency, costly maintenance, and expensive overdesign. It can also jeopardize safety and inhibit technological progress.

```
Step 1 iron + oxygen → iron oxide
Step 2 iron oxide + water → hydrated iron oxide (rust)
```

Corrosion is the atmospheric oxidation of metals. That means that oxygen combines layer can be good or bad. By far the most important form of corrosion is the rusting of iron and steel. Rusting is a process of oxidation in which iron combines with water and oxygen to form rust, the reddish-brown crust that forms on the surface of the iron. Because iron is so widely used, *e.g.*, in building construction and in tools, its protection against rusting is important. Rusting can be prevented by excluding air and water from the iron surface, *e.g.*, by painting, oiling, or greasing, or by plating the iron with a protective coating of another metal. Many alloys of iron are resistant to corrosion. Stainless steels are alloys of iron with such metals as chromium and nickel; they do not corrode because the added metals help form a hard, adherent oxide coating that resists further attack.

Although metals like aluminium, chromium, and zinc corrode more readily than iron, their oxides form a coating that protects the metal from further attack. Rust is brittle and flakes off the surface of the iron, continually exposing a fresh surface. Thus these metals might be a better selection choice for a product that will be exposed to rusting conditions, like water and air.

Sodium Bromide Brine Corrosion Testing

The general, crevice and pitting corrosion, stress corrosion cracking and galvanic corrosion behavior were studied on seven CRAs (HC, HG, IG3, 1625, 1725, 1825 and 925) and one low alloy steel (T-95), by autoclave exposure tests. The exposures were carried out in non-inhibited and inhibited NaBr packer fluid environments with 100 psia H_2S and 700 psia CO_2 at 450 F for thirty and ninety days.

Hydrogen Charging in Hf Alkylation

Measurement of corrosion rates and steady state hydrogen permeation curents in environments containing hydrofluoric acid (hydrogen flouride, HF) doped with arsenic (As) by electrochemical techniques.

Hydrogen Permeation Tests At High Pressure/temperature

High pressure/high temperature hydrogen tests in simulated sour environments. The tests were performed on two types of steel flange materials and one pipe material. One steel flange was tested for hydrogen damage in sulfinol with acid gas up to 240 F (116 C) while the other was tested in hydrocarbon/condensate up to 340 F (171 C). The pipe steel was tested in sour bring up to 176 F (80 C).

Corrosion Evaluation of Seamless Versus Welded Tubing

Electrochemical polarization measurements were performed on seamless, seamwelded, and buttwelded 316L, Inconel 825, and Monel 400 tubing in selected environments to study the general corrosion behaviour. SSR tests were conducted on the seamless and buttwelded tubing specimens to evaluate the susceptibility to environmental cracking.

Electrochemical Polarization Curves in Synthetic Seawater

Electrochemical anodic polarization measurements were performed on two AM355 materials, heat no. 8854-21 (AM355-1) and 7511-8 (Am355-2), per ASTM G-61 in deaerated synthetic seawater at 25°C. The materials were provided in the form of helicopter rotor straps. Twelve specimens from each helicopter rotor strap were tested.

SSC Resistance of Valve Materials in High Pressure H_2S/CO_2 Environments

A series of test were conducted to evaluate the sulfide stress cracking (SSC) resistance of candidate low alloy steels for application in valve components. Fracture mechanics test methods were combined with autoclave exposures to develop information on the behaviour of these materials in high pressure simulated field environments containing H_2S, CO_2 and brine.

Corrosion Behaviour of Selected Stainless and Nickel Alloys, and Anti-Galling Coatings in a Sour Environment

This report summarizes an experimental investigation conducted on the behavior of various antigalling coatings in sour environments. Laboratory tests were conducted to examine the corrosion resistance of antigalling coatings and their effect on the corrosion behavior of selected stainless steels and nickel alloys.

Behaviour of Aluminum Alloys in CO_2/NaCl Brine Environment

The present study was conducted to evaluate the performance of selected conditions of aluminum alloys in a CO_2/brine environment at 80 F. The results indicate that the most severe attack is generally observed in the weld heat affected zone area. However, Some materials were found that did not exhibit this type of attack. Generally, corrosion rates decreased with increasing exposure time. In all but one case, the corrosion was less severe on machined surfaces as compared to the as-received surface of the materials.

Evaluation of some SS and Ni Alloys in Acidizing Environments

The present study involved the evaluation of selected stainless and nickel base alloys in the following acidizing media: (1) 15 percent HCl, (2) 28 percent HCl, (3) 12 percent formic acid and (4) 13 percent HCl/2 percent HF. Tests were conducted both with H_2S and with CO_2 at 200 F and 400 F.

Elastomer Material Evaluation in High Pressure Gaseous Environment

Samples of elastomeric materials were provided for evaluation of performance in high pressure gas environments containing CO_2, CH_4 and N_2. Of particular concern are the effects of rapid decompression from test pressure and environmental degradation of the materials. The series of tests conducted in this investigation was developed to generate information that will aid in the selection of elastomeric materials for wellhead service where high pressures, temperatures and CO_2 concentrations are present.

Corrosion Tests Under Simulated Steam Generator Conditions

A series of autoclave corrosion tests were conducted to simulate conditions in the steam generator. Nal-Prep III, a commercial passivating treatment, produced only slight reduction in the pitting rate on carbon steel. Type 316 stainless steel and more highly alloyed stainless steels showed no or low pitting and no stress corrosion cracking in the steam generator conditions. Alloys with low or no molybdenum such as Type 304 showed some pitting at the liquid-vapour interface. Electrochemical tests suggest that eliminating the caustic 94D4 wash water from the bottoms water may decrease pitting in the steam generator tubes.

Evaluation of Elastomers in Methane & CO_2 Environments

Samples of Elastromeric materials were provided and evaluated to assess performance in a high pressure gas environment containing CO_2 and CH_4. The study presented herein was conducted as an extension to work, described earlier. All test methods employed presently are similar to those described in depth in the previous report and will be only briefly covered herein.

Corrosion Evaluation of Steel in LiBr Solutions

Corrosion tests conducted in inhibited 60% LiBr indicate that liquid phase corrosion rates remain low even up to 350 F. However, vapour phase and liquid-vaour interface corrosion rates are such that carbon steel will require protection above 300 F. 70-30 cupro-nickel showed low corrosion rates with no pitting at all temperatures tested. Grade 2 titanium showed significant pitting at all temperatures tested. Grade 7 titanium showed significant pitting at 300 F and above.

Fluoroshield, a fluorocarbon polymer coating, performed well in these tests and appears to be a viable method for protecting carbon steel in LiBr environments.

Electrochemical Measurements of Localized and General Corrosion in LiBr

This work is relevant to a process design project that involves high capacity heat exchangers. The heat exchangers in question are slated to use a 65 percent LiBr solution as the heat transfer agent. The engineering question of concern is the compatibility of the LiBr fluid with the materials of construction of the heat exchangers.

Electrochemical Measurement of Corrosion Rates in Acetic Acid Process

This report details results of test program to measure corrosion rate of five alloys as a function of temperature using electrochemical techniques. The test environment was a process fluid consisting of acetic acid, methyl acetate and methyl iodide.

Corrosion of Steel Alloys in CO_2 Environments

This report describes an experimental program to evaluate corrosion of carbon steels in CO_2 environment. Coupon exposure and electrochemical tests were conducted as a part of the program. The objective of the test pogram was to determine the corrosion rate of steel exposed to wet CO_2 conditions. Test variables were brine chemistry, CO_2 partial pressure, temperature, metal alloy and exposure duration.

Corrosion in Gases

In gaseous corrosion the environment is non-conductive, and the ionic processes are restricted the surface of the metal and corrosion products layers.

Corrosion is the disintegration of an engineered material into its constituent atoms due to chemical reaction with its surrounding. In the most common use of the word, this means electrochemical oxidation of metals in reaction with an oxidant such as oxygen. Formation of an oxide of iron due to oxidation of the iron atom in solid solution as a well known example of electrochemical corrosion, commonly known as rusting.

In material science environmental stress fracture or environment assisted fracture is the generic name give to premature failure under the influence of tensile stresses and harmful environments of materials such as metals and alloys, composites, plastics and ceramics.

Metals and alloys exhibit phenomena such as stress corrosion cracking, hydrogen embitterment liquid metal embattlement and corrosion fatigue all coming under this category. Environment such as moist air, sea water and corrosive liquid and glasses cause environmental stress fracture. Metal matrix composites also susceptible to many of these processes.

Plastics and plastic-based composites may suffer swelling depending and loss of strength when exposed to or geinic fluids and other corrosive environments, such as acids and alkalis. Under the influence of stress and environment many structural materials, particularly the high-specific strength ones becoming brittle and lose their resistance to fracture while their fracture toughness remains unaltered their threshold stress intensity. Factor for crock propagation may be considerably lowered consequently they becomes prone to premature fracture becomes of sub-critical of crack growth. This Corrosion is an expensive process that leas to enormous damage for modern industrial societies. Corrosion science is not only important from an economic point of view, but also due to its inter disciplinary nature combining metallurgy physics and electrochemistry. Now a days corrosion science even gets new impetus from metal physics, surface science, organic chemistry and microbiologists. It is important that scientists and engineers with different back grounds and experience collaborate and exchange their knowledge.

Ulick R. Evans the British Scientist who is considered the father of corrosion science has said that "corrosion is largely an electrochemical phenomenon [which] may be defined as destruction by electrochemical or chemical agencies....." corrosion in an aqueous environment and in an atmospheric environment is an electrochemical process because. Corrosion involves the transfer of electron between metal surface and an aq. electrolyte solution. It results from the over whelming tendency of metal to react electrochemistry with oxygen, water and other substances in the aq. environment. Fortunately most useful metals. React with the environment to form more or less protective films of corrosion reaction products that prevent the metals from going into solution as ions,

While the term corrosion has in recent years been applied to all kinds of materials in all kinds of environments, this article will only consider the electrochemistry of corrosion of metals and alloys in aq. solution at ambient temperature. Electrochemical corrosion occurring under such conditions is a major destructive process that results in such costly unsightly and destructive effects as the formation of rust and other corrosion products, the creation of the gaping holes or cracks in aircraft, automobiles boats, screens, plumbing and many other items constructed of every metals except gold.

Systems such as boiling water nuclear reaction involving aqueous solution are also examples of electrochemical corrosion but will not be covered. This article will also not cover the non-electrochemical process termed high temperature oxidation a destructive process which is the exposure of a metal or alloy to high temperature in a gaseous environments where much thicker layer of corrosion products or formed.

However, it must be pointed out that if the high temperature oxidation process results in the formation of salt layer that condensed aq. films involved in atmospheric corrosion. This technologically important corrosion process leads to the failure of such application as goes turbines, heat exchange and many others that operate at high temperature.

Consequences of Corrosion

Corrosion has many serious economic, health, safety, and technological and cultural consequences to our society.

Economic Effects

Studies in a number of countries have attempted to determined the national cost of corrosion. The most extensive of the studies was the one carried out in the united states in 1976 which found that the over all annual cost of metallic corrosion to the U.S. economy was $70 billion or 4.2% the gross national product. To get a feeling for the seriousness of this loss, we may compare it to another economic impact even one is worried about the importation of foreign crude oil, which const $45 billion in 1977.

Health Affects

Recent years have been seen an increasing use of metal prosthetic devices in the body, such as hip, joints, pacemakers and other implants. New alloys and better techniques of implantation have been developed, but corrosion continues to create problems. Examples include failures through broken connections in pacemakers, inflammation caused by corrosion products in the tissue around implant and fracture of weight-bearing prosthetic device. An example of the latter is the use of metallic hip joints, which can alleviate some of the problems of Arthritic hips. The situation has improved in recent years. So that hip joints which were was at first limited to person over 60 are now being used in younger persons, because they will last longer.

Safety Effects

An even more significant problems is corrosion of structures, which can result in severe injuries or even loss of life safety is compromised by corrosion contributing to failures of bridges, aircrafts automobile gas pipelines etc the whole complex of metal structures and devices that make up the modern world.

Technological Effects

The economic consequences of corrosion affect technology. A great deal of the development of new technology is held back by corrosion problems because metatarsals are required to with stand, in many cases simultaneously, higher temperatures, higher pressures and more highly corrosive environments. Corrosion problems that are less difficult to solve affect solar energy systems, which require alloys to with stand hot circulating that transfer fluids. For long periods of time, and geothermal systems, which require materials to with stand highly concentrated solutions of corrosive. Salts at high temperature and pressures. Another example, the drilling for oil in the sea and on land, involves overcoming such corrosion problems as sulfide stress corrosion, microbiological corrosions and the vast array of difficulties involved in working in the highly corrosive marine environment. In many these instances, corrosion is a limiting factor preventing the development of economically or even technologically workable system.

Cultural Effects

International concern was aroused by the disclosure of the serious deterioration of the artistically and culturally significant glided bronze statues in Venice Italy. Corrosive processes will accelerate the deterioration of precious artifacts such as those in Venice by the highly. Polluted environments that now are prevalent in most of the countries of the world. Likewise, inside the world's museums conservators and restores labour to protect cultural treasures against the ravages of corrosion or to remove its traces from artistically or culturally important at artifices.

Electrochemistry of Corrosion

The small metallic surface exposed to an aqueous electrolyte usually possesses sites for an oxidation (or anodic chemical reaction) that a reduction (or cathode reaction) that consumes the electrons produced by the anodic reaction. Thee sites together make up a corrosion cell". The anodic reaction is the dissolution of metal to form either soluble ionic products of an insoluble compound of the metal usually an oxide. Several cathodic reaction are possible depending on what reducible species on present in the solution. typical reaction are the reduction of dissolved oxygen gas or the reduction of the solvent (water) to produce hydrogen gas. Examples of these the anodic and cathodic reactions occur simultaneously on a metal surface they create on electrochemical cell.

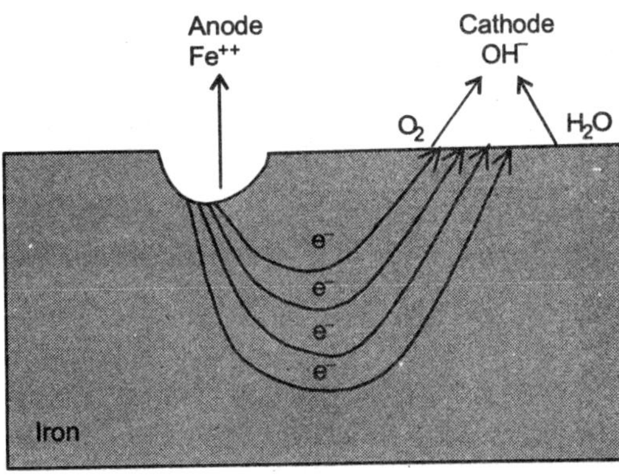

Fig. 13.2

The electrochemical cell set up between anodic and cathodic sites on an iron surface undergoing corrosion.

The sites where the anodic and cathodic reaction takes place, the anodes and the cathodes of the corrosion cells, are determined by many factor.

REFERENCES

1. Principles and prevention of corrosion (2nd edition) D.A. Jones, prentice Hall, upper saddle River NJ 1996.
2. Corrosion (two volumes) L.L. Shreir R.A. Jaman and, Oxford, Boston 1994.
3. Advances in Localised corrosion NACE-9, H.S. Isaac, U. Bertocci, J. Kruger and Smialowska (editors), National Association of corrosion Engineers, Houston TX, 1990.
4. Passivity of metals-A materials –science perspective J. Kruger, International materials Reviews "vol. 33 pp.113-129, 1988.
5. Corrosion Engineering (2nd edition), M.G. Fontana, MC Graw-Hill, New York 1986.
6. Corrosion and corrosion control An Introduction to corrosion science and Engineering (3rd edition) Uhling and R.W. Revice,Wiley, New York,1985.
7. Basic corrosion and corrosion J.M. West, Elis Harwood, Chichester UK 1980.
8. Passivity of metals,and Frankenthal Aqueous and J.Kruger (edition) The electrochemical Society, Pennington N. J1978.
9. Atlas of electrochemical equilibria in aq.solutions (2nd edition) M. Pourbaly,National Association of corrosion Engineers, TY1974.

CHAPTER 14

Photochemical Smog

Ms. Neelofar Iqbal[1]

Introduction

The word smog is derived form an emission of smoke and fog, coined by H.A. Voeux (1905). The oldest smog, which consists of a mixture of coal smoke and fog, has plagued human beings since the 14th century. The indicants of deadly smoky Fog, called Urban smog (occurs mainly over industrial areas) happened in December 1930 in Meuse Valley of Belgium, which killed about 600 people. Smog is best in a single year and the donora Pennsylvania Smog which Killed people typified by classic example as the London smog of December 1952, which killed about 5000 people in a single year and the donora Pennsylvania smog which killed 20 people and made hundreds ill. The five day health disaster of donora Pennsylvania (USA) was preceded by the formation of trick fog due to strong inversion temperature on December 26, 1948.

Los Angles in California experienced a new and serious form of air pollution during 1944, which was characterized by reduced visibility, eye irritation (i.e. lacrimation) and plant damage (e.g. metallic sheen on levels), whereas London Smog is characterized by bronchitis irritation. It is not related to smoke or fog. It is worst in the sunshine and, unlike London type of smog peaks early in the day, it peaks in the afternoon. It is also called as photochemical snog as an important reaction, the dissociation of NO_2, is caused by sunlight.

Smog arises from photochemical reaction in the lower atmosphere by the interaction of hydrocarbon and nitrogen oxide released by exhausts of automobile and some stationary sources. This interaction results in a series of complex reactions producing secondary pollutants such as ozone, aldehydes, ketones, and peroxy acyl nitrate. The reaction mechanisms are complicated and are not fully understood.

Chemical Reaction Involved in Photochemical Smog

In air, the formation of oxidants, especially ozone, including PAN and NO_x—has been indicative of smog formation. Smog formation are initiated by ultra violet light. PB_xN is produced in photochemical smog in presence of NO_2 and olefins. Hydrocarbon and olefins are especially conductive to smog formation PB_xN is 100 times more powerful than PAN and 200 times as powerful as formaldehyde.

Smoggy atmosphere shows characteristic time of day in levels of NO, NO_2, aldehydes, hydrocarbon and oxidants. For example, NO concentration is at the peak in early morning but after sunshine, NO

[1]Asian Biotech Research Centre, Bhopal-462 001 (India)

decrease and NO_2appears. The time variation in levels of O_3, NO, NO_2 are explained in terms of chemical reaction first proposed by Friedlander and Seinfield. The important steps are :

(1) Formation of Nitric Oxide (NO)—

$$N_2(g) + O_2(g) \longrightarrow 2NO(g)$$

Some of this NO reacts further with oxygen to form nitrogen dioxide (NO_2)

$$2NO(g) + O_2(g) \longrightarrow 2NO_2(g)$$

Actually man made NO_x is about 10% of the total in the atmosphere, while the rest comes from natural bacterial action. Typical levels of NO and NO_2 in the United state are 0.002 ppm and 0.004 ppm respectively. In urban areas these level often reach 100 times of these values.

(2) Nitrogen dioxide is a part of Photolytic NO_2cycle : Acoording to Haagen Smith theory, ultraviolet rays in the sunlight cause the photo dissociation of NO_2 to from NO and atomic oxygen.

$$NO_2\ hv \longrightarrow NO + O$$

NO_2 being an efficient absorber of UV light is cleared at 400 nm. For $\lambda > 400$ nm, the photons are too weak to cause dissociation the molecules are simply excitedto higher energy

(3) Formation of oxidants : formation of ozone nascent oxygen and peroxide etc. during the smog formation.

$$NO + [O] \longrightarrow NO_2$$

(4) Reaction Involving Oxygen Species :

$$O + O_2 + M \longrightarrow O_3 + M$$

Where M being a body such as N_2, O_2 Argon or CO_2 Which is capable of carrying off the larger excess energy.

(5) Combination of O_3 and NO to close cycle :

$$O_3 + NO \longrightarrow NO_2 + O_2$$

O_3 can also combine with NO_2 to form nitrogen pentaoxide

Hydrocarbons, which are emitted in automobile exhaust, are believed to react by free radical mechanism either with [O] or O_3 to from secondary pollutant such as aldehydes. Ketones and organic peroxide.

During the reaction, NO is oxidized to NO_2 and lack of NO scavenger makes the ozone content to rise. How hydrocarbon upset the photolytic NO_2 cycle is still uncertain. Most proposed schemes. go through the unsaturated hydrocarbons and have them attacked by monatomic oxygen or less probably by ozone.

(6) Production of Organic Free Radicals Form Hydrocarbons RH :

$O + RH \longrightarrow R$ + Other products $O_3 + RH \longrightarrow R$ + Other products

R' is a free radical which may or may not contain oxygen. These free radicals then react with O_2 or NO to produce other unstable intermediates and various by-products. Often a series of 20 or more consecutive reaction are needed to account for all the products.

(7) Chain Propagation and Termination :

$$NO + OR^{\bullet} \longrightarrow NO_2 + R$$

Where 'R' contains oxygen and oxidises NO

$$NO_2 + R^{\bullet} \longrightarrow \text{Products (e.g. PAN, HCHO, Acrolein)}$$

Smog Behavior

The formation of photochemical smog is a dynamic process whose nature is illustrated in Fig. 1. In the morning the NO and hydrocarbon levels increase followed quickly by increase in NO_2, which reacts with the sunlight leading to various chain reaction and ultimately to the production of ozone and other oxidation. Ozone concentration now increase until, some-time in the afternoon, it reaches a maximum, and then decreases gradually- NO_2 concentration diminishes from its peak as ozone concentration builds up and is usually low by late afternoon. The typical smog episode occurs in hot, sunny weather under low humidity conditions. The characteristic symptoms of the smog are the brown haze in the atmosphere, reduced visibility, eye irritation, respiratory distress and plant damage.

The control of photochemical smog may require a substantial reeducation in NO_x produced in urban areas. At the same time it is necessary to control the release of hydrocarbon from numerous mobile and stationary sources.

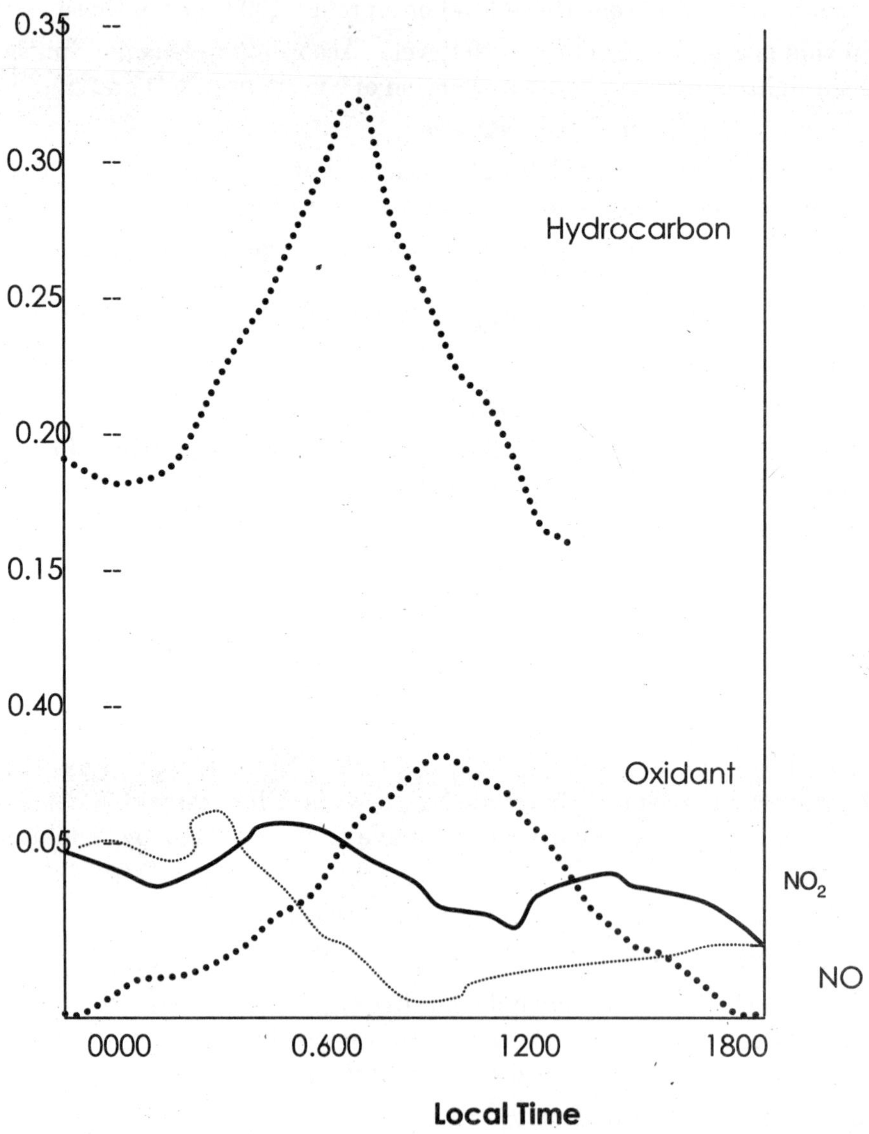

Fig. 14.1. Dynamic behavior of Photochemical Smog

Effect of Photochemical Oxidants on Humans

Both ozone and peroxy acyl nitrate (PAN) cause irritation of the eyes creating lachymation and affect severely the respiratory tract of human begins. Exposure to 50 ppm of O_3 for several hours will lead to mortality due to pulmonary edema, i.e. accumulation of blood in the lungs.

Primary photochemical pollutant, i.e. NO_2 produces a brownish haze causing nose and eye irritation and pulmonary discomfort. NO_x also cause several chronic disease of heart, lings and eyes, lower concentration of ozone irritation the nose and throat while its higher concentration causes headache, cough, dryness of the throat, chest pain, difficulty in breathing etc. Experimental studies have shown that exposure of ozone upto 0.2 ppm, creates no ill effect but a level of 0.3 ppm appears to be the threshold level at which nose and throat irritation begins. However, exposure to ozone concentration of 1.0 to 3.0 ppm. For a period of two hours produces extreme fatigue and lack of coordination in central nervous system.

Effects of Photochemical Oxidation on Plants

PAN and other member of its family (PBN, PPN etc) are photo chemically in air having olefins and NO_x. It is very damaging to plants, attacking younger leaves and bringing about "bronzing" and 'glazing' of their surfaces. Photochemical smog is characterized by brown hazy fumes which leads to cracking of rubber and extensive damage to plant life. Exposure for several hours to PAN. atmosphere at the level 0.02 to 0.05 ppm cause a great loss to vegetation. The (sulphhydryl) group present in proteins has been susceptible to damage by PAN. It acts both as an oxidizing and acetylating agent while reacting with sulph hydryl group contained in proteins.

Experimental result indicate that ozone is the most toxic photochemical oxidant. Among visible effects of ozone injury to plants are bleached or light flecks or stipples (clusters of dead cells) on the upper surface of leaves. Fully expended, mature leaves are more susceptible to damage. Leaves tip burn, a disease of white pines is mainly caused by ozone.

Control of Photochemical Pollutants

The control of primary precursors, such as NO_x and hydrocarbon will ultimately control ozone and PAN which are secondary pollutants. Today the techniques, like incineration, absorption, absorption and condensation are devised to control hydrocarbon emission from stationary sources.

(1) Incineration method : Hydrocarbon removal efficiencies are greater in flame after burner in which flame is used to complete the oxidation of hydrocarbon into CO_2 and water. The other device makes the use of catalytic after burner in which the catalyst oxidizes the hydrocarbon at a lower temperature The efficiency and fuel cost are lower in this type of incinerator. catalyst may also cause poisoning during the reaction.

(2) Adsorption method : In this method, exhust gases are passed through a bed of granulated adsorber ocnsotong of activated carbon Hydrocarbon vapours are adsorbed on the surface of activated carbon which remain there until they are periodically removed by passing stream through the system. The hydrocarbon are then condensed to liquids. are then condensed to liquid and can be used for further purpose.

(3) Absorption method : Liquid containing dissolved hydrocarbon is brought intimate contact of exhaust gases. The scrubbed exhaust then passes on, leaving the hydrocarbons trapped in the scrubbing liquid. The contact between absorbing liquid and exhaust gases is usually carried in tall scrubbing towers.

(4) Condensation method : Here the sufficiently low temperature will causes gaseous hydrocarbon i.e. smog formation pollutant to condense to liquids, which are collected and reused.

REFERENCES

1. Bowonder, B., *et. al.,* 1985. Avoiding Future Bhopals. *Environment.* 27, 6-13.
2. Carbide Considered Removal before Accident. *Economic Times,* 29th September 1985.
3. Firket, J., 1931. The Cause of the Symptoms Found in the Meuse Valley during the Fog of December, 1930. *Bill. Roy. Acad. Med.* (Belgium). 11. 683-739.
4. Greenburg, L.C., Erhardt, F. Field, J.I. Reed, and N.S. Seriff. 1962. Intermittent Air Pollution Episodes in New York City. *Public Health Repts.,* (U.S.) 78. 1061-1065, 1963.

CHAPTER 15

Industrial Hygiene, Occupational Health and Safety Management

Dr. Manju Tembhre[1]

Industrial hygiene is defined as the recognition, evaluation, and control of workplace hazards. Its origins are based on limiting personal exposures to chemicals, and have evolved to address the control of most other workplace hazards including over-exposure to noise, heat, vibration, and repetitive motion.

Industrial hygiene is "that science and art devoted to the anticipation, recognition, evaluation and control of those environmental factors or stresses arising in or from the workplace, which may cause sickness, impaired health and well-being, or significant discomfort among workers or among citizens of the community." Industrial hygiene is primarily concerned with materials, chemicals, pressures and energies encountered by man in his occupations[1].

According to the American Academy of Industrial Hygiene Code of Ethics for the professional, the practice of industrial hygiene, the primary responsibility of the industrial hygienist is as follows:

To employees:

- To protect the health of employees;
- To maintain an objective attitude toward the recognition, evaluation and control of health hazards regardless of external influences, realizing that the health and welfare of workers and others may depend upon the industrial hygienist's professional judgment;
- To counsel employees regarding the health hazards and necessary precautions to avoid adverse health effects.

To employers:

- To respect confidences, advise honestly and report findings and recommendations accurately;
- To act responsibly in the application of industrial hygiene principles toward the attainment of a healthful working environment;
- To hold responsibilities to the employer or client subservient to the ultimate responsibility to protect the health of employees.

Recognition and Control of IH's Hazards

Industrial hygienists recognize that engineering, work culture, and administrative controls are the primary means of reducing employee exposure to occupational hazards. Engineering controls minimize

[1]Principal, Sant Hirdaram Girls College, Bairagarh (B.U.) Bhopal (India).

employee exposure by either reducing or removing the hazard at the source or isolating the worker from the hazards.

Engineering controls include eliminating toxic chemicals and replacing harmful toxic materials with less hazardous ones, enclosing work processes or confining work operations, and installing general and local ventilation systems[2].

Work practice controls alter the manner in which a task is performed. Some fundamental and easily implemented work practice controls include

(1) following proper procedures that minimize exposures while operating production and control equipment;
(2) inspecting and maintaining process and control equipment on a regular basis;
(3) implementing good house-keeping procedures;
(4) providing good supervision
(5) mandating that eating, drinking, smoking, chewing tobacco or gum, and applying cosmetics in regulated areas be prohibited.

Administrative controls include controlling employees' exposure by scheduling production and workers' tasks, or both, in ways that minimize exposure levels. For example, the employer might schedule operations with the highest exposure potential during periods when very few employees are present.

When effective work practices and/or engineering controls are not feasible to achieve the permissible exposure limit, or while such controls are being instituted, and in emergencies, appropriate respiratory equipment must be used. In addition, personal protective equipment such as gloves, safety goggles, helmets, safety shoes, and protective clothing may also be required. To be effective, personal protective equipment must be individually selected, properly fitted and periodically refitted; conscientiously and properly worn; regularly maintained; and replaced as necessary(Refs. 3, 4).

Air Contaminants

These are commonly classified as either particulate or gas and vapor contaminants. The most common particulate contaminants include dusts, fumes, mists, aerosols, and fibers. Dusts are solid particles that are formed or generated from solid organic or inorganic materials by reducing their size through mechanical processes such as crushing, grinding, drilling, abrading or blasting.

Fumes are formed when material from a volatilized solid condenses in cool air. In most cases, the solid particles resulting from the condensation react with air to form an oxide.

The term mist is applied to a finely divided liquid suspended in the atmosphere. Mists are generated by liquids condensing from a vapor back to a liquid or by breaking up a liquid into a dispersed state such as by splashing, foaming or atomizing. Aerosols are also a form of a mist characterized by highly respirable, minute liquid particles.

Fibers are solid particles whose length is several times greater than their diameter.

Gases are formless fluids that expand to occupy the space or enclosure in which they are confined. Examples are welding gases such as acetylene, nitrogen, helium, and argon; and carbon monoxide generated from the operation of internal combustion engines or by its use as a reducing gas in a heat treating operation. Another example is hydrogen sulfide which is formed wherever there is decomposition of materials containing sulfur under reducing conditions.

Liquids change into vapors and mix with the surrounding atmosphere through evaporation. Vapors are the volatile form of substances that are normally in a solid or liquid state at room temperature and pressure. Vapors are the gaseous form of substances which are normally in the solid or liquid state at room temperature and pressure. They are formed by evaporation from a liquid or solid and can be found where parts cleaning and painting takes place and where solvents are used.

Chemical Hazards

Harmful chemical compounds in the form of solids, liquids, gases, mists, dusts, fumes, and vapors exert toxic effects by inhalation (breathing), absorption (through direct contact with the skin), or ingestion (eating or drinking). Airborne chemical hazards exist as concentrations of mists, vapors, gases, fumes, or solids. Some are toxic through inhalation and some of them irritate the skin on contact; some can be toxic by absorption through the skin or through ingestion, and some are corrosive to living tissue.

The degree of worker risk from exposure to any given substance depends on the nature and potency of the toxic effects and the magnitude and duration of exposure.

Information on the risk to workers from chemical hazards can be obtained from the Material Safety Data Sheet (MSDS) that Occupational Safety and Health Administration (OSHA'S) *Hazard Communication Standard requires* be supplied by the manufacturer or importer to the purchaser of all hazardous materials. The MSDS is a summary of the important health, safety, and toxicological information on the chemical or the mixture's ingredients. Other provisions of the *Hazard Communication Standard require* that all containers of hazardous substances in the workplace have appropriate warning and identification labels.

Biological Hazards

These include bacteria, viruses, fungi, and other living organisms that can cause acute and chronic infections by entering the body either directly or through breaks in the skin. Occupations that deal with plants or animals or their products or with food and food processing may expose workers to biological hazards. Laboratory and medical personnel also can be exposed to biological hazards. Any occupations that result in contact with bodily fluids pose a risk to workers from biological hazards.

In occupations where animals are involved, biological hazards are dealt with by preventing and controlling diseases in the animal population as well as proper care and handling of infected animals. Also, effective personal hygiene, particularly proper attention to minor cuts and scratches, especially those on the hands and forearms, helps keep worker risks to a minimum[5,6].

In occupations where there is potential exposure to biological hazards, workers should practice proper personal hygiene, particularly hand washing. Hospitals should provide proper ventilation, proper personal protective equipment such as gloves and respirators, adequate infectious waste disposal systems, and appropriate controls including isolation in instances of particularly contagious diseases such as tuberculosis.

Physical Hazards

There include excessive levels of ionizing and non-ionizing electromagnetic radiation, noise, vibration, illumination, and temperature.

In occupations where there is exposure to ionizing radiation, time, distance, and shielding are important tools in ensuring worker safety. Danger from radiation increases with the amount of time one is exposed to it; hence, the shorter the time of exposure the smaller the radiation danger.

Distance also is a valuable tool in controlling exposure to both ionizing and non-ionizing radiation. Radiation levels from some sources can be estimated by comparing the squares of the distances between the worker and the source. For example, at a reference point of 10 feet from a source, the radiation is 1/100 of the intensity at 1 foot from the source.

Shielding also is a way to protect against radiation. The greater the protective mass between a radioactive source and the worker, the lower the radiation exposure.

Non-ionizing radiation is also dealt with by shielding workers from the source. Sometimes limiting exposure times to non-ionizing radiation or increasing the distance is not effective. Laser radiation, for example, cannot be controlled effectively by imposing time limits. An exposure can be hazardous that is faster than the blinking of an eye. Increasing the distance from a laser source may require miles before the energy level reaches a point where the exposure would not be harmful.

Noise, another significant physical hazard, can be controlled by various measures. Noise can be reduced by installing equipment and systems that have been engineered, designed, and built to operate quietly; by enclosing or shielding noisy equipment; by making certain that equipment is in good repair and properly maintained with all worn or unbalanced parts replaced; by mounting noisy equipment on special mounts to reduce vibration; and by installing silencers, mufflers, or baffles.

Substituting quiet work methods for noisy ones is another significant way to reduce noise, for example, welding parts rather than riveting them. Also, treating floors, ceilings, and walls with acoustical material can reduce reflected or reverberant noise. In addition, erecting sound barriers at adjacent work stations around noisy operations will reduce worker exposure to noise generated at adjacent work stations.

It is also possible to reduce noise exposure by increasing the distance between the source and the receiver, by isolating workers in acoustical booths, limiting workers' exposure time to noise, and by providing hearing protection. OSHA requires that workers in noisy surroundings be periodically tested as a precaution against hearing loss[7,8].

Another physical hazard, radiant heat exposure in factories such as steel mills, can be controlled by installing reflective shields and by providing protective clothing.

Ergonomic Hazards

The science of ergonomics studies and evaluates a full range of tasks including, but not limited to, lifting, holding, pushing, walking, and reaching. Many ergonomic problems result from technological changes such as increased assembly line speeds, adding specialized tasks, and increased repetition; some problems arise from poorly designed job tasks. Any of those conditions can cause ergonomic hazards such as excessive vibration and noise, eye strain, repetitive motion, and heavy lifting problems. Improperly designed tools or work areas also can be ergonomic hazards. Repetitive motions or repeated shocks over prolonged periods of time as in jobs involving sorting, assembling, and data entry can often cause irritation and inflammation of the tendon sheath of the hands and arms, a condition known as carpal tunnel syndrome[9].

Ergonomic hazards are avoided primarily by the effective design of a job or jobsite and better designed tools or equipment that meet workers' needs in terms of physical environment and job tasks. Through thorough worksite analyses, employers can set up procedures to correct or control ergonomic hazards by using the appropriate engineering controls (e.g., designing or re-designing work stations, lighting, tools, and equipment); teaching correct work practices (e.g., proper lifting methods); employing proper administrative controls (e.g., shifting workers among several different tasks, reducing production demand, and increasing rest breaks); and, if necessary, providing and mandating personal protective equipment. Evaluating working conditions from an ergonomics standpoint involves looking at the total physiological and psychological demands of the job on the worker.

Overall, industrial hygienists point out that the benefits of a well-designed, ergonomic work environment can include increased efficiency, fewer accidents, lower operating costs and more effective use of personnel.

In sum, industrial hygiene encompasses a broad spectrum of the working environment. Early in its history OSHA recognized industrial hygiene as an integral part of a healthful work setting. OSHA places a high priority on using industrial hygiene concepts in its health standards and as a tool for effective enforcement of job safety and health regulations. By recognizing and applying the principles of industrial hygiene to the work environment, America's workplaces will become more healthful and safer[10].

Occupational Health

Improvement in the health and safety of the work environment is the result of a multidisciplinary effort involving people from several professions: industrial hygiene (recognition, evaluation and control of health hazards), occupational epidemiology (evaluation of risk and the demonstration of the occurrence of disease), industrial toxicology (evaluation of health effects), occupational medicine (diagnosis and treatment of work-related illnesses), occupational health nursing (biological monitoring of employees), occupational health administrator (workmen's compensation), safety engineer (designing and maintaining a safe environment), the environmental specialist (pollution control) and the compliance officer (regulatory controls and standards).

Occupational Disease and Injury

According to the best available estimates 100 million workers are injured and 200,000 die each year in occupational accidents and 68 million to 157 million new cases of occupational disease are attributed to hazardous exposures or workloads. Such high numbers of severe health outcomes contribute to one of the most important impacts on the health of the world's population. Occupational injuries and diseases play an even more important role in developing countries, where 70% of the working population of the world lives. By affecting the health of the working population, occupational injuries and diseases have profound effects on work productivity and on the economic and social well-being of workers, their families and dependants. According to recent estimates, the cost of work-related health loss and associated productivity loss may amount to several per cent of the total gross national product of the countries of the world.

The officially registered working population constitutes 60% - 70% of the adult male and 30%-60% of the adult female population of the world. When work at home and informal work are taken into consideration the percentage is even higher. In unfavorable cases the levels and intensities of hazardous exposures may be 10 or even 1,000 times greater at work than elsewhere. Workers in the highest risk industries, such as mining, forestry, construction and agriculture, are often at an unreasonably high risk and one-fifth to one-third may suffer occupational injury or disease annually, leading in extreme cases to high prevalence of work disability and even to premature death.

World Health Organization Target

By the year 2000, the health of workers in all Member States should be improved by making work environments more healthy, reducing work-related disease and injury, and promoting the well-being of people at work. This target can be achieved if effective measures are implemented in all Member States that:

- reduce disease, injury, disability and absence from work resulting from exposures to workplace hazards such as dust, noise, chemicals and stress;
- ensure that all employees have access to occupational health services;
- facilitate the adoption of work practices and routines that contribute to the health and well-being of workers;
- promote healthy lifestyles such as healthy nutrition, physical exercise and non-smoking;
- promote cooperation between relevant interest groups and sectors such as labour, industry, environment, education and health, and with the International Labour Office and other relevant international bodies, in the formulation and implementation of strategies.

Occupational Health Hazards

Occupational health hazards may mean (1) conditions that cause legally compensable illnesses, or it may mean (2) any conditions in the workplace that impair the health of employees enough to make them lose time from work or to cause significant discomfort. Both are undesirable. Both are preventable. Their correction is properly the responsibility of management. The basic principle applied is to "eliminate the danger at the source." This may involve:

- Elimination of a hazardous substance;
- Substitution of a less hazardous substance;
- Modification of the process;
- Control of the danger at the source (e.g., local exhaust ventilation);

- Control of the danger in the workplace (e.g., dilution ventilation);
- Use of personal protective equipment (e.g., respiratory protection);
- Administrative controls (e.g., work schedules).

Industrial Hygiene & Safety

The protection of worker health and safety presents complex challenges to business decision makers. Regulatory emphasis is shifting from enforcement toward collaboration and partnership with business and industry. Workers compensation costs continue to be a significant business concern. While many reform efforts have brought these costs under control, insurance premiums rank among the highest costs to a company after wages and benefits. In addition, efforts to circumvent workers compensation programs are leading to increased litigation. A strong health and safety program that emphasizes prevention of injuries and illness is the key to maintaining regulatory compliance, minimizing the impact of workers compensation losses, and reducing the potential for litigation to businesses as well as their key equipment and material suppliers. Worker health and safety is not just a laudable goal. Businesses have found that improvements in their health and safety programs can have a positive impact on their bottom line[12].

ENVIRON provides a wide range of specialized health and safety consulting services to international concerns ranging from multinational corporations to small businesses. Clients from the chemical, manufacturing, services, distribution, commercial, financial, and legal sectors, as well as national, state, and local agencies, rely on us to develop solutions that effectively address both business needs and health and safety program requirements. Our experienced and certified health and safety professionals offer clients comprehensive and integrated health, safety, and environmental capabilities, extensive technical expertise, and a long-standing reputation for excellence.

REFERENCES

1. Schwamm, H. World Trade Needs World-wide Standards, ISO Bulletin, September 1997.
2. International Standard Organization. ISO 9001:2000, Quality Management Systems—Requirements, Geneva, 2000.
3. International Standards Organization. ISO 14001:1996, Environmental Management Systems—Specification with Guidance for Use, Geneva, 1996.
4. British Standards Institution. BS 8800, Guide to Occupational Health and Safety Management Systems, 1996.
5. Redinger, C., BS 8800:1996—A British Occupational Health and Safety management System. The Synergist 8:32, 1997.
6. V.J. Davies and K. Thomasin "Construction Safety Hand Book" Thomas Telford Ltd., London, 1990.
7. Rao, CS, "Environmental Pollution Engineering: Wiley Eastern Limited, New Delhi, 1992.
8. Brown, D.B. System Analysis and Design for Safety, Prentice Hall, 1976.
9. The Manufacture, Storage and Import of Hazardous Chemical Rules 1989, Madras Book Agency, Chennai.
10. Encyclopaedia of "Occupational Health and Safety", Vol. I and II, published by International Labour Office, Geneva, 1985.
11. Terry Brimson, "The Health and Safety Guide", McGraw Hill Book Company, Europe-England.
12. Clayton & Clayton, Patty's "Industrial Hygiene and Toxicology", Vol. I, II and III, Wiley Inter Science, 1986.

CHAPTER 16

Water : Types, Resources, Quality and Pollution

Dr. S.A. Iqbal[1] and Ms. Vandana Magarde[2]

THE NATURE AND COMPOSITION OF WATER

Natural Characteristics

Water, (H_2O) hydrogen oxide, is an extraordinary chemical compound of fundamental environmental importance. The electronic strucutre of its mlecules impart the very special chemical and physical properties which lead to water described as 'the universal solvent' or 'the liquid of life'. Water in the gaseous state is made up of single molecules, the structure of which is usually represented as shown in Fig. 6.1. Water in the liquid state is made up of groups of molecules associated together by linkaages (hydrogen bonding) between each of the hydrogen atoms of one water molecule and the oxygen atom of an adjacent water molecules, as depicted in Fig. 7.2. In ice the water molecules are associated in tetrahedal structure formed by four molecules arranged around a central molecule. It is the nature of the molecular structure and the ways in which these molecules associate in hydrogen bonding which yield, among other chemical and physical characteristics, the general solvent power, the chemical reactant power in biological processes, and the heat storage and transfer power of water as a key environmental resource.

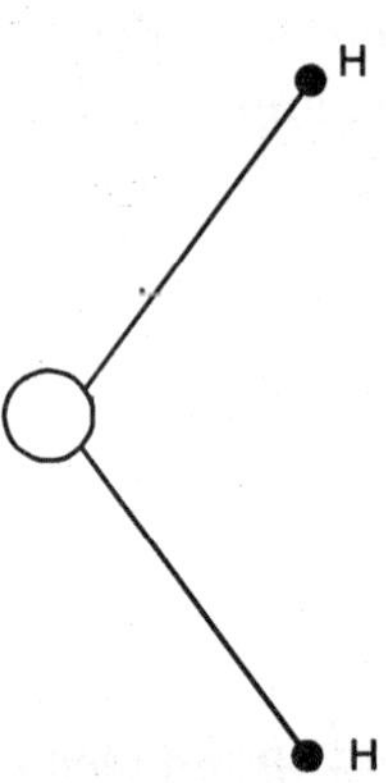

Fig. 16.1. Diagram of the water molecule

The earth's water resources, the 'hydrosphere', consists of the oceans and seas, the ice and snow of the polar regions and mountain glaciers, the water contained in surface soils and underground strata, and the water in lakes, rivers, and streams. Less than 1% of thes« resources consists of freshwater, some *2%* is freshwater ice located mainly in the polar regions, and the remaining 97% or so consists of seawater and sea ice. The annual evaporation of water from the hydrosphere, and its return as rainfall (the hydrological cycle) amounts to about $260 \times 10^{12} m^3$. The total water content of the atmosphere is about 7 × 10 m, indicating that atmospheric water is replaced on the average some 37 times a year.

For all practical purposes it can be said that the waters of the hydrosphere were of natural quality until the Industrial Revolution in Europe and North America initiated the development of technology driven by the energy of the fossil fuels, coal and oil. Now the stage has been reached where the entire

[1]Department of Chemistry, Saifia Science College, B.U., Bhopal (India)
[2]Department of Chemistry, Rajiv Gandhi College, B.U., Bhopal (India)

hydrosphere, except the polar ice which was formed before the industrial revolution began, is contaminated by the polluting activities of man. Prior to the 'Environmental Revolution' (which began to occur a mere 30 years ago) interest in water pollution in the developed world was very, limited. Now of course there is world-wide concern on pollution and all environmental matters. Neverthless the bulk of fresh water resources is not yet sufficiently contaminated by impurities derived . from the activities of man.

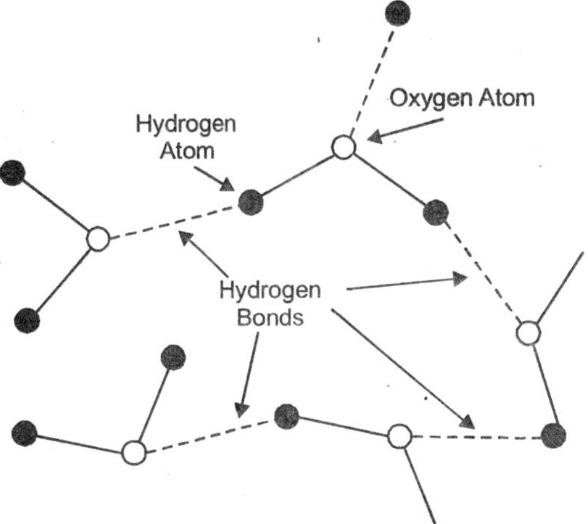

Fig. 16.2 Hydrogen bonding in water

However the pollution of water is not the only cause of man-made degradation of that environment. Water abstraction for domestic, industrial, and agricultural supply, and the lowering of near-surface groundwater levels to increase agricultural output, can bring about major depreciations of the aquatic environment." The quality of waters is defined in physical, chemical and biological (Including microbiological) terms. This definition may be lengthy and detailed, or short and broadly indicative, according to the purpose of the definition. Where the purpose is to decide whether or not the water is fit for a particular use, for example as a source of potable water supply, chemical parameters of quality will predominate in the definition. In contrast, if the purpose of the definition is to decide the extent to which a particular reach of river is subject to pollution, the occurrence of sporadic toxic chemical pollution is more likely to be deduced from observation of the biology and the physical characteristics of the river regime than by other means. Of course the proof of occurrence of any such sporadic chemical pollution will inevitably be expressed in chemical terms. Furthermore the parameters of water quality which are specified as limitations of the acceptability of a water for particular use purposes, or as a general environmental resource, are predominantly chemical and biochemical ones. Such specifications usually appear as 'quality standards', 'quality criteria', or 'quality guidelines' promulgated by international bodies and individual nations.

Rainfall, even when it was in its wholly natural state, was quite impure water in the scientific sense. Its physical characteristics in the liquid phase are that it would normally be naturally clear, colourless, and odourless with a temperature varying from ambient to colder depending on the nature and height of its precipitation. Rainfall naturally contains materials dissolved from the atmosphere. Firstly it contains the atmospheric gases—mainly nitrogen, oxygen, and carbon dioxide dissolved from the atmosphere. Secondly it contains matter dissolved from impurities in the air derived from the earth's surface—such as sea spray, volcanic emissions, wind-borne dusts, hydrogen sulphide and methane from anaerobic decomposition, and volatile organic compounds derived from land and aquatic plants. Thirdly it contains traces of ozone and other gases derived from chemical reactions, triggered by solar and cosmic radiation, of naturally occurring materials in the atmosphere. While it is essential to understand this natural chemistry of rainfall, it is the chemical consequences of the man-made pollutions of the atmosphere, which cause the greatest concern. Thus, apart from the effects of atmospheric pollution, the natural quality of rainfall is determined mainly by seawater and dust take-up, varying according to weather, locality, and major volcanic events. Rainfall naturally contains around 10 to 20 mg L^{-1} of dissolved solids according to circumstances. It is naturally somewhat acidic in character, with a pH value of around 5.5, because of its solution of atmospheric acid gases, mainly carbon dioxide.

On reaching the ground, rainfall picks up more impurities naturally from vegetation and the ground surface. If the ground is permeable, all or some of the rainfall will pass underground according to circumstances. Some will be evaporated and the balance will run off to streams and rivers and via underground strata *en route* to the sea. On the average some 70% of the rainfall evaporates directly, or indirectly via plant transpiration. Of the emainging 30% more travels through underground strata than along river and streams, and about 90% of the total run-off reaches the oceans. Natural run-off from bare, substantially impermeable, hard rock gathers material from gradual weathering of the rock to yield low concentrations of dissolved calcium, magnesium, sodium, and potassium. Traces of other materials will be present according to the chemistry of the rock. In contrast, where the rock underlies a cover of peat and associated vegetation the take up of stable organic matter will be high. The run-off from peat areas is usually acid in the p^H range 5 to 6, and has a pronounced yellow-brown colour, highly so in times of heavy rainfall. A similar situation applies to the run-off from woodland areas where ground cover of leaf mould produces a similar effect.

The run-off from clay soil is characterized by its content of calcium sulfate and calcium carbonate, compounds which are naturally abundant in clays. Turbidity in the water is often evident because of the presence of colloidal clay particles. An analysis of a typical water running off a clay area in south-east England is shown in Table 16.1. The run-off from clay soils during prolonged heavy rainfall, particularly from ground with little plant cover, is heavily loaded with suspended particles of clay and fine sand derived from the clay. While during heavy rainfall the water referred to in Table 7.1 would contain up to 0.1% suspended soil particles, the larger rivers of the world subject to tropical rainfall intensity carry relatively immense concentrations of solids in suspension.

Freshwater Rivers, Streams, and Lakes

By ard large the natural quality of rivers and streams (henceforth collectively referred to as rivers) at any point reflects the quality ot the upstream contributions of surface run-off and groundwater discharge. Similarly, but to a lesser degree according to circumstance, the natural quality of a lake reflects the quality of the inflows of water that maintain the lake level. However these waters, open to the energy of sunlight, the solution of oxygen from the air, and containing the mineral nutrients sufficient to supports plant growth, naturally become the media for the growth of aquatic biota. These biota can be divided into four categories, namely :

The phytoplankton, being plants mainly algae, floating within the water ;

The zooplankton, which include bacteria and protozoa, floating within the water ;

The rooted plants, mostly growing from the river or lake bed, but some floating on the water ;

The large animals which are either free swimming or are attached to the bed or the larger plants.

The nature, variety, and abundance of these biota—the biology of the water—is essentially determined by the physics and chemistry of the water. However there is a feed-back loop in this system which results in the growth of the biota exerting an impact on the physics and chemistry of the water—for example the effects of phytoplankton growth in increasing the oxygenation and alkalinity of the water, and in transferring some of the calcium, carbonate, and phosphate content of the water to the river or lake bed.

The natural shape of the channel of a river at any point is determined by the rate of flow (discharge) of water and the topology and geology of the terrain through which the river passes. These factors determine the gradient of the river, its velocity of flow, its depth and width, its shoals and pools, and the nature of the river bed at various points—silt, gravel, or bare rock as the case may be. Even as the growth of river biota has a secondary effect on the physical and chemical characteristics of the water, so the growth of the biota can influence the effective shape of the river channel. A good example of this is the heavy growth of rooted water plants which occurs in the shallow, calcareous waters of chalk streams. Heavy growths of these plants, if not controlled by judicious cutting and removal from the rivers, often so reduce the flow cross-section as to cause flooding during heavy summer storms. Another major cause of quality variability is the biochemical activity which proceeds within micro-organisms in the water, particularly bacteria, as key factors in the operation of the well known carbon and nitrogen cycles and other biochemical transformations. The most important of these are summarized in Table 7.2. It needs to be borne in mind of course that straight-forward chemical reactions occur also—for example the precipitation of metals from solution in the water.

Table 16.1. Main biochemical processes proceeding in water

Process	Requirements	Main and prducts
Oxidation of organic carbon	Disslved oxygen, T >0°C	CO_2, H_2O
Oxidation of ammonia	Dissolved oxygen, T>4°C	NO_3^- little NO_2^-
Reduction of nitrate	Dissolved xygen absent, T>0°C	N_2, some N_2O
Reduction of organic carbon	Dissolved oxygen absent nitrate absent, T>4°C	CH_4, CO_2

Water Pollution

It is part of the natural scheme of things that man, a saprophytic animal, should cause environmental pollution in almost all he does. Fortunately it is also part of the natural scheme of things that man, blessed with thinking ability, should recognize the need to control pollution and devise technological and administrative means of effecting this control. That this recognition is now apparently widespread does not necessarily mean that the required controlling action will be man-made pollution of water is divided into two kinds, namely into point sources and non-point sources. The most important of these are listed in Table 7.3. As might be expected, the point sources are mainly discharges of wastewaters from sewage works, factories, and farms, while the non-point sources arise mainly from specific categories of general land use (*e.g.* high intensity fanning). As regards the type of pollution which occur, these can be listed according to the effect exerted by the polluting matter, as follows :

Table 16.2. Main point and non-point sources of pollution

Point Sources	Non-point sources
Discharges from sewage treatment works to rivers	Run-off and underdrainage from agricultural land into rivers
Discharges of industrial wastewaters to rivers	General contamination of recharge rainfall to outcropping aquifers
Discharges of farm effluents to rivers	Septic tank soakaways into permeable strata
Discharges from small domestic sewage treatment plants to rivers	Wash-off of litter, dust and dry fallout, from urban roads to rivers
Discharges by means of well or borehole into underground strata	General en ry of sporadic and widespread losses of contaminants to rivers
Discharges of collected landfill leachate to rivers	Seepage of landfill leachate to underground strata and to rivers

(*a*) Substances acutely toxic to man and/or aquatic flora and fauna (*e.g.* lead, mercury, cadmium, cyanide, pesticides);

(*b*) Substances which are hazardous to man and/or to flora and fauna in causing chronic, or long dormant cumulative, harm (*e.g.* polynuclear aromatic hydrocarbons, chlorophenols, trihalomethanes);

(*c*) Substances at very low concentrations which are not highly toxic but which can either be rendered acutely toxic by biochemical transformation in the water (*e.g.* methylation of inorganic mercury) or by bioconcentration (*e.g.* triphenyl tin);

(*d*) Substances which add to the load of biochemical oxygen demand in the water or in benthal (bottom) deposits, (*e.g.* sewage effluent, food-industry wastewater, farm wastes);

(*e*) Substances which add to the eutrophication (plant-nutrient content) of the water (*e.g.* sewage, farm wastes);

(*f*) Substances which have a detrimental effect on the physical appearance of the water (*e.g.* oil, detergent foam, litter, suspended matter);

(*g*) Substances which only have a polluting effect at relatively high concentrations in water (*e.g.* mineral salts such as sodium chloride) ;

(*h*) The waterborne organisms which are pathogenic to man *(e.g. Salmonella, sp. Vibrio cholerae).*

In this list, the organic substances in categories (a), (b), and (c) exert their polluting effects at very low (μg L^{-1}) concentrations. The inorganic substances exert their effects at higher concentrations (in the range of mg L^{-1}). Those in categories (d) to (g) exert their effects at mulii mg L^{-1} concentrations. The hygienic safety of waters, through which man may be exposed to infection by the organisms in category (h), can only be high where bacteria of faecal origin cannot be detected in 100 ml of water (per litre in the case of a plaque forming unit of polio virus). dainik bhaskar

DIFFUSE POLLUTION

Pollution from non-point sources, or diffuse pollution, in its very nature presents difficult control problems. If it is known that a river, lake, or an aquifer is polluted by a particular substance or group of substances, but the point of access of this pollution to the water cannot be located and no particular person or corporate body can be proved to have caused or knowingly permitted the pollution to occur, the normal processes of penalizing polluters and making them pay for remedial and/or preventive measures are thwarted; Special research and administrative action has then to be taken, and since this is a difficult and slow process, such action can be almost guaranteed to take place too late to avoid major damage arising. There, are three main problems before us. The first is the acidification of waters caused by acid rain and the second is the high and still rising level of nitrate in many waters, particularly groundwaters. The third relates to water contamination by leachate from landfill sites and spills of chemicals.

Acidification of Waters

The acidification of rivers and lakes by acid deposition from the atmosphere is essentially a local problem. Much of the acidity is a part of the consequences of the major problem of atmospheric pollution in industrial areas.

The effects of acid deposition varies greatly according to the type of soil on which it falls. Alkaline soils based on limestone can neutralize large amounts of acid whereas soils based on peat or granite cannot do so. There are two practicable ways in which the impact of acidification can be cased. One is to reduce the emission of the atmospheric pollutants. The second is to add a neutralizing alkali (such as powdered limestone) to the acid-sensitive areas. The key link between the emissions and their ecological impact is the transfer of the acidity from deposition to run-off to rivers and lakes, and understanding of this link is essentially a matter of the chemistry of the interactions between the soil and the soil water. In essence the lightly buffered soils, thin and base poor, are the most sensitive to acid deposition and arc the main generators of acid rivers and lakes from that deposition.

Nitrate in Waters

The EC directive on the quality of water intended for human consumption sets the maximum acceptable limit for nitrate in water at 50 mgL^{-1} (11.3 mg L^{-1} nitrate nitrogen).

Nitrate problems in groundwaters arise mainly because of heavy loadings of nitrogenous matter on land through which percolating rainfall recharges the underlying aquifer. The only practicable way of reducing the rate of increase of nitrate concentration, and eventually reducing the actual concentration in the groundwater, is to reduce the intensity of agricultural activity which produces the heavy nitrogenous loading. There is plenty of evidence that nitrate levels in many groundwaters have increased greatly over recent decades, in some case to the extent that water abstractions from boreholes has had to be curtailed or even abandoned. Further, computer models indicate that this position will worsen unless remedial action is taken.

Water Pollution via the Contamination of Land

The subject of land contamination by the disposal of solid wastes and sludges, and spillages of polluting liquids, to land will be dealt with in forthcoming chapters. The sound management of water resources demands of course that the entry of foul leachates from landfill sites and other sources to waters should be kept under effective control. The greatest difficulty arises when valuable groundwater sources are rendered unfit for use, often by quite small levels of contamination, from leachates from landfill or spillages. It is usually most difficult, and most costly, to decontaminate the water source within a reasonable time, and there are many groundwater sources that have been abandoned because of the virtual irreversibility of the contamination. The modern techniques of landfill control and management are developing such that new leachate pollution can be avoided and serious existing contamination greatly reduced.

Table 16.3. Normal chemical characteristics of some public water supplies (ppm)

	Source of water			
	Chalk borehole (softened)	Upland river	Lowland river	Lowland reservoir
pH value	73	73	7.9	7.6
Colour (units)	0	2	5	4
$NH_3(N)$	0	0.10	0.02	0.03
$NO_3(N)$	3.0	1.2	4.0	1.7
Total solids	220	80	410	430
Total hardness ($CaCO_3$)	150	45	260	225
Non-CO_3 hardness	10	25	135	70
Alkalinity ($CaCO_3$)	110	20	130	160
Chloride	16	15	35	55
Iron	0.01	0	0.02	0
Manganese	< 0.01	<0.01	<0.01	<0.01

Table 16.4. Some aspects of water quality specified in the EC drinking water Directive

Contaminant	Maximum admissible concentration
Acidity/alkalinity	Maximum admissible concentration
Acidity/alkalinity	pH range 5.5 – 9.5
Colour	20 units
Turbidity	4 untis
Iron	$200\ \mu l^{-1}$
Manganese	$50\ \mu g\ 1^{-1}$
Aluminium	$200\ \mu g\ 1^{-1}$
Nitrate (nitrogen)	$11.5\ mg\ 1^{-1}$
Lead	$50\ \mu g\ 1^{-1}$
Pasticides (individual substances)	$0.1\ \mu g\ 1^{-1}$
Pesticides (total)	$0.5\ \mu g\ 1^{-1}$
E. coli	Not detectable in 100 ml of water
Coliform bacteria	95% of samples should be coliform free

Table 16.7. Algae and invertebrate animals of concern in public water supply

Type of organism	Common species or genera	Cause of presence in public supply
Small single-celled *Chlorophyta* (green algae)	*Chlamydomonas, Arkistrodesmus Scendesmus*	*Passing through water works filters*
Bacillariophyta (diatoms)	*Stephonodiscus*	Passing through waterworks filters
Cyanophyta (blue-green algae)	Various species	Passing through waterworks filters
Cyanophyta	Various species	Decay of algae causes taste and odour in water
Xanthophyta (yellow-greenalgae)	*Chryptomonas Rhodomonas*	Passing through waterworks filters
Xanthophyta	*Synura peridinium*	Causes cucumber taste Causes fishy odour
Worms	*Nias*	Can pass filters and infest water mains
Rotifera	Various species	Passing through filters
Fly larvae	*Chironomid* sp.	Can pass filters and infest water mains
Crustacea	*Daphnia, Cyclops, Asellus*	Can pass filters and infest water mains

The nature and extent of the treatment to be given to any particular raw water, to produce a public supply meeting specified quality standards at minimum cost, depends, entirely on the nature and quality of the raw water. The minimum treatment, of disinfection using chlorine, would be appropriate for a deep groundwater source in a rural area, or for water from an upland impoundment not open to public access. At the opposite extreme, water drawn from the lowland reaches of a river, containing a high proportion of sewage or industrial wastewater discharge upstream, would need extensive and very carefully monitored treatment. The conventional techniques for the treatment of raw waters for public supply are set out in Table 6.8., and Fig. 6.3 shows the layout of a typical waterworks treatment plant.

Appendix Table :

Table 16.6. Various Devices used for the elemental analysis

Element	Devices used
Na	Flame photometer
Ca	Titration EDTA
Mg	Titration EDTA
K	Flame photometer
HCO_3^-	Titration acid
Cl	Ion chromatograph
F^-	Ion chromatograph
NO_3	Ion chromatograph
SO_4^{--}	Ion chromatgraph
pH	PH meter
TDS	Oven 105°C

Table 16.7. Chemical Analysis of Tap and Zamzam water

Elements	Tap water	Zamzam well water	Zamzam pipe water
Na	37.8	133	135
Ca	75.2	96	96
Mg	6.8	38.88	38.88
K	2.7	43.3	43.2
HCO_3^-	70.2	195.4	195.4
Cl^-	73.3	163.3	159.7
F^-	0.28	0.72	0.68
NO_3^-	2.6	124.8	126.1
SO_4^{--}	107	124	123.3
pH	7.2	8	7.90
TDS	350	835	840

Note : all values except pH are in mg/*l*

Table 16.8. Concentration of Ions and Salinity in Sea Water

Chemical Ion Contributing to Seawater Salinity	Concentration in (parts per thousand) in average seawater	Proporation of Total Salinity (No matter what the salinity)
Chloride	19.345	55.03
Sodium	10.752	30.59
Sulfate	2.701	7.68
Magnesium	1.295	3.68
Calcium	0.416	1.18
Potassium	0.390	1.11
Bicarbonate	0.145	0.41
Bromide	0.066	0.19
Borate	0.027	0.08
Strontium	0.013	0.04
Fluride	0.001	0.003
Other	less than 0.01	less than 0.001

Table 16.9. Average Chemical Compositin (ppm) of Some Indian Rivers

River	HCO_3^-	Cl^-	SO_4^{--}	SiO_2	Ca	Mg	Na	K	TDS
Brahmaputra	56	11	4	7	14	5	7	3	107
Cauvery	135	20	13	23	21	9	43	4	272
Ganga	128	10	11	18	25	8	11	3	241
Godawari	105	17	8	10	22	5	12	3	181
Gomti	274	9	15	15	30	19	27	5	394
Indus	64	5	23	5	54	12	10	0.3	173
Krishna	178	38	49	24	29	8	30	2	360
Mahanadi	122	23	3	17	24	13	14	8	224
Narmada	225	20	5	9	14	20	27	2	322
Tapi	150	65	1	16	19	22	48	3	322

STANDADRS OF QUALITY OF WATER

The U.S. Public Health Service has laid down the following standards for drinking water.

(a) Physical Characteristics : The drinking water should be free from such impurities which would cause offensive taste, smell and sense of sight. Following physical limits should not be exceeded otherwise water will becme unfit or drinking (Table 16.10).

Table 16.10

Characterstic	Acceptable	Cause of rejection
(*i*) Turbidity (on J.T.U. scale)	2.5	10
(*ii*) Temperature	10°C to 15.6°C	
(*iii*) Taste	Unobjectionable	unobjectionable
(*iv*) Odour	unobjectionable	unobjectionable
(*v*) Colour (on Platinum cobalt scale)	5.0	25.0

(b) Chemical characteristics : The concentration of metals and other chemical substances in potable water should not be execeded by the amount given in the following table 16.11.

Table 16.11

Characterstic	Acceptable	Casue of rejection
1. pH value	7.0 – 8.5	6.5 – 9.2
2. Total dissolved solids	500	1500
3. Total Hardness (as $CaCO_3$) in mg/L	200	600
4. Chlorides (mg/L)	200	1000
5. Sulphates (mg/L)	200	400
6. Fluorides (mg/L)	1.0	1.5
7. Nitrates (mg/L)	45	45
8. Calcium (mg/L)	75	200
9. Magnesium (mg/L)	30	150
(If there are 250 mg/L of sulphates, Mg content can be increased to a maximum of 125 mg/L with the reduction of sulphates at the rate of one unit per every 2.5 units of sulphates).		
10. Iron	0.1	1.0
11. Manganese	0.05	0.5
12. Copper	0.05	1.5
13. Zinc	5.0	15.0
14. Phenolic compounds	0.001	0.002
15. Anionic detergents	0.2	1.0
16. Mineral oil	nil	nil
Toxic Materials		
17. Arsenic	0.05	0.05
18. Cadmium	0.01	0.01
19. Chromium	0.05	0.05
20. Cyanide	0.05	0.05
21. Lead	0.1	0.1
22. Selenium	0.01	0.01
23. Mercury	0.001	0.001
24. Poly-nuclear aromatic hydrocarbons	0.2 hg/l	0.2 hg/l
Radio Activity		
25. Gross Alpha activity	3 pcu	3 pcu
26. Gross Beta activity when pcu = pico curie unit.	30 pcu	30 pcu

Bactereological Standards :

(i) **Water entering the distribution System :** Coliform count in any sample of 100 ml should me zero. A sample of the water entering the distribution system that does not conform to this standard calls for an immediate investigation into both the efficiency of the purification process and the method of sampling.

(ii) **Water in the distribution :** Water in distribution system shall satisy all the three criteria indicated below.

(A) *E-coli* count in 100 ml sample should be zero.

(B) Colifrm organisms not more than 10 per 100 ml shall be present in any sample.

(C) Coliform organisms should not be detactable in 100 ml of any two consecutive samples or more than 50% of the samples collected for the year.

The appearance of higher numbers in any sample should necessitate the investigation and removal fo the source of pollution.

(iii) **Individual or small community supplies :** E-coli-count should be zero in sample of 100 ml and coliform organisms should not be more than 3 per 100 ml.

Virolgical standard : 0.5 ml/l of free chlorine residual for one hour is considered sufficient to inactivate virus, even in water that was originally polluted. This free chlorine residual is to be insisted in all disinfected supplies in areas suspected of endemicity of infections hapatities to take care f the safety of the supply frm wirus point of view which incidentialy take care of the supply from virus point of view which incidentially takes care of the safety from the bacteriological point of view as well. For other areas 0.2 mg/l of free chlorine residual for half an hour should be insisted.

The International standards for drinking water are given in Table 16.12.

Utilization of Available Water for Various Purposes

Efforts should be made for optimum utilization of available water in every state. Water resources should be so protected as to ensure their maximum conservation an minimum misuse and erosion. Water resources should be planned on river basins or sub-basins. Every development scheme should be so planned as to ensure its inclusion in the integrated planning framework of the basins and the best options are chosen. Quality of water should be taken full care of in these plans so that there is no adverse effect on the current utilization of water. There are three stages of development of water resource projects,

1. Construction
2. Utilization and Operation
3. Scientific study of the projects impacts on crops and social structure

These three stages should be duly taken into consideration while planning new schemes. Suitable organizations should be setup for this integrated and coordinated development and management. Units should be constituted for different institutions so that not only irrigation but other uses of water are also properly adjusted in the schemes chalked out by them.

Utilization and re-use cycle of water should be considered an important component of water resources development. Utilization of surface water should augment ground water and should be utilized for surface water schemes. These should be used as complimentary to each other. It should be made sure that purity of water is maintained in this process.

Water Utilization Priorities

Priorities of water utilization in planning and coordination of different systems should be as under:

- Drinking water

Table 16.12. International Standards for Drinking Water (Prescribed by WHO)

Substance	Maximume Allowable Limit
(*a*) Toxic Metals	
Lead	0.05 mg/L
Arsenic	0.05 mg/L
Barium	0.10 mg/L
Cadmium	0.01 mg/L
Chromium	0.05 mg/L
Selenium	0.01 mg/L
Cyanide	0.20 mg/L
(b) Substances Affecting Health	
Fluoride	1.5 mg/L
Nitrate	45 mg/L
(*c*) Substances Affecting Potability of Water	
Total Solids	1500 mg/L
Colour (Platinum cobalt scale)	50 units
Turbidity (Turbidity units)	25 units
Taste	—
Odour	—
Iron	1.0 mg/L
Managanese	0.5 mg/L
Copper	1.5 mg/L
Calcium	200 mg/L
Magnesium	150 mg/L
Chlride	600 mg/L
Sulphate	400 mg/L
Magnesium + Sodium sulphate	1000 mg/L
Phenolic cmpounds	0.002 mg/L
Carbon Chlorofrm Extract (CCE)	0.5 mg/L
Alkyl benzene sulphonates pH range	Surfactants not less than 6.5 or greater than 9.2
(*d*) Radiological Requirements	
Strontium–90	30 μ μc/L
Radium–226	10 μμ c/L
Gross beta concentraton (in the absence of Sr–90 and alpha emitters)	1000 μμ c/L

- Irrigation and land management
- Power generation/Industrial use

However, in view of the requirements of a particular area, these priorities maybe recorded. Checking of natural water flow of rivers by constructing small and big dams are barrage for different private industrial or irrigation purposes makes an adverse impact on low lying areas and sometimes it creates drinking water problem. To effectively curb this, the right of Water Resources Department on all natural water sources should be recognized by law. Utilization of natural water sources by private industries or irrigation should not be allowed without permission of Dater Resource Department so that the department can make allocation of water keeping in view the above said priorities for different purposes. After utilization of river water, the drained water should be decontaminated as per requirement so that the drinking water in rivers does not get polluted. Water should be reused for industries by rotation method. After purification water should be stored by industries in their own tanks so as to reduce the dependence on rivers and tanks. In view of the growing drinking water demand in rural as well as urban areas and the rapid industrialization, advance proposal should be invited from the PHE department in context of deciding on water planning for proposed major irrigation projects and efforts should accordingly, be made to ensure water storage capacity to facilitate fulfillment of requirements as per the priorities.

Potable Water

Drinking water in adequate quantity should be provided to the entire population in urban as well as rural areas in next five years. Drinking water supply should form integral part of the irrigation projects in areas devoid of any alternative source of drinking water. To meet the future drinking water requirements of cities, reservoirs should have to be constructed in adjacent water catchments areas to supply irrigation as well as drinking water. Thirst of humans and other beings should always be given a priority from the available water. To save the water sources from pollution, rules should be framed to prevent establishment of industries along the banks of rivers/lakes up to a certain distance in the catchments area form where drinking water is taken. The farmers should be provided adequate water as and when required for their crops. Incase of water shortage, availability of adequate water maybe ensured for drinking and irrigation purposes through efficient water management. Quality of surface and drinking water should be regularly tested by concerning departments in order to improve-water quality.

Irrigation and Land Management

In irrigation planning for various projects we should keep in mind that there should be economy in utilization of land for irrigation. The water should be provided for maximum crops for irrigation facility and therefore the benefit should go to the small farmers, at the same time the need of higher production should also be taken into consideration. There should be a close relation between the policies of water utilization and leveling and utilization of land. For irrigation projects allocation of water should be based on principles of equality and justice. If the availability of canal water is there it should be provided to the small farmers on priorities. In order to encourage drip and sprinkle irrigation, there is a need to develop a demonstration forms in different areas of every canal with the cooperation of agriculture department.

Power Generation and Industrial Use

Since last many years stress has been laid on restructuring of the price fixation system for water supply, whether it is for irrigation or domestic or for industrial supply so that the financial return from water supply can be used not only for operation and maintenance of water system, but also to make consumers aware of the importance of economizing on water. Water rates should be so fixed as may make the consumers realize the in-valuability and importance of water and inspire them to economize on water utilization. It is necessary that all the projects should be self sustainable. In view of the importance of forests for environmental production, concessional rates should be fixed for irrigation and aforestation.

NGO's and Folklores Participation

It is necessary to have a participation approach for the various uses for efficient management of water resources. Not only government agencies, officers and NGO's but also the users concerned with the various aspects of effective and decisive planning, design, development and management of water resources should be involved and to ensure realization of the above objectives necessary institutional changes would have to be effected at different levels. Constant efforts have to be made to gradually increase the participation of farmers in the different arrangements of canal irrigation system. Assistance and giving know how from various NGO's should be taken an imparting training to farmers in efficient irrigation management. This will increase the sense of responsibility among farmers in works for the distribution of water.

Utilization of Every Drop of Water

Efficiency of water utilization, be it different processes or in irrigation or in industrial. Construction, means obtaining higher production from unit quantity of water. Utilization of water can be economized by adopting amended systems of drip and sprinkle irrigation systems. Drip irrigation is especially suitable for gardens and sugarcane cultivation. Although, the installation cost of drip system is very high, yet it saves 25 to 50 percent of water and may lead to 15 to 25 percent increase in production. The sprinkle irrigation system is particularly suitable for sandy, dry and semi-dry areas. Accordingly, the land of top areas should be prepared and farmers should be motivated to take to drip or sprinkle irrigation system.

As far as possible, private sectors participation should be encouraged in different aspects of planning, development and management of water resources projects of different purposes. Participation of private sector can be helpful in introducing new modified concepts, creating financial resources and starting corporate management for improving service efficiency and accountability of the users. Different linkages of private sectors participation can be considered in creating water resources facilities, ownership implementation, lease system and transfer.

Flood Awareness and Management

A master plan should be drawn up for the flood control and safety in flood-prone areas. Such solid arrangement should be made by intensive soil-conservation, treatment of catchments areas, protection of forests, increasing forest cover and construction of check dams or to reduce the possible loss due to floods. Wherever possible, provisions should be made for additional flood water holding capacity in reservoirs to facilitate flood control. To avert loss of life and properties due to floods a well -nit communication network and flood forecast system should be developed so that the [people of flood-prone areas can be alarmed on time and to regulate the settlement of people and economic activities in the areas.

Role of Science and Technology

There is need to make all-round expansion of technical knowledge for efficient and suitable arrangement of water resources from economic point of view. Knowledge and technology has a significant role to play in development of water resources in rural areas. Intensive research is necessary in the following areas:

- Hydrometeorology
- Hydrology and ground water recharging
- Restricted use of canal water
- Conservation of water in agricultural fields through appropriate measures
- Development of water wealth in catchments area
- Economic designing of water resources projects
- Designing of dams, canals and related hydel power works by high techniques
- Continuity of water flow in rivers and purification of drinking water
- Scientific study of cropping pattern
- Study of the process of silting in reservoirs and the measures to reduce it
- Safety of water related construction works

- Research on river structure and dam construction material
- Improvement in water distribution pattern and selection of low cost alternatives
- Distribution of adequate quantity of water as per-requirement
- Study on water purification and re-use
- Effective drainage system
- Hazard analysis, disaster management

To conduct research work, it will be necessary to re-organize the jurisdiction of the director, irrigation research and to make high-tech arrangement at research laboratories. Establishment of a training institute for engineers will also be necessary for this. This will help prepare a cadre of dedicated engineers with profound technical knowledge.

Meaningful Training

To ensure standard training advanced arrangements should form part of the Water Resources Department. This should cover field planning of the information system, project designing and solid arrangements for construction. Along with the employees of all categories, farmers should also be included in the training. The main objective of the training should be to get maximum productivity from each water unit.

In view of the most important of water for the humans and other beings and to maintain ecological balance, for all types of economic and development activities and keeping in view the increasing rarity of it, now it has become all the more necessary to ensure more efficient planning and management and optimum, restricted and equitable utilization of this resource. Success of the state's water policy would depend on its wider acceptability and maintenance and on the commitment to its underlined principles and objectives.

In order to streamline the system of water management from the canals of the water resources department, it is proposed to impart special training to Sub-Engineers, Assistant Engineers and Executive Engineers in water management. In the training policy it should be necessary to provide specialized technical knowledge and water management training in their respective fields even after their promotion. With the establishment of an engineers training institute, the engineers will receive high level in the service technical training and the quality of construction works would also improve.

Conclusion

1. States are rich in water resources. A scientific exploitation of this vast wealth would correct regional imbalances and lead to optimum utilization of water in a just manner. Development will also get fresh momentum.
2. The state's water policy will give a scientific basis to the assessment of resources, integrated planning. For this, the state will be required to change its organizational structure, impart training at different levels and to go for modernization.
3. Reasonable price should be taken from the different consumers for irrigation water, so that financial resources can increase. The funds, thus received, would help in maintenance and modernization of projects.
4. It is necessary to draw up a time-bound action plan for development of water wealth in every area as per State's Water Policy.
5. The agreements entered into with neighboring states about sharing and utilization of water can be reviewed under the State's Water Policy.

CHAPTER 17

Water Pollution and Water Borne Diseases

Dr. Khalid M.S. Al-Ghamdi[1]

It is part of the nature scheme of things that man, a saprophytic animal, should cause environmental pollution in almost all he does. Fortunately it is also part of the natural scheme of things that man blessed with thinking ability, should recognize the need to control pollution and devise technologic and administrative means o effecting necessarily mean that the required controlling action will be Man-made pollution of water is divided into kinds, namely into point sources.

Table 17.1. Main point and non-point sources of pollution

Point Sources	Non-Point Sources
Discharges form sewage treatment works to rivers	Run-off and under drainage from agricultural land into rivers
Discharge of industrial wastewaters to rivers	General contamination of recharge rainfall to outcropping aquifers
Discharge of farm effluents to Rivers	Septic tank soakaways into permeable
Discharges from small domestic sewage treatment plants to rivers	Wash-off of litter, dust permeable strata urban roads to rives
Discharge by means of well or Borehole into underground strata	General of sporadic and widespread losses of contaminants to rivers
Discharge of collected landfill Leachate to rivers	Seepage of landfill leachate to underground strata and to rivers

(*a*) Substances acutely toxic to man and/or aquatic flora and fauna (*e.g.* lead mercury, Cadmium, cyanide, pesticides);

(*b*) Substance which are hazardous to man and/or to flora and fauna in causing chronic, or long dormant cumulative, harm (*e.g.* polynuclear aromatic hydrocarbons, chlorophenols, trihalmethanes);

(*c*) Substances at very low concentration which are not highly toxic but which can either be rendered acutely toxic by biochemical transformation in the water (*e.g.* methylation of Inorganic mercury) or by biococentration e.g. triphenyl tin, see case study 1);

(*d*) Substances which add to the load of biochemical oxygen demand (B.O.D.) in the water or in Benthic (bottom) deposits, (*e.g.* sewage effluent, food-industry wastewater, farm wastes;

[1]Faculty of Science, King Abdul Aziz University, Jeddah (Saudi Arabia)

(*e*) Substances which add to the eutrophication (plant-nutrient content) of the water (*e.g.* Sewage, farm wastes);

(*f*) Substances which have a detrimental effect on the physical appearance of the water (*e.g.* oil, detergent foam, litter, suspended matter);

(*g*) Substances which only have a polluting effect at relatively high concentrations in water (*e.g.* minerals salts such as sodium chloride);

(*h*) The waterborne organisms which are pathogenic to man (e.g. *Salmonella*, *Cholerae Vibrio*).

Water and Health

Clean water was just as difficult to guarantee 2,000 years ago as it is today. The Romans did not have filtration plants. chlorination facilities, or similar modern equipments. They had to rely on local streams and springs or transport the water over long distances. The Roman aqueducts, a number of which are still in use today. provided clean water to cities. where there was good water, it was planned by reputed architects and builders.

Different water quality standards are required depending on the use for which the water is intended. The major beneficial uses are listed below :

- Domestic water supply, including drinking water.
- Industrial water supply, including cooling water. Agriculture water supply for irrigation.
- Livestock and wildlife water.
- Swimming, bathing, water skiting, surfing and other water sports. Boating and aesthetic enjoyment.
- Hydroelectric power and navigation.
- Waste disposal, dispersion, and assimilation

Bacterial and Parasite Pollution

Bacterial contamination is the most common water-borne disease hazard in the world, both bacterial and other types of organisms ever-present threats. Typhoid fever and cholera periodically cause widespread illness and death. Various forms of gastroenteritis * are common. The protozoan *Entamoeba hystolitica* causes a type of dysentery that is often serious. Parasitic worms of various kinds are transmitted by water. Eggs of domestic and wild animals may be a source of contamination. Several species of flukes, all of which are parasitic worms, cause serious diseases in different parts of the world.

An early method of determining the microbial purity of water was to take a reading of its "Keeping power" water that was free of bacteria could be stored for a long time, while contaminated water would often develop growths and undergo visible changes. Increasingly widespread use of the microscope made it possible to establish standard based on the number and kinds of organisms that could be found in the water. Later, bacteriological tests were devised that were reliable and specific for the detection and identification of microorganism.

Human Waste

Part of the living process is to get rid of unwanted material. Otherwise organism die of the toxins in their own waste. The biological process for ridding the body of waste is called excretion and takes place in some manner in all living creatures. Man and other vertebrates have several mechanisms for excretion. In man, the skin eliminates unwanted water, salts, and carbon dioxide; the lungs expel carbon dioxide and water; the liver drains bile constituents into the alimentary canal; the kidneys purify the blood and expel the wastes through the urinary system; the intestinal tract gets rid of salts, minerals, fats, an indigestible material rejected as unsuitable for food or metabolic processes.

The excreta of the alimentary canal is called faeces. It consists primarily of intestinal bacteria, which comprise the most bulky portion of human waste. The urine, of which about 1.5 liters are excreted daily by a human adult, is mostly water with dissolved nitrogenous wastes and salts. The urine also carries away foreign substances such as drugs and other toxins.

Water Purification and Supply

The primary requirement regarding the quality of public water supplies is the public health requirement that the water should be 'wholesome' for drinking. Wholesome in this context is generally interpreted to mean, as stated in dictionaries, 'promoting or conducive to health'.

While the quality of public supply, like any other water, is described in physical, chemical, and biological terms, it is the microbiological quality which matters the most -simply because it is very easy for water to be contaminated by the micro-organisms causing human disease and major epidemics. The physical characteristic of public supply are measured in terms of its colour, turbidity, odour, taste, and pH value. The chemical characteristic cover a wide and complex field, embracing low concentration of the acutely toxic inorganic chemicals, very low concentration of the hazardous organic micro-pollutants such as polycyclic aromatic hydrocarbons, haloforms as produced in the chlorination of organic residuals in water, pesticides mercury, and cadmium. Also specific inorganic chemical such a fluoride, nitrite, magnesium, and sulphate which at particular levels of concentration are or may be harmful to health are included.

Table 17.2. Normal chemical characteristic of some public water supplies (ppm)

	Source of water			
Chalk borehole	**Upland (softened)**	**River**	**Low land river**	**Low land reservoir**
pH value	7.3	7.3	7.9	7.6
Colour (units)	0	2	5	4
NH_3(N)	0	0.10	0.02	0.03
NO_3(N)	3.0	1.2	4.0	1.7
Total solid	220	80	410	430
Total hardness ($CaCO_3$)	150	45	260	225
CO_3 hardness	10	25	135	70
Alkalinity ($CaCO_3$)	110	20	130	160
Chloride	16	15	35	55
Iron	0.01	0	0.02	0
Manganese	<0.01	<0.01	<0.01	<0.01

The importance of the microbiology of public supply concentrates on the organism responsible for waterborne human disease these may be classified as bacteria, virus and other organism such as worms flukes and protozoa which are known to be associate with the waterborne spread of dieses. Table 6.6 gives a very brief summary of the main aspects of these microbiology in the day to day control of the microbiology o public water supply since it is not practicable to examine waters frequently for the presence of all the organism that cause disease, Standard bacteriological examination of water in a relatively simple but effective way is carried out.This involves assessing the numbers of bacteria.

Escherichia coli (E. coli) and the coli form group is generally present in water. The formers is indicative of the faecal pollution of water while the presence of the other member of the group is indicative of pollution from animal. Sources but not necessarily of faecal pollution.

Table 17.3. Summary of Main Sewage Treatment Processes

Type of organism	Name	Source	Disease
Bacterium	*Salmonella typhi*	Man	Typhoid fever
Bacterium	*Salmonella typhi*	Man	Paratyphoid fever
Bacterium	*Vibrio cholerae*	Man	Cholera
Bacterium	*Shigella*	Man	Bacterial dysentery
Bacterium	*Salmonella group*	Man and animal	Gastro-enteritis (food poisoning)
Bacterium	*Leptospira interoha morrhagiae*	Rats	Weil's disease
Protoza	*Entamoeba histolytica*	MAN	Amoebic dysentery
Tapeworm	*Taenia saginata*	Manvia Cattle	

Wastewaters and Wastewater Treatment

The present system of wastewater disposal worldwide is based on the provision and operation of public sewerage reticulation and sewage treatment plants in urban areas, and the private provision of drains and wastewater treatment arrangement at industrial premises and stock rearing farms. Much of the industrial wastewater produced in urban areas is disposed of, with or without pre-treatment, into public sewers with the consents of the water services. This general position applies in many countries but in others, particularly developing ones industrial wastewater disposal direct to river is the norm. Wastewater produced on farms are usually too strong in the load of organic matter they carry for disposal to public sewerage systems. Instead, efforts are made to contain the farm wastes on the producing farm where the fertilizer value of the wastes can be realized.

Sewage Treatment

Modern sewage treatment employs three basic processes, namely :

(*a*) The removal of pollution matter from the sewage flow as solid, or slurries of solid in water. (sludge);

(*b*) The removal of polluting matter from the sewage flow and from separated sludge by accelerated natural process of biochemical breakdown;

(*c*) The separation of water from sludges to reduce the volume of sludge for disposal.

Table sets out the conventional description of these processes and the type and nature of the treatment given, and Fig 6.4 shows the layout of typical sewage treatment works.

Table 17.4. Summary of Main Sewage Treatment Processes

Process	Type of treatment	Basic Purpose
Preliminary	Screening and grit removal	Removal of gross of abrasive solid
Primary settlement	Settlement in tanks	Removal of solid and grease
Secondary	Activated sludge per-collating filter other bio-reactor plus settlement.	Bio-oxidation carbonaceous matter and ammonia and removal of solid
Polishing treatment	Sand filtration micro-starting or lagooning	Removal of very fine solids
Tertiary treatment	Denitriication chemical precipitation	Removal of very fine solids residuals.
Sludge treatment	Digestion thickening dewatering drying	CH_4 Production, Preparation disposal

Treatment of Industrial Wastewaters

Industrial wastes can be divided into four categories, namely :

1. Wastewaters that have been changed in chemical quality during use in or in connection with, manufacturing processes. This category consists of process waters and unclean cooling waters.
2. Contaminated run-off from roofs and yards at industrial premises.
3. Clean cooling waters.
4. Grossly contaminated wastewaters, waste chemical and liquid or semi liquid sludges of small volume kept separate, or separated from, the main wastewater streams of manufacturing processes. These wastes are usually transported away for specialist disposal.

Table 17.5. Typical limits and pre-treatments given to industrial wastewaters before discharge to sewers

Contaminant	Consent Limit	Pre-treatment (where necessary)
Suspended solids	400-1000	Screening and settlement
BOD	500-1000	High rate bio-oxidation
Oil and grease	10	Passage through oil traps or separators
Cyanide	1-5	Chlorination or enzyme treatment for CN removal
Heavy metals	1-20	Alkaline precipitation and settlement
Acidity/ alkalinity	pH 10 and 5	Neutralization
Solvents	Substantially absent	Recovery or activated carbon treatment
Strong dyestuff colour	Low colour	Bleaching with chlorine

Water-borne Diseases

Contaminated drinking water is especially dangerous to humans because of the many diseases that it often transmits. These diseases fall into three major categories: *bacterial* (caused by bacteria in water), *viral* (caused by viruses in water), and *parasitic* (caused by a parasitic protozoa or worm in water).

Bacterial Diseases

Random Quote In the struggle between the stone and the water, in time, the water wins.

Chinese Proverb

The bacterial diseases that can result from polluted water include typhoid, cholera, bacterial dysentery, and enteritis. *Typhoid* resides in the bacterium *Salmonella typhi,* and it is often fatal if untreated. The symptoms and effects of typhoid include diarrhoea, severe vomiting, an enlarged spleen, and an inflammed intestine. Another bacterial disease, *cholera,* is caused by the bacterium *Vibrio choleras.* Its symptoms consist of diarrhoea, severe vomiting, and dehydration, and, like typhoid, it is often fatal if left untreated. *Bacterial dysentery* is rarely fatal except in infants, and its major symptom include diarrhea, abdominal pain, and cramps. Bacterial dysentery results from exposure to the bacterium *Shigella dysenteriae.* The final bacterial disease is *enteritis,* which is caused by the bacterium *Clostridium perfringens.* It is characterized by severe stomach pain, nausea, loss of appetite, and vomiting, and is rarely fatal.

Viral Diseases

Hepatitis A (also known as infectious hepatitis) is a major viral disease that can be transmitted through contaminated drinking water. The Hepatitis A virus is associated with the following effects: fever, severe headache, loss of appetite, abdominal pain, muscle ache, and inflammation of the liver. It is rarely fatal but may cause permanent liver damage if untreated. *Polio,* caused by the poliovirus, entails a sore throat, fever, diarrhea, and muscle aches, and may eventually lead to muscle paralysis.

Parasitic Diseases

The remaining water-borne diseases are transmitted by parasites. *Ameobic dysentery* is caused by the parasitic amoeba *Entamoeba histolytica,* and its symptoms include severe diarrhoea, abdominal pain, headache, chills, and fever. It is more serious than bacterial dysentery and may result in death in not treated. *Giardiasis* is another disease caused by a parasitic protozoon, and it usually results in diarrhoea, abdominal cramps, belching, and fatigue. The *Ancylostoma* worm transmits the disease ancylostomiasis, which results in lung irritation, coughing, and severe anemia. Finally, the *Schistosoma* worm carries *schistosomiasis,* a tropical disease whose symptoms include diarrhoea, urinal bleeding, and abdominal pain.

ENVIRONMENT AND WATER BORNE DISEASES

Water borne diseases are caused by pathogenic microorganisms which are directly transmitted when contaminated fresh water is consumed. Contaminated fresh water used in the preparation of food can be the source of food borne disease through consumption of the same microorganisms. According to the World Health Organisation, diarrheal disease accounts for an estimated 4.1% of the total DAILY global burden of disease and is responsible for the death of 1.8 million people every year. It was estimated that 88% of the burden is attributable to unsafe water supply, sanitations and hygeine and is mostly concentrated in children in developing countries waterborne disease can be caused by protozoa, viruses, or bacteria, many of which are intestinal parasites.

Adequate supply of fresh & clean drinking water is a basic need for all human beings on the earth, yet it has been observed that millions of people worldwide are deprived of this industrial growth organisation and the increasing use of synthetic organic substances have serious and adverse impact on freshwater bodies. Many areas of groundwater and surface water are now contaminated with heavy metals, pop's (Pesticide Organic Pollutants), and nutrients that have an adverse effect on health.

Water borne diseases and water caused health problems are mostly due to in adequate and incompetent management of water re-sources. In the urban areas water get contaminated in many different ways, some of the most common reason being leaky water pipe joints in areas where the water pipe and sewage line pass close together. Sometimes the water gets polluted at source due to various reason & mainly due to inflow of sewage into the source.

Pesticides

Run-off from farms, back yards and golf sources contain pesticides such as DDT that in turn contaminate the water leech ate from land fill sites is another major contaminating source. Its effect on the ecosystem's and health are endocrine and reproductive damage in wildlife ground water is susceptible to contamination, as pesticides are mobile in the soil. It is a matter of concern as these chemicals are persistent in the soil and water.

Sewage

Untreated or inadequately treated municipal sewage is a major source of groundwater and surface water pollution in the developing countries. The organic material that is discharged with municipal waste into the water courses uses substantial oxygen for biological degradation hereby upsetting the ecological balance of rivers and lakes. Sewage also carries microbial pathogens that are the cause of the spread of disease.

Nutrients and Eutrophication

Domestic waste water agricultural run-off and industrial effluents contain phosphorus and nitrogen fertilizer run-off manure from live stock operations, which increase the level of nutrients in water bodies and cause eutrophication in the lakes and rivers and continue on the coastal areas. The nitrates come mainly from the fertilizer that is added to the fields excessive use of fertilizers cause nitrate contamination of ground water, with the result that nitrate levels in drinking water is for above the safety levels recommended good agricultural practices can help in reducing the amount of nitrates in the soil and thereby lower its content in the water.

Synthetic Organics

Many of the 100,000 synthetic compounds in use today are found in the aquatic environment and accumulate in the food chain pops or persistent organic pollutants, represent the most harmful element for the ecosystem and for human health, for example industrial chemicals and agricultural pesticides. These chemicals can accumulate in fish and cause serious damage to human health. Where pesticides are used on a large scale, groundwater gets contaminated and this leads to the chemicals contamination of drinking water.

Acidification

Acidification of surface water, mainly lakes and reservoirs, is one of the major environmental impacts of transport over long distance of air pollutants, such as sulphur dioxide from power plants, other heavy industries such as steel plants and moter vehicles.

Chemicals in drinking water can be both naturally occurring or introduced by human interference can have serious health effects. Fluorides in water are essential for protection against dental caries and weakening of the bones but higher level can have adverse effect of health India. In high fluride content is found naturally in the water in Rajasthan.

Arsenics

Arsenic occurs naturally or is possibly aggravated by over powering aquifers and by phosphorus from fertilizers. High concentration of arsenic in water can have an adverse effect health. A few years back high concentrations of this element was found in drinking water in six districts in West Bengal. A majority of people in the areas was found suffering from arsenic skin lesions. It was felt that arsenic contamination in the groundwater was due to natural causes. The government is trying to provide an alternative drinking water source and a method through which the arsenic content from water can be removed.

Lead

Lead pipes fitting solder and the service connections of some plumbing systems contain lead that contaminates the drinking water source.

Recreational use of water, untreated sewage, industrial effluents, and agricultural waste are often discharged into the water bodies such as the lakes, coastal areas and rivers dangering their use for recreational purposes such as swimming and canoeing.

Petrochemicals

Petrochemicals contaminate the groundwater from underground petroleum storage tanks.

Other Heavy Metals

These contaminants come from mining waste and tailings, landfils or hazardous waste dumps.

Chlorinated Solvents

Metals and plastic effluents, fabric cleaning, electronic and aircraft manufacturing are often discharged and contaminate groundwater.

Water-Borne Disease

Pollution in water cause the following water-borne disease.

1. Bacterial infections, typhoid
2. Cholera
3. Para typhoid fever
4. Bacilliary dysentery
5. Viral infections infectious hepatitis (Jaundice)
6. Poliomyelitis
7. Protozoal infectious amoebic dysentery

Water-borne diseases are infectious disease spread primarily through contaminated water. Though these diseases are spread either directly or through files or fish water is the chief medium for spread of these diseases and hence they are formed as water-borne diseases.

Most intestinal (enteric) diseases are infectious and are transmitted through faeces waste pathogens which include virus, bacteria, protozoa and parasitic worms which are disease-producing agents found in the faeces of infected person. These diseases are more prevalent in areas with poor sanitary condition. These pathogens travel through water sources and directly through persons handling of food and water.

Since these disease are highly infectious, extreme hygiene should be maintained by people looking after an infected patient. Hepatitis, Cholera, dysentery and typhoid are the more common water-borne disease that affect large populations in the tropical regions.

A large number of chemicals that either exist naturally in the land or are added due to human activity dissolve in the water, thereby contaminating it and leading to various diseases. Pesticides, the oranophosphates and the carbonates present in pesticides affect and damage the nervous system and can cause cancer some of the pesticides contain carcinogens that exceeds recommended levels. They contain chlorides that cause reproductive and endocrinal damage. Lead is hazardous to health as it accumulates in the body and affects the central nervous system; children and pregnant woman are most at risk.

Fluoride

Excess fluorides can cause yellowing of the teeth and damage to the spinal cord and other crippling disease. In India, the most common cause of fluorosis is fluoride – laden water derived from bore wells dug deep into the earth of India's 32 states, 17 have been identified as endemic areas for fluorosis with an estimated 25 million people impacted and other 66 million at risk.

In a news "Increased threat of fluorosis in city-Times of India, June 29, 2004 NEW DELHI." with one-third of Delhis ground-water laced with excessive fluorides, the number of people falling prey to fluoride poisoning is increasing.

India's largest state suffers from fluorosis a debilitating disease that damages, bones with teeth, research by a voluntary body shows. "The incidence of fluorosis, caused by an excess of fluoride compounds in drinking water, has been rising at an alarming rate in the state" says Mahitosh Bagoria of Health Environment and Development consortium" It is estimated that around 25% of the rural population in the state is affected he said.

Villagers who consume non-potable water suffer form yellow, cracked teeth joint pains and crippled limbs and also age rapidly.

Nitrate

Drinking water that gets contaminated with nitrate can prove fatal especially to infants that drink of oxygen that reaches the brain causing the blue body syndrome. It is also linked to digestive tract cancers. It cause algae to bloom resulting in eutrophication in surface water.

Petrochemicals

Benzene and other petrochemicals can cause cancer even at low exposure levels. Chlorinated solvents. These are linked to reproduction disorders and to same cancers.

Arsenic

Arsenic poisoning through water can cause liver and nervous system damage, vascular disease and also skin cancer.

Other Heavy Metals

Heavy metals cause damage to the nervous system and the kidney and other metabolic disruptions.

Salts

It makes the fresh water unusable for drinking and irrigation purposes. Exposure to polluted water can cause diarrhea, skin irritation, respiratory problems and other disease depending on the pollutant that is in the water body stagnant water and other untreated water provide a habitat for the mosquito and a host of other parasites and insects that cause a large number of disease especially in the tropical regions. Among these malaria is undoubtedly the most widely distributed and cause most damage to the health.

Table 17.6. I. Protozoal Infections

Sl. No.	Disease and Transmission	Microbial Agent	Sources of Agent in water supply	General symptoms
1.	Amoebiasis (hand to mouth)	Protozoan (*Entamoeba histolyticacyst*-like appearance	Sewage, non-treated drinking water, flies in water supply	Abdominal discomfort, fatigue light loss, diarrhea, bloating, fever
2.	Cryptospori-chiosis (oral)	Protozoan (*Cryptospor-dium parvum*)	Collects, on water filters and membrane that can not be disinfected, animal manure, seasonal runoff of water	Flue-like symptoms, watery diarrhea, loss of appetite substantial loss of weight, bloating increased gas, nausea
3.	Cycloporasis	Protozoan parasite (Cyclospora Cayetaensis)	Sewage, non-treated drinking water	Cramps,m nausea, vomiting, muscle pain, fever and fatigue.
4.	Giardiasis (Oral-fecal) (Hand-to-Mouth)	Protozoan (*Giardia lamblia*) most common intestinal parasite	Untreated water, poor dis-infections, pipe breaks, leaks, groundwater conta-mination, campground where humans and wildlife use same source of water, beavers and muskrats create ponds that acts as reservoirs for Giardia	Diarrhea, abdominal discomfort
5.	Microsporidiosis	Protozoan phylum (microspordia)	The genera of encephalito-zoom intestinal has been detected in ground water the origin of drinking water [2]	Diarrhoea and wasting in immune compromised individuals.

II. Parasitic Infections (Kingdom Animalia) :

6.	Schistosmiasis (immerson)	Members of the genus (schistosoma)	Fresh water contaminated with certain types of snails the carry schistosomer.	Rash or itchy skin, fever, chills, cough and muscles aches.
7.	Dracunculiasis (Guinea worm disease)	Dracunculus (medinensis)	Stagnant water containing, larve	Allergic reaction, urticaria rash, nausea, vomiting, diarrhea, asthmatic attack.
8.	Taeniasis	Tapeworm of the genus (Taenia)	Drinking water contaminated with eggs	Intestinal disturbances, neurologic manifestations, loss of weight cysticercosis.
9.	Fascioloopsiasis	Fasciolopsis buski	Drinking water contaminated with encysted metacercaria	GIT disturbance, diarrhea, cholecystitis, obstructive jaundice.
10.	Hymenolepiasis (Dwarf Tapeworm Infection)	Hymenolepsis nana	Drinking water contaminated with eggs	Abdominal pain, anorexia, itching around the anus, nervous manifestation
11.	Echinococosis (Hydatid disease)	Echinococcus granulosus	Drinking water contaminated with feces (usually canidy) containing eggs.	Liver enlargement, hydatid GSTs press on bile duct and blood vessles, if cyst rupture they can cause anaphylactic shock.
12.	Coenurosis	Multiceps multiceps	Contaminated drinking water with eegs.	Increase interracial tension
13.	Ascariasis	Ascaris lumbricoides	Drinking water contaminated with feces (usually canid) containing eggs.	Mostly disease is asymptomatic or accompanied by inflammation, fever and diarrhea severs cases involve loffer's syndrome in lungs, nausea, vomiting malnutrition and under development.
14.	Enterobiasis	Enterobius Vermicularis	Drinking water, contaminated with eggs.	Peri-anal itch, nervous irritability, hyper acting and insomnia.

Disease	Morbidity (cases per year)	Mortality (death's per year)
—	1,500,000,000	100,000
Schistosomiasis	200,000,000	200,000

III. Bacterial Infections :

15.	Botulism	Clostridium botulinum	Bacteria can enter a wound from contaminated water sources can enter the gastrointestinal tract by consuming contaminated drinking water (more commonly) good.	Dry mouth, blurred and or double vision, difficulty swallowing, muscle weakness, difficulty breathing, slurred speech, vomiting and sometimes diarrhea, death is usually caused by respiratory failure.
16.	Camphylobacteriosis	Most commonly caused by (camphylbacter jejani)	Drinking water contaminated with feces	Produce dysentery like symptoms along with high fever, usually lasts 2-10 days.

...(Contd.)

17.	Cholera	Spread by the bacterium vibrio cholerae	Drinking water contaminated with the bacterium	In sever forms it is known to be one of the most rapidly fatal illnesses known symptoms include very watery diarrhea, rapid pulse, vomiting and hypovolemic shock (in severe cases) at which point death can occur in 12-18 hours.
18.	E. coli, Infection	Mycobacterium marinum certain strains of escherichia coli. (commonly E. coli)	Water contaminated with the bacteria	Mostly diarrhea, can cause death in immuno compromised indivuals, the very young, and the elderly due to dehydration from prolonged illness.
19.	M. Marinum infection	Mycobacterium marinum	Naturally occurs in water, most cases from exposure in swimming polls or more frequently aquariums, rare infection. Since it mostly infects immuno compromised individuals.	Symptoms include lesions typically located on the elbows, kness, and feet (from swimming pools) or lesions on the hands (aquariums). Lesions may be painless painful.
20.	Dysentery	Caused by a number of sps. the genera shigella and salmonella with the most common beign shisella dysenteries	Water contaminated with bacterium	Frequent passage of feces with blood and / or muscus and in same cases vomiting of blood.
21.	Legionellosis (two distinct forms: Legionnaries disease and Pontiac fever)	Caused by bacteria belonging to genus Legionella (90% of cases caused by Legionella pneumophila)	Contaminated water the organism thrives in warm aquatic environment.	Influenza without pneumonia, Legionnaires disease has severe symptoms such as fever, chills, pneumonia (with cough that sometimes produces sputum) ataxia, anorexia, muscle aches, malaise and occasionally diarrhea and vomiting.
22.	Leptospirosis	Caused by bacterium of genus Leptospira	Water contaminated by the animal urine carrying bacteria	Begins with flue-like symptoms then resolves. The second phase then occurs involving meningitis, liver damage (causes jaundice and rena) failure.
23.	Otitis externa (swimmer's ear)	Caused by a number of bacterial and fungal species	Swimming in water contaminated by the responsible pathogens.	Ear canal swells causing pain and tenderness to the touch.
24.	Salmonellosis	Caused by many bacteria of genus salmonella	Drinking water contaminated with the bacteria more common as a food borne illness.	Symptoms include diarrhea fever, vomiting and abdominal cramps
25.	Typhoid fever	Salmonella typhi	Ingestion of water contaminated with feces of an infection person.	Characterised by sustained fever upto 40°C (104°F) profuse seating, diarrhea less commonly a rash may occur symptoms progress to delirium and the spleen and liver enlarge if untreated. In this case it can last up to four weeks and cause death.

...(Contd.)

26.	Vibro illness	Vibrio vulnificus, Vibrio, alginolyticus and vibrio parahae-mylytices	Can enter mounds from contaminated water also got by drinking contaminated water or eating under cooked oysters.	Symptoms include explosive, watery diarrhea, nausea, vomiting abdominal cramps and occasionally fever.
27.	Adenovirus infection	Adenovirus	Manifests itself in improperly treated water	Symptoms include common cold, symptom, pneumonia, croup and bronchitis.
28.	Gastroenteritis	Astrovirus, calicivirus enteric adenovirus and parvovirus	Manifests itself improperly treated water	Symptoms include diarrhea, nausea, vomiting fever, malaise and abdominal pain
29.	SARS (Severe Acure Respiratory Symdrome)	Coronavirus	Manifests itself in improperly treated water	Symptoms include fever myalgia, letharsy, gastrointestinal symptoms cough and sore throat
30.	Hepatitis-A	Hepatitis-A virus (HIV)	Can manifests itself in water and food	Symptoms are only acute (no. chronic stage to the virus) and include fatigue, fever, abdominal pain, nausea, diarrhea, weight loss, itching jaundice and depression.
31.	Polio mavirus (polio)	Poliovirus	Enters water through the feces of infected individuals	90-95% apatients show no symptoms, 5-8% have miner symptoms (comparatively) with delirium, headache, fever and occasional seizures and spastic paralysis 1% have symptoms of non-paralytic aseptic meningitis. The rest have serious symptoms resulting in paralysis of death.
32.	Poly mavirus infection	Two of polyomavirus JC virus and BK virus	Very widespread, can manifest itself in water 80% of the population has antibodies polyo mavirus	BK virus produces and mild respiratory infection and can infect the kidney of immuno suppressed transplant patients JC virus infects the respiratory system, kidney or can cause progressive multi focal leukoencephalopathy in the brain (which is fatal).

REFERENCES

1. "WHO I Burden of disease and cost-effectiveness estimates.
2. "ab Mwachcuka N. Gerba CP (June 2004) curing water borne pathogens, can we kill them all curvopin Biotechnol. 15(3); 175-80 PMID 15193323.
3. "Dziaban EJ, Liang JL. Craun GF, Hill V, Yu PA et. al., (22 Dec-2006) surveillance for waterborne disease and outbreaks associated with recreational water united states 2003-2004. MMWR surveill Summ. 55(12); 1-30 PMID, 17183230.
4. "Petrini B. (October 2006)" Mycobacterium marinum, Ubiquitous agent of waterborne glaucomatous skin infections. Eur J. Clin Microbiol infect. Dis-25(10); 609-13 PMID 17047903.
5. Nwachuka N. Gerba CP, Oswald A, Mashadi FD (Sept. 2005). Comparative inactivation of Adenovirus serotypes by UV light disinfections Appl. Environ Microbiol 71(9) 5633-6 :PMID 16151167 Pinc 12144670.

CHAPTER 18

Assessment of Water Quality : Prevailing Practices and Implications

Dr. Subrata Pani[1] and Dr. S.A. Iqbal

Introduction

Water, which is vital to life, may contain many impurities both physical and chemical. It can absorb chemical wastes and toxic refuse from various sources including anthropogenic origin. Therefore, the source of any water determines the kinds and amounts of its impurities if any. For example, groundwater, which is being used as potable sources in absence of surface water sources usually, contains high concentrations of dissolved minerals. This water is usually clear, colorless due to its filtration through rock, and sand. However, it also may contain various types of pollutants, including detergents and industrial wastes. Deep wells and large lakes provide water, which is more or less consistent from season to season. Smaller bodies of water, shallow wells and springs often reflect seasonal-even daily variations in their mineral content. Thus to understand why water from different sources varies in quality, it is necessary to know something about basic water chemistry.

Today it is almost impossible to find a source of water that will meet basic requirements for a public water supply without requiring some form of treatment. Thus to understand the quality of water we have today, it is important to know the various methods used for analysis of the water quality so that it can be compared with the prevailing water quality standards for designating its best uses.

Although various methods are available for analysis of different parameters but the techniques to be used varies with the sensitiveness of the analysis and prevailing resources. Therefore, selection of methods should carefully be done to achieve the objectives with fair degree of accuracy.

Some common methods that are being discussed in this chapter are usually being practised for determination of various physical, chemical, biological and microbiological characteristics of different types of water. The methods that are being discussed here are simplified on the basis of day-to-day experiences. However as each method have its advantages and disadvantages, therefore selection of method for specific analysis should be done meticulously. Some of the important steps that are required for analysis of environmental parameters are discussed below.

Collection of Sample

The first and foremost thing that is required for analysis of water sample is collection of sample or sampling. The main objective of sampling is to collect a portion small enough volume to be transported to the

[1]Lake Conservation Authority of Madhya Pradesh, Bhopal (India)
[2]Department of Chemistry, Saifia Science College, (B.U.), Bhopal (India)

laboratory yet accurately and true representative of the material being sampled. The sample should be collected in such a way that the relative concentration of all the components must be same in both the samples and the material being sampled. The sample must be handled in such a way so as to preserve its composition before analysis. The date, time & site of collection should always be recorded on diary as well as on the can in which the samples are collected.

There are various methods for sample collection. The following sampling techniques are generally applied for collection of sample:

***(i)* Grab Sample :** Also known as random or catch sample. It is the most commonly applied sampling technique used in the collection of water samples from natural water bodies, lakes, streams, rivers etc for assessing the water quality of the water body. This type of sampling is preferred when the source is known to be fairly constant in composition over a considerable period of time.

***(ii)* Composite Sample :** It refers to a mixture of grab samples collected at the same point over a period in proportion to flow rate. The composite samples are more useful in evaluating average concentration.

***(iii)* Integrated Samples :** For certain purposes, best information is provided by a mixture of grab samples collected from different points simultaneously or as early as possible. Such mixtures are called integrated samples.

Precautions to be Adopted

***(i)* General sample :** The samples should be collected in plastic cans for analysis of general parameters like hardness, chloride, phosphate etc. The cans used for this purpose should be properly washed and free from any contamination.

Prime care should be taken while collecting sample for Dissolved Oxygen. The sample should be collected in 125 ml glass D.O bottle without bubbling and should be fixed at the spot and analyzed as soon as possible.

***(ii)* Bacteriology :** Utmost care should be taken while collecting the sample for microbiological analysis. The glass bottles used for sample collection should be pre-sterilized and care should be taken to avoid any contamination while sampling. The samples should be brought back to the laboratory in ice packs.

***(iii)* Heavy metals :** The sample should be collected in small plastic cans & the heavy metals should be preserved by adding 5 ml of 1N HNO_3 and bringing down the pH to near about 4.

***(iv)* Pesticides :** The samples for pesticides should be collected in dark bottles and should be kept at a low temperature.

***(v)* B.O.D. :** Samples for BOD should be collected in 300 ml BOD bottles & should be kept at low temperature.

***(vi)* Plankton :** Known amount of sample (1 to 20 liters preferably) should be filtered through plankton net (Nylo bolt 25: mesh size 40µm preferably) and should be preserved by adding Lugol's iodine or 4% formaline solution.

Analysis of Parameters

After collection of samples analysis is done either on the site (for field parameters) or in the laboratory.

I. Analysis of field parameters : The value of certain physico-chemical parameters changes very rapidly on standing & so they must be analyzed on the field. In addition, the observation regarding climate, light, temperature etc. should also be made on the field.

The observations /parameters that should be analyzed on the field are –

Light

Light is the source of energy for autotrophs and is therefore affects various physico chemical parameters. Light intensity is measured using the Lux-meter. The unit of light is LUX.

Ambient Temperature

Climatic conditions, time of collection, climate, solar radiation and topography of water bodies have an impact on the ambient temperature. The ambient temperature is measured by using a probe and also using a mercury thermometer graduated up to 100°C with an accuracy of 0.1 – 0.2°C.

Water Temperature

Temperature of water depends upon water depth besides solar radiation, climate and topography. It may be mentioned that no other single factor has so much profound direct or indirect influence on physico-chemical, biological, metabolic and physiological behaviour of aquatic ecosystem than temperature.

In general, it can be stated that temperature plays an important role in treatment of water supplies; in the aquatic life of water reservoirs (biochemical reactions may double the reaction rate for a 10°C increase in temperature); in the activity of organisms producing bad taste and odour; and in the rate of corrosion in the distribution system.

In sampling, temperature readings must be done immediately due to changes caused by air temperature. The thermometer must be left in water long enough to get a final, constant reading.

Odours and Tastes

Various odours and tastes may be present in water. They can be traced to many conditions. Unfortunately, the causes of bad taste and odor problems in water are so many, it is impossible to suggest a single treatment that would be universally effective in controlling these problems,

In many cases it is difficult to differentiate between tastes and odors

pH

The hydrogen ion concentration or pH is the logarithm in base 10 of the reciprocal of the hydrogen ion concentration given in moles per litre. It is the measure of the relative acidity or alkalinity and represents the negative logarithm of the concentration of the free hydrogen ions in the solution.

The pH scale of values extends from 0 (very acidic) to 14 (very alkaline), with the middle value (pH=7) corresponding to exact neutrality at 25°C. The pH value represents the hydrogen ion activity, while values of "alkalinity" and "acidity" represent the buffering capacity of the sampled water.

$$pH = -\log[H^+] \text{ or } pH = \log 1/[H^+]$$

The logarithmic function is selected not only for the simplicity of value representation but also because electrochemical potentials vary with the logarithm of the concentration and the wide range of ionic activities is better included in the logarithmic graphical representation.

Hydrogen ion concentration is an important environmental factor the variation of which, among other causes are linked with the chemical changes, species composition and life process of plant and animal communities inhabiting in them. pH changes with change in temperature & on standing. So pH should be determined on the site of using pH meter or pH indicator solution or strips on the spot as soon as possible after the sample collection. The pH of water is measured with the help of pH meter. The 'p' of pH on the other hand denotes power of the hydrogen ion activity in mole per litre.

Total Suspended Solids

The suspended particles in water that remain in suspension due to water currents, wind action and lower density than water are mainly responsible for turbidity in the water quality. Suspended solids hinder the light penetration to the bottom layers of the water body thus preventing primary production in the lower strata of the lake.

The suspended particles are measured by filtering a known amount of water through pre weighed filter paper and then weighing the filter paper after drying at 105° C (APHA, 19th Ed. 1998). The difference in the weight of the filter paper after drying and initial weight of the filter paper represents the weight of the suspended solids present in the water sample.

Total Dissolved Solids (TDS)

A large number of salts are found dissolved in natural waters, the common ones are carbonates, bicarbonates, chlorides, sulfates, phosphates, and nitrates of calcium, magnesium, sodium, potassium, iron, and manganese etc. A high content of dissolved solids elevates the density of water influences osmoregulation of freshwater organisms, reduces solubility of gases (like oxygen) and utility of water for drinking, irrigational, and industrial purposes. It is especially an important parameter in the analysis of saline lake, coastal, estuarine, and marine waters. This factor, having high value in such waters, is often expressed as g/L or ppt (parts per thousand) instead of mg/L or ppm (parts per million).

TDS can be determined by TDS meter, which gives reading directly in mg/l.

Conductivity

The ability of water to conduct an electric current is known as conductivity or specific conductance and depends on the concentration of ions in solution. Conductivity is measured in milli Siemens per centimeter. The measurement we made *in situ*, or in the field immediately after water sample has been obtained, because conductivity changes with storage time. Conductivity is also temperature-dependent; thus, if the conductivity meter is used for measuring conductivity is not equipped with automatic temperature correction; the temperature of the sample should be measured and recorded.

The conductivity meter consists of a conductivity cell containing two rigidly attached electrodes, which are connected by cables to the body of the meter. The meter contains a source of electric current (a battery in the case of portable models), a Whetstone Bridge (a device for measuring electrical resistance) and a small indicator (usually a galvanometer). Some meters are arranged to provide a reading in units of conductance (mhos). The conductivity cell forms one arm of the Whetstone Bridge. The design of the electrodes, *i.e.* shape, size and relative position, determines the value of the cell constant, K_C, which is usually in the range 0 to 2.0. A cell with a constant of 2.0 is suitable for measuring conductivity from 20 to 1000 mS / cm. It is measured by conductivity meter, which gives reading in mS/cm.

Turbidity

Turbidity and suspended matter are not synonymous terms, although most of us use the terms more or less interchangeably. Correctly, speaking, suspended matter is that material which can be removed from water through filtration or the coagulation through a filtration process.

Turbidity, on the other hand, is a measure of the amount of light scattered and absorbed by water because of the suspended matter in the water.

There is also some danger of confusion regarding turbidity and colour. Turbidity is the lack of clarity or brilliance in water. Water may have a great deal of colour — it may even be dark brown — and still be clear and without suspended matter.

Turbidity is the measure of the fine suspended matter in water, mostly caused by colloidal matter, silica of diatomaceous earth that could cause the optical effect. The suspended matter causing turbidity is expected to be clay, silt, non-living organic particulate, plankton, and other microscopic organisms, in addition to suspended organic or inorganic matter. Turbidity represents light scattering and absorbing properties of suspended matter in water.

Guideline Value:

BIS has laid down 5-10 NTU Maximum turbidity for drinking water.

Methodology: The turbidity determined by Jackson turbidity tube results expressed as Jackson Turbidity Unit (JTU).

Colour

Ordinarily we think of water as being blue in color. Ideally, water from the tap is not blue or blue-green. If such is the case, there are certain foreign substances in the water. Infinitely small microscopic particles add color to water. Colloidal suspensions and no colloidal organic acids as well as neutral salts also affect the color of water.

Dissolved Oxygen

The occurrence of dissolved oxygen in water may be attributed to two distinct phenomenons:

- Direct diffusion from air
- Photosynthetic evolution by aquatic autotrophs.

The first one is purely a physical phenomenon and depends upon various abiotic factors like temperature, salinity, water movements etc. whereas the later is a biological process and depend upon both biotic and abiotic factors like availability of light, duration of light hours etc among the abiotic and distribution of aquatic autotrophs among the biotic factors. Eutrophic water bodies have a wide range of oxygen and depict a clinograde curve whereas the oligotrophic water bodies have a narrow range of dissolved oxygen vertically and depict an orthograde curve of oxygen distribution. Concentration of all the gases in the water depends upon temperature, pressure humidity etc. therefore gases like Dissolved Oxygen (DO) should be fixed on the field and should be analyzed immediately as per the Wrinkle's method with azide modification.

Dissolved oxygen is measured by a number of methods including electronic oxygen meters but the most commonly used and preferred method for DO determination is the Winkler's method with azide modification. The water samples are collected in 125 ml. glass bottles avoiding any agitation in the water column without bubbling and are immediately fixed by adding 1 ml each of manganese sulphate and alkaline iodine azide respectively. Adding 1ml concentrated sulphuric acid and tilting the bottle upside down at least 5-6 times dissolves the resultant brown precipitate. 50ml aliquot is titrated against 0.025 N solution of sodium thiosulphate titrant upto the disappearance of the blue color, using starch as an indicator (APHA, 1976) the dissolved oxygen content was calculated by using the formula.

$$\text{Dissolved oxygen mg/mL} = \frac{\text{mL of titrant} \times N \times E \times 1000}{\text{ml. of sample}}$$

Where, N = Normality of titrant

E = Equivalent weight of oxygen

Free Carbon Dioxide

It refers to carbon dioxide gas dissolved in water. The term is used to distinguish a solution of the gas from the combined carbon dioxide present in bicarbonate and carbonate ions.

On reaching the earth, the rainwater now slightly acid will absorb additional amounts of carbon dioxide if it flows through decaying vegetation. At the same time, the carbon dioxide becomes carbonic acid. If the water now passes through limestone formations, its carbonic acid content will react with the limestone to form soluble calcium bicarbonate. In this process the carbonic acid is partially neutralized.

The value of Free CO_2 changes on standing. So it must be analyzed on the field. For analysis take 50 ml of sample add few drops of phenolphthalein indicator, if pink color doesn't appear, Free CO_2 is present. Titrate with 0.02 N NaOH till pink appears. Calculate Free CO_2 by formula –

$$\text{Free } CO_2 \text{ in mg/L} = \frac{\text{mL of titrant} \times 1000}{\text{mL of sample}}$$

Free carbon dioxide dissolved in water is essentially the only source of carbon that can be assimilated and incorporated in the skeleton of the living matter of all the aquatic autotrophs; once fixed it can be further utilized by the organisms of other categories. In the absence of Free CO_2 plants utilize the bicarbonates. Carbon dioxide dissolved in natural water actively participates in the carbonate system.

Free CO_2 is estimated by adding two drops of phenolphthalein indicator to 50 ml. of water sample. If pink color develops then Free CO_2 is absent and if the solution remains colorless Free CO_2 is present and is titrated against a standard NaOH (0.02 N) to a slight pink end point. The final result may be calculated as

$$\text{Free } CO_2 \text{ mgL} = \frac{\text{ml of titrant} \times 1000}{\text{ml of sample}}$$

Total Alkalinity

The alkalinity of water is a measure of how much acid it can neutralize. If any changes are made to the water that could raise or lower the pH value, alkalinity acts as a buffer, protecting the water and its life forms from sudden shifts in pH. This ability to neutralize acid, or H+ ions, is particularly important in regions affected by acid rain.

Most alkalinity in surface water comes from calcium carbonate, $CaCO_3$, being leached from rocks and soil. This process is enhanced if the rocks and soil have been broken up for any reason, such as mining or urban development. Limestone contains especially high levels of calcium carbonate. Alkalinity is significant in the treatment of wastewater and drinking water because it will influence treatment processes such as anaerobic digestion. Water may also be unsuitable for use in irrigation if the alkalinity level in the water is higher than the natural level of alkalinity in the soil.

Carbonate Alkalinity

It is known that if Free CO_2 is present, carbonate alkalinity is absent & vice versa. If the sample is brought to the lab CO_2 may defuse into the sample in which it was absent in the field. So carbonate alkalinity should also be analyzed on the filed.

For the analysis of Carbonate alkalinity 50 ml of water sample is taken and few drops of phenolphthalein indicator is added as. If pink color appears, the sample is titrated with 0.02 N H_2SO_4 to a colorless end point.

Calculate Carbonate alkalinity by the formula–

$$\text{Carbonate Alkalinity in mg/L} = \frac{\text{ml of titrant} \times 1000}{\text{ml of sample}}$$

Total Hardness

Hard water is a serious problem, and it is a common one. There are only a few areas where water is sufficiently soft to be satisfactory for most home-making needs. No natural water supply is completely free of hardness.

50 ml of sample was taken in a conical flask and 1ml of Ammonia Buffer solution with solochrom (Eiochrome) black T as indicator was also added to develop pink color. It is then titrated with N/50 solution of di-sodium salt of Ethylene Diamine Tetra acetic Acid (EDTA). Appearance of blue color indicates end point. Values were computed by following formula:

$$\text{Total hardness mg/L} = \frac{\text{ml of titrant} \times 1000}{\text{ml of sample}}$$

Calcium Hardness

50 ml of sample is taken and added 1 ml of sodium hydroxide as an intermediate solution using ammonium purpurate (Meroxide) as an indicator, pink color appears which is then titrated with N/50 solution of di sodium salt of ethylene diamine tetra acetic acid (EDTA) upto violet end point. Values were computed by following formula:

$$\text{Calcium hardness mg/L} = \frac{\text{ml of titrant} \times 1000 \times 1.05}{\text{ml of sample}}$$

Chloride

Chloride concentration can be estimated by titrating 50 ml. of water sample with standard silver nitrate (0.0141 N) titrant up to a brick red end point, using potassium chromate as an indicator.

The value of Chloride can be calculated as:

Chloride (mg/L) = ml of titrant × N × 35.46 × 1000

Biochemical Oxygen Demand

For determining Biochemical oxygen demand (BOD) incubation method is adopted. It is the decrease in oxygen concentration after incubation in the dark at a constant temperature for a period of time.

One set of bottles was kept in BOD incubator at 20° C for five days and dissolved oxygen in another set was determined immediately. After completion for 5 days of incubation D.O was determined according to modified Winkler's method. The value of BOD was computed as: B.O.D.(mg/lit.) =(Do-D5)x Dilution factor.

Where, DO = Initial D.O. in sample D5 = D.O after five days.

Chemical Oxygen Demand (COD)

As mentioned earlier that BOD is measured for assessment of biodegradable organic compound that forms a reasonable fraction of organic matter in lakes. In recent times, with the increase of pollution by large amount of various chemically oxidisable organic substances of different nature entering in the aquatic system, BOD alone does not give a clear picture of the organic matter contents of the sample. Further more, the presence of various toxicants in the sample may severally affect the validity of the BOD test. Hence, chemical oxygen demand is a better estimate of the organic matter, which needs no sophistication and is time saving.

However COD i.e. the oxygen consumed (OC) does not differentiate the stable organic matter from the unstable form. Therefore, the COD values are not directly comparable to that of BOD. Further more, some cyclic organic compounds are not oxidized; whereas, on the other hand, many inorganic compounds like nitrites, sulphites and reduced metal ions get oxidized. Samples containing chlorides more than 2 gm/ l, the chloride ions are oxidized to chlorine giving erratic results. Despite of these limitations COD is still an important parameter for estimating the carbonaceous fraction of the organic matter much closer to the actual amount.

Chemical Oxygen Demand (COD) is determined by potassium dichromate open reflux method (NEERI, 1988) 20 ml of water sample was taken in a 100 ml flask, then 10 ml of potassium dichromate (0.25N) and 30 ml of COD reagent (concentrated sulphuric acid and pinch of silver sulphate) and 0.4 gm. of mercuric sulphate was added & refluxed for two hours on a hot plate. After two hours it is cooled down and distilled water was added to make the volume up to 140 ml. And 2 or 3 drops of ferroin indicator were added to refluxed sample, mix thoroughly and titrated with. 0.1N ferrous ammonium sulphate to a brick red color end point. A blank is run with distilled water.

$$\text{COD mg/L} = \frac{(B - A) \times N \times 1000 \times 8}{\text{ml of sample}}$$

Where, A = ml of titrant used with sample

B = ml of titrant used with blank, N = Normality of Titrant

FLUORIDE

Fluoride occurs in all natural water supplies and inchemical waste from industries, Fluorides, if present in small concentration up to 1 ppm, are generally considered to be beneficial in water. Excessive fluorides in drinking water may cause mottling of teeth or *dental fluorosis or cruppling effects* are observed in case the connectration of fluorides exceeds 1.0 mg/L.

Spectrophotometric Method

Alizarin-S Visual Method : Fluoride reacts with Zr Alizarin-S lake to form colourless ZrF_6^{2-} and the dye. The colour of the dye lake becomes progressively weak with increase in amount of F.

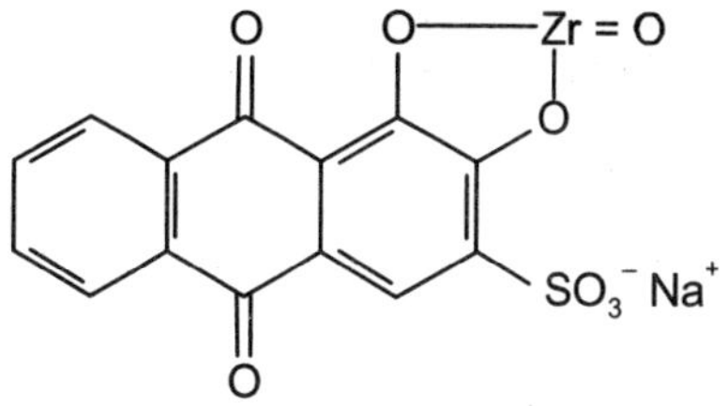

Zr-Ahzarun red-S lake

Zr-Alizarin red-S lake

To a 100 mL sample add 1 drop of $NaAsO_2$ solution (5g/L) to remove residual Cl, if any. Add 5 mL acid-zircony-alizarin reagent (300 mg $ZrOCl_2.8H_2O$/50 mL + 70 mg alizarin red S/50 mL + 800 mL of 1.5 N HCl – 1.2 NH_2SO_4 made up to 1 litre). Mix thoroughly and compare the samples and standards after 1 hour.

Hydrogen Sulfide

Hydrogen sulfide is a gas present in some waters. There is never any doubt as to when it is present due to its offensive "rotten egg" odor. This characteristic odor is sometimes apparent in concentrations below 1 mg/l. Obnoxious, as are the taste and odor of hydrogen sulfide, these are only two of the problems it presents. Hydrogen sulfide promotes corrosion due to its activity as a weak acid. Further, its presence in the air causes silver to tarnish in a matter of seconds. High concentrations of hydrogen sulfide gas are both flammable and poisonous.

Nitrate and Nitrite

Nitrate and nitrite are usually monitored regularly in water supplies as they are deemed to be potentially hazardous to halth if their maximum admissible concentrations of 50 mg/L and 0.1 ml/L, respectively, are exceeded. NO_3^- is particularly dangerous to infants less than six months old, causing child disease mathemoglobinaemia. A limit of 10 mg/L for NO_3^- has been imposed on drinking water.

Nitrates generally occur in traces in surface water and ground water. Natrites are an intermediate product, both in the oxidation of NH_3 to NO_2 and in the reduction of NO_3, which occurs in waste-water treatment plants, water-distribution systems and natural waters.

Spectrophotometric Method for Total Nitrate and Nitrite

Nitrite and nitrites are reduced to NH_3 by Deverda's alloy (50 Cu, 45 Al, 5 Zn) in strongly alkaline solutions, the NH_3 is distilled into excess standard acid and finally estimated spectrophotometrically.

- Take 500 mL sample in NH_3 distillation apparatus. Add 50 mL of 10% NaOH and Evaporate to about 200 mL.
- Cool the solution, Add Immediately connect the flask. With a vartical condenser whose outlet dips into a receiver containing 200 mL of 0.2 NH_2SO_4.
- Distil at 50 to 80°C for one hour.
- Disconnect the receiver. Make up the volume of the solution in the receiver to 250 mL. Take 5 to 40 m Lalquot in a 50 mL volumetric flask and neutralise to pH 4.5.
- Add 2 mL Nessler's reagent and estimate the absorbance at 424 nm as described under ammonia.
- This spectrophotometric method is valid for NO_3^- concentrations greater than 0.5 ppm.

Titrimetric method may be followed for NO_3^- contant exceeding 5 ppm. Here the distillate can be directly back titrated with standard alkali (0.2 N NaOH) using methyl red as indicator.

$$1 \text{ mL } NH_2SO_4 \equiv 0.06201 \text{ g } NO_3^-$$

NITRITE

Method Based on Diazotisation Reaction

A reddish purple azo dye is produced at pH 2.0 to 2.5 by the coupling of diazotised sulphanilamide with N-(1-naphythy) ethylenediamine dihydrochloride.

- Take a 40 mL sample in a volumetric flask and adjust pH to 7.0.
- Add 2 mL of sulphanilamide solution (50 g in 500 mL of 1.2 N HCl), shake and allow to stand for 10 minutes.
- Add 2 mL of N-(naphythl) ethylene dimine dihydrochloride (0.83 g in 200 mL warm water, cooled, filtered and diluted to 250 mL with glacial cetic acid), dilute to 50 mL and mix thoroughly.
- Measure the resulting purple azo dye at 543 nm within two hours, also against standards covering the range of NO_2^- from 1 to 25 μg/L.

Silica

Many water supplies contain silica. This is not surprising since silicon is the second most abundant chemical element in the earth. Silica. (silicon dioxide) A compound of silicon and oxygen (SiO_2). It is a hard, glassy mineral substance, which occurs in a variety of forms such as sand, quartz, sandstone, and granite. It also is found in the skeletal parts of various animals and plants. Silicon. (Si) One of the nonmetallic elements in abundant supply as part of various compounds in the crust of the earth.

The solid crust of the earth contains 80% to 90% silicates or other compounds of silicon. Water passing through or over the earth dissolves silica from sands, rocks and minerals as one of the impurities it collects. Silicates compounds, which contain silicon and oxygen in combination with such metals as aluminum, calcium, magnesium, iron, potassium, sodium and others. Silicates are classed as salts. Silicates are widely distributed in such minerals as asbestos, mica, talc, lava, etc.

The silica content of water ranges from a few parts per million in surface supplies to well over 100 ppm in certain well waters. In its colloidal form, it consists of very fine particles in suspension. These can usually be removed by coagulation and settling or filtering.

Sodium

Sodium salts are present to a greater or lesser degree in all natural waters. Their concentrations vary from a few parts per million in some surface supplies to several hundred grains per gallon in certain well supplies. Sodium is extremely soluble and increases its solubility as the temperature of water rises. Because of this characteristic, sodium salts do not form scale when water is heated. Likewise, sodium salts do not produce curd when combined with soap. In fact, ordinary soap is an organic sodium compound. As such it does not react with the sodium in water.

Soap may be made from a fatty acid and a strong alkali:

$$C_{17}H_{35}COOH + NaOH \longrightarrow C_{17}H_{35}COONa + H_2O$$

Sodium content in water is determined by Flame Photometer method.

PHOSPHATE

Industrial effuents, domestic sewage and agricultural run off are the major contributor of phosphorus in water. Phosphates are largely used for laundry purposes and treatment of boiler waters. Phosphorus occurs both in inorganic and organic (85%) forms.

Spectrophotometric Method

Principle: Orthophosphates form heteropoly acid when they react with ammoniummolybdate and potassium antimony tartrate in acid medium. The phosphomolybdic acid reduce to molydenum blue by asorbic acid.

REAGENTS

Standard Phosphate Solutions

Dissolve 2.194 g of an hydrous potassium hydrogen phosphates in deionised water and make up the volume to 500 mL. Take 10 mL of this solution and add deionised water to make 1 L of stock solution containing 1 Mg P/L.

Prepare standard phosphorus solutions of various strengths from 0.0 to 1.1 mg P/L at intervals of 0.1 mg P/L by diluting the stock solution with distilled water.

Reagent A : Take 1 g of ammonium molybdate and 0.02 g of potassium antimony tartrate in 1000 mL volumetric flask. Add 16 mL of connectrated H_2SO_4 slowly. Dilute with distilled water to the mark.

Reagent B : Weight 0.88 g of ascorbic acid and dissolve in 1L of reagent A.

Procedure

Take the flask gently so that the content becomes colourless. Cool and add 10 mL distilled water and two drops of 1% phenoplhthalein indicator.

- Heat the flask gently so that the content becomes colourless. Cool and add 10 mL distilled water and two drops of 1% phenolphthalein indicator.
- Titrate against 1 N NaOH solution until pink colour appears. Make up the volume to 25 mL with distilled water.
- Transfer this solution into 50 mL volumetric flask and add 10 mL of reagent B.
- Make the volume to 50 mL with water till the blue colur develops.
- Record the absorbance on spectrophotometer at 660 nm.
- Run simultaneously a distilled water blank in the same manner.
- Process the standard phosphorus solutions of various strengths in a similar way.
- Plot a curve between absorbance and concentration of standard phosphorus solution.
- Deduce the phosphorus content of the sample by comparing its absorbance with standard curve.

Calculation

$$P\ (\text{mg/L}) = \frac{\text{mg } P \text{ in 50 mL}}{\text{Volume of sample}} \times 1000$$

Potassium

Potassium content in water is determined by Flame Photometer method.

Methane

Wells that contain methane are generally located in areas where gas and oil wells are common sights. Amounts run from 0.1 to 11.6 cubic feet per 1,000 gallons. This is roughly equivalent to 0.8 to 87 milliliters of methane per liter of water. Methane is objectionable in drinking water because of the odor and flammability. When water contains methane gas, it is advisable to aerate it prior to use for either industrial or household purposes. This is necessary to avoid the dangers of fire or explosion. The aerator must be vented to the open air to permit the gas to escape into the atmosphere.

Phenol

There is a growing trend by both government and private groups to control pollution of water due to the discharge of industrial waste materials. One of the offensive wastes is phenol. Phenol (C_sH_50H) imparts a medicinal taste and odor to water when the latter is chlorinated. This objectionable taste in a chlorinated water occurs in concentrations as low as one part per billion due to the formation of chlorophenols, which may be removed by activated carbon filtration.

Total Organic Nitrogen in Water

For determination of Total Organic Nitrogen, 200 ml of sample is taken in an evaporating dish and evaporate to dryness. Thereafter 4 ml of digestion mixture is added to the residue and dissolve it in about 20 ml of distilled water. The solution is heated to fuming for over 15 minutes and cool. Then the digest is transferred to micro-kjeldahl distillation assembly through way and about 3.5 ml of hypo solution is added.

25 ml of boric acid solution containing 2-3 drops of mixed indicator is taken in a conical flask placed below the condenser so that the tip of outlet of condenser is dipped in contents of conical flask. The boiling flask is heated so that the steam passes into sample through chamber. The distillation is continued for about 10 minutes. The conical flask having distillate is then removed. The boiling flask is then cooled so that all wastes are sucked into chamber and removed through tap. The distillate is titrated against hydrochloric acid. Turning of blue Colour to pink indicates end point. The distilled water blank is run in similar way.

Oil and Grease

250 ml of sample is collected in a wide mouth bottle and 10 ml of sulfuric acid is added [reagent A]. Thereafter 50 ml of Petroleum ether is added to it [reagent B] with little amount of ethyl alcohol [reagent C]. The whole suspension is shaken well and transferred to a separator funnel. The standing is allowed for some time so that the two layers upper one of petroleum ether and lowering one of water become distinct.

The lower layer of water is discarded from the separator funnel, through a filter paper soaked in petroleum ether [reagent b], in a pre weighed glass beaker. Some more petroleum ether is passed through filter paper so that no oil and grease remain stuck to the paper.

The beaker with contents is kept in a hot water both [never on flame] so as to evaporate the petroleum ether. The weight of beaker with residue is recorded in it.

Calculation

$$\text{Oil and Grease (mg/mL)} = \frac{\text{Wf} - \text{Wi} \times 100}{\text{V}}$$

Where, Wf = Final weight of beaker with residue (g)

Wi = Initial weight of empty beaker (g)

V = Volume of the sample (l)

MICROBIAL ANALYSIS

Introduction

The microbial populations in water tend to change and fluctuate, in response o the source of the water in which they occur. The origin of microorganism has been generally the result of surface runoff, precipitation that carries microorganisms attached to the dust particles to earth or the addition of wastewater to waterways, bays, lakes, and so on. The microorganisms derived from wastewater may, on the other hand, be harmful. Generally, wastewater is having both pathogenic and non-pathogenic forms. The danger arises when wastewater enters the water supplies, intended for human use.

Diseases commonly transferred in this manner have been those of enteric origin (those caused by microorganisms found in the intestinal tract). Microbial pathogens found in the excretory matter include, Shigella, the agent for bacillary dysentery; Salmonella, the bacterial agent for typhoid fever; Vibrio, the agent of cholera; Entamoeba, a protozoan responsible for amoebic dysentery; viral agents, polio and hepatitis.

An early method of determining the microbial purity by examining the various kinds of pathogenic microorganism was been slow, unwieldy and costly. Later on bacteriological tests got devised and are used to detect the presence of bacteria of fecal origin and thus indicative of fecal contamination.

Bacteriological Evidence of Pollution

The evidence of pollution in a given water body is identified by the presence of coliofrm group as one of the normal inhabitant of human intestine and hence consequently present in human feces.

The presence of Escherichia coli, fecal streptococci and clostridium perfringes in water is evidence of fecal pollution of human and animal origin. If these organisms are present in water, the way is open to the entrance of intestinal pathogens, sine they, too occur in feces. The enumeration of pathogen is difficult because-

- They gain entrance into water sporadically
- Do not survive for long periods
- Small numbers and escape the detection of laboratory procedures

It takes minimum 24h for detection, if pathogens containing water is consumed during the detection period then people might be exposed to severe infection.

As far as pollution indicator species *i.e. Escherichia coli* they are the inhabitants of intestinal tract and they survive for longer duration in water samples. Their detection is also possible within 24h. Thus the presence of coliforms in water is regarded as a warning signal for consumption without prior treatment. All bacteriological testing of water is based on the identification of these indicator species.

The enumeration of coliforms is done as per the standard method given in APHA. Generally three methods are followed-

Standard Plate Count

This is the procedure for estimating the number of heterotrophic bacteria in water and measuring the changes during treatment and distribution.

Multiple Tube Fermentation Technique (MPN)

This technique involves the quantification of coliforms in the given water sample. Coliforms consist of several genera of bacteria belonging to the family enterobacteriaceae. This group is defined as all facultative anaerobic, gram negative, non-spore forming, rod shaped bacteria that ferment lactose with gas and acid formation within 48 hrs at 35°C.

Coliform density, together with other information regarding sanitary survey, provides the best assessment of water treatment effectiveness and the sanitary quality of source water.

Analysis of salt or brackish water

Sediments and sludge

Membrane Filtration Technique

This test is highly reproducible but has limitations particularly when testing water is highly turbid or (non-coliform) background. Densities above 10^4 can interfere during the analysis.

MULTIPLE TUBE FERMENTATION TECHNIQUE FOR MEMBERS OF THE COLIFORM GROUP

Introduction

The coliform group consists of several genera of bacteria belonging to the family Enterobacteriaceae. The historical definition of this group has been based on the method used for detection of (lactose fermentation) rather than on the trends of systematic bacteriology. Accordingly, when the fermentation technique is used. This group is defined as all facultative anaerobic, Gram negative, non-spore forming, rod shaped bacteria that ferment lactose with gas and acid formation within 48 hours at 35°C. The standard test for coliform group may be carried out by multiple-tube fermentation technique or presence absence test.

When multiple tubes are used in the fermentation technique, results of the examination of replicate tubes and dilutions are reported in terms of the Most Probable Number (MPN) of organisms present. This number, based on certain probability formulas, is an estimate of the mean density of coliforms in the sample.

QUANTIFICATION OF TOTAL COLIFORM BY STANDARD FERMENTATION TECHNIQUE

Presumptive Phase

Media used-Mac Conkey

Procedure:

Arrange fermentation tubes in rows of five or ten tubes each in a test tube rack. The number of rows and the sample volumes selected depend upon the quality and character of the water to be examined. For portable water use five 20-ml portions, ten 10 ml portions, or a single bottle of 100 ml, portion, for non-potable water use five tubes per dilution (of 10, 1,0.1 ml etc.).

Incubate inoculated tubes or bottles at 35°C. After 24h swirl each tube or bottle gently and examine for growth, gas, and acidic reaction (shades of yellow color) and , if no gas or acidic reaction is evident, re incubate and re-examine at the end of 48 = 3 h. Record presence or absence of growth, gas, and acid production. If the inner vial is omitted, growth with acidity signifies a positive presumptive reaction.

Interpretation

Production of an acidic reaction or gas in the tubes or bottles within 48h constitutes a positive presumptive reaction. Submit tubes with a positive presumptive reaction to the confirmed phase (9221 B.2)

Confirmed Phase

Media used - Brilliant green lactose bite broth

Procedure

Submit all presumptive tubes or bottles showing growth, any amount of gas, or acidic reaction within 24h of incubation to the confirmed phase. If active fermentation or acidic reaction appears in the presumptive tube earlier than 24h, transfer to the confirmatory medium; preferably examine tubes at 18 h. With a sterile loop transfer one or more loopfuls of culture to fermentation tube containing brilliant green lactose bile broth. Incubate the inoculated tube at 35°C. Formation of gas in any amount in the inverted vial of the Brilliant green lactose bile broth fermentation to at any time within 48 hours constitutes a positive confirmed phase.

Complete Test

Media used - Endo Agar

The positive tubes of Brilliant green lactose bile are streaked on the endo agar plates and incubate at 35°C for two days. Growth with metallic sheen confirms the growth of Escherichia coli.

Staining

From the endo agar plates take a loopful of culture and streak on nutrient agar and then prepare slides from the growth on the nutrient media. Air dry and fix the by passing the slide on the flame and stain with crystal violet, iodine and safranin. Observe under the microscope for Gram-negative, nonspore-forming, rod shaped bacteria.

QUANTIFICATION OF FECAL COLIFORMS

Introduction

Elevated - temperature tests for distinguishing organisms of the total coliform group that also belong to the fecal coliform group. The fecal coliform test (using EC medium) is applicable to investigation of drinking water, lake water, wastewater treatment systems and general water quality monitoring. Prior enrichment in presumptive media is required for optimum recovery of fecal coliforms when using EC medium.

Media used - EC Medium

The fecal coliform test is used to distinguish from total coliforms. Submit all presumptive fermentation tubes showing any amount of gas, growth, and acidity with 48h of the incubation to the fecal coliform test. Using a sterile loop transfer the growth of the TC broth into another set of EC broth tubes and incubates at 44.5^{o} for 24h. Place all EC tubes in water bath within 30 min after inoculation.

Interpretation

Gas production with growth in an EC broth culture within 24h or less is considered as positive fecal coliform reaction. Failure to produce little or no gas constitutes a negative reaction. The results for the multiple tube fermentation technique are calculated as MPN index from the number of positive EC broth tubes.

QUANTIFICATION OF FECAL STREPTOCOCCUS

Fecal Streptococcus Group

The fecal streptococcus group consists of a number of species of the genus streptococcus, such as S.Faecal; is, S. Faecium, S.avium, S.boivs, S.equinus, and S.gallinarum. They all give a positive reation with Lancefields Group D. antisera and have been isolated from the feces of warm-blooded animals. S.Faecalis and S.Faecium once were thought to be more human specific than other Streptococcus. The fecal streptococcus has been used with fecal coliforms to differentiate human fecal contamination from that of other war-blooded animals. Editions of Standard Methods previous to the 17th suggested that the ratio of fecal coliforms (FC) to fecal streptococci (FS) could provide information about the source of contamination. A ratio greater than four was considered indicators of human fecal contamination, whereas ratio of less than 0.7 was suggestive of contamination by non-human sources.

Multiple Tube Technique

The multiple tube technique is applicable to lake water and sediments as the membrane filter technique is unsuitable for high turbid water.

Presumptive Test

Media used - Azide dextrose broth

Innoculate as series of tubes of Azide dextrose broth with appropriate graduated quantities of sample Use sample of 10 ml. portions of less. Use double-strength broth for 10 ml. inocula. The portions used w vary in size and number with the sample character. Use only decimal multiples of 1 ml.

Incubate inoculated tubes at 35°C. Examine each tube for turbidity at the end of 24 h. If no definite turbidity is present reincubate, and read again at the end of 48 h.

Confirmed Test Procedure

Media used - Pfizer selective Enterococcus (PSE) agar

Subject all azide dextrose broth tubes showing turbidity after 24 or 48 h incubation to the confirmed test. Streak a portion of growth from each positive azide dextrose broth tube on PSE agar. Incubate the inverted dish at 35=0.5°C for 24 = 2h. Brownish black colonies with brown halos confirm the presence of fecal streptococci.

Brownish black colonies with brown halos may be transferred to a tube of brain heart infusion broth containing 6.5% NaCI. Growth in 6.5% NaCI broth and at 45°C indicates that the colony belongs to the enterococcus group.

Computing and Recording of MPN

Estimate fecal streptococci densities from the number of tubes in each dilution series that are positive on PSE agar. Compute the combination of positives and record as the most probable number (MPN).

ANALYSIS OF BASIC PHYSICO-CHEMICAL PARAMETERS IN SOIL SAMPLES

Physical parameters

pH

pH is the scale of intensity of acidity or alkalinity and measures the concentration of hydrogen ions. pH 7 indicates neutral sludge , pH 7 to 14 alkaline , and pH below 7 acidic.

10 gm of air dried soil /sediment is taken and 50 ml of distilled water is added to make a suspension of 1:5 dilutions. The pH of suspension is determined with the help of pH meter.

Conductivity

10 gms of air-dried soil /sediment is taken and 50 ml of distilled water is added to make a suspension of 1:5 dilution. The conductivity of suspension is thereafter determine with the help of conductivity meter.

Moisture Content

The water in soil is not only important as a solvent and transporting agent but it maintains texture and compactness of oil and make it habitable for plants and animals. The moisture in soil is gained mainly from infiltration of precipitated water and irrigation, and its content in soil at any time depends upon the water holding capacity of soil and drains through percolation, evaporation, and uptake by plans etc.

For determination of moisture content of soil a fresh homogenized sample is taken for initial weight. It is then dried in an oven at 105°C until a constant weight is attained. The sample is cooled in a desiccator for recording final weight of sample.

Moisture Content % = (I-F/I) × 100

I = Initial weight of Soil

F = Weight of soil after oven drying

Bulk Density

The bulk density of the soil is defined as the dry weight of a unit volume of it, and it is expressed as g/cm^3. Fine textured soil has lowest bulk –density and slightly higher in coarse textured soil and maximum in case of alkaline soil. The bulk density is inversely proportional to pore space.

The soil sample is dried in an oven at 105°C until a constant weight is attained. A little dried soil is transferred to a measuring cylinder and the volume is noted. The weight of this volume of soil on a balance is recorded.

Bulk Density (g/cm^3) = Weight of Soil (g) / Volume of Soil (cm^3)

Preparation of Soil Suspension

Approximately 50 gms soil and 250 ml distilled water is taken in a beaker. The suspension is then shaken manually or on shaker for half an hour. The suspension is filtered through Whatmann No. 50. Filtrate obtained is used for the analysis of Chloride, Ca Content, Mg Content, Sodium, Potassium etc.

Analysis: All the above mentioned parameters are analyzed in soil filtrate as in water.

ANALYSIS OF NITROGEN BY KJELDAHL NITROGAN METHOD (WATER AND SOIL)

1.Total Organic Nitrogen in Soil

10 gms of air - dry soil/sediment is taken in a 300ml round bottom flask and 20gms of catalyst mixture is added with 35ml of sulphuric acid. The contents from bottom of the flask are heated for about 2 hours. The contents (digest) is cooled, about 100 ml. of-distilled water is added and the supernatant is delivered into 1L distillation flask of Kjeldahl distillation assembly. A little of distilled water is washed several times and the supernatant transferred each time to the same distillation flask.100 ml of sodium hydroxide solution is added with a few granules of Zn, 25 ml of Boric acid cum indicator solution is take in a 500 ml Erlenmeyer flask and is placed below the condenser of distillation so that the lower open end of the condenser is dipped in solution. The distillation flask is heated on a hot plate and about 150 ml of distillate is collected in a flask. The flask with distillate is removed and titrated (which has turned blue due to dissolution of ammonia) against 0.1N hydrochloric acid. Turning of blue colour to light brown-pink indicates the end point. Distilled water blank is run in the same manner.

Calculation

$$\text{Kjeldahl Nitrogen (mg/g)} = \frac{T_1 - T_2 \times N \times 14}{W}$$

$$\text{Kjeldahl Nitrogen (\%)} = \frac{T_1 - T_2 \times N \times 1.4}{W}$$

T_1 = Volume of the titrant against sample (ml)

T_2 = Volume of the titrant against distilled water blank (ml)

N = Normality of the titrant (0.1)

W = Weight of soil/sediment used (g)

ANALYSIS OF BIOLOGICAL PARAMETERS

Phytoplanktons

Phytoplankton are chlorophyll bearing suspended microscopic organisms consisting of algae with representatives from all major taxonomic phyla; the majority of members belong to Chlorophyceae, Cyanophyceae and Bacillariophyceae. Their unique ability to fix inorganic carbon to build up organic matter through primary production makes their study a subject of prime importance. The quality and quantity of phytoplankton and their seasonal succession patterns have been successfully utilized to assess the quality of water and its capacity to sustain heterotrophic communities.

Quali-Quantitative Analysis

Among the several method known for the collection of plankton , the use of 'plankton net' is most common even for algal work; which however should be abandoned because many minute algae (e.g. nannoplankton, mainly diatoms) pass through the pores of the net resulting in under estimation of the population density. Therefore, the direct concentration of the phytoplankton of the water sample by sedimentation (and occasionally by centrifugation) is a pre-requisite for accurate quali-quantitative analysis. The sample concentrated as above is then subjected to qualitative analysis of algal forms (use the keys and monographs). Out of the several method used for quantitative enumeration, the drop-method and Sedgwick-Rafter cell method are described here; the former, despite its simplicity and in expensiveness, is comparatively more accurate.

Preparation of the Sample

One litre of the water sample is taken in a glass bottle. 10 ml Lugol's iodine is added and allowed to stand for at least 24 hours to ensure complete sedimentation (centrifuge if necessary).The supernatant (clear) liquid is taken with the help of pipette. Further the remaining sample is concentrated upto 10 to 100 ml depending on the number of plankton.

Plankton Enumeration (Drop count Method)

This method involves the plankton enumeration in one drop of the concentrated sample taken on a slide using a standard calibrated dropper.

One drop of concentrated sample is taken on a clean microslide with the help of a standard dropper holding it vertically. The whole drop is carefully covered with a cover glass of suitable size so that the sample does not run out. The micro slide is placed under the microscope and one edge of the cover glass is focused. The phytoplankton is counted species-wise, while moving the slide with the help of movable stage to the other edge. The slide is shifted to the next 'field'. The process is repeated on the path parallel to that of earlier one, but in the reverse direction (to save time).

In this way the whole cover glass is observed and planktonic estimations is worked out at least for five to ten drops depending on the density of plankton.

Sedgwick-Rafter Cell

It has been named after the inventor of the cell. The rectangular cavity in the slide (50x20x1 mm) contains exactly 1 ml (1000 mm^3) of water sample. This method is better suited for the enumeration of larger organisms only (e.g. zooplankton).

Requirement

Sedgwick-Rafter cell and other requirements as given in earlier procedure.

Method

The width of the field of microscope is measured with the help of ocular and stage micrometers. The cover slip is placed diagonally on the cell cavity. The sample is then shaken gently and 1 ml of it is transferred quickly into the cavity with the help of a graduated pipette. The cover slip is adjusted properly to cover the cavity without air bubbles. One edge of the cavity is focused and the slide is moved horizontally, simultaneously counting the plankton till the other edge.

Zooplankton

Zooplankton, the microscopic free swimming animal components of aquatic systems, are represented by a wide array of taxonomic groups; of which the members belonging to Protozoa, Rotifera, Cladocera and

Copepods are most common and often dominate the entire consumer communities. They are endowed with many remarkable features and are often armoured with spines, which hamper their predation by higher organisms. The ability of movement not only provide them an effective defense measure but also enable them to actively search and feed upon the phytoplankton. Their high and rapid rate of parthenogenetic reproduction usually overcomes the predation losses and enables them to exploit algal blooms. They constitute an important link between primary producers and consumers of higher order in aquatic food webs. Therefore, the population dynamics of zooplankton with reference to system energetics provides key information for the management practices.

Preparation of the Sample

Known amount of water sample e.g. 25 litres is collected and filtered through a plankton net of bolting silk No.25. The net plankton is transferred in 50 ml bottle and preserved in 5% formaline. few drops of glycerin is added to it. If further concentration is required the sample is allowed to stand for a day. Practically all the zooplankton will settle down at the bottom of the bottle. The supernatant plankton-free water is removed with the help of pipette and the sample is reduced to the described volume.

Quali-Quantitative Analysis

The zooplankton is identified in the sample using keys and monographs, and then the quali-quantitative analysis is done using the methods described for the phytoplankton.

Periphyton

All microscopic organisms (both plants and animals) which grow attached on materials submerged in water are known as periphyton. At times they get represented in plankton samples either due to fragmentation or detachment of the filaments or colonies. The periphyton form a heterogeneous community and are broadly classified by the nature of their substratum as epipelic (grow on sediments), epilithic (on rocks and stones), epipsammic (in or on sand), epiphytic (on plants) or epizoic (on animal surfaces). In shallow lakes, heavily infested with macrophytes, contribution of algal periphyton to the total primary production may assume significant proportions and cannot be overlooked. Furthermore, many periphytonic organisms show a remarkably distinct response to the pollution of water bodies, due to their simple structure and lack f mobility. This property could be exploited for pollution assessment. However, due to a lack of realization of their significance for a long time, the study of their population dynamics is still in its early stages.

Collection of the Periphyton

They can be collected directly from the natural substrata

A known volume (ml) of sample is taken. To preserve every 10 ml of the sample 2-4 drops of Lugol's iodine or 5% solution of mercuric chloride is added.

Qualitative-Quantitative Analysis

The preserved sample is shaken well, and a drop on the slide is taken. It is then examined under the microscope and the Periphytonic organisms are identified with the help of keys, monograph etc. For quantitative estimations the drop count method is followed.

Benthos

The term benthos denotes the whole assemblage of organism dwelling at the bottom, which shows a marked diversity depending upon the change in the depths and properties of the sediments. They are represented from almost at the phyla and provide important information which is useful for pisci culture practices. Most of the benthic organisms are detritivores and form an important link in the food chain on account of their ability to convert low quality and low energy detritus into better quality food for higher organisms in the food web. By virtue of being relatively stationary, they are constantly exposed to change undergoing in the mud-water interface and respond very well to it. Therefore, several pollution indices have been proposed using qualitative and quantitative change in benthic populations.

Collection and Preparation of Sample

Since benthos dwell at or below the surface of the bottom sediments of littoral and profundal zones, they are collected with the help of mud samplers.

The sediment sample is collected by a suitable mud sampler. The sample is taken in a bucket, its volume is noted and water is added to it. The suspension is sieved out through 2 mm and 0.5 mm mesh size sieves, one after the other. The micro and macro invertebrates are collected from the sieve with the help of forceps and transfer them to a wide mouth bottle containing 70% alcohol. The residual organic matter retained in the sieve is transported to the laboratory and 50 to 100 ml of water is added to it. 5 g. of sucrose is dissolved in it. (This helps in easy collection of benthic fauna. Because they now float on the surface due to the change in the density of the medium). Alternatively, the dyeing of the organisms (by treating residual matter with 0.1 – 0.2% cosin, erythrosin or acid fuchsin in 70% alcohol) helps in easy collection of smaller organisms like oligochactes, chironomids, mematodes etc. which pose a greater problem in comparison to sponges, molluscs, arthropods and others.

The organisms are fixed in 70% alcohol.

Quali-Quantitative Analysis

The organisms are identified using keys and monographs. The number are counted species-wise in the sample.

REFERENCES

1. F. Moriarty, 'Ecotoxicology : The Study of Pollutants in Ecosystems', 2nd Edn., Academic Press, London, 1988.
2. E.D. Schulze and P.H. Freer-Smith, 'Acid Deposition—Ist Natureand Impacts', eds. F.T. Last and B. Watling, Proc. Roy. Soc. Edinburgh, 1991, Section B, 97, p. 155.
3. P.R. Miller and J.R. McBride, 'Air Pollution and Forest Decline', eds. J.B. Bucher and I. Bucher-Wallin, 1989, p. 61.
4. J.N.B. Bell, 'Gaseous Air Pollutants and Plant Metabolism', eds. M.J. Koziol and F.R. Whatley, Butterworth, 1984, Chpt. 1, p. 3.
5. M.W. Holdgate, 'A Perspective of Environmental Pollution', Cambridge University Press, Cambridge, 1979.
6. J. Widdows and D. Johnson, 'Biological Effects of Pollutants', ed. B.L. Bayne, K.R. Clarke, and J.S. Gray. Marine Ecology Progress Series Special Issue, Amelinghausen : Inter Research, 1988, 46.
7. T.W. Clarkson, L. Friberg, G.F. Nordberg, and P.R. Sagar (ed), 'Biological Monitoring of Toxic Metals', Plenum, New York and London, 1988.
8. L. Friberg, G.F. Nordberg, and V.B. Vouk (ed.s), 'Handbook on the Toxicology of Metals', VII, 2nd End., Elsevier, 1986.

CHAPTER 19

Conservation of Water

Syed Akhtar Husain[1]

"Water is Life" and every thing on earth contains water. This quotation is not just said by some philosopher, its, importance is described in holy books. The Quran says "Every thing is made of water". So saving water is a noble affair and wasting water in any form in any religion is always forbidden. It is mentioned in holy books that every body has to give account of every drop of water wasted. Not only saving water is necessary but to keep it form pollution is also equally important. It is a universal fact that the quantity and quality of ground water is adversely going down year by year. The day by day increase of consumption of water for drinking, domestic, agricultural, sanitation and industrial use, is for more than, charging ground water level and water bodies. Consequently this grave shortage of water is causing crop failure, poor yield of food grain, adverse salt balance and intrusion of sea water in costal areas.

It is now extremely necessary to improve the quantity and quality of ground water. For this purpose different methods are adopted by the Government, NGO' s and some individuals.

Natural Recharge of Ground Water

Centuries old rivers, streams, lakes and other water bodies are charged by natural rains. Just fifty year before the situation was quite satisfactory.

Wells, rives, lakes and other water bodies had enough water for all purposes.

Pollution was not so drastic as it is today. Rain was enough to drain out surface and bottom pollution. Increase in rural and urban and population directly effected the natural ground water charging and tremendous pollution in wells, streams, lakes, ponds and all sorts of water bodies. Poor drainage and sewage system and increasing number of toilets, discharge of poisonous chemicals from industries directly soused severe pollution to very high danger levels.

Just half century before the quantity of water consumed was almost equal to input of natural rains. Wastage of water in rural and urban areas was minimum as water was mostly drawn manually and with the help of "Persian Wheels" and leather "Charas" operated by bullocks.

Few villages had electric supply lines and so there were very few electric pumps and so consequently there was minimum wastage of water.

Artificial Recharge of Ground Water

There are many methods for artificial recharging of ground water such as recharging through wells, pits, spreading water in the agricultural field manually or by pumping and other methods. Choice of method depends on local topographic, geological, soil condition, climatic condition etc. Presently due to global

[1]Ex. Advisor, Indira Gandhi National Museum of Man, Bhopal-462001 (India)

warming and drastic change in weather conditions, artificial recharge of ground water is serious and extremely challenging issue for government, NGO's and individual level.

Recharge Methods

1. Irrigation of agricultural land : This way water is spread in crop-land during no crop season. This is possible in such areas where huge dams are existing and water may be flown through canals. Such areas are very few and may not be enough for more than very short percentage area of that region. The other method of spreading water by pumping is, now a days, very difficult due to power crisis.

2. Tank or Pit method : Digging of pits or construction of small dams on small streams is a very successful method for recharging ground water. Excavation of a pit on some vacant government or public land is very easy and economical method to collect natural rain water. This increases water level in nearby wells and tube wells. Such work is easily possible if some farmers contribute fund with the help of Government. Such digging is possible with heavy earth moving equipment without the need of expensive boulder or RCC construction. Author's son Dr. S. Arham Husain has made such pit of about 250 × 250 feet at his farmland at Shampur Tehsil of district Sehore, M.P. Rainwater collected in this pit has increased water level in surrounding wells and tube wells.

Rain Water Harvesting

Conservation of rain water is called rainwater harvesting. This method is useful for domestic use which otherwise goes waste through drainage. Rain water from roof top or any other type of RCC storage is passed to a pit dug out near the pipe coming from roof top. The rain water passing through the small pit goes down the earth and increases ground water level. During the last few years the Govt. is trying best to adopt this technique. But contractors and builders are forced to make such system on govt. buildings. A common man does not take interest in making such water harvesting system because he thinks that why he should spend atleast Rs. 5,000/- for making such fitting in his house because such expenditure does not directly gives him any benefit.

Author has designed a System of Rainwater Harvesting-3 in One

This device gives direct utility of pure rainwater for drinking and domestic purposes. It is a simple device and extremely cheap and free of maintenance. Before rainy season roof top is cleaned and if necessary a coat of white cement is done on the roof. A bucket of about 50 litres is kept under the outlet of pipe coming from top. A small piece of fine nylon net is covered on bucket to stop any thick particles etc. On one side of the bucket a hose nipple is fitted with a 3/4 inch pipe, delivering water to the ground reservoir of the building. If due to heavy rains water overflows from the ground tank than it is passed to the nearby tubewell. If still there is heavy rain and even the tube well overflows, than it is passed to the pit mostly under a tree, to go down the earth, to increase ground level.

The main advantage and interest of this system is that the house owner gets "PUREST WATER" for drinking and domestic purposes. The water from municipal tapes is always polluted due to leakages in streets where dirty drainage water mixes with the incoming tape water. The water collected with this system can be used at least for about 2-3 months during rainy season. This way lot of water supplied from any dam or river is saved and to be utilised during hot summer season. If calculated it will give saving of millions of litres of water in a small town. See diagram.

Fog Water Harvesting

Most part of the world is suffering from severe water crisis. Population growth has caused a acute water shortage. Fog water and dew water harvesting is done by specially designed collection system. This system is useful specially in North Indian areas where fog and dew drops are abundant. This system is expensive and needs special fittings.

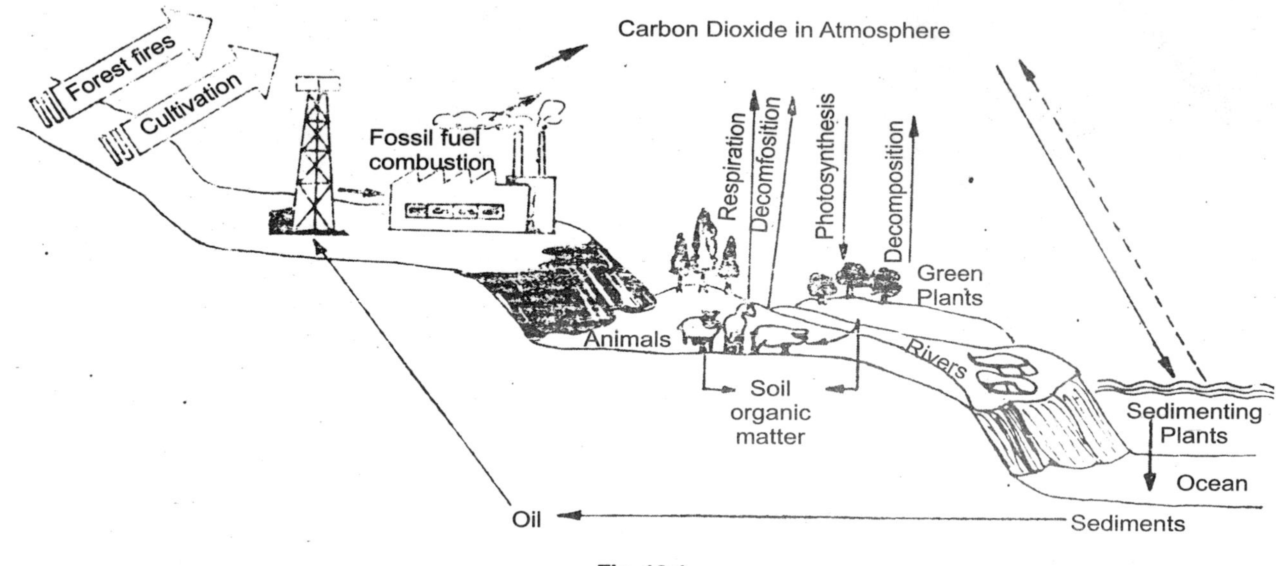

Fig. 19.1.

Rainwater Harvesting and Ground Water Recharge the only tool to fight with the alarming threat of Groundwater Depletion

Harvesting rainwater and recharge groundwater—You can make a difference for a better tomorrow.

Benefits of Rainwater Harvesting

(1) Environment friendly and easy approach for water requirements. (2) RWH is the ideal solution for all water requirements. (2) Increase in ground water level. (3) Mitigates the effects of drought. (4) Reduces the runoff, which other wise flood storm water drains.

Reduces flooding of roads and low-lying areas. (5) Reduces soil erosion. (6) Improves the ground water quality. (7) Low cost and easy to maintain. (8) Reduces water and electricity bills.

Who can Harvest Rainwater and Where

People planning construction of House, Modification of house existing house etc. from rooftops Govt. Buildings, Institutions, Hospitals, Hotels, Shopping malls etc. from rooftops and open areas Farmlands, Public Parks, Playground, etc. Paved and unpaved areas of a layout/city/town/village.

Need for Rainwater Harvesting: (1) As water is becoming scarce, it is the need of the day to attain self-sufficiency to fulfill the water needs. (2) As urban water supply system is under tremendous pressure for supplying water. (3) to ever increasing population. (4) Groundwater is getting depleted and polluted. (5) Soil erosion resulting from the unchecked runoff. (6) Health hazards due to consumption of polluted water.

Methods of Rainwater Harvesting

1. Rainwater stored for direct use in above ground or underground sumps/overhead tanks and used directly for flushing, gardening, washing etc. (Rainwater Harvesting).
2. Recharged to ground through recharge pits, dug wells, bore wells, soak pits, recharge trenches, etc. (Ground water recharge).

Rainwater Harvesting Potential

In an are of 2,400 Sq. ft. or 223 Sq. mt. (40 ft × 60 ft. site) around 2,23,000 liters of rainwater can be harvested (Local Annual rainfall in is around 100 mm/ 39.4 inches). The amount of rainwater that can be harvested depends on the catchment area, rainfall in that area and collection efficiency etc.

Fig. 19.2. A view of Roof rain water harvesting from harvested rain water.

Fig. 19.3. A view of groundwater recharge from Harvested rain water.

FOG WATER HARVESTING

Fog water harvesting can make a significant impact on the quality of life of the people living in the water scarce area and, if sufficient collectors will be erected, the water can be used to start various small scale agricultural projects. With sufficient water, the abundance of sunshine and the absence of other natural hazards such as hail and frost, the selection of appropriate crops can generate a substantial income and significantly improve the quality of life of the community.

Introduction

Many countries in the world are facing water security and scarcity problems. The rapid population growth and increasing pollution of rivers and ground water have put much stress on the water resources to meet the needs of the teaming millions. India is blessed with vast water resources. The sustainability of the water resources has gained increasing importance, as scarcity has already started being felt in some parts of the country. Therefore, the sustainable management of water resources has been emphasized equally with creation of new facilities.

There is an urgent need to conserve fresh water, either being polluted or wasted to the rivers, oceans or on land. Collecting rain-water on roofs (urban areas) and in well planned watersheds (rural areas) are very common and ancient practices in India. But, in the time of new challenges, it's necessary to collect water from its each and every source, even from smaller sources like fog and dew.

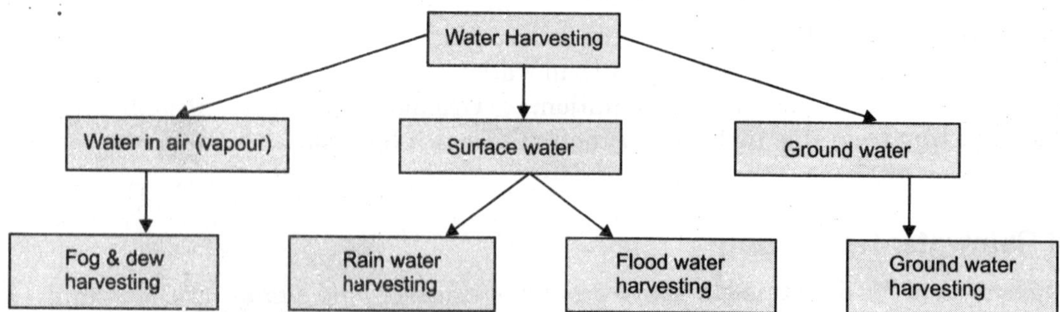

Fig. 19.4. Types of water harvesting.

Fog water harvesting (FWH) offers an untapped resource of water to the society for various applications, particularly over the plains in northern India, which experience dense, widespread and prolonged foggy days in winter. Under specific atmospheric conditions, fog and dew can be captured and yield substantial amount of water which can be used for domestic purposes, livestock, establishment of trees, or for the growth of crops. Small and simdple installations or the condensation of fog and dew can yield several liters of water per day.

What is Fog ?

Fog is defined as almost microscopically small droplets of water (condensed form) and suspended in air near the surface of earth in sufficient number to reduce the horizontal visibility to less than 1 km. In simple words it's a cloud with its base at or very near the ground is provoked by the following conditions :

- Substantial heating during the daytime.
- Clear skies or very light high clouds at night.
- No or very light wind.
- Thermal inversion at moderate heights.
- A sufficiently high atmospheric humidity (so it's more common in the vicinity of ocean where there is an abundant supply of moisture near the earth surface than inland).

Fog formation takes place on a synophilic scale : however its characteristic may change over regional to localized scales depending upon terrain effects, urban and non-urban regional availability of water bodies, wet surfaces, etc.

History of Fog Water Harvesting

Fog though an alternative source of potable water is largely ignored, it was used extensively in the ancient times. Palestinians for instance, build small low circular honeycombed walls around their vines. So that, mist and dew can precicipate in immediate variety of plants. Historically, both dew and fog were collected in the Atacama and other deserts from piles of stones, arranged so that the condensation would drop to the inside of the base of pile where it shielded from the sunshine. In the Canary Islands, fog drip from trees was a sole source of surface water for man and animals for many years.

A number of projects have been mutated which specifically aimed at supplying fog water to communities. In this direction, the efforts are already on in many countires like Chilly, Peru, Oman and Nepal. The first was implemented at Mariepstop in Mpumalanaga, South Africa during 1969-70. The second and largest fog collection project to date was initiated by researchers at the National Catholic University of Chile and the International Development Research Center in Canada, at a small fishing village in northern Chile in 1987.

Though, plains in northern India, experiences dense, widespread and prolonged foggy days in winter, no such study has been done. Only a few efforts have been made in the country. In a study it has been estimated that during January, 2003, the maximum suspended water available in fog was about 12.5 billion liters over India.

Fog Characteristics

Occurrence of fog is not uniform. It varies over space and time. Annual fog incidence decreases with increasing latitude and longitude. It means fog occurs more frequently at sites with a slightly higher elevation than sea level.

Fog occurs more often during the cooler nocturnal hours peaking between midnight and 8.00 a.m. and confined to the period of midnight. Fog intensity is the most considerable character of fog. It can be determined by the degree of visibility in fog, which further depends upon the opacity of the air, resulting from the number of particles held in the suspension. It is defined as, "the greatest distance at which a dark object can be seen".

The mean annual fog day frequencies were recorded as 66 days/year at Dassen Island; 111 days/ years at Cape Columtine and 148 days/year at Port Molloh. In India, not only the hilly regions, most of the northern plain of the country like Delhi, U.P., M.P. and Bihar faces more than 40 dense foggy days/ year, which offers the scientists to collect water.

Fog Collectors

Fog droplets are much small than both raindrops and drizzle drops with diameters varying from 01 to 40 microns and fall at velocities ranging from less than 01 cm/s to about 05 cm/s. These low velocities result in fog droplets being influenced even by light winds that can cause the droplets to travel almost horizontally. Therefore, such surfaces are constructed as vertical mesh panels on which fog droplets are intercepted and collected.

A full-scale fog collection system consists of simple, rectangular nets supported by a post at either and, arranged perpendicular to the direction of wind (Fig. 2). These surface-nets are usually made up of nylon or polypropylene netting (usually has an extraction of 30%). An anemometer is attached with the pannel to know the wind direction (Fig. 3). As water condenses, it falls to a collector, which conveys the water to a storage tank (Fig. 4) where the water can be filtered, chlorinated and distributed.

Fig. 19.5. A fog harvesting arranged in perpendicular direction of wind can condense fog into water.

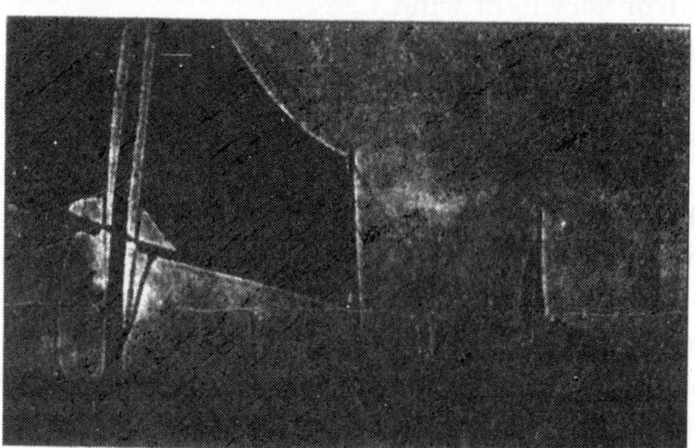

Fig. 19.6. Fog collector with anemometer to detect the wind direction.

Fig. 19.7. A storage tank attached with the fog harvesters.

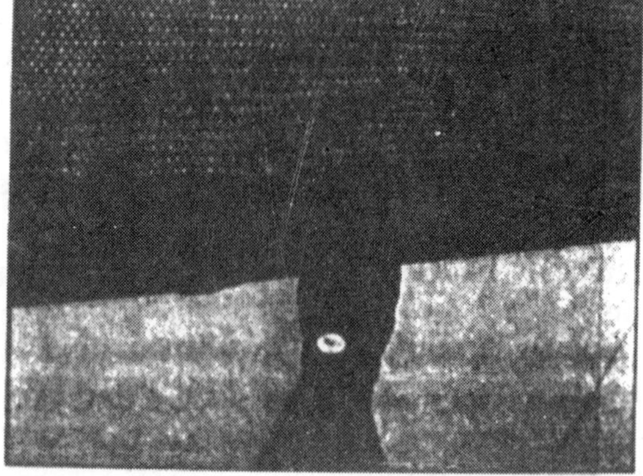

Fig. 19.8. A close view of nylon made fog collector.

Cost and Operation

Cost of the fog-collector panel is not very high. Virtually all of the input materials to construct, operate and maintain the systems are available locally, thus, cost differs from place to place. However, an automatic weather station costs about $50,000 to purchase and install. Operations requirement include the measurement of volume collected and the recording of meteorological data, either manually or by automatic weather station since changes in weather conditions may change the operational design of harvesters.

Effectiveness of Technology

Also no intensive, permanent structure are necessary to implement this technology and the drrived water is normally potable. The technology can easily canter to the water need of coasts or desert settlements or camps currently relying on saline water sources or some other expensive options such as desalinization and tanker delivery. The water is available within the demand area and therefore, requires little or no pumping. The water source is also sustainable for many years. The collectors are simple and require no energy other than the wind. Thus, there will be no pollution during its operation.

A surface area of about 50 m^2 can harvest a significant amount of got and convert it into water. Fog harvesting is suitable in regions which have hills or mountains close to potential users, on a coastline with an old current offshore. The quantity of derived water is a function of scale of project and fog unavailable. There is a chance of contamination of dust, microbes and other pollutants present in atmosphere. In Namibia, the resulting water is slightly salty as a result of the inclusion of some marine aerosols and contains some dust. Due to highly stable atmosphere, the pollutants released locally, cannot be transported to other place by wind easily. However, many of such pollutants can be removed using a simple sand-filter. Their complete removal needs more sophisticated technical tools which further increase the cost of the project.

PATTERN OF DESIGNING SOME RECHARGE STRUCTURES

Settlement Tank

For designing the optimum capacity of the tank, the following parameters need to be considered:

(1) Size of the catchment

(2) Intensity of rain falI

(3) Rate of recharge, which depends on the geology of the site

The capacity of the tank should be enough to retain the run off occurring from conditions of peak rainfall intensity. The rate of recharge in comparison to run off is a critical factor. However, since accurate recharge rates are not available without detailed geo-hydrological studies, the rates have to be assumed. The capacity of reacharge tank is designed to rtainrunfoo from at least 15 minutes rainfall of peak intensity. Illustration :

For an area of 100 sq. m

Suppose: Area of rooftop catchment (A) is 100 sq.m.

Peak rainfall in 15 min (r) is 25 mm (0.25m)

Runoff coefficient (C) is 0.85

Then the capacity of deslting tank = A × R 0.025 × 0.85

= 0.125 cu. m. (2.125 litres)

= 2.125 cu. m. (2,125 liters)

Reacharge Trench

The methodology of design of a recharge trench is similar to that for a settlement tank. The difference is that the water-holding capacity of a recharge trench is less than its gross volume because it is filed with porous material. A factor of loose densit of the media (void ratio) has to be applied to the equation. The void ratio of the filler material varies with the king of material used, but for commonly used materials like brickbats, pebbles and gravel, a void ratio of 0.5 may be assumed.

Using a void ratio of 0.5, the required capacity of a recharge tank = A × r × C/0.5

= (100 × 0.025 × 0.85)/0.5

= 4.25 cu.m. (4,250 litres)

In designing a racharge trench, the length of the trench is an important factor. Once the required capacity is calculated, length can be calculated by considering a fixed depth and width.

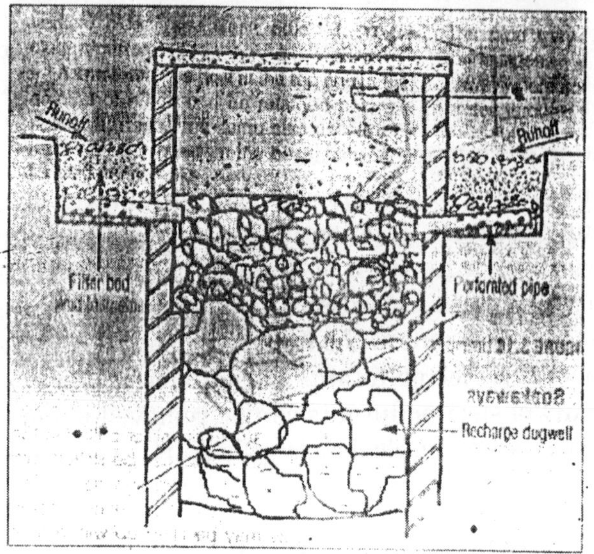

Fig. 19.9. Recharge Well.

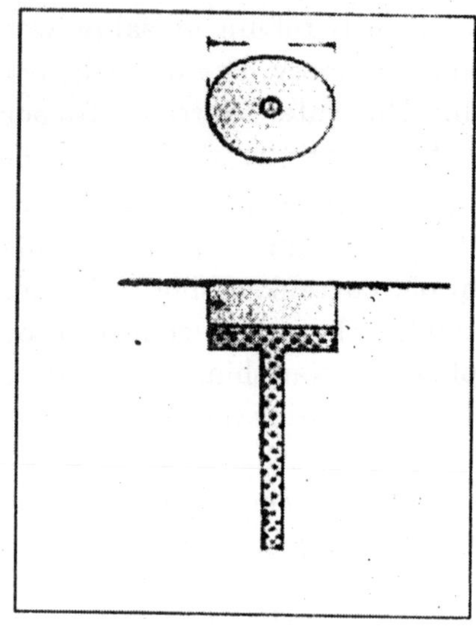

Fig. 19.10. Recharge Pit.

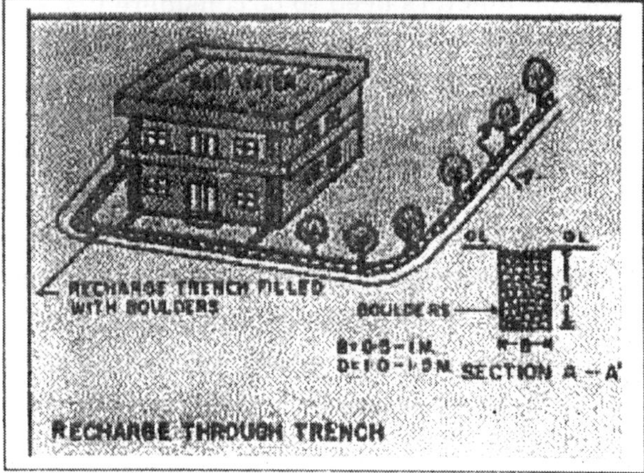

Fig. 19.11. Recharge Trench.

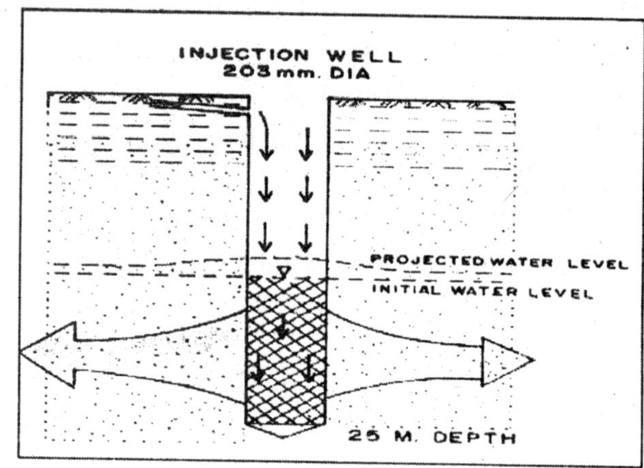

Fig. 19.12. Injection Tubewell.

Conclusion

All countries need to join their scientific and financial resources to help fight the world water crises before it encompasses the entire world. It is hard to imagine that water, something so basic, so simple and so necessary is at risk of elimination.

Thus, there is an urgent need to study the benefits and drawbacks seriously in Indian context before installation of fog-collectors. Due to sufficient foggy days and other atmospheric conditions, this technique will certainly be suited to us.

REFERENCES

1. Raghunath, H.M., 1997. " Ground Water", 2nd ed. New Age International (P) Ltd. New Delhi.
2. Brown, R.F., and D.C. Signor 1974. Artificial Recharge-State of Art. Ground Water, V. 12.
3. Artificial Ground Water Recharge. Design and Operation of Recharge Besins, Journal., Amer. Water Works Association, V-55, pp. 697-704, 2001.
4. Dangi, R.S. *et al.,* Project on Rainwater Harvesting, Indian Railways Institute of Civil Engineers, Pune, 2007.
5. Karanth, K.R., 2003, Ground Water Assessment Development and Management Tata McGraw-Hill Publishing Company Limited, New Delhi.

6. Karanth, K.R., 1980, "Hydrogeology", Tata McGraw Hill Publishing Company Limited, New Delhi.
7. Goldman, J.C. *et al.,* 1972, Water Research, 6:637.
8. Langelier, W.F., 1946, "Effect of Temperature on the pH of Natural Waters", journal of the American Water Works Association, 38:179.
9. O'Connor, J.T. *et al.,* 1975, "Deterioration of Water Quality in Distribution Systems", Journal of the American Water Works Association, 67: 113.
10. Tayler, F.B. *et al.,* 1996, "The Case of Water Borne Infectious Hepatitis". American Journal of Public Health, 56 : 2093.
11. Sodium Chloride and Conductivity in Drinking Water Supplies. Copenhagen. WHO Regional Office for Europe, 1979 (Euro Reports and Studies), No. 21.
12. Ridgway, J. *et al.,* "Water Quality Changes, Chemical and Microbiological Studies, In: Water distribution System Maintenance of Water Quality and Pipeline Integrity", Medmenham, England Water Research Centre.
13. Thomas, J.F.L., 1953, "Industrial Water Resources of Canada, Water Survey Report No. 1, Scope, Procedure and Interpretation of Survey Studies". Ottawa Queen's Printer.
14. Nitrate, Nitrites and N-Nitroso Compounds Geneva, World Health Organization, 1978 (Environmental Health Criteria 5).
15. Guidelines for Canadian Drinking Water Quality, 1978, Quebec, Ministry of Supply and Services, 1980 (supporting documentation).
16. Rrazer, P. Et.Al., 1980, "Nitrate and Human Cancer : A Review of the Evidence. " International Journal of Epidemiology, 9:3.
17. National Research Council, Drinking Water and Health, Washington, DC, National Academy of Sciences, 1977.
18. Frazek, P. & Chilveks, C., 1980. Health Aspects of Nitrate in Drinking Water". Netherlands, (paper presented at the International Symposium on Water Supply and Health).
19. Schbneider, W.J. Hydrologic Implications of Solid-waste disposal. Washington, D.C., Department of the Interior, 1970, pp. F1-F10 (US Geological Survey Circular 601-F).
20. Petiyjohn, W.A., 1971, "Water Pollution by Oil-field Brines and Related Industrial Waters in Ohio", Ohio Journal of Science, 71: 257.
21. Ralston. I.G. Deicing Salts as a Source of Water Pollution, Toronto, Ministry of the Environment, 1971.
22. Kirkpatrick, D.C. & Coffin, D.E., 1976, "The Trace Metal Content of Representative Canadian Diet in 1970 and 1971". Canadian Institute of Food Science and Technology Journal, 7:56 (1974).
23. Meranger, J.C. & Smith, D.C., 1972, "The Heavy Metal Content of a Typical Canadian Diet", Canadian Journal of Public Health, 63: 51.
24. Zoeteman, B.C. J. & Brinkman, F, J.J., In: Hardness of Drinking Water and Public Health. Proceeding of the European Scientific Colloquium, Luxembourg.1975. Oxford. Pergamon Press 1976, p. 173.
25. Ehipple, G.C., The value of Pure Water, New York, John Wiley and Sons, 1907.
26. WHO Technical Report Series, No., 505, 1972 (Evolution of Certain Food Additions and the Contaminates : Mercury, Lead and Cadmium).
27. Guidelines for Canadian Drinking Water Quality, 1978, Quebee, Supply and Services, 1979 (supporting documentation).
28. Commission of the European Communities, 1978, "Criteria does effects Relationships for Cadmium", Oxford, Pergamon Press.
29. Schroeder, H.A., 1968, The Role of Chromium in Mammalian Nutrition, American Journal of Clinical Disease Nutrition, 21 : 230.
30. Toxicology of Water. Vol.II Washington, DC. US Environmental Protection Agency, 1977 (Environmental Health Effects Research Series).
31. Schramel., P. *et al.,* 1973. "Some Determination of Hg. As. Se. Sb. Sn. and Br. in Water, Plants, Sediments, and Fishes in Bavarian Rivers". International Journal of Environmental Studies, 5:37.
32. Voege, F.A., 1971, "Levels of Mercury Contamination of Water and its Boundaries", In: Proceedings of the Symposium on Mercury in Man's Environment. Ottawa, Royal Society of Canada, p. 107.
33. Ermakov. V.V. & Kovalskij, V.V., 1974, Biological Significance of Selenium, Moscow, Nauka Publishing House.
34. Underwood, E.J., Trace Elements in Human and Animal Nutrition, New York, Academic Press, 1977.

CHAPTER 20

Marine or Sea Pollution

Dr. E.H.El-Mossalamy[1]

Introduction

A survey of literature shows that, about 13000 tonnes of lead, 18500 tones of copper, 78000 tonnes of zinc. 8500 ionnes of arsenic, 950 tonnes of Barium, 78,000 tonnes of manganese, 7,000 tonnes of chromium, 4000 tonnes of antimony, 18,000 tonnes of iron, 7,000 tonnes of mercury and 4,800 tonnes of tin are discharged per year into the sea without any dilution. The industrial wastes loaded with toxic substances such as acids, alkalies, pesticides, oils, varnishes, plastics, petrochemicals, rubber, paints, and wastes of paper, soap, sugar, distillery, mine drainage, tannery, cyanides and radioactive substances pollute heavily causing fish killing in different coasts of different nations. The fast industrialization has caused danger for sea life and posed a serious threat to fishery resources. The dyes and other chemicals have changed the colour of the sea at many places and transparency has completely vanished.

Oil Pollution

The most important pollution of sea is oil. Hence we will discuss it here. Every year, approximately 3.5 million metric tons of the oil are released into the world's oceans. That represents one metric ton of oil spilled for every thousand metric tons of oil extracted—a rather high figure.

These oil releases can be divided about equally between land-based sources and sea-based (see table 1). Land base sources are in the hand of coastal states and they have the capacity to control them through national action if they wish to do so. Sea-based pollution, however requires international action.

What are the major sources of sea-based pollution? Table 1 indicates that production platforms are a negligible source world-wide, although they may have significant impacts in estuaries, bays or in a regional sea like the North Sea.

Sea-based pollution is linked mostly to maritime transport. Approximaately 1.5 million metric tons of the oil are released into the sea every year by maritime transport (i.e., one metric ton for every thousand metric tons are transported by sea). Of this, about 1.1 million metric tons are not accidental stemming from the frequent discharge of oil by tankers at sea. The balance about 0.4 million metric tons, results from tanker at sea. The balance about 0.4 million metric tons, results from tanker accidents and many of these too, are avoidable.

[1]Faculty of Science, King Abdul Aziz University, Jeddah (Saudi Arabia)

Table 20.1. Source of Oil Pollution of the Sea
(Millions of metric tons per year)

Sea based (52%)		
Maritime transport		1.49
Non accidental	–1.08	
Accidental	–0.41	
Production plateforms		0.05
Natural seeps, erosion		0.30
Land-based (48%)		
Municipal industrial wastes run-off		1.40
Atmospheric		0.30
Total		3.54

Source : J. B. Baker : "Impact of Oil Pollution on Living Resources". The Environmentalist IUCN 1983 (Provisional figures from the US National Academy of Science Report. 1981).

Concerted international action started in 1954 with the *"Convention of the Prevention of Pollution of the Sea by Oil"* and this was followed by an impressive series of conventions negotiated mostly within the framework of the IMO.

International action has been successful because these conventionwere ratified ultimately by most of the major shipping nations although there are a few outstanding exceptions such as Greece, Panama, and the Unhited States. It is a slow process, however, it took 10 years for the 1973 Convention on the Prevention of Pollution of the sea to enter into force and when it did only two of its parts came into force and they affected only 68 percent of the gross tonnage of the world fleet. Nevertheless, this Convention is very significant, because it deals with concrete techniques and measures to reduce oil releases to the sea. In particular, it provides for the creation of reception facilities, the operation tankers with segregated ballast crude oil washing and the use of oil separators. With its entry into force in 1983, one can expect to see a significant reduction in oil pollution of the sea stemming from routine operations.

Indian Ocean

Data for dissolved/dispersed petroleum residues in the India Ocean were clustered around the coastline of India and along a single transect between the Strait of Malacca and South Africa. This set of data contained very few values below 10 μg/l. While some values exceeded 300 μg/1 (GM = 8-9 μg/1). This suggests either that this region was much more highly polluted than any other area of the world ocean or that the data are suspect. Nevertheless, there was a tremendous difference between the concentration of μg/l) and those along the east coast (GM = 0.7 μg/l) of India. This, presumably, is a consequence of the ranker lane that passes along the west coast of India and across the Bay of Bengal enroute to the Strait of Malacca and of the monsoonal circulation of surface water is this area.

Global Assessment

MAPMOPP data for dissolved/dispersed petroleum residues were highly regional in character and so sparse over such enormous expanses of the world ocean that it is not possible to obtain a complete, assessment of the levels of these substances on a global scale. Nevertheless, thedata suggest some general trends. Analysis of the data for the eastern hemisphere by 20° × 20° squares of latitude and longitude not only pointed out the paucity of the data but indicated an extensive area in the southewest Pacific where the level of contamination by dissolved/dispersed petroleum residuces was around 0.1μg/l, while somewhat higher concentrations were present in the waters adjacent to Japan. Concentrations seemed to be remarkably low around Japan considering the amount of oil consumed in that country and were only slightly higher in the Strait of Malacca where tanker and other ship traffic converges while passing into the Pacific from the Indian Ocean. Concentrations in the northern part of the Indian Ocean were very much higher and reached 0.7µg/1 in the Bay of Bengal and along the east coast of India.

The highest concentrations of all were found along the west coast of India where the geometric mean was 86-4(µg/l. Such high concentrations suggest that the water contained disperse particles or droplets of oil from tankers that passed through the area.

Types of Marine Pollution

Before examining in detail the major types of marine pollution and the specific problems which they cause, or to which they contribute, let us see how they are grouped according to their basic properties.

Table 20.2. Sources and Effects of Marine Pollution

Type	Primary Source/Cause	Effect
Nutrients	Runoff approximately 50% sewage, 50% from forestry, farming, and other land use. Also airbone nitrogen oxides from power plants, cars etc.	Feed algal blooms in coastal waters. Decomposing algae depletes water of oxygen, killing other marine life. Can spur algal blooms (red tides), releasing toxins that can kill fish and poison people.
Sediments	Erosion from mining, forestry, farming, and other land-use; coastal dredging and mining	Cloud water; impede photosynthesis below surface waters. Clog gills of fish. Smother and bury coastal ecosystems. Carry toxins and excess nutrients.
Pathogens	Sewage, livestock.	Contaminate coastal swining areas and seafood, spreading cholera, typhoid and other diseases.
Alien Species	Several thousand per day transported in ballast water; also spread through canals linking bodies of water and fishery enhancement projects.	Outcompete native species and reduce biological diversity. Introduce new marine diceases. Associated with increased incidence of red tides and other algal blooms. Problem in major ports.
Persistent Toxins (PCBs, Heavy metals, DDT etc.)	Industrial discharge; wastewater discharge from cities; pesticides from cities; pesticides from farms, forests, home use etc.; seepage from landfills.	Poison or cause disease in coastal marine life, especially near major cities or industry. Contaminate seafood. Fatsoluble toxins that bio-accumulate in predators can cause disease and reproductive failure.
Oil	46% from cars, heavy machinery, industry, other land-basde sources; 32% from oil tanker operations and other shipping; 13% from accidents at sea; also offshore oil drilling and natural seepage.	Low level contamination can kill larvae and cause disease in marine life. Oil slicks kill marine life, especially in coastal habitats. Tar balls from coagulated oil litter beaches and coastal habitat. Oil pollution is down 60% from 1981.
Plastics	Fishing nets; cargo and cruise ships; beach litter; wastes from plastics industry and landfills.	Discard fishing gear continues to catch fish. Other plastic debris entangles marine life or is mistaken for food, Plastics litter beaches and coasts and may persist for 200 to 400 years.
Radioactive substances	Discarded nuclear submarine and military waste; atmospheric fallout; also industrial wastes.	Hot spots of radio activity. Can enter food chain and cause disease in marine life. Concentrate in top predators and shellfish, which are eaten by people.
Thermal	Cooling water from power plants and industrial sites	Kill off corals and other temperature sensitive sedentary species. Displace other marine life.
Noise	Super tankers, other large vessels and machinery	Can be heard thousands of kilometers away under water. May stress and disrupt marine life.

Source: Compiled by World Watch Institute.

Biodegradable and Non-biodegradable Materials

Most of the materials which reach the sea disintegrate either through simple chemical reactions or because of the activities of bacteria and some larger organisms. There are some substances nevertheless which are either extremely stable or else they have a very slow rate of degradation. The organic compounds in domestic sewage, wastes from the food industry and agricultural fertilizers, all belong to the first category while plastics, heavy metals and nuclear wastes belong to the second a substance is necessary for the ecosystem in small or moderate quantities, i.e. both absence and excessive concentration inhibit vital biological functions, while in the second case the substance is anything but necessary for the environment and every increase in its concentration leads to deterioration of the environmental conditions.

Bioaccumulated and Non-bioaccumulated Substances

Many of these long-lasting materials demonstrate the phenomenon of bioaccumulation i.e. they accumulate in the tissues of living organisms in concentrations much higher than those in the sediment or in the water. Such concentrations may not be lethal for these organisms either because they have developed a certain resistance or perhaps because the toxic material is gradually deposited in their fat tissues. Severe effects stemming from such bioaccumulation appear in those predators (including human consumers) who are exposed without previous warning to high concentrations of the toxic substances in question.

Solid wastes - Plastics

Pollution caused by solid wastes, in particular plastic wastes, displays some of the most conspicuous effects from human activities, very often in the form of used bottles, cups, plastic bags and other objects scattered on the seashore. Though the end result is undoubtedly aesthetically unpleasant, there are other more objectionable features. Plastics decompose very slowly, and therefore they are very difficult to eliminate.

It has to be said however, that the biological consequences of this type of pollution are negligible or extremely limited in comparison with other types of pollution. There have been cases reported offish having swallowed plastic objects which caused a blockage in the digestive tube; there are other cases of fish having been trapped inside plastic bags" or other plastic objects and dying of suffocation; still other cases have been reported offish getting entangled in plastic fishing nets which have been torn off fishing boats and drift in the water column, causing a phenomenon known as "ghost fishing".

The discharge of plastic sheets, coming for instance from greenhouses, and being subsequently deposited on the sea bottom does cause local problems in respect of the oxygenation of the sediments, giving rise to anoxic conditions underneath the plastic sheet. However, mortality related to plastic solid waste is not very important and does not affect the faunal composition of marine areas to any notable extent. Of course the discharge of large quantities of solid material into the sea and especially on areas with sandy or silty bottoms will result in the alteration of the texture of the substrate and hence will be the agent of changes in the composition of the fauna. This process though is not always undesirable (i.e. if the solid material does not contain toxic compounds), and in some cases the development of artificial reefs has been considered as a very good way to increase the environmental heterogeneity in areas which function as nursery or feeding grounds for important fishing stocks.

Organic Enrichment

Organic pollution or organic enrichment is the most common and the oldest type of pollution on the planet. Urbanization i.e. the high concentration of human beings in a limited space, created a need for the construction of networks of sewage outfalls which usually end up in the sea. The major characteristic of this type of pollution is the presence of high quantities of organic compounds of biological origin, i.e. produced by plants or animals. An equivalent quality of pollution is caused by certain industrial discharges such as t food industry, paper mills etc., the raw materials of which are also of biological origin.

The organic load concentration is expressed by means of a quantity called Biochemical Oxygen Demand (BOD), i.e., the amount of oxygen required by the micro-organisms for the decomposition of the organic compounds dissolved in a given volume of sewage. Consequently high BOD values indicate high concentrations of organic material and therefore low quantities of dissolved oxygen (DO).

The oxygen consumption for the decomposition of the organic material which results in anoxic conditions near the discharge point is the major consequence of organic enrichment. Since most mar organisms cannot survive under these conditions, the faunal composition changes dramatically within a short distance of the sewage discharge point.

Four sub-areas are distinguished, with sediments being more and more organically loaded and thus less oxygenated:

1. a heavily polluted area completely unsuitable for all macrofaunal organisms
2. a polluted area occupied by so-called opportunistic species, i.e. small species with a high reproduction potential, easily able to colonise azoic areas, whose major representative is the polychaete species *Capitella capitata*
3. a transitional area with species from both polluted and unpolluted areas
4. an unpolluted area where the oxygenation conditions are natural and the fauna is diverse, and include; species from various taxonomic groups as well as a wide range of sizes.

Macrofauna

Macrofauna (and especially the infaunal component) reflects environmental changes cause organic enrichment with remarkable sensitivity. Because of this sensitivity, macrofauna is often used in pollution studies. On the other hand, nekton species (and epifauna to a certain extent) can easily emigrate from a polluted area but can also return from time to time in order to take advantage of the large quantities of food which can be found there.

Bacteria

Besides the effects on concentrations of oxygen, an important hazard caused by domestic sewage is the existence of pathogenic bacteria, viruses and all sorts of parasites. These organisms do not die when they come into contact with sea water (as was previously held to be the case) but they do suffer a considerable drop in population size due to the ultraviolet radiation effect. A high proportion, however, do change into tolerant forms not easily detected by the usual microbiological methods. Any swallowing of contaminated sea water during a swim, or any consumption of shellfish caught in polluted areas may give rise to severe public healt problems.

Purification Plants

Difficulties caused by organic enrichment can be overcome up to a point by making use of biological purification plant technology. Such treatment plants fall into three categories in respect of the proportion of sewage undergoing treatment.

5. Primary treatment consists of the elimination of suspended solids using mechanical systems (racks);
6. Secondary treatment: consists of the reduction of BOD using aerobic decomposition by bacteria. This process usually needs energy for the continuous aeration of the fermentation tanks.
7. Tertiary treatment consists of the elimination of microbes, heavy metals and other chemical compounds by employing special techniques e.g. filtration, ultra-violet radiation, chemical sedimentation etc.

Sewage treatment technology has made a good deal of progress over the last decades so that by the end phase the water is considered to be pure enough to be used for agricultural irrigation instead of being discharged into the sea. Of course this technology is not cheap (both in terms of construction expenses and function costs) and therefore in most cases it is primary and secondary treatments which are used.

Pesticides

The so called "green revolution" led to the establishment of intensive agriculture, or, to put it another way, to an increase of the areas taken up by cultivated organisms. This concentration of many plants in a small space had certain advantages but also certain side-effects, namely, an increased susceptibility of the plants to parasites for those organisms which took advantage of these high concentrations or merely took advantage of the general change in the conditions of cultivation. These parasites belong to different taxonomic groups (plants, insects, fungi, acarea etc.) and nowadays they are kept under control by the use (sometimes excessive) of chemical compounds. These chemical compounds belong to four main categories.

8. Organochlorine compounds, such as DDT): most of these do not decompose or else decompose very slowly (half-life period from 9 to 116 years), and they usually accumulate in the fat tissues of certain organisms, in concentrations much higher than those encountered in the environment. It is noteworthy that these compounds have been detected in the fat tissues of Antarctic penguins who live in areas where such compounds have never been used.
9. Organophosphate esters, such as Parathion : these are very toxic compounds, much more so than the organochlorines
10. Carbamid compounds, such as Baygon: they decompose rapidly through hydrolysis and have a half-life period of approximately one week.
11. Chlorophenyl-acids such as 2,4,5-T used during the Vietnam war for defoliation: these are hazardous compounds whose use has been banned.

These chemical compounds usually end up in the sea, either through runoff into rivers or having been transported by the wind (dust - aerial spray). Once in the sea they enter the biological cycle of certain organisms and, through the bioaccumulation mechanism, some amounts are stored in the fat tissues of particular marine organisms. This storage has no visible effects on these organisms except when a shortage of food occurs, and the fat deposits are used up. It is then that large quantities of the chemical compounds enter the metabolic cycle with possible lethal consequences. This time-lag in the physiological activity of these pesticides facilitates the biomagnification phenomenon, that is to say, the multiplication of the concentrations of pesticides in the higher levels of the food web with its concomitant dramatic effects on marine birds and mammals.

Fertilisers

Fertilisers are industrial products rich in nutrient salts, especially nitrates and phosphates, and are used in order to increase agricultural yield. These salts are soluble in water and are often carried through the runoff of rainwater to the sea via streams and rivers. Another major pathway of nutrient input to the sea (and the land) is through wet and dry atmospheric deposition. The atmospheric nitrogen input in northern Europe may reach 10-20 kg Nitrogen per hectare and year. The origin is not only the car traffic but also intense culture of cattle were cows are fed like pigs. It may be noteworthy that one cow fed like this releases the same amount of nitrogen as 1-2 cars without catalstors. Nutrients, as has been mentioned are necessary for primary production both in the terrestrial and the marine environment. In this snese they cannot be considered as pollution factors except when occurring in very large quantities, i.e. when they stimulate excessive vegetal production, at levels which disturb the equilibrium of the ecosystem capacity.

The Efects of Oil Pollution in Sea Water

The effects of oil pollution in water can be summarised as follows :

(a) Physical Effects of Oil in Marine Water : The three main effects caused by oil coating are as follows :

(*i*) **Smothering :** Smothering coats of oil have killed lichens and algae along the shore lines of the sea.

(*ii*) **Reduction in Dissolved Oxygen :** Oil films reduce the rate of oxygen uptake by water resulting in death of aquatic animals.

(*iii*) **Reduction in Light Penetration :** Oil slick decreases the light intensity upto 80-90% or 2 metres below the surface of water than in clean water. This diffused light may reduce the photosynthesis of aquatic life in marine water.

(*b*) **Effect of Oil Pollution on Marine Ecosystem :**

(*i*) Oil pollutants may block the taste receptors of organisms and may mimic the natural stimulation.

(*ii*) Continuous inhalation of aromatics result in acute intoxication. Benzene, even in minute traces is particularly toxic. Its long exposure can cause anaemia and leukopenia in fishman and others.

(*iii*) The oil may reach the bottom of sea hampering the aquatic , animals and plants.

(*iv*) Extensive spreading of oil affects the floating plantation and marine life continuously.

(*v*) It has been reported that hydrocarbons cause anaesthia and necrosis in a wide variety of lower aniamals while their higher doses result in cell damage and death.

(*vi*) Hydrocarbons in oil get incorporated in body tissues of marine animals. These are stable like heavy metals and pesticides and damage tissues permanently.

(*vii*) Oil spilling causes lethal toxicity on aquatic flora. Fingler in 1997 reported large scale fish kill in 1996 on sea banks of Europe.

(*viii*) According to scientists waste from oil refineries and discharged petroleum from ships and other carriers cause heavy damage to fishery. Recently heaps of dead fish were washed ashore between Dabolim and Velcao coast in Goa. It created a big scare especially .among fisherman all over India.

(*ix*) Aromatic hydrocarbons in oil may accumulate in aquatic plant tissues even at low concentration of 10 ppb. These are carcinogenic and adversely affect plant metabolism.

(*x*) It has been reported that soluble aromatics (C_{10} or less) present in oil affect the aquatic organisms by disrupting their biological, physiological and behavioural activities and ultimately make them inactive.

(*xi*) Sublethal dose *i.e.* 10 to 100 ppb of oil component distrubs *Y* communication system of marine organisms permanently.

(*xi*) The marine organisms can not survive on exposure to 1 to 100 ppm. of soluble aromatics contained in oil, while even 0-1 ppm. acts as a lethal dose for fish larvae.

(*xii*) Naphthalene and phenanthrene have been found as extremely toxic to marine biota.

(*xiv*) Benzopyrene accumulates in food chain in fish which may cause cancer in human beings.

Some Measures to Check Oil Spills

Some methods have been suggested to check oil spill in the sea. The methods are as follows:

Simple Methods

(1) A powder of high density is spread over the oil patch, the oil patch can be sunk to the bottom in the ocean.

(2) Some chemicals can be used to coagulate the oil.

(3) The floating oil can be absorbed using a suitable absorbing material like polyurethane foam, saw dust can be used to absorb oil from i he marine water.

(4) The chalk or $CaCO_3$ treated with stearate and 10% sand in water slurry is a good method to remove oil from sea.

(5) Skimming the oil off the surface with a suction device has been suggested the simplest method to remove oil from marine.

Physical and Chemical Methods

1. Absorbents : An oil spill can be cleaned up by using ahsorbetits such as moss, saw peat, dust, pine bark and Straw etc.s Synthetic absorbents include polystyrene, polyethylene, polyurethane and polypropylene which can also be used. If the absorbent material U removed from the water, the oil can be removed easily and used.

2. Emulsification : It is simple process in which water is incorporated into the floating oil forming a water-in-oil emulsion. Such emulsions, containing 20 to 80% water are often viscous and .ire called "mousse".

3. Dispersion : Natural dispersion also removes some oil from the water. During this process small droplets of oil which are bigger than disolved molecules get incorporated into water in the form of a dilute oil-in water suspension.

The crude oil consists of little nitrogen, sulphur and oxygen bearing compounds which can act as natural surfactants. The surfactant decreased the oil water interfacial tension dissociating the oil into tiny droplets. A dispersant contains a surfactant, a solvent and a stabilizer which cause the oil to spiead farther and disperse. The *solvent* enables the surface-active agents to mix with it and penetrate into the oil slick forming emulsions. The *stabilizer* fixes the emulsion and prevents it from coalescing once it is formed.

4. Evaporation : Evaporation removes about 50% of oil during an oil slick's life time. Low boiling hydrocarbons such as *benzene, toluene and xylenes* are lost rapidly from the water surface. Much of the oil that evaporates is photo-oxidized in air while some of it may return to the seas as atmospheric fall out. *Photolysis and physico-chemical changes* in oil cause it to coalesce forming tar balls which would sink in water decreasing its toxicity.

5. Floating Booms : The floating booms are in common use in harbours and areas where transfer of petroleum products occur. Make shift booms are useful in rough water situations that generally accompany accidental spills in marine water.

6. Burning the Oil Slick : Burning oil slicks on the open seas is comparatively less successful because the more volatile light fractions evaporate quickly from the oil slick.

7. Using Chemical Additives : Chemical additives are generally used to solidify oil from water surface. Mechanical methods are also used to remove oil slicks from marine water.

French used small amounts of sorbents and sinking agents on the oil slick from Torry Canyon accident to prevent oily drops in water bodies.

8. Improved Navigation Aids : The offshore drilling operations can effectively protect the water from oil pollution.

Role of Micro-organisms in Oil Clean-up in the Ocean

In the clean up operations either genetically engineered organisms or biotechnological approaches have been adopted to solve vexed pollution problems. Certain bacteria like *"Pseudomonas"* are also being employed in the recovery of oil.

Micro-organisms that utilize hydrocarbons are found universally on soil, water and oil rich areas where they use free available hydrocarbon as a source of energy.

According to scientists oil eating bacteria are being used to remove oil pollution problems in sea. Accelerated bacterial seeding and fertilization of oil slicks account as the best effective measure against marine oil pollution all over the world.

Now a days new biotechnologies can be applied to tackle the hazard Y/created by oil spills are :

Microbes can be deployed as Voracious scavengers removing all kind of oil pollutants. Various varieties of Pseudomonas can consume esteric compounds and hydrocarbons from the oil. *The gene secreting enzymes* are found on plasmids, small and semi-autonomous rings of DNA. Some microbes can ingest dispersed oil droplets and subsequently deposit them as faecal pellets.

Recently some American Scier tists used have developed new strains of oil-eating bacteria at the *University of Illinois Medical Centre Chicago, USA.* The *"Superbug"* created by experts was the first ever made living organism to be patented in the history of USA.

Recently *(1997) USA Government* allowed the use of natural oil-eating microbes to be used in cleaning up an oil spill in the waters off of the state of Texas. Even in the Exxon Valdez disaster in *Alaska,* nitrogen and phosphorus containing *oleophilic fertilizers* were sprayed over waters to bolster the population of native oil eating bacteria so that the oil could degrade much faster. The new experiments are going in America to find new bacterias. Which may be safe for the environment.

Some Oil Disasters

1. *The National Academy of Sciences, Washington* has inform that between 1-7 to 8-8 million metric tonnes of oil enter the oceans every year.
2. *Collision in Maltese Tanker :* On *June 1989*, an oil spill threatened the *Indian coast* when a Maltese tanker *M.T. Puppy* collided with British vessel. It spilled over 5,500 tonnes of furnace oil into the open seas off Mumbai. The fear of oil contamination in fish brought the fish industry to a virtual halt as people feared that fish eating might cause cancer in Mumbai people.
3. A major oil spilling occurred on *July 30, 1987* when a large oil tanker of Union Oil Company leaked 117,000 tonnes of crude oil into the sea.
4. One such oil disaster also occurred in *Bombay High on July 30, 1982*, when ONGC suffered a loss of Rs. 1000 million including damage of drilling rigs and crude oils.
5. In *1978*, in the much publicized *Amoco Cadiz* disaster 68 million gallons of oil was dumped all along the *French coast* (A report).
6. An American *Navy Super Tanker Sealift Mediterranean* on *April 1978* ran a ground at *Rondo* islandoff northern tip of Sumatra spilling 1000 tonnes of black oil. This oil moved for several days very close to the Island of *Great Nicobar 1980-82.*
7. In *July 1976*, a Greek tinker disappeared in rough sea off Veraval. Large slicks of oil were observed for several days floating on the surface close to the Indian coast.
8. The leakage of 100,000 tonnes of crude oil from the wreckage of tanker vessel *Urquiola* near the *Spanish coasts La coruna on May 12, 1976* revealed the nature of oil spilling damaging the aquatic flora in marine sea.
9. In *September 1974*, an American oil tanker Transhuron crilided with one of the atolls of the *Kilton* Island in The Laccadives sea spilling 5000 tonnes of navy special furnance oil on the beaches causing extensive damage to the inter-tidal fauna.
10. *On March 24, 1989 :* The 30,000 tonnes supertanker spilled over 11 million gallons of oil into the clean waters of *Alaska's Prince Willifm Sound.* As the oil hit 1930 km. of shoreline, 100,(KX) seabirds including 150 rare species of bald eagles died.

REFERENCES

1. E. Goldberg, M. Koide, J.S. Yang, and K.K. Bertine, 'Metal Speciation: Theory, Analysis and Application', ed. J.R. Kramer and H.E. Allen, Lewis Publishers, Chelsea, 1988, p. 201.
2. E. Goldberg, Environment, 1986, 28, 17.
3. C. Stewart and S.J. de Mora, Environ. Techno., 1990, 11, 565.
4. S. Murray, 'Pollutant Transfer and Transport in the Sea', ed. G. Kullenberg, CRC Press, Boca Raton, 1982, Vol. 2, p. 169.

CHAPTER 21

Water Quality Parameters and their Variation Effects on Human Health

Ishrat Alim[1], Mohammad Rafi[2] and Neelofar Iqbal[3]

INTRODUCTION

Water is life! It is a precondition for human, animal and plant life as well as an indispensable resource for the economy. Water also plays a fundamental role in the climate regulation cycle. This ordinary liquid: amorphous, colourless and odourless and makes up over two-thirds of our own body mass, and is the essential constituent life as "Water is the material cause of everything" said Thales of Millet - 6th century B.C. Its genesis is related to that of the universe and our genesis to that of the water.

It is a well-known fact that clean water is absolutely essential for healthy living. Adequate supply of fresh and clean drinking water is a basic need for all human beings on the earth, yet it has been observed that millions of people worldwide are deprived of this.

Freshwater resources all over the world are threatened not only by over exploitation and poor management but also by ecological degradation. The main source of freshwater pollution can be attributed to discharge of untreated waste, dumping of industrial effluent, and run-off from agricultural fields. Industrial growth, urbanization and the increasing use of synthetic organic substances have serious and adverse impacts on freshwater bodies. It is a generally accepted fact that the developed countries suffer from problems of chemical discharge into the water sources mainly groundwater, while developing countries face problems of agricultural run-off in water sources. Pollutants like chemicals in drinking water causes problem to health and leads to water-borne diseases which can be prevented by taking measures even at the household level.

Water quality is the physical, chemical and biological characteristic of water. It is most frequently used by reference to a set of standards against which compliance can be assessed. The most common standards used to assess water quality relate to drinking water, safety of human contact and for the health of ecosystems.

The parameters for water quality are determined by the intended use. Work in the area of water quality tends to be focused on water that is treated for human consumption or in the environment.

[1] Quality Assurance Laboratory M.P. Council of Science and Technology, Bhopal, (India)
[2] ICFAI, Institute of Science and Technology, Medchal, Hydrabad, A.P. (India)
[3] Asian Biotech Research Centre, Idgah Hills, Bhopal-462001 (India)

Contaminants that may be in untreated water include microorganisms such as viruses and bacteria; inorganic contaminants such as salts and metals; pesticides and herbicides; organic chemical contaminants from industrial processes and petroleum use; and radioactive contaminants. Water quality depends on the local geology and ecosystem, as well as human uses such as sewage dispersion, industrial pollution, use of water bodies as a heat sink, and overuse (which may lower the level of the water).

In the United States, the U.S. Environmental Protection Agency (EPA) limits the amounts of certain contaminants in tap water provided by public water systems. The Safe Drinking Water Act authorizes EPA to issue two types of standards: primary standards regulate substances that potentially affect human health, and secondary standards prescribe aesthetic qualities, those that affect taste, odour or appearance. The U.S. Food and Drug Administration regulations establish limits for contaminants in bottled water that must provide the same protection for public health. Drinking water, including bottled water, may reasonably be expected to contain at least small amounts of some contaminants. The presence of these contaminants does not necessarily indicate that the water poses a health risk.

Some people use water purification technology to remove contaminants from the municipal water supply they get in their homes, or from local pumps or bodies of water. For people who get water from a local stream, lake, or aquifer (well), their drinking water is not filtered by the local government.

Environmental water quality, also called ambient water quality, relates to water bodies such as lakes, rivers, and oceans. Water quality standards vary significantly due to different environmental conditions, ecosystems, and intended human uses. Toxic substances and high populations of certain microorganisms can present a health hazard for non-drinking purposes such as irrigation, swimming, fishing, rafting, boating, and industrial uses. These conditions may also affect wildlife which uses the water for drinking or as a habitat. Modern water quality laws generally specify protection of fisheries and recreational use and require as a minimum, retention of current quality standards.

There is some desire among the public to return water bodies to pristine or pre-industrial conditions. Most current environmental laws focus of the designation of uses. In some countries these allow for some water contamination as long as the particular type of contamination is not harmful to the designated uses. Given the landscape changes in the watersheds of many freshwater bodies, returning to pristine conditions would be a significant challenge. In these cases, environmental scientists focus on achieving goals for maintaining healthy eco-systems and may concentrate of the protection of populations of endangered species and protecting human health.

The complexity of water quality as a subject is reflected in the many types of measurements of water quality indicators. Some of the simple measurements can be made on-site i.e. temperature, pH, dissolved oxygen, conductivity, Oxygen Reduction Potential (ORP). More complex measurements that must be made in a lab setting require a water sample to be collected, preserved, and analyzed at another location. Making these complex measurements can be expensive. Because direct measurements of water quality can be expensive, ongoing monitoring programs are typically conducted by government agencies. However, there are local volunteer programs and resources available for some general assessment. Tools available to the general public are on-site test kits commonly used for home fish tanks and biological assessments.

The Importance of Water and Human Health

"I'm dying of thirst!" Well. We just might. It sounds so simple. H_2O, two parts hydrogen, one part oxygen. But this boon of Nature, better known as water, is the most essential, next to air, to our survival. Water truly is every where; still most of us take it for granted.

Water makes up more than two thirds of the weight of the human body, and without it. humans would die in a few days. The human brain is made up of 95% water; blood is 82% and lungs 90%. A mere 2% drop in our body's water supply can trigger signs of dehydration: fuzzy short-term memory, trouble with basic math, and difficulty focusing on smaller print, such as a computer screen. Mild dehydration is also one of the most common causes of daytime fatigue. An estimated seventy-five percent of peoples have mild, chronic dehydration. (Pretty scary statistics for a developed country, where water is readily available through the tap or bottle).

Water is important to the mechanics of the human body. The body cannot work without it, just as a car cannot run without gas and oil. In fact, all the cell and organ functions made up in our entire anatomy and physiology depend on water for their functioning as :

- Water serves as a lubricant
- Water forms the base for saliva
- Water forms the fluids that surround the joints.
- Water regulates the body temperature, as the cooling and heating is distributed through perspiration.
- Water helps to alleviate constipation by moving food through the intestinal tract and thereby eliminating waste- the best detox agent.
- Regulates metabolism

In addition to the daily maintenance of our bodies, water also plays a key role in the prevention of disease. Drinking eight glasses of water daily can decrease the risk of colon cancer by 45%, bladder cancer by 50% and it can potentially even reduce the risk of breast cancer. Since water is such an important component to our physiology, it would make sense that the quality of the water should be just as important as the quantity. Drinking water should always be clean and free of contaminants to ensure proper health and wellness.

The following is a list of indicators generally measured by situational category for drinking water:

- ➢ Colour of water
- ➢ pH
- ➢ Taste and odour
- ➢ Alkalinity
- ➢ Dissolved metals and salts (Sodium, Chloride, Potassium, Calcium, Manganese, Magnesium)
- ➢ Microorganisms such as fecal coliform bacteria (Escherichia coli), Cryptosporidium, and Giardia lamblia.
- ➢ Dissolved toxic metals and metalloids (Lead, Mercury, Arsenic, etc.)
- ➢ Dissolved organics: Coloured Dissolved Organic matter (CDOM), Dissolved Organic Carbon (DOC)
- ➢ Heavy metals

Essential Parameters for Water Quality Monitoring

1. Physical assessment

- ➢ pH
- ➢ Temperature
- ➢ Total suspended solids (TSS)
- ➢ Turbidity
- ➢ Secchi Depth

2. Chemical assessment

- ➢ Chlorophyll
- ➢ Alkalinity
- ➢ Ammonia
- ➢ Carbondioxide
- ➢ Chlorine
- ➢ Nitrates and Nitrites
- ➢ Phosphates
- ➢ Biochemical Oxigen Demand (BOD)

3. Biological parameters

- Benthic Macro Invertebrates
- Submerged aquatic vegetation

4. Bacteriological parameters

- Fecal coliform

5. Heavy Metals

How Water Quality is measured

Water quality measurements include physical, chemical and biological parameters. The following is a brief description of some commonly used parameters, which have been cited as an essential parameters for water quality monitoring.

Physical Parameters

pH

pH is a term used to indicate the alkalinity or acidity of a substance as ranked on a scale from 1 .0 to 14.0, or the pH of water is the measure of how acidic or basic the water is on a scale of 0 – 14. It is a measure of hydrogen ion concentration. Acidity increases as the pH gets lower. A pH of 7.0 is neutral. Aquatic organisms differ as to the range of pH in which they flourish.

In a lake or pond, the water's pH is affected by its age and the chemicals discharged by communities and industries. Most lakes are basic (alkaline) when they are first formed and become more acidic with time due to the build-up of organic materials. As organic substances decay, carbon dioxide (CO_2) forms and combines with water to produce a weak acid, called "carbonic" acid—the same stuff that's in carbonated soft drinks. Large amounts of carbonic acid lower water's pH.

Most fish can tolerate pH values of about 5.0 to 9.0, but serious anglers look for waters between pH 6.5 and 8.2. The vast majority of rivers, lakes and streams fall within this range, though acid rain has compromised many bodies of water in our environment.

Water's acidity can be increased by acid rain but is kept in check by the buffer limestone. Extremes in pH can make water sources inhospitable to life. Low pH is especially harmful to immature fish and insects. Acidic water also speeds the leaching of heavy metals harmful to fish.

Temperature

Water temperature affects the ability of water to hold oxygen, the rate of photosynthesis by aquatic plants and the metabolic rates of aquatic organisms. Causes of temperature change include weather, removal of shading stream bank vegetation, impoundments, discharge of cooling water, urban storm water, and groundwater inflows to the stream.

Temperature impacts the rates of metabolism and growth of aquatic organisms, rate of plants' photosynthesis, solubility of oxygen in river water, and organisms' sensitivity to disease, parasites, and toxic materials. At a higher temperature, plants grow and die faster, leaving behind matter that requires oxygen for decomposition.

Variables that affect a waterway's temperature include:

1. The colour of the water. Most heat warming surface waters comes from the sun, so waterways with dark-Colored water, or those with dark muddy bottoms, absorb heat best.
2. The depth of the water. Deep waters usually are colder than shallow waters simply because they require more time to warm up.

3. The amount of shade received from shoreline vegetation. Trees overhanging a lake shore or river bank shade the water from sunlight. Some narrow creeks and streams are almost completely covered with overhanging vegetation during certain times of the year. The shade prevents water temperatures from rising too fast on bright sunny days.
4. The latitude of the waterway. Lakes and rivers in cold climates are naturally colder than those in warm climates.
5. The time of year. The temperature of waterways varies with the seasons.
6. The temperature of the water supplying the waterways. Some lakes and rivers are fed by cold mountain streams or underground springs. Others are supplied by rain and/or surface run-off. The temperature of the water flowing into a lake, river or stream helps determine its temperature.
7. The volume of the water, the more water there is, the longer it takes to heat up or cool down.
8. The temperature of effluents dumped into the water. When people dump heated effluents into waterways, the effluents raise the temperature of the water.

Suspended Minerals or Total suspended solids

Water is a good solvent and picks up impurities easily. Pure water - tasteless, colourless, and odourless—is often called the universal solvent. Dissolved solids" refer to any minerals, salts, metals, cations or anions dissolved in water. Total dissolved solids (TDS) comprise inorganic salts (principally calcium, magnesium, potassium, sodium, bicarbonates, chlorides and sulphates) and some small amounts of organic matter that are dissolved in water.

TDS in drinking-water originate from natural sources, sewage, urban run-off, industrial wastewater, and chemicals used in the water treatment process, and the nature of the piping or hardware used to convey the water, *i.e.*, the plumbing. In most of the country, elevated TDS has been due to natural environmental features such as: mineral springs, carbonate deposits, salt deposits, and sea water intrusion, but other sources may include: drinking water treatment chemicals, storm water and agricultural runoff, and point/non-point wastewater discharges.

In general, the total dissolved solids concentration is the sum of the cations (positively charged) and anions (negatively charged) ions in the water. Therefore, the total dissolved solids test provides a qualitative measure of the amount of dissolved ions, but does not tell us the nature or ion relationships. In addition, the test does not provide us insight into the specific water quality issues, such as: Elevated Hardness, Salty Taste, or Corrosiveness. Therefore, the total dissolved solids test is used as an indicator test to determine the general quality of the water, the sources of total dissolved solids can include all of the dissolved cations and anions.

Suspended minerals are a measure of the amount of sediment moving along in a stream. It is highly dependent on the flow of water and usually increases during and immediately after rain events. As the sediment settles out of the water, aquatic habitats are often destroyed. A known volume of water is filtered through a pre-weighed filter paper (0.45 μm Whatman GF/C). The filter paper and retained sediment is then oven-dried at 105°C for one hour, cooled to room temperature in a Desiccator, then weighed. The drying, desiccation and weighing steps are then repeated, to give an average final weight (filter paper plus retained sediment) based on two measurements. A final concentration in mg/L is calculated by subtracting the initial weight of the filter paper from the final weight (filter paper plus retained sediment), and dividing by the volume of water that was filtered.

Potential Health Effects

An elevated total dissolved solids (TDS) concentration is not a health hazard. The TDS concentration is a secondary drinking water standard and therefore is regulated because it is more of an aesthetic rather than a health hazard. An elevated TDS indicates the following:

1. The concentration of the dissolved ions may cause the water to be corrosive, salty or brackish taste, result in scale formation, and interfere and decrease efficiency of hot water heaters; and
2. Many contain elevated levels of ions that are above the Primary or Secondary Drinking Water Standards, such as: an elevated level of nitrate, arsenic, aluminium, copper, lead, etc.

Hence an elevated total dissolved solids concentration does not mean that the water is a health hazard, but it does mean the water may have aesthetic problems or cause nuisance problems. These problems may be associated with staining, taste, or precipitation. With respect to trace metals, elevated total dissolved solids may suggest that toxic metals may be present at an elevated level. It is important to keep in mind that water with a very lower TDS concentration may be corrosive and corrosive waters may leak toxic metals such as: copper and lead from the household plumbing. This also means that trace metals could be present at levels that may pose a health risk. Water is not a health hazard, but dealing with hard water in the home can be a nuisance.

Turbidity

Turbidity is the condition resulting from suspended solids in the water, including silts, clays, industrial wastes, sewage and plankton. Such particles absorb heat in the sunlight, thus raising water temperature, which in turn lowers dissolved oxygen levels. They also prevent sunlight from reaching plants below the surface. This decreases the rate of photosynthesis, so less oxygen is produced by plants. Turbidity may harm fish and their larvae. It is caused by soil erosion, excess nutrients, various wastes and pollutants, and the action of bottom feeding organisms which stir sediments up into the water.

Turbidity is a measure of the amount of particulate matter that is suspended in water. Water that has high turbidity appears cloudy or opaque. High turbidity can cause increased water temperatures because suspended particles absorb more heat and can also reduce the amount of light penetrating the water.

- Turbidity is a measure of cloudiness in water. The more turbid the water, the murkier it is.
- Turbidity can be caused by soil erosion, waste discharge, urban runoff, bottom feeders like carp that stir up sediments, household pets playing in the water, and algal growth.
- Turbid waters become warmer as suspended particles absorb heat from sunlight, causing oxygen levels to fall. (Warm water holds less oxygen than cooler water.) Photosynthesis decreases with lesser light, resulting in even lower oxygen levels.
- Suspended solids in turbid water can clog fish gills, reduce growth rates, decrease resistance to disease, and prevent egg and larval development. Settled particles smother eggs offish and aquatic insects.

Secchi Depth

Secchi Disk Transparency is a measure of the clarity of the water, and a quick, simple, and accurate method for estimating lake water quality. Secchi depth is the depth to which one can see into a lake and is an indication of water clarity. A black and white disk (called a Secchi disk) is lowered into the water until it just disappears from sight—this depth measurement is recorded. The deeper the measurement, the clearer the water.

Secchi disk measurements give a general indication of problems with algae, zooplankton, water color and silt. This measurement is obtained by lowering a black and white disk into the water and recording the depth at which it is no longer visible.

Chemical Parameters

Chlorophyll

Chlorophyll is the pigment that allows plants, including algae, to convert sunlight into organic compounds through photosynthesis. Measuring chlorophyll a concentrations in water is a surrogate for an actual measurement of algae biomass. Excessive amounts of chlorophyll-a-indicate the presence of blooms, which usually consist of a single species of algae - typically one that is not desirable for consumption by fish and other predators. Unconsumed algae sink to the bottom and decay, depleting deeper water of oxygen.

Water samples are filtered through 0.45μm Whatman GF/C filter papers in known volumes. Each filter paper with the retentate (*i.e.*, the material retained on the paper after filtration) is stored in a plastic tube with preservative, covered in aluminum foil to prevent light entering, and transported on ice to the laboratory. The amount of chlorophyll-a in the retentate is determined using a spectrophotomeler and the concentration of chlorophyll-a in the original sample calculated in μg/L.

Alkalinity

Alkalinity refers to the capability of water to neutralize acid. This is really an expression of buffering capacity. A buffer is a solution to which an acid can be added without changing the concentration of available H^+ ions (without changing the pH) appreciably. It essentially absorbs the excess H^+ ions and protects the water body from fluctuations in pH. The presence of calcium carbonate or other compounds such as magnesium carbonate contribute carbonate ions to the buffering system (Carbonate-bicarbonate). Alkalinity is often related to hardness because the main source of alkalinity is usually from carbonate rocks (limestone) which,are mostly $CaCO_3$. If $CaCO_3$ actually accounts for most of the alkalinity, hardness in $CaCO_3$ equal to alkalinity. Since hard water contains metal carbonates (mostly $CaCO_3$) it is high in alkalinity. Conversely, unless carbonate is associated with sodium or potassium which don't contribute to hardness, soft water usually has low alkalinity and little buffering capacity. So, generally, soft water is much more susceptible to fluctuations in pH from acid rains or acid contamination. Some Recommended Alkalinity Values has been given in Table 21.1.

Table 21.1. Some Recommended Alkalinity Values

Industry and Process	Recommended Value Total Alkalinity (Maximum in mg/L $CaCO_3$)
Carbonated beverages	85
Food products (canning)	300
Fruit juiece	100
Washing dispers	60
Pulp and paper making (ground-wood process)	150
Rayon manufacture	50
Tanning hides	135
Textile mill products	52-200
Petroleum refining	500
Reference 1,2	

Environmental Impact

Alkalinity is important for fish and aquatic life because it protects or buffers against rapid pH changes. Living organisms, especially aquatic life, function best in a pH range of 6.0 to 9.0. Alkalinity is a measure of how much acid can be added to a liquid without causing a large change in pH. Higher alkalinity levels in surface waters will buffer acid rain and other acid wastes and prevent pH changes that are harmful to aquatic life.

Ammonia

Pure ammonia is a strong-smelling, colourless gas. It is produced from nitrogen and hydrogen or is produced from coal gas. In nature, ammonia is formed by the action of bacteria on proteins and urea. Ammonia makes a powerful cleaning agent when mixed with water. For this reason, it is one of the most common industrial and household chemicals. The formula for ammonia, NH_3, means it consists of one atom of nitrogen and three atoms of hydrogen. Ammonia is rich in nitrogen so it makes an excellent fertilizer. In fact, ammonium salts are a major source of nitrogen for fertilizers. Like nitrates, ammonia may speed the process of eutrophication in waterways.

Ammonia is toxic to fish and aquatic organisms, even in very low concentrations. When levels reach 0.06 mg/L, fish can suffer gill damage. When levels reach 0.2 mg/L, sensitive fish like trout and salmon begin to die. As levels near 2.0 mg/L, even ammonia-tolerant fish like carp begin to die. Ammonia levels greater than approximately 0.1 mg/L usually indicate polluted waters.

The danger ammonia poses for fish depends on the water's temperature and pH, along with the dissolved oxygen and carbon dioxide levels. Remember, the higher the pH and the warmer the temperature, the more toxic the ammonia. Also, ammonia is much more toxic to fish and aquatic life when water contains very little dissolved oxygen and carbon dioxide.

Carbon dioxide

Carbon dioxide is an odourless, colourless gas produced during the respiration cycle of animals, plants and bacteria. All animals and many bacteria use oxygen and release carbon dioxide. Green plants, in turn, absorb the carbon dioxide and, by the process of photosynthesis, produce oxygen and carbon-rich foods. The general formulas for plant photosynthesis and respiration are summarized below.

Photosynthesis (in the presence of light and chlorophyll):

Carbon dioxide + Water + Oxygen + Carbon-rich foods

$$6CO_2 + 6H_2O = 6O_2 + C_6H_{12}O_6$$

Respiration

+ Oxygen + Carbon dioxide + Water

Carbon-rich foods + Oxygen + Carbon Dioxide + Water

$C_6H_{12}O_6$ O_2 CO_2 H_2O

Green plants carry on photosynthesis only in the presence of light. At night, they respire and burn the food they made during the day. Consequently, more oxygen is used and more carbon dioxide enters waterways at night than during, the daytime. When carbon dioxide levels are high and oxygen levels are low, fish have trouble respiring (taking up oxygen), and their problems become worse as water temperatures rise. As you can see from the table, even small amounts of carbon dioxide can affect fish.

It's lucky for fish that "free" carbon dioxide (by "free" we mean it is not combined with anything) levels rarely exceed 20 mg/L (milligrams per liter), because most fish are able to tolerate this carbon dioxide level without bad effects.

When several days of heavy cloud cover occur, plants' ability to photosynthesize is reduced. When that happens in a pond containing lots of plant life, fish can be hurt in two ways: by low dissolved oxygen and by high carbon dioxide levels.

Carbon dioxide quickly combines in water to form carbonic acid, a weak acid. The presence of carbonic acid in waterways may be good or bad depending on the water's pH and alkalinity. If the water is alkaline (high pH), the carbonic acid will act to neutralize it. But if the water is already quite acid (low pH), the carbonic acid will only make things worse by making it even more acid.

Chlorine

Chlorine is a greenish-yellow gas that dissolves easily in water. It has a pungent, noxious odour that some people can smell at concentrations above 0.3 parts per million. Because chlorine is an excellent disinfectant, it is commonly added to most drinking water supplies in the India. In parts of the world where chlorine is not added to drinking water, thousands of people die each day from waterborne diseases like typhoid and cholera.

Chlorine is also used as a disinfectant in wastewater treatment plants and swimming pools. It is widely used as a bleaching agent in textile factories and paper mills, and it's an important ingredient in many laundry bleaches.

Free chlorine (chlorine gas dissolved in water) is toxic to fish and aquatic organisms, even in very small amounts. However, its dangers are relatively short-lived compared to the dangers of most other highly poisonous substances. That is because chlorine reacts quickly with other substances in water (and forms combined chlorine) or dissipates as a gas into the atmosphere. The free chlorine test measures only the amount of free or dissolved chlorine in water. The total chlorine test measures both free and combined forms of chlorine.

If water contains a lot of decaying materials, free chlorine can combine with them to form compounds called trihalomethanes or THMs. Some THMs in high concentrations are carcinogenic to people. Unlike free chlorine, THMs are persistent and can pose a health threat to living things for a long time.

People who are adding chlorine to water for disinfection must be careful for two reasons: (1) Chlorine gas even at low concentrations can irritate eyes, nasal passages and lungs; it can even kill in a few breaths; and (2) The formation of THM compounds must be minimized because of the long-term health effects.

Less than one-half (0.5) mg/L of free chlorine is needed to kill bacteria without causing water to smell or taste unpleasant. Most people can't detect the presence of chlorine in water at double (1.0 mg/L) that amount. Although 1.0 mg/L chlorine is not harmful to people, it does cause problems for fish if they are exposed to it over a long period of time.

It is important to realize chlorine becomes more toxic as the pH level of the water drops. And it becomes even more toxic when it is combined with other toxic substances such as cyanides, phenols and ammonia.

Nitrates and Nitrites

Nitrite and Nitrate are forms of the element Nitrogen, which makes up about 80 percent of the air we breathe. As an essential component of life, nitrogen is recycled continually by plants and animals, and is found in the cells of all living things. Organic nitrogen (nitrogen combined with carbon) is found in proteins and other compounds. Inorganic nitrogen may exist in the Free State as a gas, as ammonia (when combined with hydrogen), or as nitrite or nitrate (when combined with oxygen). Nitrites and nitrates are produced naturally as part of the nitrogen cycle, when a bacteria 'production line' breaks down toxic ammonia wastes first into nitrite, and then into nitrate. Sources of nitrites and nitrates

Nitrites are relatively short-lived because they're quickly converted to nitrates by bacteria. Nitrites produce a serious illness (brown blood disease) in fish, even though they don't exist for very long in the environment. Nitrites also react directly with haemoglobin in human blood to produce methemoglobin, which destroys the ability of blood cells to transport oxygen. This condition is especially serious in babies under three months of age as it causes a condition known as methemoglobinemia or "blue baby" disease. Water with nitrite levels exceeding 1.0 mg/L should not be given to babies. Nitrite concentrations in drinking water seldom exceed 0.1 mg/L.

Nitrate is a major ingredient of farm fertilizer and is necessary for crop production. When it rains, varying nitrate amounts wash from farmland into nearby waterways. Nitrates also get into waterways from lawn fertilizer run-off, leaking septic tanks and cesspools, manure from farm livestock, animal wastes (including fish and birds), and discharges from car exhausts.

Nitrates stimulate the growth of plankton and water weeds that provide food for fish. This may increase the fish population. However, if algae grow too wildly, oxygen levels will be reduced and fish will die.

Nitrates can be reduced to toxic nitrites in the human intestine, and many babies have been seriously poisoned by well water containing high levels of nitrate-nitrogen. The Public Health Service has established 10 mg/L of nitrate-nitrogen as the maximum contamination level allowed in public drinking water.

Phosphates

The element phosphorus is necessary for plant and animal growth. Nearly all fertilizers contain phosphates (chemical compounds containing the element, phosphorous). When it rains, varying amounts of phosphates wash from farm soils into nearby waterways. Phosphates stimulate the growth of plankton and.water plants that provide food for fish. This may increase the fish population and improve the waterway's quality of life. If too much phosphate is present, algae and water weeds grow wildly, choke the waterway, and use up large amounts of oxygen. Many fish and aquatic organisms may die.

The Phosphorus Cycle is said to be "imperfect" because not all phosphates are recycled. Some simply drain off into lakes and oceans and become lost in sediments. Phosphate loss is not serious because new phosphates continually enter the environment from other sources (Table 2).

Table 21.2. Phosphate-Phosphorus Levels and Effects

Total phosphate/ /phosphorus	Effects
0.01-0.03 mg/L	Amount of phosphate-phosphorus in most uncontaminated lakes
0.025 mg/L	Accelerates the eutrophication process in lakes
0.1 mg/L	Recommended maximum for rivers and streams

Effects on Humans

Phosphates won't hurt people or animals unless they are present in very high concentrations. Even then, they will probably do little more than interfere with digestion. It is doubtful that humans or animals will encounter enough phosphate in natural waters to cause any health problems.

Dissolved Oxygen

An adequate supply of dissolved oxygen gas is essential for the survival of aquatic organisms. A deficiency in this area is a sign of an unhealthy river. There are a variety of factors affecting levels of dissolved oxygen. The atmosphere is a major source of dissolved oxygen in river water. Waves and tumbling water mix atmospheric oxygen with river water. Oxygen is also produced by rooted aquatic plants and algae as a product of photosynthesis.

Bacteria which decompose plant material and animal waste consume dissolved oxygen, thus decreasing the quantity available to support life. Ironically, it is life in the form of plants and algae that grow uncontrolled due to fertilizer that leads to the masses of decaying plant matter.

Too much dissolved oxygen is not healthy, either. Extremely high levels of dissolved oxygen usually result from photosynthesis by a large amount of plants. Great uncontrolled plant growth, especially algal blooms, is often the result of fertilizer runoff. This phenomenon is called cultural eutrophication.

Dissolved oxygen analysis measures the amount of gaseous oxygen (O_2) dissolved in an aqueous solution. Oxygen gets into water by diffusion from the surrounding air, by aeration (rapid movement), and as a waste product of photosynthesis.

When performing the dissolved oxygen test, only grab samples should be used, and the analysis should be performed immediately. Therefore, this is a field test that should be performed on site.

How Dissolved Oxygen Affects Water Supplies

A high DO level in a community water supply is good because it makes drinking water taste better. However, high DO levels speed up corrosion in water pipes. For this reason, industries use water with the least possible amount of dissolved oxygen. Water used in very low pressure boilers have no more than 2.0 ppm of DO, but most boiler plant operators try to keep oxygen levels to 0.007 ppm or less.

Environmental Impact

In a nutrient-rich water body the dissolved oxygen is quite high in the surface water due to increased photosynthesis by the large quantities of algae. However, dissolved. oxygen tends to be depleted in deeper waters because photosynthesis is reduced due to poor light penetration and due to the fact that dead phytoplankton (algae) falls toward the bottom using up the oxygen as it decomposes. In a nutrient-poor water body there is usually less difference in dissolved oxygen from surface to bottom. This difference between surface and bottom waters is exaggerated in the summer in reservoirs, stream-pools, and embayment when thermal layering occurs which prevents mixing. The surface may become supersaturated with oxygen (> 100%) and the bottom anoxic (virtually no oxygen). Shallower reservoirs and actively flowing shallow streams generally are kept mixed due to wind action in the shallow reservoirs and physical turbulence created by rocks in the stream beds.

Adequate dissolved oxygen is needed and necessary for good water quality. Oxygen is a necessary element to all forms of life. Adequate oxygen levels are necessary to provide for aerobic life forms which carry on natural stream purification processes. As dissolved oxygen levels in water drop below 5.0 mg/L, aquatic life is put under stress. The lower the concentration, the greater the stress. Oxygen levels that remain below 1-2 mg/L for a few hours can result in large fish kills. Total dissolved oxygen concentrations in water should not exceed 110 percent. Concentrations above this level can be harmful to aquatic life. Fish in waters containing excessive dissolved gases may suffer "from "gas bubble disease"; however, this is a very rare occurrence. The bubbles or emboli block the flow of blood through blood vessels causing death. Aquatic invertebrates are also affected by gas bubble disease but at levels higher than those lethal to fish.

Dissolved oxygen analysis measures the amount of gaseous oxygen (O_2) dissolved in an aqueous solution. Dissolved oxygen is one of the most important parameters in aquatic systems. This gas is an absolute requirement for the metabolism of aerobic organisms and also influences inorganic chemical reactions. Therefore, knowledge of the solubility and dynamics of oxygen distribution is essential to interpreting both biological and chemical processes within water bodies. Oxygen gets into water by diffusion from the surrounding air, by aeration (rapid movement) and as a waste product of photosynthesis. The amount of dissolved oxygen gas is highly dependent on temperature. Atmospheric pressure also has an effect on dissolved oxygen. The amount of oxygen (or any gas) that can dissolve in pure water (saturation point) is inversely proportional to the temperature of water. The warmer the water, the less dissolved oxygen.

Though water molecules contain an oxygen atom, aquatic organisms rely upon a small amount of oxygen that is actually dissolved in the water. In general, rapidly moving water contains more dissolved oxygen than slow or stagnant water and colder water contains more dissolved oxygen than warmer water. Bacteria consume oxygen as organic matter decays. As a result, an oxygen-deficient environment can develop in lakes and rivers with excess organic material. These conditions can eventually lead to fish kills.

Biochemical Oxygen Demand

Biochemical oxygen demand or BOD is a chemical procedure for determining the rate of uptake of dissolved oxygen by the rate biological organisms in a body of water use up oxygen. It is not a precise quantitative test, although it is widely used as an indication of the quality of water.

BOD can be used as a gauge of the effectiveness of wastewater treatment plants, and it is listed as a conventional pollutant because :

- Biochemical oxygen demand is a measure of the quantity of oxygen used by microorganims (*e.g.*, aerobic bacteria) in the oxidation of organic matter.
- Natural sources of organic matter include plant decay and leaf fall. However, plant growth and decay may be unnaturally accelerated when nutrients and sunlight are overly abundant due to human influence.
- Urban runoff carries pet wastes from streets and sidewalks; nutrients from lawn fertilizers; leaves, grass clippings, and paper from residential areas, which increase oxygen demand.
- Oxygen consumed in the decomposition process robs other aquatic organisms of the oxygen they need to live. Organisms that are more tolerant of lower dissolved oxygen levels may replace a diversity of more sensitive organisms.

Biological Parameters

Water samples should be monitored for the followings using standard methods.

Benthic Macro invertebrates

Macro invertebrates are organisms that are large (macro) enough to be seen with the naked eye and lack a backbone (invertebrate). Benthic refers to the bottom of a waterway. Examples of benthic macro invertebrates include insects in their larval or nymph form, crayfish, clams, snails, and worms. Most live part or most of their life cycle attached to submerged rocks, logs, and vegetation. The basic principle behind the study of macro invertebrates is that some are more sensitive to pollution than others. Therefore, if a stream site is inhabited by organisms that can tolerate pollution and the more pollution-sensitive organisms are missing a pollution problem is likely.

Submerged Aquatic Vegetation

Submerged aquatic vegetation (SAV) provides invaluable benefits to aquatic ecosystems. It not only provides food and shelter to fish and invertebrates but also produces oxygen, traps sediment and absorbs nutrients such as nitrogen and phosphorus. Whereas SAV are dependent upon the transmission of sunlight through the water, the location of individual species depends upon a variety of factors such as salinity, depth and bottom sediment.

Sediment samples are taken for the assessment of concentrations of trace elements (metals and metalloids), organic compounds (pesticides, polychlorinated biphenyls and organophosphates) and nutrients (total nitrogen and phosphorus). Sediment sample collection, preparation and storage are undertaken according to methods outlined in the EPA Water Quality Sampling Manual 3rd Edition (1999).

Bacteriological Parameters

Faecal Coliforms

Samples are collected from 20cm below the water surface, in 250mL sterilised containers. They are stored on ice or in a refrigerator and laboratory quality assessment tests should be carried out for analysis within 24 hours of collection. At the laboratory, faecal coliform concentrations are determined using a standard membrane filtration method. A measured volume of water sample is passed through a membrane filter, retaining any bacteria on the membrane. The membrane is placed onto the surface of a growth medium (plate) and incubated at 44.0 – 44.5 °C; the number of colonies on the plate is then counted. The final concentration of faecal coliforms is calculated in colony-forming units per 100mL, by dividing the number of colonies counted by the volume of water that was filtered.

Water entering the distribution system in piped supply chlorinated or otherwise disinfected shall satisfy the following criteria: Coliforms count in any sample of 100 ml should be zero. A sample of the water entering the distribution system that does not conform to this standard calls for an immediate investigation into both the efficacy of the purification process and the method of sampling.

Heavy Metals

Infect heavy metals that can be very harmful to our health if found in the drinking water. Severe effects include reduced growth and development, cancer, organ damage, nervous system damage, and in extreme cases, death. Exposure to some metals, such as mercury and lead, may also cause development of autoimmunity, in which a person's immune system attacks its own cells. This can lead to joint diseases such as rheumatoid arthritis, and diseases of the kidneys, circulatory system, and nervous system. The young are more prone to the toxic effects of heavy metals, as the rapidly developing body systems in the foetus, infants and young children are far more sensitive. Childhood exposure to some metals can result in learning difficulties, memory impairment, damage to the nervous system, and behavioural problems such as aggressiveness and hyperactivity. At higher doses, heavy metals can cause irreversible brain damage. Children may receive higher doses of metals from food than adults, since they consume more food for their body weight than adults.

Toxic metals can be present in industrial, municipal, and urban runoff, which can be harmful to humans and aquatic life. Increased urbanization and industrialization are to blame for an increased level of trace metals, especially heavy metals, in our waterways. There are over 50 elements that can be classified as heavy metals, 17 of which are considered to be both very toxic and relatively accessible. Toxicity levels depend oh the type of metal, it's biological role, and the type of organisms that are exposed to it.

The heavy metals linked most often to human poisoning are lead, mercury, arsenic and cadmium. Other heavy metals, including copper, zinc, and chromium, are actually required by the body in small amounts, but can also be toxic in larger doses.

Heavy metals in the environment are caused by air emissions from coal-burning plants, smelters, and other industrial facilities; waste incinerators; process wastes from mining and industry; and lead in household plumbing and old house paints. Industry is not totally to blame, as heavy metals can sometimes enter the environment through natural processes. For example, in some parts of the India, naturally occurring geologic deposits of arsenic can dissolve into groundwater, potentially resulting in unsafe levels of this heavy metal in drinking water supplies in the area. Once released to the environment, metals can remain for decades or centuries, increasing the likelihood of human exposure.

In addition to drinking water, we can be exposed to heavy metals through inhalation of air pollutants, exposure to contaminated soils or industrial waste, or consumption of contaminated food. Because of contaminated water, food sources such as vegetables, grains, fruits, fish and shellfish can also become contaminated by accumulating metals from the very soil and water it grows from.

BACTERIOLOGICAL STANDADS

- Water in the distribution system shall satisfy these three criteria
  1. E. Coli count in 100 ml of any sample should be zero.
  2. Coliform organisms not more than 10 per 100 ml shall be present in any sample
  3. Coliform organisms should not be detectable in 100 ml of any two consecutive samples or more than 50 percent of the samples collected for the year.
- In individual or small community supplies E. coli count should be zero in any sample of 100 ml and coliforms organisms should not be more than 3 per 100 ml. If coliform exceed 3 per 100 ml, the supply should be disinfected.

VIROLOGICAL STANDARDS

0.5 mg/l of free residual chlorine for one hour is sufficient to inactivate virus, event in water that was originally polluted. This free chlorine residual is to be insisted in all disinfected supplies in areas suspected of endemicity of infectious hepatities to inactivate virus and also bacteria 0.2 mg/l of free residual chlorine for half an hour should be insisted for other areas.

- The figures indicated under the column 'acceptable' are the limits up to which the water is genrally acceptable to the consumers.
- Figures in excess of those mentioned under 'acceptable' render the water not acceptable, but still may be tolerated in the absence of alternative and better source up to the limits indicated under column 'case for rejection' above which the supply will have to be rejected (Table 3).

Table 21.3. Water Quality Standards: The Water Quality Standards as set by Union Health Ministry and followed by APHED are :

PHYSICAL STANDARDS			
S.No.	**Characteristics**	**Acceptable***	**Case for Rejection***
1.	Turbidity (units on J.T.U. scale)	2.5	10
2.	Colour (units on Platinum cobalt scale)	5.0	25
3.	Taste and odour	Unobjectionable	Unobjectionable
CHEMICAL STANDARDS			
S.No.	**Characteristics**	**Acceptable***	**Case for Rejection***
1.	pH	7.0-8.5	6.5-9.2
2.	Total dissolved solids (mg/l)	500	1500
3.	Total hardness (as $CaCO_3$) (mg/l)	200	600
4.	Chlorides (as Cl) (Mg/l)	200	1000
5.	Sulphates (as SO_4) (mg/l)	200	400
6.	Fluorides (as F) (mg/l)	1.0	1.5
7.	Nitrates (as NO_3) (mg/l)	45	45
8.	Calcium (as Ca) (mg/l)	75	200
9.	Magnesium (as Mg) (mg/l)	>30	150
		(If there are 250 mg/l of sulphates, Mg content can be increased to a maximum of 125 mg/l with the reduction of sulphates at the rate of 1 unit per every 2.5 units of sulphates)	
10.	Iron (as Fe) (mg/l)	0.1	1.0
11.	Manganese (as Mn) (mg/l)	0.05	0.5
12.	Copper (as Cu) (mg/l)	0.05	1.5
13	Zinc (as Zn) (mg/l)	5.0	15.0
14.	Phenolic compounds (as phenol) (mg/l)	0.001	0.002
15.	Anionic detergents (as MBAS) (mg/l)	0.2	1.0
16.	Mineal oil (mg/l)	0.01	0.3
17.	Arsenic (as As) (mg/l)	0.05	0.05
18.	Cadmium (as Cd) (mg/l)	0.01	0.01
19.	Chromium (as hexavalent Cr) (mg/l)	0.05	0.05
20.	Cynides (as CN) (mg/l)	0.05	0.05
21.	Lead (as Pb) (mg/l)	0.01	0.1
22.	Selenium (as Se) (mg/l)	0.01	0.01
23.	Mercury (total as Hg) (mg/l)	0.001	0.001
24.	Polynuclear aromatic (hydrocarbons (PAH) (mg/l)	0.2	0.2
25.	Gross alpha activity (pCi/l)	3	3
26.	Gross beta activity (pCi/l)	30	30

REFERENCES

1. Quality Criteria for Water, *U.S. Environmental Protection Agency,* July (1976).
2. Water Quality Criteria, *California Water Quality Resources Board,* Publication No. A, (1963).
3. Water Quality Criteria, *Environmental Studies Board,* National Academy of Sciences (1972).
4. Study and Interpretation of the Chemical Characteristics of Natural Water, United States Geological Survey, Water Supply Paper 1473 (1970).
5. Water Pollution Microbilogy, Ralph Mitchell ed., Wiley-Interscience (1972).
6. Quality Criteria for Water, U.S. Environmental Protection Agency, EPA#440/5-86-001 (1986).
7. Ammonia Toxicity levels and nitrate tolerance of Channel Catfish, The Progressive Fish-Culturist, 35: 221, Knepp and Arkin (1973).
8. Bartram J., *et. al.*, eds. (2003) Heterotrophic plate counts and drinking-water safety: the significance of HPCs for water quality and human health. WHO Emerging Issues in Water and Infectious Disease Series. London, IWA Publishing.
9. Bartram J., *et. al.*, eds. (2004) *Pathogenic mycobacteria in water:* A guide to public health consequences, monitoring and management. Geneva, World Health Organization.
10. Chorus I, Bartram J., eds. (1999) Toxic cyanobacteria in water: A guide to their public health consequences, monitoring and management. Published by E & FN Spon, London, on behalf of the World Health Organization, Geneva.
11. Dufour A., *et. al.*, (2003) Assessing microbial safety of drinking water: Improving approaches and methods. Geneva, Organisation for Economic Co-operation and Development/World Health Organization.
12. FAO/WHO (2003) Hazard characterization for pathogens in food and water: guidelines. Geneva, Food and Agriculture Organization of the United Nations/World Health Organization (Microbiological Risk Assessment Series No. 3). Available at http://www.who.int/foodsafety/publications/micro/en/pathogen.pdf.
13. Havelaar AH, Melse JM (2003) Quantifying public health risks in the WHO Guidelines for drinking-water quality: A burden of disase approach. Bilthoven, National Institute for Public Health and the Environment (RIVM Report 734301022/2003).
14. Davison A., *et. al.*, (2004) *Water safety plans*. Geneva, World Health Organization.
15. Howard G., Bartram J., (2003) *Domestic water quantity*, service level and health. Geneva, World Health Organization.
16. Le Chevallier MW., Au K-K (2004) *Water treatment and pathogen control:* Process efficiency in achieving safe drinking-water. Geneva, World Health Organization and IWA.
17. Sobsey M., (2002) *Managing water in the home:* Accelerated health gains from improved water supply, Geneva, Wrold Health Organization (WHO/SDE/WSH/02.07).
18. Sobsey MD., Pfaender FK (2002) Evaluation of the H_2S *Method for detection of fecal contamination of drinking water*. Geneva, World Health Organization (WHO/SDE/WSH/02.08).
19. Wagner EG., Pinheiro RG (2001) *Upgrading water treatment plants*. Published by E & FN Spon, London, on behalf of the World Health Organization, Geneva.
20. WHO (in preparation) *The arsenic monograph*. Geneva, World Health Organization.
21. WHO (in preparation) *Desalination for safe drinking-water supply*. Geneva, World Health Organization.
22. WHO (in revision) *Guide to hygiene and sanitation in aviation*. Geneva, World Health Organization.
23. WHO (in revision) *Guide to ship sanitation*. Geneva, World Health Organization WHO (in preparation) Health aspects of poumbing. Geneva, World Health Organization.
24. WHO (in preparation) Legionella and the prevention of legionellosis. Geneva, World Health Organization.
25. WHO (in preparation) Managing the safety of materials and chemicals used in the production and distribution of drinking-water. Geneva, World Health Organization.
26. WHO (in preparation) *Protecting ground waters for health*—Managing the quality of drinking-water sources. Geneva, World Health Organization.
27. WHO (in preparation) *Protecing surface waters for health*—managing the quality of drinking-water sources. Geneva, World Health Organization.
28. WHO (in preparation) *Rapid assessment of drinking-water quality*: a handbook for implementation. Geneva, World Health Organization.

CHAPTER 22

Environment, Biodiversity and Exotics—An Overview

Dr. Shaukat Saeed Khan[1]

Environment plays an important role in the type of vegetation of a geographic area, various environmental factors synergistically influence both flora and fauna. For instance low rainfall, high temperature and a long dry period between two rainy seasons encourage the growth of scrub and desert vegetation. Incessant rainfall, high humidity & temperature which prevails in the equatorial region results in evergreen and broadleaved forests with a variety of lianas. Places with rainfall around 1000 mm, temperature from 10°C in winter and above 40°C in summer and with about 8 months dry period between two rainy season, support the growth of deciduous forests. High rainfall. plus snowfall and low temperature in most part of the year results in the development of temperate vegetation with coniferous forests, *Platanus, Acer, Aesculus, Juglans, Pyrus, Prunus* and *Rhododendron* species.

The coastal areas with high humidity, high temperature and saline soils support the growth of Mangrove vegetation with characteristic vivipary & pneumatophores. Similarly drastic change in environmental factors adversely affect the flora of a place. For instance hailstorm or snowfall in subtropical areas results in leaf fall of trees and death of herbaceous plants.

Biodiversity is the totality of genes, species and ecosystems in a region and can be divided in three main categories, viz, genetic diversity, species diversity and cultural diversity. Genetic diversity is the variation of genes within species. Species diversity refers to the potentially inter-breeding natural populations that are reproductively isolated from other such groups, cultural diversity of human society includes diversity in language, religious beliefs, social structure, food habits etc.

The term Biodiversity is very vast and encompasses all the life forms inhabiting this crust of earth. Variation is the rule of nature, which further results in diversity. For the sake of convenience, the Biodiversity may be divided as under:

The loss of biological diversity and the degradation of habitats and ecosystems may at its most be irreversible, ultimately resulting in extinction of species. The loss may be due to hunting, collection, persecution, habitat destruction and modification due to natural calamities such as volcano eruptions, floods, Tsunamis, Katrinas, Ritas and earthquakes as well as human activities as deforestation, industrialization and urbanization. There is environmental, genetic and also demographic uncertainty of the 3,00,000 estimated angiospermic species of the world out of which about 2,50,000 have been described till now. India is one of the mega diversity region of the world and is reported to have 18,000 + species of flowering plants occupying Ca 7% of the world species, out of the total twelve biodiversity hot

[1] Department of Microbiology, Saifia Science College, Bhopal-462001 (India)
e-mail: drkhanss_micro@yahoo.co.uk

spots (Table 22.1) in the world, India has two, one is the north east region and other is Western Ghats. The geographical area of India is about 329 million ha. and its coastline stretches to over 7000 km.

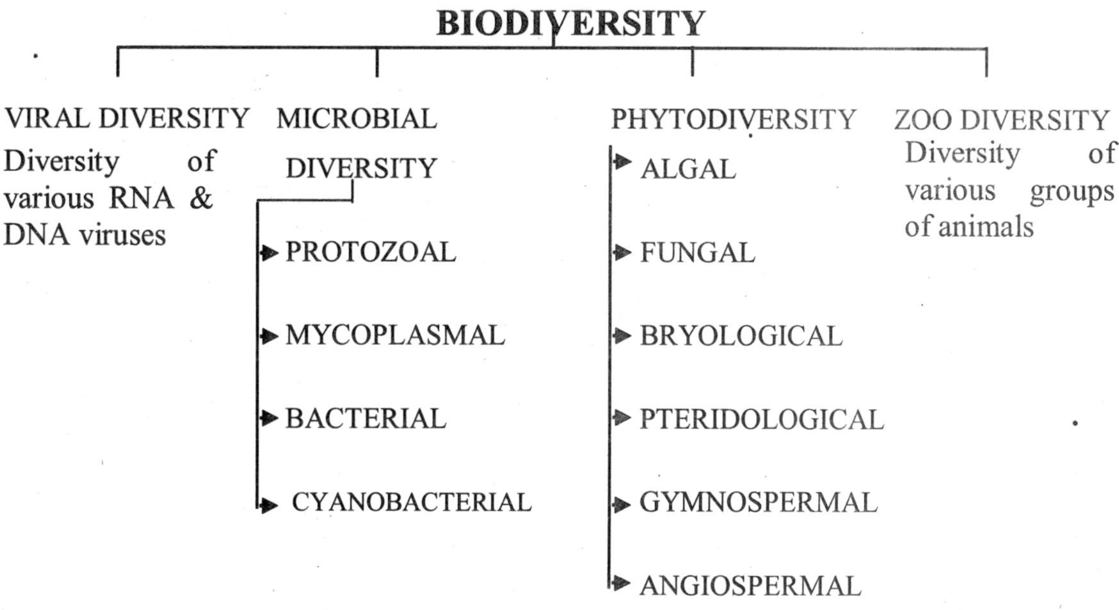

Angiospermic Diversity in India

The rich and diverse vegetation of India is due to its varying climatic and geographical conditions with varied ecological habitats. For instance we have temperate climate in Himachal and Kashmir with coniferous forests and broad leaved trees in the ever green forests of Meghalaya with tropical climate, Deciduous forests in M.P., Maharashtra, Desert vegetation in the arid zones of Rajasthan and mangrove plants in the coastal areas. Apart from naturally occurring native plants, a number of plant species have come accidentally (Table 22.4) with items of trade or introduced intentionally for the sake of aesthetic value or commercial exploitation (Table 22.5). Thus the exotic species coming from far off places as Tropical, North and Central America, Africa, Brazil, Eurasia, China, Japan, Mexico and many other countries have further enriched our country floristically. The description of exotic species in local floras is not given due weightage, which they deserve, as they have adapted themselves to Indian conditions and are interspersed with local vegetation in an inseparable manner. Not only this but a number of species have entered into Indian systems of medicine (Table 22.6). In majority of Floras, the exotic species are not described as the native ones and are merely reported/ignored.

Apart from this a variety of vegetables, pulses, oil yielding, ornamental and fruit yielding species, so commonly grown in our country are natives of other parts of the world. They, in fact have become part and parcel of our life. In the present paper only some of the exotic plants found or grown in various parts of India are mentioned, otherwise there is a long list of these plants which have acclimatized & naturalized in different regions of the country.

It is evident from table 22.1 that compared to U.S.A. and Australia, India has more number of species of flowering plants despite its smaller area. India presents a panorama of climatic conditions which permit higher biodiversity, especially phytodiversity.

Owing to various reasons such as population explosion, industrialization, urbanization, loss of specific pollinators, seed germination capacity, habitat destruction and genetic variability, many plant species are becoming extinct or on the verge of extinction. About 70 Indian plant species are possibly/presumably extinct.

Table 22.1. Mega diversity countries of the world

Country	No. of species (Flowering plants)
Brazil	55,000
Columbia	35,000
China	30,000
Mexico	25,000
South Africa	23,000
Former USSR	22,000
Indonesia	20,000
Venezuela	20,000
U.S.A.	18,000
Equador	18,000
India	18,000+
Australia	15,000

India - A primary / secondary centre for Domestication

So far as the number of agricultural species are concerned, India ranks 7th in the world and has been a primary centre for domestication for rice, sugarcane, banana, tea, mango, cucumber, citrus, jute, minor millets, *Brassicas, Alocasia, Colocasia,* cardamom, black pepper, ginger, turmeric, cucurbits, bamboos and *Vignas* etc. It is secondary centre of domestication for maize, potato, tomato, sesame and soybean etc. About 167 species of flowering plants are said to have originated in India. It is also estimated that 334 species of wild relatives of crop plants have originated in India of which about 60 are endemic. Table 22.2 summarises the plant groups / species occurring in India.

Table 22.2. Plant groups/species in India

Category	%	No. of species	Area
Possibly/ presumed extinct	Ca 0.37	Ca 70	Indian subcontinent
Rare, endangered and threatened	Ca 9.50	Ca 1700	N.E. India, W. Himalayas, Western Ghats, Andaman & Nicobar Islands
Endemic	Ca 32	Ca 5725	Himalaya (20%), Peninsular region (12%), Andaman & Nicobar Islands (1.5%)
Medicinal	Ca 17	Ca 3000 of which about 1300 are extensively exploited	Various parts of the country
Mangroves	Ca 0.40	Ca 69	Distributed over 7000 Km stretch of the coastline of the country
Wetlands	Ca 1.70	Ca 300	Distributed over 4.1 million ha of wetlands, (minus paddy fields & mangroves) of which 1.5 mi ha. are natural & 2.6 mi ha are man made.
Wild relatives of crop plants	Ca 1.80	Ca 334	Indian subcontinent

Species Diversity

India is bestowed by nature with rich wealth of biological diversity in its forests, wetlands and in its marine areas. The richness is depicted in absolute numbers of species and the proportion they represent of the world total (table 22.3).

Table 22.3. Comparison between the number of species in India and the world. (Vertebrates and Angiosperms)

Group	Number of species		SI/SW (%)
	In India (SI)	In World (SW)	
Mammals	350	4,629	7.6
Birds	1224	9,702	12.6
Reptiles	408	6,550	6.2
Amphibians	197	4,522	4.4
Fishes	2546	21,730	11.7
Flowering plants	18000	2,50,000	7.2

SI = species in India; SW = species in the world

Table 22.4. Some exotic weeds which have become naturalised in various parts of India

S.No.	Name of the species	Family	Nativity
1.	*Acanthospermum hispidum* DC	Asteraceae	S. America
2.	*Alternanthera pungens* H.B. & K.	Amaranthaceae	Tropical America
3.	*Argemone mexicana* L.	Papa veraceae	Mexico
4.	*Cichorium intybus* L.	Asteraceae	Europe
5.	*Coronopus didymus* (L.) Sm.	Brassicaceae	Tropical America
6.	*Crotolaria juncea* L.	Fabaceae	Malay islands/Australia
7.	*Croton bonplandianum* Baill.	Euphorbiaceae	C. America
8.	*Datura innoxia* Mill.	Solanaceae	America / Mexico
9.	*Desmodium spirale* DC.	Fabaceae	Africa
10.	*Euphorbia geniculata* Orteg.	Euphorbiaceae	Tropical America
11.	*Flaveria repanda* Lag.	Asteraceae	Peru
12.	*Gomphrena celosioides* Mart.	Amaranthaceae	S. Brazil
13.	*Hymenantherum tenuifolium* Cass.	Asteraceae	C. America
14.	*Ipomoea hederifolia* L.	Convolvulaceae	Tropical America
15.	*I. nil* (L.) Roth	Convolvulaceae	Tropical America
16.	*I. fistulosa* Mart. & Choisy	Convolvulaceae	S. America
17.	*Jatropha gossypifolia* L.	Euphorbiaceae	Brazil
18.	*Lagascea mollis* Cav.	Asteraceae	C. America
19.	*Lantana camara* L.	Verbenaceae	Tropical America
20.	*Martynia annua* L.	Pedaliaceae	Mexico
21.	*Nicotiana plumbaginifolia* Viv.	Solanaceae	Mexico & West Indies
22.	*Oxalis latifolia* HB & K.	Oxalidaceae	Mexico
23.	*Parthenium hysterophorus* L.	Asteraceae	Tropical America
24.	*Prosopis juliflora* (SW.) DC	Mimosaceae	Mexico & C. America
25.	*Ruellia tuberosa* L.	Acanthaceae	Tropical America
26.	*Scoparia dulcis* G. Don	Scrophulariaceae	America
27.	*Triumfetta rhomboidea* Jacq.	Tiliaceae	Malaya & China
28.	*Vicia hirsuta* (L.) S.F. Gray	Fabaceae	Europe

Table 22.5. Exotic fruits, vegetables and pulses which are cultivated in India

S.No.	Name of the species	Vernacular name	Family	Nativity
1.	*Abelmoschus esculentus* (L.) Moench.	Bhindi, Lady finger	Malvaceae	Africa
2.	*Anacardium occidentale* L.	Kaaju, Cashewnut	Anacardiaceae	Tropical America
3.	*Ananas comosus* Merr.	Anannas, Pine apple	Bromeliaceae	Brazil
4.	*Annona squamosa* L.	Sitaphal, Shareefa, Custard apple	Annonaceae	West Indies
5.	*Arachis hypogoea* L.	Moongphali, Ground nut	Fabaceae	S. America/ Philippines
6.	*Brassica oleracea* Var. *botrytis* L.	Phoolgobhi, Cauliflower	Brassicaceae	W. Europe
7.	*B. oleracea* Var. *capitata* L.	Band Gobi, Cabbage	Brassicaceae	Europe & Asia
8.	*Cajanus cajan* (L.) Millsp.	Arhar	Fabaceae	Tropical Africa
9.	*Carica papaya* L.	Papeeta, Papaw, Papaya	Caricaceae	W. Indies, C. & Tropical America
10.	*Cicca acida* (L.) Merr.	Harfarori, Harfareori	Euphorbiaceae	Malay Island & Madagascar
11.	*Citrullus vulgaris* Schrad	Tarbooz, Tarmus	Cucurbitaceae	Tropical Africa
12.	*Coriandrum sativum* L.	Dhania, Kothmeer, Coriander	Apiaceae	Greece
13.	*Daucas carota* L.	Gajar Gazar, Carrot	Apiaceae	Europe
14.	*Ficus carica* L.	Anjeer, Fig, Anjera	Moraceae	Mediterranean region
15.	*Grewia asiatica* L.	Faalsa, Parusha	Tiliaceae	East Tropical Africa
16.	*Glycine max* (L.) Merr.	Soybean, Soyabean	Fabaceae	Cochin China, Japan & Java
17.	*Lycopersicum esculentum* Mill.	Tamatar, Tomato	Solanaceae	Tropical America
18.	*Litchi chinensis* Sonner Syn. *Nephelium litchi* Camb.	Leechi, lichi	Sapindaceae	China
19.	*Manihot esculentus* Crantz.	Cassarva Cassava	Euphorbiaceae	S. America
20.	*Manilkara achras* (Mill.) Fosberg. Syn. *Achras sapota* L., *A. zapota L.*	Sapota, Cheeku	Sapotaceae	S. America
21.	*Pithecellobium dulce* (Roxb.) Benth.	Jangal jalebi, Vilayati imli	Fabaceae	Mexico
22.	*Prunus persica* (L.) Stokes	Aru, Peach	Rosaceae	China
23.	*P. amygdalus* Batsch	Badam, Almond	Rosaceae	E. Mediterranean region
24.	*Pyrus communis* L.	Naspati, pear	Rosaceae	Eurasia
25.	*P. malus* L.	Seb, apple	Rosaceae	Europe & W. Temp. Asia
26.	*Psidium guajava* L.	Amrood, Bihi, Guava	Myrtaceae	Mexico
27.	*Punica granatum* L.	Anar, pomegranate	Punicaceae	Iran
28.	*Phoenix dactylifera* L.	Khajoor	Arecaceae	Arabia / Persia
29.	*Solanum tuberosum* L.	Aaloo, potato	Solanaceae	America
30.	*Phaseolus lunatus* L.	Lobiya	Fabaceae	Brazil
31.	*Physalis peruviana* L.	Tipari, Raspberry	Solanaceae	Tropical America
32.	*Pistacia vera*	Pista, Pistacia	Anacardiaceae	Syria, Mesopotamia

Table 22.6. Exotics used as drugs by practitioners of medicine in India

S.No.	Name of the species	Vernacular name	Family	Nativity
1.	*Aconitum napellus* L.	Visha, Katbish, Meetha Zahar	Ranunculaceae	Temp. & Arctic Europe
2.	*Aframonium melegueta* (Rosc.) K. SchumSyn. *Amomum melegueta* Ross.		Zingiberaceae	Tropical Africa
3.	*Aganosma calycina* A.DC.	Malati	Apocynaceae	Burma
4.	*Agave americana* Haw.	Kantala, Khetki	Agavaceae	America
5.	*Aglaia odorata* Lour	Priyangu	Meliaceae	Malaya
6.	*Alocasia denudata* Engl.	Keladi	Araceae	Malay Peninsula
7.	*Aloe abyssinica* Lam.	Gunwar Patha	Liliaceae	Abyssynia and C. Africa
8.	*Aloe barbadensis* MillSyn. *A. vera* Tourn. ex Linn.	Ghikwar, Ganwarpatha	Liliaceae	Barbados, W. Indies
9.	*Alpinia officinarum* Hance	Kulinjan, Khulinjan	Zingiberaceae	China
10.	*Anacyclus pyrethrum* DC.	Aqarqara	Asteraceae	S. Europe
11.	*Ananas comosus* Merr.Syn. *A. sativus*	Anannaas	Bromeliaceae	Brazil
12.	*Anastatica hierochuntica* L.	Garvaphul	Brassicaceae	Arabia, Palestine
13.	*Anthemis nobilis* L.	Babuni Ka Phool	Asteraceae	England, France, Spain
14.	*Arctostaphylos uva-ursi Spreng.*	—	Ericaceae	N. America, Europe
15.	*Aristolochia longa*	Zarwandi-i-tawil	Aristolochiaceae	S. Europe
16.	*A. reticulata* Nuttal	—	Aristolochiaceae	Senegal
17.	*A. rotunda* L.	Zarwand-i-gard	Aristolochiaceae	S. Europe
18.	*A. serpentina* L.	—	Aristolochiaceae	N. America
19.	*Arnica montana* L.	—	Asteraceae	W. & C. Europe
20.	*Artemisia persica* Boiss.	Pardesi Parwano	Asteraceae	W. Tibet
21.	*Asarum europaeum* L.	Upana, Taggar	Aristolochiaceae	Temp. Europe & N. Asia
22.	*Berberis vulgaris* L.	Kashmal	Berberidaceae	Europe
23.	*Camellia sinensis* (L.) Kuntze	Cha, Chai	Theaceae	China
24.	*Carapa granatum* (Koenig) Alston Syn. *C. obovata* Blume; *C. moluccensis* W.P. Hiern.	Dhundal, Pussur	Meliaceae	Ceylon, Burma & Malaya
25.	*Catha edulis* Forsk.	—	Celastraceae	Tropical Africa
26.	*Centaurea behen* L.	Safed Bahman	Asteraceae	Persia
27.	*Ceratonia siliqua* L.	Kharnub	Fabaceae	E. mediterranean
28.	*Chieranthus cheiri*	Todri surkh	Brassicacea	S. Europe
29.	*Chlorophora excelsa* Benth. & Hook. F.	—	Moraceae	Tropical America
30.	*Chrysobalanus icaco* L.		Rosaceae	Tropical & subtropical America, W. Africa
31.	*Cocoloba uvifera* L.		Polygonaceae	West Indies
32.	*Cochlearia armoracia* L.		Brassicaceae (Cruciferae)	E. Europe
33.	*Commiphora myrrha* (Nees) Engl.	Rasagandhi	Burseraceae	Arabia & of the African coast of Red Sea
34.	*C. opobalsamum* (L.) Engl. syn. *Balsamodendron opobalasamum* Kunth.	Balasan, Habbul balasan	Burseraceae	Arabia, on both sides of Red sea
35.	*Conium maculatum* L.	Kurdumana	Apiaceae (Umbelliferae)	Europe & Temp. Asia

...(Contd.)

S.No.	Name of the species	Vernacular name	Family	Nativity
36.	*Convolvulus scammonia* L.	Sak-munia, Saqmoonia	Convolvulaceae	Mediterranean
37.	*Corylus avellana* L.	Bindak, Findak	Betulaceae	England, France
38.	*Cyclamen europaeum* L.	Hathajooree	Primulaceae	Europe and Caucasia
39.	*Dorema ammoniacum* D. Don	Ushak	Apiaceae	Persia
40.	*Dryobalanops aromatica* Gaertn. f.	—	Dipterocarpaceae	Sumatra & Borneo
41.	*Ecballuim elaterium* A. Rich	Kateri Indryan	Cucurbitaceae	S. Europe
42.	*Elaeis guinensis* Jacq.	African oil palm	Arecaceae (Palmae)	Africa
43.	*Euphorbia lathyrus* L.	Barg-sadab, Sudab	Euphorbiaceae	C. & S. Europe
44.	*E. resinifera* Berg	Farfiyum	Euphorbiaceae	Morocco
45.	*E. tirucalli* L.	Konpal, Sehund	Euphorbiaceae	Africa
46.	*Ferula alliacea* Boiss	Heeng, Hingu	Apiaceae (Umbelliferae)	Eastern Persia
47.	*F. foetida* Regela	Hingu, Hing	Apiaceae (Umbelliferae)	E. Persian & Western Afghanistan
48.	*F. galbaniflua Coiss* et Buhse	Gandabiroza, Jawashir	Apiaceae (Umbelliferae)	Persia
49.	*Fumaria parviflora* Lam.	Pitpapra	Fumariaceae	Baluchistan
50.	*Guazuma tomentosa* Kunth.	Nipatunth	Sterculiaceae	Tropical America
51.	*Guizotia abyssinica* Cass.	Ramtil, Kala Til	Asteraceae	Tropical Africa
52.	*Haematoxylon campechianum* L.	Bakkan, Partanga	Caesalpiniaceae	Tropical America
53.	*Hagenia abyssinica* Gmel Syn. *Brayera anthelmintica* Kunth.	Cusso	Rosaceae	Abyssinia
54.	*Helianthus annuus* L.	Surajmukhi	Asteraceae	America
55.	*Helleborus niger* L.	Kalurchini, Khorasani Kutki	Ranunculaceae	Europe
56.	*H. viridis*	Krishnabhedi Kalikatuki	Ranunculaceae	Europe
57.	*Hibiscus cannabinus* L.	Patsan, Nali	Malvaceae	Africa
58.	*Hippomane mancinella* L.	Manchincal tree	Euphorbiaceae	Tropical America
59.	*Humulus lupulus* L.	Hop	Moraceae	N. America
60.	*Hura capitans*	Mullarasanam	Euphorbiaceae	Tropical America
61.	*Ilex paraguariensis* St. Hillaire Syn. *I. paraguayensis* Hook.		Aquifoliaceae	Brazil & Paraguay
62.	*Illicium anisatum* L. Syn. *I. religiosum* Sieb & Zucc.	Anasphal, Badiyan	Magnoliaceae	Japan
63.	*Inula helenium* L.	Rasan	Asteraceae	Europe & N. Asia
64.	*Ipomoea tuberosa* L.	Woodrose	Convolvulaceae	Tropical America
65.	*Iris foetidissima* L.	Dadmari	Iridaceae	W. & S. Europe, N. Africa
66.	*Jatropha curcas* L.	Bagbherenda, Ratanjot	Euphorbiaceae	Tropical America
67.	*J. gossypifolia* L.	Karituru Kaharalu (Kan.)	Euphorbiaceae	Brazil
68.	*Liquidamber orientalis* Sonner	Silaras, Silhaka	Hamamelidaceae	S.W. Asia Minor
69.	*Lupinus albus* L.	Turmuz, Turmas	Fabaceae	Levant
70.	*Mandragora officinarum* Bertol	Lakshmana	Solanaceae	Mediterranean region
71.	*Mangifera caesia* Jack	Binjai	Anacardiaceae	Malay Peninsula
72.	*Manilkara kauki* Dub Syn. *Mimusops kauki* L.	Khirni	Sapotaceae	Malay Peninsula & Archipelago
73.	*Maranta arundinacea* L.	Ararot, Arrow root	Marantaceae	Tropical America
74.	*Martynia annua* L.	Bichu Hathajori	Pedaliaceae	Mexico
75.	*Memecylon amplexicaule* Roxb.	Vachi (Tamil)	Melastomaceae	Malay Peninsula
76.	*Opuntia coccinellifera* Mill.	Pachikkali (Tamil)	Cactaceae	Mexico & Peru
77.	*O. dillenii* Haw.	Nagphani	Cactaceae	Mexico

...(Contd.)

S.No.	Name of the species	Vernacular name	Family	Nativity
78.	*Orchis laxiflora* Lam.	Salap Misri	Orchidaceae	Asia Minor
79.	*O. mascula* L.	Salep Misri, Salum	Orchidaceae	C. & S. Europe
80.	*Pistacia lentiscus*	Rumi mastaki	Anacardiaceae	Mediterranean region
81.	*P. terebinthus* L.	Kabuli mustaki	Anacardiaceae	Mediterranean, Asia Minor
82.	*Physalis angulata* L.	Jangli Rasbhari	Solanceae	Tropical America
83.	*P. alkekengi* L.	Kaknaj	Solanceae	S.E. Europe
84.	*Polianthes tuberosa* L.	Gulshabbo, Rajanigandha	Amaryllidaceae	Mexico
85.	*Quercus infectoria* Oliv.	Majuphal	Fagaceae	Greece, Asia minor & Syria
86.	*Rheum palmatum* L.	Rewand Chini	Polygonacea	China
87.	*Rhododendron javanicum*	Rhododendron	Ericaceae	Malay peninsula
88.	*Rhus coriaria* L.	Tatrak, sumak	Anacardiaceae	Mediterranean region, Persia
89.	*Ricinus communis* L.	Arand, Castor	Euphorbiaceae	Africa
90.	*Rumex crispus* L.	Amla-betase	Polygonaceae	Europe & N. Asia
91.	*Rosmarinus officinalis* L.	Rusmari	Lamiaceae	S. Europe
92.	*Ruta graveolens*	Sadab	Rutaceae	Westward to the Canary Island
93.	*Salvia officinalis* L.	Salbia sefakuss	Lamiaceae	S. Europe
94.	*Sambucus nigra* L.	Uti-Khaman (Arabic)	Caprifoliaceae	Europe & Asia minor
95.	*Styrax benzoin* Dryand	Luban Shambirani	Styraceae	Malacca & Malaya
96.	*S. officinalis*	Silajit, Usturak	Styraceae	Levant, Asia Minor & Syria
97.	*Syzygium aromaticum* (L.) Merr.	Laung, clove	Myrtaceae	Moluccas
98.	*Teucricum chamaedrys* L.	Kamazariyns (Arabic)	Lamiaceae	Europe & some parts of Asia
99.	*T. polium* L.	Buliun	Lamiaceae	Mediterranean region
100.	*Vinca rosea* L. Syn. *Catharanthus roseus*	Sadabahar, Periwinkle	Apocynaceae	W. Indies

An Example of Conservation and Exploration

Efforts in the last 10 years in south western India by MS Swaminathan Research foundation, a centre for Research on sustainable Agriculture & Rural development, have resulted in regeneration of many Rare, Endemic and Threatened (RET) species of Wayanad district in Kerala. A conservation garden has been established with an orchidarium consisting of 175 orchids, an arboretum with about 200 tree species, shade houses with a collection of over 50 critically endangered species and about 400 species of medicinal plants and a herbarium with 2000 flowering plant species distributed in Wayanad district.

Wayanad, district, a part of the Nilgiri Biosphere reserve, is a hot spot of rare flora in Kerala. A study conducted by scientists of M.S. Swaminathan Research Foundation (MSSRF) at Puthurvayal, near here, identified 2034 flowering plants including 3 new species - *Miliusa wayandica, M. gokhale*, belonging to the *Annonaceae* (Custard apple family) and *Oberonia swaminathanii* of the orchid family.

The rare and endangered species considered, possibly extinct - *Eugenia argentea* and *Hedyotis wayandensis* were collected after 130 years (Manoj, 2009).

Western Ghats—A Highly Threatened Biodiversity Hot Spot

Faster rate of species extinction is recorded in floristically rich sites like the tropical forest. The Western Ghats in India is one of the most threatened Biodiversity hot spots of the world. Among the 600 taxa considered to be rare or threatened in the flora of Peninsular India, about 90% species are in the western Ghats (Sastry & Sharma, 1991). Western Ghats, the 1600 Km long hill chain that run parallel to the west coast of India between the river Tapti in Gujarat and Kanyakumari in Tamil Nadu covering approximately 1,60,000 sq km area is a treasure house of plants and animals next only to Himalayan tracts in terms of diversity of unique species, of the total recorded species 38% flowering plants, 11% butterflies and 53% fishes.

Similarly a number of species which are in cultivation today never existed before, and owe their origin to natural hybridization followed by polyploidy and establishment of new species by natural selection. Thus apart from wheat, rice and sugarcane long - staple cotton has similar origin.

Further variation and diversity comes from minor genetic changes leading to the development of varieties differing in some characteristics, which may be morphological, anatomical, physiological, or phytochemical. For instance, the Asian rice *Oryza sativa* has more than 8,000 known varieties.

Thus on one hand we incur losses due to extinction of species and on the other nature provides us with new species and varieties through natural hybridization, mutation, natural selection and adaptability of the new species to specific environments. Some species have come into being through natural hybridization. Many of them have established themselves. Some examples given below testify it.

Origin of Bread Wheat

The grasses which fall in the category of wheats comprise of 3 categories - *Einkorn* (diploid with Chromosome no. 2n = 14) *eg. Triticum monococcum*; *Emmer* (Tetraploids with 4n = 28) *eg. T. dicoccum* and *vulgare* (Hexaploid with 6n = 42), eg. *T. aestivum.*

The bread wheat is hexaploid and has come into being by natural hybridization between *Triticum urartu* and a wild grass *Aegilops speltoides*. It resulted in the formation of emmer or tetraploid group. The hexaploid wheat originated by the hybridization of *emmer* wheat, with another wild diploid grass *Aegilops tauschii* and subsequent chromosome doubling. Thus hexaploid wheat enriched the biodiversity by its origin.

Origin of *Primula kewensis*

Pellew (1927) reported that natural hybridization between *Primula floribunda* and *P. verticillata* resulted in the origin of a new species which was named *P. kewensis*, because of its origin at Royal Botanic Garden, Kew, England.

Raphanobrassica is a synthetic species and is the result of a cross between *Raphanus sativus* and *Brassica oleracea* var. *botrytis*. Another such species is *Triticale*, a hybrid between *Triticum vulgare* and a grass, *Secale cereale*.

Moreover, our forests still have enormous floristic wealth which has either been underexplored or unexplored. Plant explorers, every now and then report new species from various parts of the country. By extensive explorations, not only new species shall come into light, but those species which have been declared extinct, may also be discovered as has been found at Wayanad district of Kerala.

Phytodiversity of Angiosperms in India and its Sources of Information

After discussing the biodiversity in general my emphasis shall be focussed on Angiospermic diversity of India. India is a vast country with a rich floristic heritage and more than 18000 species of angiosperms. The information is based on various regional floras which encompass the botanical description of all those species which inhabit a particular geographical area within the country. In the lines to follow, few important and monumental works are being highlighted.

Stewart (1869) published the Flora of undivided Punjab under the title, "Punjab Plants". Stewart & Brandis (1874) published, "The Forest Flora of North-West and Central India". Talbot (1894) explored and published, "The trees shrubs and woody climbers of the Bombay Presidency". A very exhaustive compilation of flora of India was done as early as 1872-1897 by Hooker, who published 7 volumes under the title. "The Flora of British India".

Cook (1901 - 1908) published the Flora of Presidency of Bombay in 2 volumes. Wood (1902) published. Plants of Chhota Nagpur including Jaspur and Surguja. Prain (1903) published Bengal plants in 2 volumes. Duthie (1903-1920) explored Flora of the Upper Gangetic plain and the adjacent Siwalik and Sub-Himalayan tracts. Originally his work was published in 3 volumes however its reprinted edition (1960) was published by B.S.I. in two volumes.

Hole (1904) published a taxonomic work entitled, "A contribution to the Forest flora of Jubbulpore Division of Central Provinces". Indian trees was brought out by Brandis (1906). Talbot (1909-1911) also published forest flora of the Bombay presidency and Sind. Haines (1910) published a Forest Flora of Chhota Nagpur. A list of trees and shrubs was published by Biscoe (1910).

A 3 volume publication pertaining to the Flora of the Nilgiri and Pulney Hilltops (above 6,500 feet) being the wild and commoner introduced flowering plants around the Hill stations of Ootacamund, Kotagiri and Kodaikanal was brought out by Fyson (1915-1920). At the same time Gamble and Fischer (1915-1936) extensively explored the Flora of the Presidency of Madras and got it published. Haines (1921-24) published the flora entitled, "The Botany of Bihar & Orissa" in 6 volumes, and its reprinted edition (1961) was published by B.S.I. in 3 volumes.

A small booklet under the title "Weed Manual of Gwalior State" with line diagrams was published by Kenoyer (1924). A two volumed publication entitled, "The Flora of the South Indian Hill stations" was authored by Fyson (1932). Kanjilal (1933) published 'A forest flora of Pilibhit, Oudh, Gorakhpur and Bundelkhand'. Based on his extensive work, Champion (1936) published a preliminary survey of the Forest Types of India. Chatterjee (1939) studied the endemic flora of India and Burma. Bor (1947) published 'Common Grasses of the United Provinces' and also, The 'Grasses of Burma, Ceylon, India and Pakistan' in 1960. Cowen (1950) published photographs of Flowering Trees and Shrubs in India alongwith description of the species, however from horticultural point of view. Tiwari published a series of books and papers mainly based on his exploration of Madhya Pradesh. His publications include "The Grasses of Madhya Pradesh" (1955); "Supplement to the Grasses of Madhya Pradesh" (1963); "Flora of Bandhavgarh" (1968), Flora of Bansapur (1972); The Orchids of Madhya Pradesh (1963) in collaboration with Prof. J.K. Maheshwari and The Cyperaceae of Madhya Pradesh (1964).

Subramanyam (1962) published a monograph entitled, "Aquatic Angiosperms of India'. The book has description as well as illustrations and few photographs. However, some of the flora such as that of J.K. Maheshwari's (1963). The Flora of Delhi has a separate volume entitled, Illustrations to the Flora of Delhi published in 1966. Similar is the case with the 'Weed Manual of Gwalior State' by Kenoyer (*loc. cit*). Panigrahi *et al.*, (1966) published a paper on Botany of Madhya Pradesh and listed the plants from Dilleniaceae to Moringaceae. Panigrahi and Arora (1965) listed the plants from Rosaceae to Rubiaceae of Madhya Pradesh. Panigrahi *et al.*, (1965) reported the plants of the families from Ebenaceae to Convolvulaceae, followed by Euphorbiaceae and Urticaceae in 1967. Oommachan (1971) published The Flora of Bhopal, which was based on his Ph.D. work. Waheed Khan (1973) published 'Madhya Pradesh plants'. Thus one can get glimpses of Indian Flora by going through these masterpieces, but it is a fact that almost all these treatises and research papers deal with the description and diagnostic features of a species only and lack photographs or line sketches. The Flora of Bhopal was reexplored by Khan (1993), who added 178 new species to Oommachan's Flora of Bhopal.

The information on medicinal plants of India can be gathered from Chopra's (1933), Indigenous Drugs of India; Indigenous Drugs of India (Dey, 1896); Kirtikar & Basu's (1935) Indian medicinal plants (4 volumes of description & 4 of illustrations); Nadkarnis (1927), The Indian Materia Medica, with description only. Chopra *et al.*, (1956) Glossary of Indian medicinal plants and supplement to Glossary of Indian Medicinal plants (1969) are based on rigorous & extensive surveys and research work.

Global Species Estimates

According to an estimate, the number of global species range from 2 million to 100 million. Ten million is probably nearer the mark. Only 1.4 million species have been named and identified. Of these, approximately 2,50,000 are flowering plant (Angiospermic) species and amazingly 750,000 are insects. As a result of explorations new species are continually being discovered every year.

The number of species present in little known ecosystems such as the soil beneath our feet, and the deep oceans can only be guessed at. It is presumed that the deep sea floor may harbour as many as a million unexplored and undescribed species. In other words, we really have absolutely no idea of how many species exist on this crust of earth.

Extinction Rate

Variation is rule of nature and extinction is a fact of life. As a dynamic process, species have been evolving and dying out ever since the origin of life on this crust of earth. One only has to look at the fossil record to appreciate this. It has been estimated that surviving species are only tip of iceberg and constitute about 1% of the species that have ever existed.

There are hypothetical estimates and predictions about extinction rates. According to one such estimate, a quarter of all species on earth are likely to be extinct, or on the way of extinction within 30 years. Another predicts that within a span of 100 years, three quarters of all species will either be extinct or in populations so small that they can be described as "the living dead". This is exemplified by species of *Cycas* and *Ginkgo biloba,* the Maiden hair tree.

In fact, these are only predictions and most are based on computer models and as such, need to be taken with a very generous pinch of salt. To begin with, we really have no idea of how many species, there are on which to base our initial premise. The reason is very clear. The persons, who venture to find out the flora or fauna of a thick forest or a city, confine themselves to the areas which are easily approachable and remote forest areas go unexplored. A university professor wrote flora of a capital city of our country and it is surmised that he included only those species of plants in his flora which were found on his way from his residence to the university campus. This is true because, I have witnessed a number of species which are not included in his flora for that particular city.

Vulnerable Species

(*i*) **Species at the top of food chain :** Large carnivores usually require extensive territories in order to provide them with sufficient prey. But owing to deforestation activities taking place at an alarming rate and human encroachments may lead to their extinction because of loss/shinkage of habitat. Gigantism leads to extinction, as said by Chamberlain, (1966) is true and exemplified by giant *Dinosaurs* and *Lepidodendrons*.

(*ii*) **Endemic local species :** These species are confined only in one geographical area. These species are very sensitive to habitat disturbance because of their limited adaptability and often with less degree of reproductive capacity.

(*iii*) **Migratory species :** Propagules of plants, which are washed away by sea water, on reaching a new habitat may fail to develop. Similarly migratory birds to a new destination may lose their life after they are hunted.

(*iv*) **Species with exceptionally complex life cycles :** If completion of a particular life cycle requires few or several different elements to be in place at very specific times, then the species is vulnerable, if there is disruption of any single element in the cycle. For instance, the rust fungus, *Puccinia graminis tritici* completes its life cycle on wheat and barberi bush, which is found on hills. To save the wheat from the attack of this fungus, barberi bushes are destroyed. This disrupts its life cycle, and if this continues, the fungus is likely to become extinct.

(*v*) **Species with chronically small populations :** If populations become too small, then simply finding a mate, or interbreeding, can become serious problem. Such plant populations may fail to continue if the specific pollinator is not available due to various reasons and there is no formation of runners, suckers or bulbils.

(*vi*) **Specialist species :** Such species have very narrow requirements, such as a single specific food source, *e.g.* a particular plant species. The saffron or *Crocus sativus* is cultivated in a small area of Kashmir, where microclimatic and edaphic conditions favour its growth, but outside this area it fails to develop. Any drastic change of soil condition or climate may lead to the extinction of this species from the area.

Loss of an individual species can have various different impacts on the surviving species in an ecosystem. These effects depend upon the how importnat the species is in the ecosystem? Removal of some species has no apparent effect whereas removal of others may have enormous effects on the remaining species. For instance, some species such as *Lantana indica* and *Baliospermum montanum* exercise teletoxic or allelopathic effect and do not permit other species to grow in their close proximity and hence their removal leads to the arrival and establishment of other species & vice-versa. Species, such as these are termed “Keystone” species. Similarly for the growth of Sandalwood tree (*Santalum album*) vicinity of grasses from the roots of which it derives mineral nutrition and water is essential otherwise it will fail to develop.

Species which Deserve Immediate Conservation Efforts

Some species need urgent conservation otherwise these will be thing of past. These include - *Catharanthus pusillus, Exacum petioliare, Hoppea dichotoma, Enicostema hyssopifolium, Indigofera hirsuta, I. trita, I. linifolia, Grangea maderaspatana. Polycarpea corymbosa Potentilla supinum, Schrebra swietenoides, Saponaria vaccaria, Swertia chirata, Bacopa monnieri, Centella asiatica, Tribulus terrestris, Anisomeles malabarica, Anisochilus carnosus, Anotis lancifolia, Gymnema hirsuta, Entada gigas* and *Zeuxine sulcata.*

All the works cited in the list of references have not been discussed in the text, but all of them represent biodiversity of our country that is present in various regions.

REFERENCES

1. Anderson, E. (1956) Man as a maker of New Plants and New Plant communities. Ed. W.L. Thomas Jr. The Univ. of Chicago Press, Illinois, pp. 763-777.
2. Bailey, L.H. (1956). Manual of Cultivated plants, New York.
3. Bates, M. (1956) Man as an agent in the spread of organisms. Man's role in changing the face of the Earth Ed. W.L. Thomas Jr. Illinois, pp. 788-804.
4. Biscoe, W.F. (1910) A list of trees and shrubs of Indore state, Bombay.
5. Blatter. E. & F. Hallberg (1918-1921). The Flora of Indian deserts (Jodhpur & Jaisalmer) JBNHS 26: 218-246, 525-551, 811-818, 968-987 *Ibid., 27*: 40-47, 270-279, 506-519.
6. Bor, N.L. (1947) Common Grasses of the United provinces, *Indian For. Rec.* (n.s. Bot.) *2*:1-220.
7. Bor, N.L. (1960). The Grasses of Burma, Ceylon, India and Pakistan (Excluding Bambuseae) Pergamon Press, London.
8. Brandis, D. (1874) The forest flora of North-West and Central India, London.
9. Brandis, D. (1906) Indian Trees, London (Reptd. 1921).
10. Chamberlain, C.J. (1966) Gymnosperms, structure & Evolution.
11. Champion H.G. (1936). A preliminary survey of the forest types of India & Burma. *Indian For, Rec.* (n.s. Bot.) *1*:1-286.
12. Chapman, A.D. (2005) “Numbers of living species in Australia and the world, Australian Biological Resources study - 04 - 23.
13. Chatterjee, D. (1939) Studies on the endemic flora of India & Burma *Jour. Royal As. Soc. Bengal Sci. 5*: 19-67.
14. Chopra, R.N. (1933). Indigenous Drugs of India.
15. Chopra, R.N., S.L. Nayar and I.C. Chopra (1956). Glossary of Indian medicinal plants, CSIR, N. Delhi.

16. Chopra, R.N. S.L. Nayar & I.C. Chopra (1969). Supplement to Glossary of Indian Medicinal Plants CSIR, New Delhi.
17. Cook, T. (1901 - 1908), The Flora of Presidency of Bombay, London 2 Vols (Reptd. Edn. *Bot. Surv. India* Calcutta, 3 Vols. 1958.
18. Cowen, D.V. (1950) Flowering Trees and Shrubs of India, Bombay.
19. Dey, K.L. (1296) Indigenous Drugs of India.
20. Duthie, J.F. (1903 - 1929). Flora of the Upper Gangetic Plain and the adjacent Siwalik and Sub-Himalayan Tracts (ed.1) Calcutta, 3 Vols. (Reptd. Edn. 1960 B.S.I. Calcutta 2 vols.)
21. Fyson, P.F. (1915-1920). The flora of the Nilgiri and Pulney Hilltops (above 6,500 feet) being the wild and commoner Introduced flowering plants around the Hill stations of Ootacamund, Kotagiri and Kodaikanal Madras, 3 Vols.
22. Fyson, P.F. (1932). The flora of the South Indian Hill stations. 2 vols. Madras.
23. Gamble, J.S. & C.E.C. Fischer (1915-1936). Flora of the Presidency of Madras London 11 parts.
24. Graham, E.H. (1956). The Recreative power of Plant communities. Man's Role in Changing the Face of the earth. Illinois pp. 677-691.
25. Haines, H.H. (1910). A forest Flora of Chota Nagpur, Calcutta.
26. Haines (1921 - 24). The Botany of Bihar and Orissa (ed.1) London 6 parts (Reptd. edn. 1961 *Bot. Surv. India.* Calcutta 3 Vols.).
27. Hole, R.S. (1904) A contribution to the Forest Flora of Jabbulpore Division, Central Provinces, *Indian For. 30*: 499-514, 566-592.
28. Hooker, J.D. (1872 - 1897). The Flora of British India, London 7 vols.
29. Hutchinson, J. (1959). The families of flowering plants (ed.2) Oxford 2 vols.
30. Jindal, S.L. (1970) Flowering shrubs in India, N. Delhi.
31. Joseph, J. (1963). A contribution to the Flora of Bori Reserve Forests. Hoshangabad District, Madhya Pradesh. *Bull Bot. Surv. India 5*(3 & 4): 281 - 299.
32. Kanjilal, P.C. (1933). A forest flora of Pilibhit Oudh, Gorakhpur and Bundelkhand.
33. Kanjilal, U.N. *et. al.*, (1934-1940). Flora of Assam Calcutta 5 Vols.
34. Kenoyer, L.A. (1924). Weed Manual of Gwalior State and adjacent parts of India, Calcutta.
35. Khan S.S. (1993). Studies on Ethnomedicinal plants of Bhopal District. Ph.D. thesis submitted to Barkatullah University, Bhopal (Unpublished).
36. Kirtikar, K.R. & B.D. Basu (1935) Indian Medicinal plants 4 Vols. of description + 4 Vols. of illustration.
37. Maheshwari, J.K. (1963). The flora of Delhi CSIR, New Delhi.
38. Manoj, E.M. (2009) Wayanad, hot spot of rare flora, shows up 3 new species. The Hindu, Sunday, Nov. 8, p. 9.
39. Nadkarni, K.M. (1927) The Indian Materia Medica.
40. Narayanaswamy, V. & R.S. Rao (1969). A contribution to our knowledge of the vegetation and Flora of Pachmarhi Plateau and the adjacent region. *J. Indian Bot. Soc. 39*: 222-242.
41. Oommachan, M. & V.K. Billore (1969). Asteraceae of Bhopal (M.P.) A systematic study *Bull Bot. Surv. India 11*(62): 35-40 (Published in June 1971).
42. Pandeya, S.C. (1952) Grasslands of Sagar, M.P. *Indian for 78*: 638 - 654.
43. Panigrahi, G., C.M. Arora., D.M. Verma & V.N. Singh (1966) Contribution to the Botany of Madhya Pradesh - I (Dilleniaceae to Moringaceae) *Bull Bot. Surv. India 8*:117-125.
44. Panigrahi, G. and C.M. Arora (1965) Contribution to the Botany of Madhya Pradesh-II (Rosaceae to Rubiaceae). *Nat. Acad. Sci. India Sec.B. 35*, Pt. I : 87-98.
45. Panigrahi G., C.M. Arora and D.M. Verma (1965). Contribution to the Botany of Madhya Pradesh - III (Ebenaceae to Convolvulaceae). *Ibid.,. B-35*, Pt.I : 99-109.
46. Panigrahi G. & R. Prasad (1967) *Ibid.,.* IV. Euphorbiaceae & Urticaceae *B-37* Pt. 4: 553-564.
47. Pellow, Caroline (1927) The Genetic behaviour of *Primula kewensis*. Molecular and General Genetics. *M.G.G. 45*(1), 402 - 403.
48. Prain, D. (1903). Bengal Plants (ed.1). Calcutta, 2 Vols. (Rep. edn. 1963) Bot. Surv. India, Calcutta, 2 Vols.)
49. Santapau, H. (1967). The Flora of Khandala & Western Ghats of India. *Rec. Bot. Surv. India. 16*(1): 1-372, ed.3 Calcutta (1st edn. 1953 & 2nd edn. 1960, Delhi).
50. Takacs, D. (1996). The idea of biodiversity: philosophies of paradise, Baltimore: The Johns Hopkins University Press.

51. Talbot, W.A. (1894) The Trees, Shrubs and Woody Climbers of the Bombay Presidency, Bombay.
52. Talbot, W.A. (1910-1911) Forest flora of the Bombay Presidency and Sind 2 Vols. Poona.
53. Tiwari, S.D.N. (1954). The Grasses of Madhya Pradesh. *Indian For 80* : 601-611, 681-689.
54. Tiwari, S.D.N. (1963). Supplement to the Grasses of Madhya Pradesh. *Indian For 89*: 593 - 602.
55. Tiwari, S.D.N. (1968) Flora of Bandhavgarh. *Ibid., 94*:5
56. Tiwari, S.D.N. (1972) Flora of Bansapur, Madhya Pradesh. *Botanique 3*(2): 73-84
57. Tiwari, S.D.N.(J.K. Maheshwari (1963). The Orchids of Madhya Pradesh. *Ibid., 89*: 426-444.
58. Tiwari, S.D.N. (1964). The Cyperaceae of Madhya Pradesh *Ibid., 98*(3): 147-159. *Ibid., 90*(9): 616-629.
59. Waheed Khan, M.A. (1973) Madhya Pradesh Plants. Govt. Regional Press Rewa.
60. Wilson, E.O. (1992). The diversity of Life, Cambridge: Belknap press.
61. Wilson, E.O. (ed.) (1988). Biodiversity, Washington DC. National Academy of Sciences / Smithsomian Institution.
62. Wood, J.J. (1902), Plants of Chota Nagpur including Jaspur and Surguja. *Rec. Bot. Surv. India 2*(1) : 1 - 170.

CHAPTER 23

Municipal Solid Wastes and Recycling

Dr. S.A. Iqbal[1] and Dr. Siddiq Qureshi[2]

Solid Wastes

Litter along the roadsides, trash on the beaches, heaps of refuse in the streets, the pungent piles of rotting garbage are unpleasant reminders of untidy human habits. Thousands of backyards, vacant plots contain the remains of discarded machines, parts of vehicles, food wastes and so on. Open dumps scar the landscape. In some countries solid wastes are dumped into the ocean. The growing mass of solid waste produced annually the world over includes millions of tons of paper and paper products, plastics, billions of bottles, cans, tyres, junked machines and discarded automobiles and millions of other appliances of different sizes and kinds. Probably the United States produces the largest solid waste. The city of New York produces the largest waste around one tonne each household per year. Around 15 per cent of solid waste from every household in an affluent society consists of food wastes. There is considerable loss of raw material and energy which goes into the manufacturing of the material thrown away as solid waste.

Disposal of Solid Wastes

The most commonly used methods for final disposal of solid wastes are sanitary landfill and open dumping. These methods are cheaper than the other disposal methods. The other advantage is that we can avoid the acute pollution problems associated with discharging wastes into waterways or polluting the air from incineration.

If properly planned, landfills can be used later for construction sites or recreational facilities. However, there are major disadvantages to open dumps, which are a public health problem, breeding flies, rats etc. Burning combustible wastes produces mixtures of particulate and gaseous matter that are obnoxious. Land disposal of solid wastes requires large parts of land. The increasing demand of land for human habitations and the ever-growing mountain of solid waste make the problem complex.

The adverse economics of this type of the disposal of solid wastes requires large parts of land. The increasing demand of land for human habitations and the ever-growing mountain of solid waste make the problem complex.

The adverse economics of this type of the disposal of solid wastes are far-reaching. Mineral resources such as copper, zinc, lead and tin are limited, and the ores are non-renewable. These metals are not destroyed by disposal but are often scattered beyond recovery, The squandering of these resources that occur when wastes are dumped indiscriminately, will produce a disruptive economic impact at some time in the future unless improvements are adopted. *Recycling* will alleviate some problems associated with the increasing density of population.

[1]Department of Chemistry, Saifia Science College, (B.U.), Bhopal-462001 (India)
[2]Department of Chemistry, Sant Hirdaram Girls College, (B.U.), Bhopal-462030 (India)

Incineration

In most developed countries the residents collect the wastes in polythene bags, and put them at prescribed places at a certain time of the day. Then onwards it becomes the responsibility of the municpal or other local bodies to collect and carry them to appropriate disposal sites. In well organised waste disposal systems articles such as cans, bottles etc. are separated and then the remaining garbage subjected to compaction in special trucks. These are transported to selected garbage disposal sites, like landfills, composting pits, open dumps or to incinerators. Once the garbage is collected, useful or recycable articles are separated. In India, some people earn or their livelihood by selling renewable resources handipicked from garbage. The solid wastes to be incinerated are *highly compbustible* trash, like paper, carboard, plastics or rubber scrap, *combustible* wastes like wood scraps, cartons, floor sweepings, food *wastes* and other combustible articles. The incinerators produce toxic ash which must be landfilled.

The incinerators are so designed that the heat generated during incineration may be usefully utilised. The most common use is to generate steam. This partly recovers the cost of waste collection and disposal. Some waste materials produce corrosive fumes. Hence a careful segregation of such materials is desirable before compaction and incineration.

Recycling

Recycling Through Composting

One of the best places to start recycling is in the home garden with a compost pile. Compost, by definition, represents a mixture of plant material, soil eggshells, straw, manure, and other materials that can be broken down by microrganisms (bacteria and fungi). Its main value for use in the home garden, after decomposition has converted it into rich humus, is to improve soil texture, increase soil water-holding capacity, serve as a much, prevent or related weed growth, and protect plants from excess winter injury due to low temperatures or desication. It may also serve as source of nutrients, especially if ashes and manure have been added to the compost pile. However, the fertilizer value of most compost humus is low; 0.5 per cent nitrogen, 0.4 per cent phosphorus, and 0.2 per cent potassium.

To make good humus we can add the following to a compost pile: eggshells, citrus rinds, coffee grounds, kitchen vegetable garbage, tops of cut perennials, leaves, grass clippings, weedy annuals, ashes, straw, manure, crushed granitie rock, soil, inorganic fertilizer, seaweed, overripe fruit, water, and earthworms.

Such a mixture of garden and home by-products will contain many types of fungi and bacteria which act to decompose the compost substrates. However, to make sure that decomposition occurs satisfactorily, many compost makers add *compost starter,* which is an aqueous suspension of bacteria and/or fungi that are adept at breaking down vegetable matter. The decomposition action by these organisms works best under aerobic conditions in warmer weather where there is a favourable supply of moisture. Like secondary treatment of sewage, the microbial action breaks down the vegetable and other matter to humus plus CO_2 and, in the process, liberates much heat, generating temperatures typically between 50 and 80°C.

A compost pile may be senstructed in many ways but basically, one makes alternating layers of vegetable matter and soil. To simulate microbial action, some fertilizer and water should be added periodically to the compost pile, especially during periods when the air temperature is warm. The soil is primarily added to provide a source of microorgaisms and to hasten the microbial decomposition action by these microrganisms. It also retains moisture needed by these organisms to grow.

Energy Production Through Recycling

Another approach to recycling involves the use of animal wastes (manure) and plant debris (including algae) to produce methane (CH_4) and organic fertilizer for urban, farm, and garden use, Methane (biogas) generators are in widespread use in Germany, France, India, Japan and the United States. They can be used in both northern temperate and in tropical regions, Several of the big advantages to their use are that they utilize readily available substrate; they help to solve energy needs especially when used in combination with wind generators and solar panels; they provide useful fertilizer by products.

In contrast to composting, an aerobic operation, methane generators involve breakdown of organic matter in the absence of oxygen (anaerobic conditions). The latter, therefore, involves a well-known process called fermentation. The main products of this fermentation are CH_4 and CO_2, with trace amounts of hydrogen sulphide and nitrogen gas given off. The conversion occurs at or around normal atmospheric pressure and at temperatures ranging between 15 and 50°C.

In general, conversion of 450 g of dry organic matter yields between 4.5 to 6.5 cubic feet of methane. In India methane gas generators are called *biogas plants.*

Recycling Paper and Plastics

The United States recycles about one-fifth of its paper. In contrast, Japan recycles about one-half of its paper. Harvesting to make pulp for paper manufacture costs less than the recycling process. Removal of ink is now feasible but expensive, and recycled paper is generally not useful in making white bond or slick paper for magazines. Recycling will become more profitable as pulp becomes more scarcre, labour costs increase, and recycled paper comes into greater use.

One of the most emphatic arguments for recycling paper is that it results in a reduction of solid waste and a reduction of the demands placed on a nation's forests and woodlots. The rate at which timber is being cut in forests for lumber and pulp is indeed alarming.

Non-degradable plastics are put in disposal sites. They make a relatively stable material for landfil because they do not breakdown and pollute underground water systems. Some plastics are however degradable. In developing and under-developed countries plastics are generally recycled. However, the quality of recycled plastics is not as good.

Various methods have been devised to separate non ferrous from ferrous metals; glass from metal and different colours of glass. Of course the more economical method of achieving this separation is in individual homes, restaurants and other places where the materials are used. Broken glass is used in the manufacture of more glass bottles. This saves energy—melted glass uses only 2/3rd of heat energy required to produce fresh glass from silica and lime. Metal is easily recycled. Used tyre casing are suitable for reuse in the manufacture of synthetic rubber. Shreaded tyres mixed with rubber-base stock is used for playground surfacing.

Developing Countries

The hi-tech methods used for solid waste management and sewage treatment are highly expensive and beyond the reach of developing nations. In most of these countries sewage is discharged untreated into open drains and then on to rivers or open seas. Some countries including India and Thailand use a low-tech system of sewage treatment. The firststep of this system is the same as the hi-tech system. The liquid waste is then left in open pond for conversion of organic waste through phtosynthesis. The low-tech system requires three times more space than the high-tech one. There are no programmes for solid-waste management. Most of solid waste is dumped on to open ground and used for landfil. Renewable resources like plastics, glass, metal etc. are however, recycled.

PROSPECTS FOR UTILIZATION OF MUNICIPAL SOLID WASTE (MSW) IN INDIA

Municipal solid waste generation in India is increasing day by day. The average per capita generation of waste was about 500 g/day in 2007 and it will increase to 925 g/day by 2047. The most important step to be used for management of this waste would be minimising its generation but this is not really possible. Then to manage waste sustainably, thewaste to wealth route remains a viable solution although in India it is not common practice. Maximum recycling, composting with organic municipal solid waste and waste to energy generation should be utilised for MSW management. This approach of sutainable waste management can solve the problem of land required for waste disposal and resulting pollution problems of air, ground, surface water, soil etc.

India has second largest population in the world (16.7 per cent of the world population) and accounts for a meager and 2.4 per cent of the world surface area. With the changing pattern of life style and consumer economy waste generation is increasing rapidly in India and the annual municipal waste generation. (MSW) is 50 million tonnes/year (5 million tonnes/year is hazardous waste) already. This is estimate to increase to approximately 310 million tonne annually by 2047 (Table 1).

Table 23.1. Increase in Municipal Solid Waste in India.

Years	1950	2000	2050
Uraban Population (millions)	57.8	256	—
Daily Per Capita Waste Generation (Grams)	300	550	1050
Total Waste Generated (Million tonnes)	7	56	330
Area Under Land Fills (Thousand of ha)	0.15	25.4	160
Annual Methane Emissions (million Tonnes)	1.25	10.2	58

Municiapl solid waste (MSW) generation is around 500 g/day/person. This is estimated to further increase to 925 g/capita/day by 2047. The MSW contains largely food/organic waste, papers, plastics textiles, rubber, leather, wood, glasses, batteries, ferrous and nonferrous metals containing components etc. Indian MSW has 40-60% of compostable material or organic material with high moisture contents (Table 23.2 and 23.3). The rising inconie levels and in turn greater purchasing power of individuals especially in big cities have given rise to consumer preferences that make goods obsolete quickly and this has led to a sharp rise in minds of policy makers, regulators and concerned citizenry. One recent example of tragic impact of improper management of municipal solid waste was seen as epidemic in Surat. Gujarat in 1994. Land scarcity in the cities has made waste disposal more difficult. The use of landfill is no longer cosidered to be a satisfactory environmental solution due to many health hazards linked with it. Therefore, generation of wastes. Wastes pose a severe challenge for Municipal Authorities for its sound disposal and the problem has been agitating the new methods have to be found to produce wealth from the waste. Obviously, the impact on environment can be reduced by reducing waste generation itself and failing this, waste should be either recycled or reused. When these options are unsuitable, waste must be incinerated for energy recovery. The last resort should be used in landfills because it requires space and run the risk of leakage.

Table 23.2. Physcial Characteristics of MSW in India Cities.

Population Range (in million)	No. of cities surveyed	Paper	Rubber, Leatherand synthetics	Glass	Metals	Total Compostable metals	Inert
0.1 to 0.5	12	2.91	0.78	0.56	0.33	44.57	43.59
0.5 to 1.0	15	2.95	0.73	0.35	0.32	40.01	48.38
1.0 to 2.0	9	4.71	0.71	0.46	0.49	38.95	44.73
2.0 to 5.0	3	3.18	0.48	0.59	56.67	49.07	
>5	4	6.43	0.28	0.94	0.80	30.84	53.90

All values in % and are calculated on dry weight basis.

(**Source:** Manual on MSW management for GOL Ministry Urban Development Ministry, 2000)

Table 23.3. Chemical Characteristics of MSW in India Cities.

Population Range (in million)	No. of cities surveyed	Moisture %	Organic Matter	N as (TN)	P as P_2O_5	K as K_2O	C/N ratio	Calorific Value in K cal/Kg
0.1 to 0.5	12	25.18	37.09	0.71	0.63	0.82	30.94	100/89
0.5 to 1.0	15	23.52	25.14	0.66	0.56	0.69	21.13	900.61
1.0 to 2.0	9	26.98	26.89	0.64	0.82	0.72	23.68	980.05
2.0 to 5.0	3	21.03	25.06	0.56	0.69	0.78	22.45	907.45
>5	4	38.72	39.07	0.56	0.52	0.52	30.11	800.70

All Values except moisture are on dry weight basis.
(**Source:** Manual of MSW Management for GOL Urban Development Ministry, 2000)

Neoative Impacts by Unplanned Dumping of MSW

Some neatuve influencers are ground water contamination through leachate generation, surface water contamination by run-off from waste dumps, fire menace, bird meance, bad odours from dumping sites, epidemic through stray animals, global warming due to release of greenhouse gases (methane and CO_2), acidification of surrounding soils, stratospheric ozone deletion, photo-oxidant formation, etc. These can be rid of if MSW is put to some use. It should be noted that much of MSW may be rich in nutrients.

Thus, it is need of the hour to bring technological interventions for achieveing waste reduction and its utilisation. Waste should be regarded as resource/wealth lying at a wrong place. Thus in view that it has immense potential for resource recovery, it is required that MSW is looked from cradle to grave, properly collected, kept transported segregated and made ready to be disposed. Only those wastes should be sent for dumping which cannot be reused or recycled. The organic waste can be used as compost. The approach in waste management should be such that maximum possible waste generated could be used in one way or the other.

Proper management of MSW requires sensitizing the community about the importance of environmentally sound waste management practices, as well as concerted efforts should be made for adoption of a scientific approach by the concerned authorities. The biggest challenge is to ensure that different kinds of wastes are stored and segregated from the very beginning when individuals or families reject them.

Waste to Wealth Potential

Decrease in waste generation should be the first priority in MSW management hierarchy and then comes its maximum recycling possiblities. The 3R principle Reduce, Reuse and Recylce should be used to minimize waste generation. Now there is a fourth R which implies Rebuy. Thus, encouraging to rebuy the products made from the recycling of MSW, recycling industries can be promoted. There is no process in the world which has zero or no waste generation.

Some of the steps that can be taken for waste to generate 'wealth' from MSW are as follows:

Municipal Solid Waste Composting

Composting of MSW is recognized as a cost-effective method for waste management that results in an end product that can be used as a soil conditioner with beneficial effects on soil productivity.

In India MSW has 40-60% (or sometime more) of compostable material with high moisture contents. The high moisture contents and low calorific value makes it unsuitable for incineration. Biomethanation and composting are perhaps better alternatives for treatment of wastes.

Energy Generation from Waste

In the waste management hierarchy, waste to energy (WTE) has been considered as a mode for the recovery of resources that must be considered before ultimate disposal of the final inert materials.

It is a fact that incineration plants are expensive. Because of this reason burning technologies are considered as inappropriate method of waste management by environmentalists. Several incineration plants that were installed in Delhi and Lucknow have failed to deliver rated energy outputs and have since been closed. However, there have been success stories elsewhere and the Govt of India provides subsidy for installation of increaters which not only generate power but reduce the bulk of wastes to a small volume of ash.

Biomethanation Anaerobic Digestion

Methane is produced by anaerobic decomposition of landfill solid waste and it contributes towards global warming. Methane is 21 times more potent green house gas (GHG) than CO_2. Calorific value of landfill gases is app. 4500 Kcal/m^3. Biomethanation technique of using methane gas for combustion or electricity generation could also provide compost for soil conditioning. This technique can be effectively used in a developing country like India. The methane value estimated for wastes generated in six selected metropolitant cities in India is 2 million tonnes/annum to several times more. It is expected to increase to 39 million tonnes/annum by 2047.

Pelletisation of Refuse Derived Fuel (RDF)

This proven and tested technology has been widely implemented in Europe for disposal of MSW and this RDF is suitable for Indian cities. The pellets are a good coal substitute. The NOX and SO_2 emissions are less than what are emitted from coal burning and the calorific value of these pellets are 2500-3000 Kcal/Kg. A 6.6 MW electricity generation plant from incinerated cellets is satisfactorily running at Mahbobnagar H; derabad.

Incineration of MSW

Calorific value of MSW in India ranges from 600-1100 kcal/m^3 and 100 tons of raw solid waste can potentially produce 1-1.5 megawatt power. In Malaysia, where 80% of MSW contains food, papers and plastics with 55% of moisture contents incineration plants operate successfully. By incinerating 1500 tons of MSW day with an averageof calorific value of 2200 Kcal/kg one can produce 640 kW/day[7]. In Delhi Municipal Corporation of Delhi (MCD) has planned to produce 16-20 MW of electricity by first converting MSW to RDF and then incinerating it to produce electricity.

Gasification Pyrolysis

This is a proccess of destructive distillation. In this process, MSW is heated to 900-1000°C so that pyroligenous liquid/water gas can be used in internal combustion engine to produce electricity. This is a bit expensive technique but can be used in more effective manner in small scale production of electricity.

VERMICOMPOST MAY BE USED AS A POTENTIAL BIOFERTILIZER

For sustainable development in agriculture, it is essential to improve the soil organic carbon status that ha deteriorated under intensive chemical framing. Among the several bio-fertilizers, farmers may use vermicompost on account of its simple preparation technology, favourable climate support and easy availability raw materials. Farmers can use vermicompost by mixing with chemical fertilizers almost in every agro production for increasing the effectiveness of the chemical fertilizer. It performs better in floriculture, horticulture, houseplants, potting soil, fruits and vegetables productions. However, for any application, better result can be obtained if farmers apply the estimated amount after knowing its per cent of organic carbon and C: N ratio.

In spite of this remarkable achievement, recently, farmers are facing problems in harvesting crops for stagnant productivity and associated hazards of the technology. As it requires a huge amount of chemical fertilizer, pesticide, insecticide and even water, chemicals accumulate gradually into the soil that causes the deterioration of soil fertility and environmental quality. A review of over 300 published reports showed that most of the environmental impact indicators which have decreasing trends are floral diversity, faunal diversity, habital diversity, landscape, soil organic matter, soil biological activity, soil structure, soil erosion, nitrate leaching, pesticide residue. CO_2, N_2O, CH_3, NH_3, nutrient use, water use and energy use. However, organic farming system performed significantly better in all the above factors and performed worse in none[1]. For eradicating these ill effects of modern agriculture, peasants are considering growing food without chemical fertilizers and pesticides throughout the world. Meanwhile, the global organic-food market has already crossed 31 billion US $ by 2005 with an average annual growth rate of 20-25%. Though the production cost in the organic farming is slightly higher, yet it will be sustainable alternative on account of its biological, ecological and environmental supportive natural approaches. Government of India has already constituted a 'Task Force' to popularise the organic farming. According to the opinion of the task force, the application of organic manure is the only option to improve the soil organic carbon for sustenance of soil quality and future agricultural productivity.

Vermicompost as a Biofertilizer

Even earthworms that decompose the agricultural wastes naturally and supply organic manure in the form of cast, have also left from our fields now. Hence, the percentage of organic carbon in the harvested soil gradually decreases to such a minimum level that hampers the exchange of plant macronutrients. At this juncture we can reintroduce the earthworms into our field by preparing and using vermicast as a source of bio-fertilizer. It is biologically active mound containing thousands of bacteria, enzymes, remnants of plant materials and animal manures which were not digested by the earthworm. An important component of this dark mass is humus which is a complicated material formed during the breakdown of organic matter. It also provides many binding sites for plant nutrients e.g. calcium, iron, potassium, sulfur and phosphorous. It dissolves slowly rather than allowing immediate nutrient leaching. It has excellent soil structure, porosity, aeration and water retention capabilities, which can insulate plant roots from extreme temperatures, reduce erosion and control weeds.

Fig. 23.1. Preparation of Vermicompost by women folklores

Fig. 23.2.

Chemical Quality and Precautions

Casting contain 5 times the available nitrogen, 7 times the available potash and 5.5 times more calcium than that found in 15 cm of good top soil. Moreover, vermicomposting adds valuable attributes viz. water retention, texture, nutrient availability and an ability to fight soil-borne plant disease like root rotten[6]. In spite of these advantages, the use of vermicomposed either in pure or mixed with chemical fertilizer as a farm manure, is still rare because the technology has not gone into farmers' concept. Aforesaid activists grop in 'Matar-Benay Trust' campus has been experimenting the use of vermicompost in different crop production for the last five years. Out of curiosity the author has collected eight different vermicompost samples from different farmers,

farmers, co-operative and NGO's and estimated its percentage of carbon content and NPK values in *Shibaprasad Bandyopadhyay maati Parikshagar'* (soil testing laboratory) at 'Matar-Benay Trust campus, Hatsimul. Burdwan West Bengal. The results obtained are shown in the Table-1 below.

Table 23.4. Percentage of Different Macronutrients in Eight Vermicompost Samples

Sample	pII Organic Carbon	% of Nitrogen	% of	& of P_2O_5	% of K_2O
1.	6.60	3.83	0.40	0.074	0.301
2.	6.56	3.25	0.33	0.220	0.220
3.	6.54	2.66	0.27	0.160	0.280
4.	–	13.06	1.41	1.502	0.258
5.	7.22	15.30	1.60	0.405	0.960
6.	8.75	6.85	0.68	0.040	1.400
7.	8.53	3.64	0.36	0.100	0.570
8.	6.98	2.78	0.30	0.095	0.210

Table 23.1 relects the estimated results of eight different vermicompost samples which show a very high variation of carbon content (2.66 to 15.3%) and also the percentage of nitrogen content in the vermicompost sample, which is obvious. But the problem is whether farmers know the percentage o carbon, content and C: N ratio of the used vermicompost sample or not. Therefore, the nutrent content, particularly the C:N ratio of each vermicompost sample should be measured before applying it into the field even if the farmers themselves. Values of C:N ratio can be improved by using different animal manure as a feeding material to the eartworms, instead of using cow-dung only. In order to project this idea, some relevant data[7] regarding the animal manure are shown in the Table 23.2.

Table 23.5. Approximate Dry Matter and Fertilizer Nutrient Composition of Various Types of Animal Manure

Type of livestock	% of Dry matter	N NH_4–N	Nutrient (kg ton^{-1} in raw waste) Total-N	P_2O_5	K_2O
Swine	18	2.72	4.54	4.08	3.63
Beef cattle	15	1.81	4.99	3.18	4.54
Dairy cattle	18	1.81	4.08	1.81	4.54
Poultry	45	11.79	14.97	21.77	15.42

Table 23.2 indicates that certainly the percent carbon content, C:N ratio and NPK values of the vermicast will be higher by using the poultry litter as feeding material. This trend of higher NPK values is also reflected in the vermicast NPK values (Table-1). Intensive care is required for protecting the vermicompost pit from ant, birds, rat and direct sunlight. Shortly a large number of vermicompost products are coming into the market packed within beautiful containers without mentioning the percentage of organic matter and C:N ratio. Where as the farmers are now being advised to use vermicast as an organic manure to increase the effectiveness of the chemical fertilzer also, hence for the optimum benefit from the organic manure, information regarding bio-fertizer should be authenticated or verified thoroughly.

Vermicompost Application

It has been established that the earthworms are one of the most useful and active agents in introducing suitable chemical, physical and microbiological changes in the soil, thereby directly increasing the fertility of the soil. Vermicompost should be used, after checking its nutrient content and mixing with the chemical fertilizers in suitable proportions. Institute of National Organic Agriculture (INORA) has applied the vermicompost in different prportions with chemical fertilizer in sugarcance production at different places of Maharastra, Gujarat and Karnataka. It has been established that 1:1 vermicompost and chemical fertilizer ratio is most effective in sugarcane production. In a certain region of New Zealand, grass production became doubled after introducing some European earthworm species. Hidago *et. al.* of Mississppi State University have shown that the mixture casting with peat moss at the ratio of 1:2, 2:1 and 3:1 performed better in all growth parameters of Marigold producition. They have also shown that earthworm casting increases the germination rate in seeding development of Cucumber nicely. (Annexure-IB) In Bangalore. India, earthworms successfully decomposed residuals from sugar factory and turned them into a soil nutrient which enabled 50% reduction in the use of chemical fertilizer. Earthworm castings are the best imaginable potting soil for houseplants as well as gardening and farming. It can also be used as a planting soil for trees, vegetables, shrubs and flowers. It was also observed at the Matar-Benay' Thrust campus that vermicompost performed better in flowers, vegetables, fruits and nursery productions. It can also be used, by mixing with chemical fertilizers, in any agro-production for rasing the effectiveness of the chemical fertilizer.

ELECTRONIC WASTE

In the past few years, technological advances in electronics have boosted the economy and improved the general lifestyle of a common man. The ever growing dependence on electronic products has paved the way for an emerging environment concern, called "Electronic Waste". Every year, an estimated 100 million computers and other electronic devices break or become obsolete and are discarded. Electronic waste, or e-waste, is an emerging problem as well as a business opportunity of increasing significance, given the volumes of e-waste being generated and the content of both toxic and valuable materials in them. E-waste is a popular informal name for electronic products nearing the end of their "useful life". Computers, televisions, VCRs, stereos, copiers, mobile phones and fax machines are common electronic products. Many of these products can be reused, refurbished or recycled. Unfortunately, electronic discards is one of the fastest growing segments of our nations waste stream. E-waste has become a problem of crisis proportions because of two reasons. Firstly, E-waste is hazardous as the vast amount of computers, televisions, mobile phones and other electronic products that are disposed of every year all contain a variety of toxic substances. When discarded electronics are dumped in landfills, or when the waste is incinerated, contaminants and toxic chemicals are generated and released into the ground or air risking pollution of the environment and toxins entering the food chain are astronomical. Another reason is E-waste being generated at an alarming rate, due to fast obsolescence along with rapidly evolving technology.

Constituents of E-Waste

Electronic and Electrical equipment, consist of multiple components, some having toxic substances that can affect human health and environment, if not properly managed. Generally, these hazards occur on account of improper recycling and disposal methods adopted. There are number of harmful substances (metals) found in e-waste. *"Antimony"* is used primarly in flame proofing, paints, ceramics, alloys, electronics and rubber. Antimony is increasingly used as an alloy that greatly increases lead's hardness and strength. Its most important use is as a hardener in lead for storage batteries.

Antimony and its compound in small does cause headaches, dizziness and depression, while larger doses can cause violent and frequent vomiting *"Arsenic"* is found in small quantities in the form of gallium arsenide in light emitting diodes. Arsenic is a poisonous metallic element and chronic exposure to it can lead to skin diseases and lung cancer. *"Barium"* is a metallic element used in sparkplugs, fluorescent lamps and in vacuum tubes. It generates poisonous oxides in contact with air. Exposure to barium can lead to muscle weakness. liver and heart problems. *"Brominated Flame Retardants (BFRs)"* is used in the plastic housings of electronic equipment and in circuit boards to prevent flammability. More than 50% of BFR usage in the electronics industry consists of tetrabromo-bis-phenol—(TBBPA). 10% is polybrominated diphenyl ethers (PBDEs) and less than 1% is polybrominated biphenyls (PBB). *"Beryllium"* found in power supply boxes contain cancer causing agent for lungs. *"Cadmium"* is present in rechargeable NiCd-batteries, CRT screens, printer inks and toners. Cadmium components, absorbed through respiration, seriously affect the kidneys. *"Chromium"*, found in data tapes and floppy disks, is used because of high conductivity and anti corrosive properties. Chromium (VI) compounds are irritating to eyes, skin and mucous membranes. It can also result in DNA damage. *"Lead"* is found in CRT screens, batteries and printed circuit boards. It is also used in solder, lead acid batteries, electronic components and cable sheathing. Exposure to high amount of lead can result in vomiting, diarrohea, appetite loss, abdominal pain, constipation, fatigue and sleeplessness. *"Mercury"* is used extensively in fluorescent lamps, LCDs, alkaline batteries and mercury wetted switches. It is a toxic heavy metal, that bioaccumulates, causing brain and liver damage, if inhaled. Excessive exposure to *"Selenium"* used in photocopying machines, can cause selenosis, leading to hair loss, nail brittleness, and neurological abnormalities. *"Polyvinyl chloride"* PVC is an extensively used plastic in electronic appliances. PVC is harmful as it contains about 55% chloride, which when burned gives rise to hydrogen chloride gas. This combines with water to produce hydrogen chloride gas. This combines with water to produce hydrochloric acid and if inhaled can lead to respiratory problems.

There are a number of valuable substances in electronic wastes Gold Silver, Aluminium plastic etc. are prestigious materials, which recyclers recover from e-waste. *"Silver"* is used mainly in connectors and PWBs to provide conductivity. In mobile phones, it is typically used in the electronics and keypad contracts in the elemental form. *"Gold"* is used in marginal amounts in connectors and PWBs. It is used to provide conductivity and connectivity. Gold has no harmful effects on environment and humans and it can be obtained again after recycling process. *"Aluminium"* is present in almost all electronic products in large quantities. It also provides conductivity. Its less cost, compared to gold & silver and because of its easy recyclability, aluminium is used extensively.

Disposal of E-Waste

The e-waste that is generated can be recycled, reused, and disposed off in landfills or incinerators. The reuse and recycling of outdated electronic products minimize the hazardous effects of electronic waste on the environment. Reuse and recycling also boost energy and resource conservation. After all possibilities for reuse have been exhausted and a computer is slated for disposal, is sent for recycling. By this it is meant that the old raw materials are reclaimed for use in making new products. However, the costs of recycling are still very high due to which most recyclers are not very much willing to take computers for recycling. If the waste cannot be even recycled then it is either sent to landfills or is burnt in incinerators.

Dumping waste in to landfills contaminates the ground water and soil. Toxic chemicals in electronics products leach into the land over time or are released in the atmosphere. These toxic substances can migrate into ground waters, and eventually into lakes, streams, or wells, and raise a potential exposure to humans and other species by entering the food chain.

Burning electronic products into incinerators leads to the formation of toxic gases due to the presence of heavy metals such as lead, cadmium and mercury. Mercury released into the atmosphere can bioaccumulate in the food chain.

E-Waste and Industry

It is a proven fact that, all who produce, distribute, use and dispose of electronic products, have a moral responsibility towards managing electronic waste. Electronic equipment manufacturers should ensure that their products contain lesser toxic constituents, more recycled content and are designed for easy upgradation and disassembly. Initiatives have been taken by some leading companies in this regard.

- A new packaging technology is already in the offering at Anadigics, which will enhance its product's moisture sensitivity level.
- National Semiconductor reveals that most of its products are now lead free (except solder). It has also banned the use of cadmium, mercury and chromium in its products.

- Nokia has created a list of hazardous substance which will not be used in their products and has sent the same to its suppliers.
- The design of Hewlett-Packard's Office Jet 500 multi-purpose printer has elminated the need for plastic flame retardants by using a metal chasis and power supply enclosure, utilizes light-emitting diodes (LEDs) instead of a mercury lamp for the scanner, and eliminates the need for batteries by using flash memory technology.
- Apple's Macintosh Power Books have used longer-life, less toxic rechargeable lithium-ion batteries for the last three products generations, in place of nickel cadmium batteries.
- The primary plastic resin used in Intel's PCs and servers (ABS + Polycarbonate) have no flame retardants containing PBBs or PBDEs. None of their products contain asbestos, or include lead or cadmium as plastic additives.
- Panasonic is the first company to apply reflow type lead-free soldering to compact portable minidisc player PC boards. its video equipment division has been developing a low-cost tin-copper base solder.
- At Motorola, housings are made of standard engineering plastics. Several Motorola phone models have eliminated the use of brass inserts in their plastic housings.

Finding methods to keep electronic waste out of landfills is a challenge now for all the electronic product manufacturs, recycling and waste management organizations, government agencies and environment management organizations. Indian Govt. should draft legislation in this direction. If all consumers plan to phase out their obsolete computers and other electronic products at the same time, the country may face a tsunami of e-scrap. Ultimately, it is the balance of nature, which always holds supreme.

REFERENCES

1. S. Singhal. S. Pandey: *Teri Information Monitor on Environment Sciences.* 6.1: 1-4, (2001).
2. A. Messineo and D. Panno: Municipal waste management in Sicily: Practices and challenges. Waste Management doi: 10.1016/j.wasman.2007.05.003, (2007).
3. W. Chaya S. H. Gheewala: *Journal of Cleaner production.* 15: 1463-1468, (2007).
4. S. M. Aggelides, P.A. Londra, *Bioresource Technology,* 71, 253-259, (2000).
5. Gopal Krishna: Waste-to-energy or waste-to-pollution? 2005 http://ww.infochangeindia.org/agenda5-14.jsp.

   Government Bureau. Power generation from municipal waste. (2007).

   http://www.igovernment.in/site/power-generation-from-municipal-waste/
6. M. Rawat. K.U. Singh, K. A. Mishra, V. Subramanian: *Environment Monitoring and Assessment.* 137. 13, 67-74, (2008).
7. S. Kathirvale, MNM Yunus, K.S. Sopian, AH: *Renewable Energy.* 29. (559-567).
8. R. Basin MCD looks to cash in on biocompost Demand, *Times of India* 24 March p. 9. (2008).
9. M. Stolze, A. Piorr. A Haring and S. Dabbert, *Organic Farming in Europe: Economics and Policy,* University of Hohenheim. Hohentheim, Germany, (2000).
10. R. Prasad. *Current Sci.* 89. 252-254, (2005).
11. S.A. Ali. I. Khan and A.S. Ali, *Science Reporter,* 43, 1, 28-30, (2006).
12. M. Appelhof, *Worms Eat Garbage,* p.-68, (1982).
13. K. E. Lee, *Earthworms. Their Ecology and Relationship with Soils and Land Use.* New York: Academic Press, (1985).
14. C. A. Edwards. *Bio Cycle,* 36, 56-58, (1995).
15. S. L. Tisdale. W.L. Nelson, J. D. Beaton and J. L. Havlin, *Soil Fertility and Fertilizers,* 5th Edition p-598, Prentice Hall of India Pvt. Ltd. New Delhi-110001, (1995).
16. N. V. Joshi and B. Kelkar, *Ind. J. Agril. Sci.* 22, 189-196, (1951).
17. P. Hidalgo, F. B. Matta and R. Harkess.
18. R. E. Gaddie and D. E. Douglas. *Scientific Earthworm Farming.* Vol. I, 175, (1975).

CHAPTER 24

Thermal Pollution and Control

Dr. Suman Malik[1]

INTRODUCTION

The term Thermal Pollution has been used to indicate the detrimental effect of heated effluents discharged by various industrial plants. The heated effluents are discharged at a temperature 8 to 10°C higher than the temperature of intake waters which reduces the concentration of D.O. (Dissolved Oxygen) in water bodies.

Thermal Pollution can be defined as:

1. The warning up of aquatic system to the point where desirable organism are adversely affected.
2. Addition of excess of undesirable heat to water that makes it harmful to man, animal, plant or aquatic life or otherwise causes significant danger to the normal activities of aquatic communication in water.
3. Heated effluents either from natural or man made sources, contaminated with water supplies may be harmful to life because of their toxicity, reduction in Dissolved Oxygen (D.O.) aesthetically unsuitable and spread disease.
4. It reduces the number of aquatic species and destroys the life in streams as in evidenced by the biological indices of community and diversity.
5. It is a by-product of rapid and unplanned industrial progress and over-pollution.

What is Thermal Pollution?

Thermal pollution is a temperature change in natural water bodies caused by human influence. The temperature change can be upwards or downwards. In the Northern Hemisphere, a common cause of thermal pollution is the use of water as a coolant, especially in power plants. Water used as a coolant is returned to the natural environment at a higher temperature. Increases in water temperature leave impact on aquatic organisms by (*a*) decreasing oxygen supply, (*b*) killing fish juveniles which are vulnerable to small increases in temperature, and (*c*) affecting ecosystem composition. In the Southern Hemisphere, thermal pollution is commonly caused by the release of very cold water from the base of reservoirs, with severe affects on fish (particularly eggs and larvae), macro invertebrates and river productivity.

[1]Department of Chemistry, Sadhu Vaswani College, (Barkatullah University), Bhopal, (India)

The *combustion of fossil fuels* always produces heat, sometimes as a primary, desired product, and sometimes as a secondary, less-desired by product. For example, families burn *coal,* oil, *natural gas,* or some other fuel to heat their homes. In such cases, the production of heat is the object of burning a fuel. Heat is also produced when fossil fuels are burned to generate electricity. In this case, heat is a byproduct, not the main reason that fuels are burned.

Heat is produced in a number of other common processes. For example, electricity is also generated in nuclear *power plants,* where no combustion occurs. The decay of organic matter in landfills also releases heat to the *atmosphere.*

It is clear, therefore, that a vast array of human activities result in the release of heat to the *environment.* As those activities increase in number and extent, so does the amount of heat released. In many cases, heat added to the environment begins to cause problems for plants, humans, or other animals. This effect is then known as thermal *pollution.*

One example of thermal pollution is the development of urban heat islands. An *urban heat island* consists of a dome of warm air over an urban area caused by the release of heat in the region. Since more human activity occurs in an urban area than in the surrounding rural areas, the atmosphere over the urban area becomes warmer ... Soil erosion and deforestation are two other major factors that can lead to unnatural increases in temperature. Shore plants and trees help shade water, keeping temperatures in check by providing protection from sunlight. When plants are removed, not only does a body of water absorb more sunlight, soil held in place by root structures falls into the water. Soil erosion can cause river and lake beds to rise, creating shallower pools of water which heat more quickly.

One of the primary causes for concern with thermal pollution is the depletion of oxygen in the water. Fish rely on a certain amount of oxygen to survive underwater; they do not easily adapt to oxygen level shifts. The warmer water is, the less oxygen it can hold, making a flourishing ecosystem turn deadly for inhabitants in a short time. Additionally, shallow and warmer waters increase the production of oxygen consuming plants such as algae, which decrease oxygen levels even farther as they decompose. There are several ways to help mitigate the effects of thermal pollution in order to stabilize and sustain aquatic environments. Factories that use water as coolant can install cooling towers and ponds which return water to original temperature before releasing it back into the natural source. Ensuring that banks and shorelines have adequate vegetation to provide shade and retain soil can also help prevent thermal pollution.

The Nature and Origin of the Pollutant

Thermal pollution describes the introduction of waste heat into water bodies such as rivers, estuaries and coastal waters. Although other industries abstract water and then discharge it, the coastal power plants produce the greatest amount of water and thus waste heat. Coastal power stations of up to 1000MW capacity discharge up to SOcumecs ($m^3 s^{-1}$) at elevated temperatures. The effects of that discharge depend on the time and place of discharge and the characteristics of the receiving area. In particular, the effects depend on the ability to disperse and absorb the additional energy. Power stations with a ready supply of cold water, such as those on the coast and in estuaries, use this to condense the steam used to turn the turbines. This flow of direct cooling, as opposed to the use of cooling towers, returns water to an adjacent area up to 10 degrees centigrade above ambient. The fossil and nuclear fuel power plants require this cooling. The high cost-effective value of direct cooling explains the large number of directly cooled estuarine and coastal power plants worldwide.

Sources of Thermal Pollution

The accelerated development, rapid industrialization extensive population density have increased demand of thermal power plants. Human activities, today, are constantly adding pollutants to air and water at an alarming rate. The following sources contribute to thermal pollution:

1. **Nuclear Power Plants:** Nuclear power plants institute nuclear experiments and explosions, discharge a lot of unutilized trapped radio nuclides in to nearby water streams. Emission from nuclear reactors and processing instruments are also responsible for increasing the temperature of water bodies. Heated effluents from power plants are discharged at 10° C higher than the coolant receptor and severely affect the aquatic flora and fauna.
2. **Coal-Fired Power Plants:** Some thermal power plants ultimately discharge effluents having temperature difference of 15°C between effluent and water body. The thermal power plants utilize coal as fuel and they constitute the major source of thermal pollutants. The heated coils are cooled with water from nearby lake or river and discharge the hot water back to the receptor water body and thereby increasing the temperature of the nearby water. The heated effluents decrease the dissolved oxygen contents of water, it results into killing offish and other marine organisms.
3. **Industrial Effluents:** Industries engaged in generating electricity, such as coal powered and nuclear powered plants, need huge amount of cooling water for heat removal. Other industries such as textile, paper and pulp as well as sugar also release heat in water but to a much lesser extent. Normally the discharged water from stream - electric power industry using turbo generators, will have a higher temperature ranging from 6 °C to 9°C than the receiving water. In order to cope with the tremendous demand for electricity, the generating power of the installation will be raised which then require a large proportion of stream flow. To increase turbine efficiency, a partial vacuum is created at the turbine exhaust by cooling and condensation of the turbine stream. This causes an increase in the stream temperature to a level at which nature dissipation of heat will be inefficient.
4. **Domestic sewage:** It is usually a discharge into river, lake, canals or streams with or without waste treatment, the municipal sewage is normally having a higher temperature than the receiving water. The discharged sewage increases the stream temperature to a measurable extent and also results in numerous deleterious effects on aquatic biota with the increase in temperature of the receiving water, the dissolved oxygen content (D.O.) decreases and the demand of oxygen increase. Hence the anaerobic condition will be set up causing the release of foul and offensive gases in water. This rise in temperature will have a profound effect on water quality and aquatic life.

Harmful Effects of Thermal Pollution

A huge amount of information is available to show how temperature affects the life process offish. Most of the effects are attributed to the impact of temperature on the rate of metabolism, which is speeded up by heat according to the Van't Hoff principle stating the rate of chemical reaction increases with rising temperature. In general, the metabolic rate becomes almost double with each increase of 10 °C.

1. **Life Span:** All organisms are having temperature limits for their survival ability, but for most, acclimation would be possibly lethal. Hence a rapid change in temperature or sudden transference of fish to warm water may cause death at a temperature well below that regarded as lethal for the species.
2. **Respiration:** All increase in temperature gives rise to the disappearance of dissolved oxygen because of the lower solubility of oxygen at higher temperature as well as to an increase in the rate of respiration of fish. The oxygen consumption of most fish becomes four- fold as the temperature of water is raised to the maximum at which they are able to survive. At higher temperature, the hemoglobin in the blood of the fish is having a reduced affinity and hence becomes less efficient in delivering oxygen to the tissue. The combination of increased need and reduced efficiency at higher temperature can exert severe stress even on fish that are normally capable of living in a limited supply of oxygen. For instance, the hardy carp, which at water temperature of 1°C can survive at an oxygen concentration of 0.5 ppm requires a minimum of 1-5 ppm when the temperature gets raised to 35°C.
3. **Reproduction:** Temperature increase influences the reproduction of aquatic animals. An increase in temperature induces seasonal development of gonads and triggers females to deposit eggs. A rise

in temperature also decreases the time taken by the fish's eggs to hatch. For example fertilized eggs of Atlantic salmon are known to hatch in 114 days in water at 2 °C, but the period gets shortened to 90 days at 7 °C. Trout eggs hatch in 165 days at 3 °C while in 32 days at 12 °C. Excessive temperature, however, retards the normal development of eggs. Further, critical temperatures are known above 11 °C and 22 °C respectively.

4. **Heart Beat:** From experiments with crayfish, it can be seen that its heart beat tends to increase from 30 beats per minute at 22 °C and then tends to decrease to 65 beats per minutes as water temperature approaches 35 °C which is lethal temperature for the crustacean. The final decrease in heart beat will show the weakening of the fish under thermal stress.
5. **Swimming:** In general, the fish tend to increase their swimming speed and usually show more spontaneous movement with increase in temperature. Stockeye salmon in known to cruise twice as fast in water at 16 °C as it does in water at 2 °C. Above 16 °C, however its speed decreases. The decrease in speed at higher temperature affects their ability to obtain food. They slow down in pursuing minnows and at still higher temperature they are almost not capable of catching minnows.
6. **Food Assimilation:** The temperature of water exerts a pronounced effect on the assimilation of food in fish. It has been well established that digestion becomes fast at higher temperature. Tracer experiments with young carp have proved beyond doubt that at 10 °C the food needs 18 h to pass through the alimentary canal, whereas at 26 °C it needs only 4-5 h. since at higher temperatures the metabolic activity of the fish increases, a substantial part of the intake in warm waters is used mainly for maintaining the vital functions of body.
7. **Morphology:** Temperature difference during growth and development gives rise to morphological changes in aquatic organism. A variety offish, under the influence of thermal pollution, show agape jaws, eye abnormalities and twinning, Above 30 °C the incidence of spinal deformation becomes appreciably high.
8. **Synergism:** The toxicity attributed to insecticides, detergents, paper-mill effluents and other chemicals increases with increasing temperature. For example, the lethal effects of 5 and 10 ppm zinc sulfate to rainbow trout almost becomes doubled on raising the temperature from 13.5 to 21.5°C. At 1°C, carp is known to tolerate a carbon dioxide concentration of 120 ppm but at 36°C a concentration of 60 ppm is almost lethal. In the presence of cadmium salts the LC50 for galaxies maculates at 21°C has been 5-7 times lower than that at 15°C.

Thermal Effect on Marine Life

Temperature plays an important role in effecting the physiology, metabolism, growth and development of marine animals, for example, the American oyster C. Virginica ceases to feed at temperature below 7°C, the European oyster ostrea lurida tends to close its shells at low temperature; the oyster shell remains closed at 4-6°C and open for about 6 hours per day at 6-8°C while they remain open for 23 hours a day at 15°C. Ciliary activity which is responsible for their movement in water decreases at 32 °C and almost every body-function ceases at 42°C. Long exposure to high temperature (34°C) may impede oyster's normal rate of water circulation. Temperature change may cause acute toxicity in water so that oyster can not feed and subsequently lose weight and die.

Thermal Effect on Bacteria

Heated effluents from domestic, industrial and installation process pose following deleterious effect of bacterial growth:

- Coagulation of their body cell proteins
- Accelerated enzymatic reaction leading to exhaustion
- Melting of cells fats
- Reduction in permeability of cell membrane

- Increase in degradation rates, and
- Toxic action of metabolic product inhibiting cell division

Effect on Water Quality

Purification of water stream is actually an aerobic oxidation process. During this process, complex organic matter is converted into innocuous substances by bacteria.

At higher temperature, the rapid bacteria activity causes a burden on the dissolved oxygen resources of the water when the waste to disposed in water has a higher demand of oxygen than the supplies, then it may cause putrefaction in water bodies developing noxious conditions. It has been reported that water purification occurs more rapidly at high temperature. Colder the water entering a purification plant, higher will be the treatment cost. Moreover, warm water is best suited for laundering purpose. But warm water promotes undesirable growth of organism refracting in chocking the balance of entire aquatic system.

Factors Responsible for the Damage of Biotic Organisms

The discharge of thermal effluents affects severely the biotic life and micro-organisms present in water. The biotic damage is caused because of the following factors:

1. Aquatic organism like Juvenile fish, Planktons, fish eggs, larvae, algae and protozoa which pass through screens and condenser cooling system are extremely sensitive to abrupt temperature changes.
2. Biotic damage is caused by buffeting with condenser surface (commonly known as entrainment) in addition to exposed heat. The cooling system may acutely affect the concentration of dissolved oxygen, pH, the rate of biochemical reaction metabolism and physiological activities of marine organisms.
3. Mechanical buffeting in the narrow tubes is another cause of biotic damage.
4. The discharge of heated effluents results in massive fish killing. This could be due to synergism because of toxic biocides like copper, nickel, chromium and chlorine used to remove slimes.
5. Fish mortality may be observed due to bio-fouling agents used for condenser coolants, corrosion products like ferric oxide and hydroxide as well as use of preservatives.
6. When higher aquatic creatures of large fishes are drawn on to the screens, they Get suffocated due to burden on gill chambers and they die as a result of impingement.

Thermal Pollution Causes Global Warming

Over longer time-scales there is no net heat inflow to Earth since incoming solar energy is re-emitted at exactly the same rate. To maintain Earth's thermal equilibrium, however, there must be a net outflow equal to the geothermal heat flow. Performed calculations show that the net heat outflow in 1880 was equal to the geothermal heat flow, which is the only natural net heat source on Earth. Since then, heat dissipation from the global use of nonrenewable energy sources has resulted in additional net heating. In, e.g. Sweden, which is a sparsely populated country, this net heating is about three times greater than the geothermal heat flow. Such thermal pollution contributes to global warming until the global temperature has reached a level where this heat is also emitted to space. Heat dissipation from the global use of fossil fuels and nuclear power is the main source of thermal pollution. Here, it was found that one third of current thermal pollution is emitted to space and that a further global temperature increase of 1.8 jC is required until Earth is again in thermal equilibrium.

Prevention of Thermal Pollution

Thermal pollution from power plants and factories is relatively easy to control. Instead of discharging heated water into lakes and streams, power plants and factories can pass the heated water through cooling towers or cooling ponds, where evaporation cools the water before it is discharged. Alternatively, power plants can be designed or refitted to be more efficient and to produce less waste heat in the first place. In a process called *cogeneration,* the excess heat energy from generating electricity can be used in another manufacturing process that needs such energy. Where homes or other buildings are located near industrial plants, waste hot water can be used for heating—an arrangement often found in Scandinavian towns and cities, and proposed for use in China.

How to Control Thermal Pollution

Following are the means to reduce Thermal pollution :

1. Theoretically, when efficiency of any heat engine is equal to 1.0 then it will convert 100% of heat energy to mechanical energy. So there will be no loss of heat to the environment. This is practically impossible. Rather, we should aim at maximizing the efficiency of heat engines (steam, 1C, nuclear etc) so that heat loss is minimum.
2. Reduce mechanical friction in any rotating parts.
3. Avoid consuming energy more than necessity. Burn less coal, oil or gas.
4. Temperature signal conditioners accept outputs from temperature measurement devices such as resistance temperature detectors (RTDs), thermocouples, and thermostats. They then filter, amplify, and/or convert these outputs to digital signals, or to levels suitable for digitization
5. Industrial fans and industrial blowers and commercial fans and blowers are designed to move air and/or powders in industrial and commercial settings. Typical applications include air circulation for personnel, exhaust or material handling
6. If we can follow these, certainly we shall be successful in reducing thermal pollution and will be able to prevent the glaciers from melting and rising of sea levels. Thermal can be prevented very easily. Most of the people who cause thermal pollution are big factories which use the water then pour it back in the ocean. To prevent this all they have to do is just cool the water before they put it back in the ocean.
7. One of the major cause of Global warming is increasing concentration of Carbon dioxide, leading to more green house effect. On the other hand, green plants have got the capacity to absorb Carbon dioxide. In the photo synthesis plants take water, sunlight and carbon dioxide to produce their food. So, plant as many trees as possible. Massive plantation is the only solution for reducing global carbon dioxide level. Indirect effect of plantation is, It will reduce soil temperature, cause more rains, some of the carbon dioxide shall be dissolved in rain and shall go to the sea - which will.

Cooling tower at Gustav Knepper Power Station. Dortmund, Germany.

Industrial Wastewater

In the United States, thermal pollution from industrial sources is generated mostly by power plants, petroleum refineries, pulp and paper mills, chemical plants, steel mills and smelters. Heated water from these sources may be controlled with:

- Cooling ponds, man-made bodies of water designed for cooling by evaporation, convection, and radiation.
- Cooling towers, which transfer waste heat to the atmosphere through evaporation and/or heat transfer.
- Cogeneration, a process where waste heat is recycled for domestic and/or industrial heating purposes.

Some facilities use *once-through cooling* (OTC) systems which do not reduce temperature as effectively as the above systems. For example, the Potrero Generating Station in San Francisco, which uses OTC, discharges water to San Francisco Bay approximately 10° C (20° F) above the ambient bay temperature.

Urban Runoff

During warm weather, urban runoff can have significant thermal impacts on small streams, as storm water passes over hot parking lots, roads and sidewalks. Storm water management facilities that absorb runoff or direct it into groundwater, such as bioretention systems and infiltration basins, can reduce these thermal effects. Retention basins tend to be less effective at reducing temperature, as the water may be heated by the sun before being discharged to a receiving stream.

Productive use of Heated Effluents

The heated effluents discharge from the frost chemical industries and thermal power plants can be put into certain beneficial uses such as:

1. **Green Houses:** The air of the green house can be warmed by using it as mechanism of cooling. The water is passed through vertical tower in the green house through which the air passes.
2. **Agriculture:** The use of heated water can be made in agriculture particularly for frost protection during cold season, plant cooling during hot period's soil heating and irrigation soil heating is normally performed by passing the water through underground networking of pipe system near the root zone which brings optimum temperature for the crop growth. The application of heated water can be made for keeping the temperature optimum in animal houses for maximum growth.
3. **Aquaculture:** Thermal discharge has been successfully used in aquaculture to increase the yield offish in many countries. The optimum temperatures for growth of many fishes/species are in the range of the thermal discharge.
4. **Heating:** The thermal discharge from an industry or power plants is also used for heating the home building or other such structures. This is achieved by circulating the hot water through pipes in the structure.

REFERENCES

1. Science Progress, March 22, 2003, Russell, A.D.
2. Environmental Encyclopedia, 2003, Newton, David E.
3. Sci. Prog. 86:115-137 [c] 2003 Science Reviews.
4. Impacts of climate change on Australian marine life report, Dr Alistair Hobday.
5. Thermal Pollution August 5, 2009. By John Smith.
6. Brown, Richard D.; Ouellette, Robert P.; and Chermisinoff, Paul N. (1983). *Pollution Control at Electric Power Stations: Comparisons for U.S. and Europe.* Boston: Butterworth-Heinemann.
7. Henry, J. Glenn, and Heinke, Gary W. (1996). *Environmental Science and Engineering.* Upper Saddle River, NJ: Prentice-Hall.

8. Hinrichs, Roger A., and Kleinbach, Merlin. (2001). *Energy: Its Use and the Environment,* 3rd edition. Monterey, CA: Brooks/Cole Publishing Company.
9. Langford, Terry E. (1990). *Ecological Effects of Thermal Discharges.* New York: Elsevier Applied Science.
10. Larminie, James, and Dicks, Andrew. (2000). *Fuel Cell Systems Explained.* New York: John Wiley & Sons.
11. Liu, Paul Ih-fei. (1997). *Introduction to Energy and the Environment.* New York: John Wiley & Sons.
12. Ristinen, Robert A., and Kraushaar, Jack J. (1998). *Energy and the Environment.* New York: John Wiley & Sons.
13. Slovic, Paul. (2000). *The Perception of Risk.* London: Earthscan Publications Ltd. Selna, Robert (2009). "Power plant has no plans to stop killing fish." *San Francisco Chronicle,* January 2, 2009.
14. U.S. Environmental Protection Agency (EPA). Washington, D.C. "Cooling Water Intake Structures - Basic Information." June 2, 2008.
15. EPA. "Technical Development Document for the Final Section 316(b) Phase III Rule." June 2006. Chapter 2.
16. EPA (1997) Profile of the Fossil Fuel Electric Power Generation Industry. (Report). Document No. EPA/310-R-97-007. p. 24.
17. California Environmental Protection Agency. San Francisco Bay Regional Water Quality Control Board. "Waste Discharge Requirements for Mirant Potrero, LLC, Potrero Power Plant." Order No. R2-2006-0032; NPDES Permit No. CA0005657. May 10, 2006.
18. EPA. "Preliminary Data Summary of Urban Storm Water Best Management Practices." August 1999. Document No. EPA-821-R-99-012. p. 5-58.
19. Michael Hogan, Leda C. Patmore and Harry Seidman, *Statistical* Prediction of Dynamic *Thermal Equilibrium Temperatures* using Standard *Meteorological* Data Bases, U.S. Environmental Protection Agency Office of Research and Development EPA-660/2-73-003, August, 1973.
20. E.L. Thackston and F.L. Parker, Effect of *Geographical* Location on *Cooling Pond* Requirements Vanderbilt University, for Water Quality Office, U.S. Environmental Protection Agency, Project no. 16130 FDQ, March 1971.
21. Edinger, J.E.; Geyer, J.C (1965). *Heat Exchange in the Environment.* Edison Electric Institute, New York City, N.Y.
22. Edward A. Laws, *Aquatic Pollution: An Introductory Text,* John Wiley and Sons (2000) ISBN 0-471-34875-9.
23. Nuclear And Thermal Pollution, by P. R. Trivedi.
24. Environmental Radiation And Thermal Pollution and their Control, by Chhatwal.
25. Environmental Pollution Control by Linda Goldstein.
26. Technology Of Environmental Pollution Control by Esber Saheen.
27. Environmental Pollution Management And Control by Kamlesh Mahanto.

CHAPTER 25

Soil, Composition, Pollution and Management

Dr. Charanjit Kaur[1]

The word soil is derived from the Latin word "solum" which means floor or ground. Soil is a three phase system of liquid, solid and air phases with an indefinite number of components- mineral matter, organic matter, soil water and soil air etc.

Soil water and soil air are dependent variables and are controlled by amount and character of mineral matter and organic matter in soil. Solid and liquid phases are always in equilibrium. The equilibrium does not remain constant but always changes due to variation of temperature, water content, removal of nutrients by plant roots and the activities of micro-organisms.

"Soil is a natural body, differentiated into horizons of mineral and organic constituents, usually unconsolidated, of variable depth, which differs from parent material in morphology, physical properties and constitution, chemical properties and composition and biological characteristics."

Composition of Soil

Soil is a complex biological system comprising of organic matter, inorganic matter, water and air in various proportions. Approximately 50 per cent of the total volume of the surface horizon of many soils is made up of inorganic matter and organic matter and the remaining volume is pore space between the soil particles. Water and air occupy these pore spaces in various proportions. The proportion of air and water varies from one season to another.

Mineral Matter

The principal constituents of soil are sand, silt and clay. The distinction between these components is rather arbitrary and is based on the particle size:

Table 25.1.

Type of the mineral particle	Particle diameter, mm
Stone and gravel	2 and above
Coarse sand	0.2 – 2.0
Fine sand	.02 – .2
Silt	.002 – .02
Clay and colloids	Less than .002

[1]Department of Chemistry, Sri Sathya Sai College for Women, (B.U.), Bhopal (India)

Particles larger than 2 mm diameter are called gravel sand particles in the range .02-2.0 mm in diameter are mainly composed of quartz and silica. Particles of silt in the diameter range .002-.02 mm are mainly made up of quartz and other silicate minerals. Particles of diameter less than .002 mm consist of a variety of silicates, aluminosilicates and colloidal substances. Depending on the proportions of sand, silt and clay present, a soil is described as sandy, sandy loam, clayey loam, clayey etc.

Organic Matter

The organic matter content of a soil is generally 2-5 per cent. Partially decomposed organic matter present in a soil is referred to as "humus" and the process of formation of humus from plant and animal remains after decomposition by micro-organisms is called "humification". During this process, cellulosic materials disappear, lignins are modified while proteins are retained. In spite of small quantity present in the soil, humus plays a very vital role in maintaining the porosity of the soil, its nutrients and water retaining capacity and the fertility of a soil. Humus also helps to bind the soil particles in aggregates of smaller or larger size which helps in restricting soil erosion. The top layer of soil consists of organic matter such as bacteria, algae, fungi, protozoa, nematodes. worms, moluscs, arthopods etc. The organisms present in the soil are called "Edaphons". The amount of organic material in the soil depends on climatic conditions, the type of inorganic constituents present and the topography. In temperate regions of the world where decomposition is slow, plenty of organic debris collects in the soil. However, in tropical or subtropical regions, warm and humid conditions cause quick decomposition and mineralization of organic matter.

Soil Water and Soil Atmosphere

Soil water and soil air fill up the space in between the soil particles. The quantity and the movement of water and gases in the soil is determined by the size of the soil particles. Soils with a higher percentage of smaller particles (e.g. Clayey soils) have more inter-particle space, in spite of the smaller pore size. Soils with a higher proportion of larger particles (e.g. Sandy soils) have little interstitial space, in spite of the larger pore size. The water-holding capacity of sandy soils is low although most of the water present is available to the plants. On the other hand, the water-holding capacity of clayey soils is higher but mostly it is held tightly by the minerals present in the microscopic pores and is unavailable to the plants. The water present in the soil acts as a solvent for dissolving nutrients and depends upon the pH of the soil solution. It is due to this reason that pollutants that affect pH of the soil water indirectly affect the availability of nutrients to the plants. Soil water also helps in maintaining soil texture, arrangement and compactness of soil particles and makes the soil as a perfect habitat for plants and animals. Soil solution is dilute solution of various dissolved solids, liquids and gases. It is held in the soil by capillary or absorptive forces between soils/layers and soil particles. The soil water may be gravitational water, capillary water, and hygroscopic water or combined water.

Some space between soil particles is also occupied by the "Soil air" which is very vital for the oxidation of organic materials in the soil. Roots of higher plants and some soil microbes also required for their growth and activity. In case of water-logged soils, absence of oxygen promotes anaerobic activity. Soil atmosphere consists of more carbon dioxide and moisture and less oxygen content, which is due to decay of organic materials. A loam soil containing 66 per cent of water and 34 per cent of air is preferred for several crops.

Texture

Soil texture refers to the relative proportions of various particle sizes found in a soil as *gravel, sand, silt and clay.* Texture should not be confused with structure. While texture denotes the size of individual soil particles, structure refers to the manner in which these individual soil particles are grouped together to form clusters of particles called aggregates. Gravel in soil is separated by screening before determining the texture. Sand, silt and clay are most commonly found in the solid state and are called soil separates. Clay has finest grade of particles and sand the coarsest.

Sand : Sand particles consist of small pieces of primary unweathered rock fragments. They are the heaviest and coarsest of the mineral particles. They do not hold water well, due to large pore spaces and do not stick together. Sand and gravel particles do not supply any nutrient for crop growth as they do not store nutrients.

Silt : Silt particles are bigger than clay particles but smaller than sand particles. They can hardly be seen by the naked eye. They hold plant nutrients better than sand but not as well as clay. Silt particles allow water and air to pass through readily, yet retain moisture for crop growth. However, silty soils are low in organic matter and surface crusts are easily formed after rains.

Clay : Clay particles are the smallest of all particles found in soil. They cannot be seen by the naked eye. Many of them are too small to be seen under the microscope. However, they are visible under the ultramicroscope. They are the most active part of the soil and hold plant nutrients well. Clay has a high moisture retention capacity. However, a soil with large amount of clay may present difficulties and turn out to be impermeable to air, water and plant roots. Clay is sticky when wet, dries out slowly and is hard to work with and may become cloddy when cultivated.

The percentage compositions of sand, silt and clay decide the basic soil textural class.The various textural classes that are found in soil are given in the following table.

Table 25.2

Group	Soil Textural Classes
Fine textured	Clay Silty clay Sandy clay
Moderately fine textured	Silty clay loam Clay loam Sandy clay loam
Medium textured	Loam Silty loam Silt
Coarse textured	Sandy loam Loamy sand
Very coarse textured	Sand

*If gravel is an important part of the soil, its presence is indicated by adding the word 'gravelly' to the textural class name, e.g., gravelly loam.

Soils contain a good mixture of all soil separates. The amount of different soil separates can be determined in two ways:

1. By separating them from each other in the laboratory by mechanical analysis.
2. By estimating the texture by feeling the soil between the thumb and the forefinger.

This is not as accurate a method as the laboratory method, but an experienced worker can find out the texture by the feel method pretty accurately. The feel method is commonly used by soil surveyors, agriculture teachers, farmers and students. Moisten the sample of soil to the consistency of a firm workable putty for determining the texture by the feel method. Make a ball about one cm in diameter from this sample. Keep the ball between the thumb and the forefinger, gradually press the thumb forward, forming the soil into a ribbon. .

If the ribbon forms easily and remains long and flexible, the sample is probably clay or silty clay, that is fine textured. These soils are sticky and plastic when wet and hard when dry.

If the ribbon forms but tends to break into pieces of not more than two cm in length, the sample is probably a silty clay-moderately fine textured. These soils are moderately sticky and plastic when wet.

If the ribbon is not formed, the sample is perhaps a loam, silty loam, sandy loam, loamy sand or sand-this group could be medium textured or very coarse textured depending upon whether it is silt or sand that is dominant in the sample. If the soil feels smooth and is like talc and has no grittiness, silt predominates and the soil is called medium textured. while if the soil feels smooth and is like talc but is slightly gritty, it is perhaps loam or silty loam and is also included in the medium textured group. If there is no smoothness and the feeling is gritty, it is an indication that sand predominates and the soil is coarse textured.

If the soil consists of almost entirely of gritty materials with little or no fine material in it. it is sand and is called very coarse textured. Soils containing much gravel and very little fine material also fall in this category.

When the soil separates are determined in the laboratory, gravel is screened and the amounts of sand, silt and clay are found by mechanical analysis of soil. After determining the relative amounts of sand, silt and clay, the soil textural class be determined by using the texture triangle.(fig. 17.1).

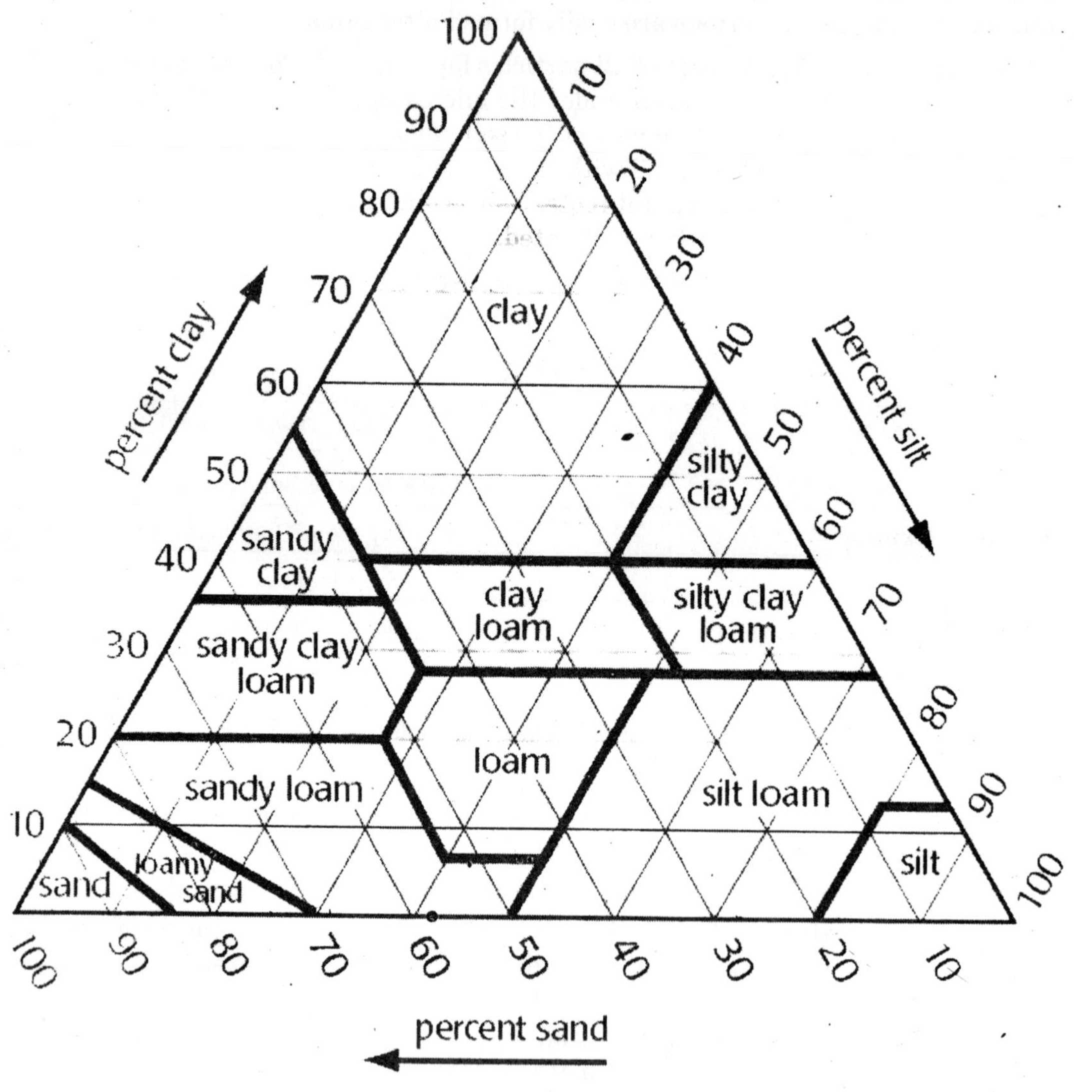

Fig. 25.1

Suppose the amounts of sand, silt and clay are 40 per cent, 40 per cent and 20 per cent respectively. Read the figures of 40, 40 and 20 on the texture triangle against the scales of sand, silt and clay respectively and see where the three lines emanating from these figures meet inside the triangle. They meet in the area of LOAM. Hence the soil is loam.

A knowledge of the texture helps in the management of soils. Coarse textured soils are very permeable and allow the movement of air and water between the soil particles very readily, but they do not hold and store plant nutrients or water.very well and tend to be droughty and infertile. Fine clay particles hold most of the plant nutrients, but a soil consisting predominantly of clay is not permeable to air, water and plant roots. When plant roots are not able to grow well, the yield will be reduced. Medium textured soils are most desirable because they allow the movement of air and water readily and hold the plant nutrients and moisture well. If the soil is coarse textured, more fertilizers will be needed and it would be necessary to increase the organic matter content to help hold water and nutrients. In coarse textured soils frequent light irrigation may be more useful as water is not held for long periods. In heavy clay soils with poor aeration, it may be necessary to plan a proper drainage system. Fine textured soils should not be worked upon when they are too wet, as they may become compact and cloddy, whereas the sandy soils do not present this problem. The texture of surface and subsurface soils should both be taken into consideration while judging the value of land for crop production.

Major Types of Soil in India

The major types of soil occurring in different parts of our country are given in the table

Table 25.3

S. No.	Types of Soil	Remarks
1.	Alluvial Soil	Occurs in the great northern planes of India and deltas of rivers in peninsular India.
	(a) Khadar	Very fine and new alluviam. Very fertile.
	(b) Bangar	Relatively coarse and old alluviam. Relatively less fertile.
2.	Mountain or hill soils and forest soils.	Rich in organic matter (humus), occurs in eastern and western ghats, hills in central India and northern hilly regions of Himalayas.
3.	Black soils	Originate from volcanic rocks. Highly fertile clayey soils. Occurs in parts of M.P., Tamilnadu, Gujarat and Deccan trap regions of Maharashtra.
4.	Red soils	Relatively less fertile. Deficient in organic and nitrogenous matter. Occurs in plateaus of Kerala. Karnataka, A. P., Tamilnadu, Orissa and Southern parts of Bihar.
5.	Lateritic soils	These are poorly fertile soils but can support pastures and scrub forests. These soils occur in regions of tropical rainy climate such as western ghats, chota Nagpur plateau, Assam, Orissa, A. P., Tamilnadu and Kerala.
6.	Desert soils	These are arid sandy soils with low moisture and low humus content. They occur mostly in Rann of Cach, Southern Punjab, Western Rajasthan and Hryana.

Soil Pollution

Soil pollution is defined as the contamination of the soil system by considerable quantities of chemical or other substances, resulting in the reduction of its fertility or productivity with respect to the qualitative and quantitative yield of the crops. However, if some of the contaminants are such that if they are taken up by the plants and enter into the food chain and impart detrimental or toxic effect on the consumers, then that also should be treated as soil pollution.

Sources of Soil Pollution

Soil pollution is receiving greater attention as soil pollutants remain in direct contact with the soil for relatively longer and hence alter the chemical and biological properties of the soil. The hazardous chemicals can also enter the human food chain from soil or water plants.

Soil pollution mainly results from the following sources:

1. Industrial wastes
2. Urban wastes
3. Radioactive pollutants

4. Agricultural practices
5. Chemical and metallic pollutants
6. Biological agents
7. Mining
8. Resistant objects
9. Soil sediments. .

Effects of Soil Pollutants

Soil pollution is receiving greater and greater attention due to its direct impact on public health. The major effects of various types of pollutants are given below:

1. Effect of Modern Agricultural Practices

Synthetic Fertilizers: Synthetic fertilizers are employed to increase the soil fertility and crop productivity. These fertilizers concentrate the essential nutrients in layer of top soil. However, the soil enriched by chemical fertilizers cannot support the microbial flora which is so essential to enrich the humus that helps in plant growth. Excessive and indiscriminate use of chemical fertilizers may result in the following undesirable effects:

(*i*) Wheat, maize, corn, etc, grown on soils fertilized with NPK fertilizers may result in considerable reduction in protein content of the crop,

(*ii*) Excessive use of nitrogenous fertilizers leads to the accumulation of nitrates in the soil which may contaminate the ground water. Nitrate concentrations exceeding 90 ppm in drinking water may lead to diarrhea, blue Jaundice in children, "methemoglobinemia" (or blue baby syndrome) in infants. Further, the nitrates and nitrites entering the human body may be eventually converted into nitroso amines and other nitroso-compounds which are suspected to cause stomach cancer. Surveys in Rajasthan and other parts of the country indicate much higher nitrate levels than the permissible 45 ppm levels,

(*iii*) Vegetation growth in nitrate-rich soils may exert toxic effects in cattle,

(*iv*) Excessive use of chemical fertilizers may enter the water bodies and contribute to "eutrophication". (Eutrophication is the excessive growth of algae and aquatic plants to undesirable levels,

(*v*) Excessive use of chemical fertilizers may reduce the ability of plants to fix nitrogen,

(*vi*) Excessive quantities of potassium fertilizers in soils may reduce the quantities of valuable ascorbic acid (Vitamin C) and carotene in fruits and vegetables grown in such soils,

(*vii*) The large-sized fruits and vegetables grown in highly fertilized soils may be more vulnerable to attacks by pests and insects.

Pesticides: As per the reports of the World Health Organization (WHO), about 50,000 people in developing countries are poisoned and about 5,000 people die because of improper use of pesticides and other chemicals in modern agricultural practices.

Pesticides pose potential hazard to animals, humans and aquatic life. They also cause deleterious effect on soil fertility and crop productivity. Pesticides applied to crops are retained in the soil in considerable quantities. They enter into cyclic environmental processes such as absorption by soil, leaching by water, etc and contaminate both lithosphere and biosphere. Pesticides including herbicides, fungicides and rodenticides, are persistent pollutants. Owing to interactions between lithosphere and biosphere, pesticides may enter the food chain and pose serious health hazards. Some of them undergo metabolic transformation and bio-degradation. The degradation products of some of the pesticides are more dangerous than their respective parent compounds. Some of the pesticides residues are carcinogenic while their metabolic products too are toxic. The rate of degradation of pesticides depends upon their properties and structural characteristics.

The following types of pesticides are commonly used:

(*i*) Chlorinated hydrocarbons (eg. DDT, Aldrin, Dieldrin, Lindane. BHC etc.)

(*ii*) Carbamate (eg. Carboryl or Sevin, Zectrion etc.)

(*iii*) Organo-Phosphorous compounds (eg. Methyl or ethyl parathion, melathion, guthion etc.)

(*iv*) Inorganic compounds (eg. As_2O_3, PbO_2, $NiCl_2$, $CuSO_4$ etc.)

(*v*) Miscellaneous compounds (eg. Organic mercurials, 2,4D; 2,4,5T etc.)

Some of of the adverse effects of pesticides are given below:

(*i*) Some arsenic pesticides may render the soil permanently infertile.

(*ii*) Pesticide residues in soil may be taken up by plants and cause phytotoxicity. They may enter the aquatic environment and enter the food chain.

(*iii*) Pesticides such as, endrin, dieldrin, DDT, heptachlor etc may seep through the soil and contaminate drinking water supplies.

(*iv*) Fruit, vegetables, rice, wheat, barley, maize etc are known to contain considerable quantities of toxic pesticide residue such as of DDT, BHC and other organochloro pesticides.

(*v*) Polychlorinated biphenyls (PCB) having half-life periods of about 25 years in soil are among the most hazardous soil pollutants. They may accumulate in soil and plants and when they eventually enter the animal or human body, they may cause severe health disorders including eye damage, skin problems, nervous disorders, foetus deformities and liver or stomach cancer.

(*vi*) Irrigated water from pesticide contaminated soils may evaporate and spread the toxic pesticide vapors in the atmosphere.

(*vii*) DDT can enter the food chain and accumulate in human fats and may lead to disorders such as impotency.

(*viii*) Persistent pesticides can damage human tissues and interfere with the normal metabolic activities by disturbing enzymatic functioning.

(*ix*) Chlorinated pesticides and herbicides are hazardous soil pollutants which can affect the soil texture and damage the ecosystem.

(*x*) Herbicides such as dioxin may cause congential birth defects in offsprings.

(*xi*) Hunting birds feeding on grains contaminated with DDT are threatened of extinction.

(*xii*) Organophosphate pesticides may cause muscular disabilities, tremors and dizziness.

(*xiii*) Excessive use of synthetic pesticides may lead to defoliation of forests and adverse effect on fauna and flora.

(*xiv*) Farm animals drinking stagnant water in fields sprayed by pesticides developed toxic symptoms and some mortalities were reported.

(*xv*) Farmers and farm workers are particularly prone to pesticide poisoning because of greater exposure while handling and spraying.

(*xvi*) Accidental spillages and leakages in pesticides manufacturing industries cause disastrous effects on the people residing in nearby areas due to pollution of air, water and soil. The Bhopal tragedy on 3rd December, 1994 is a lingering example.

(*xvii*) Contaminated soils may act as potential carries of pathogenic bacteria and other dangerous organisms which may endanger human health.

(*xviii*) Volatile pesticides may cause pollution of air in the surrounding areas.

2. Effects of Industrial Effluents

Solid, liquid and gaseous chemicals from various industries such as paper and pulp, iron and steel, fertilizers, dyes, automobiles, pesticides, tanneries, coal-based thermal power plants etc contain a variety of pollutants such as toxic heavy metals, solvents, detergents, plastics, suspended particulates, and refractory/non bio-degradable/ recalcitrant chemicals. If they are not properly treated at source, they give rise to water, air and soil pollution. Fly ash resulting from coal-based thermal power plants is one of the alarming and continuously increasing sources of soil pollution leading to degradation of soil, apart from water and air pollution in the nearby areas. Some trade wastes such as tannery wastes may contain pathogenic bacteria.

Indiscriminate dumping of untreated or inadequately treated domestic, mining and industrial wastes on land is an important source of soil pollution. Fall out of gaseous and particulate matter from mining and smelting operations, smoke-stacks, etc are major source of soil pollution in nearby areas.

3. Effects of Urban Wastes

Millions of tones of urban waste are produced every year from critically polluted cities. The inadequately treated or untreated sewage sludge not only pose serious health hazards but also pollute soil and decrease its fertility and productivity. Other waste materials such as rubbish used plastic bags, garbage, sludge, dead animals, waste medicines, hospital wastes, skins, tiers, shoes, cans, etc also cause land and soil pollution. Some solid wastes may cause clogging of ground water filters. Suspended matter present in sewage can act as a blanket on the soil and interfere with its productivity.

Apart from the above major sources, radioactive waste dumped in the soil from natural and man-made sources, soil erosion due to deforestation, unplanned irrigation and unscientific agricultural practices also result in land and soil pollution.

Remedies and Management

The major sources of soil pollution are the domestic wastes, industrial wastes, and agricultural wastes including those toxic chemicals arising from modern agricultural practices. The various approaches to control and manage the soil pollution are as follows:

1. Implementing stringent and proactive population control programmes.
2. Launching extensive afforestation and community forestry programmes.
3. Implementing deterrent measures against deforestation.
4. Formulation of stringent pollution control legislation and effective implementation with powerful administrative machinery.
5. Imparting informal and formal public awareness programmes to educate people at large regarding the health hazards and undesirable effects due to environmental pollution. Mass media, educational institutions and voluntary agencies should be involved to achieve these objectives.
6. Banning the use of highly toxic and resistant synthetic chemical pesticides or at least regulating their use only for special purposes under thorough monitoring.
7. Encouraging the use of bio-pesticides in place of toxic chemical pesticides.
8. Conservation of soil to prevent the loss of precious top soil from erosion and to maintain it in a fertile state for agricultural purposes.
9. Transforming intensive agriculture into a sustainable system by measures such as
   - Maintaining a healthy soil community in order to regenerate soil fertility by providing organic manures, increasing fallow periods, avoiding excessive use of chemical fertilizers and pesticides.
   - Infusing bio-diversity in agriculture by sowing mixed crops, crop rotation etc.
10. Effective treatment of domestic sewage by suitable biological and chemical methods and adopting modern methods of sludge disposal.
11. Municipal wastes have to be properly collected by segregation, treated and disposed scientifically in land fills. Recycling and reuse of materials should be done wherever possible, such as recycling of glass, plastics, paper and production of biogas.
12. Industrial wastes have to be properly treated at source, by segregation of wastes and adopting integrated waste treatment methods. Proper care should be taken in treating heavy metal wastes and other obnoxious waste materials.

13. Security land-fills have to be constructed for permanent disposal of hazardous and recalcitrant industrial wastes.
14. Sponsoring more intensive R & D efforts on bio-fertilizers, bio-pesticides, utilization of wastes by recovery, reuse and recycling processes and safer treatment and disposal of hazardous waste.
15. Enforcing environmental audit for industries and promoting eco-labeled products.
16. Avoiding excessive use of chemical fertilizers and insecticides and providing more organic manures to the fields and thereby maintaining healthy biota. This in turn regenerates soil fertility. A soil rich in organic matter also helps in controlling soil erosion.

REFERENCES

1. A. M. Dix, Envtal Pollution, John Wiley, New York (1980).
2. D' Itri, P.M. (Ed.). A Systematic approach to conservation Tillage (Chelsea, MI: New Publishers, Inc) 1985.
3. F.E. Bear, Chemistry of the Soil, Reinhold Publishing Corp, New York (1955).
4. G.H. Bolt and M.G.M. Bruggenwart (Eds.) Soil chem., Elsevier, New York (1976).
5. Larson, W.E., S. C. Gupta and R.A. Useche. Compression of agricultural soil from eight soil orders. "Soil Sc. Soc. Ame J., 44:450-57. (1980)
6. Lyon, T.L. and H.O. Buckman. The Nature and Properties of Soils. 5th edition (N.Y. Macmillan) 1952.
7. P.R. Hesse, A Textbook of chemical Analysis, John Murray, London (1971).
8. S. E. Manahan, Envtal chemistry, 8th edition, CRC Press, Boca Raton, Fl, (2005).

CHAPTER 26

Soil Reactions and Fertility with Reference to N, P, K

Dr. Charanjit Kaur[1]

Soil Reactions

One of the most important physiological characteristics of the soil solution is its reaction. The presence and development of micro-organisms and higher plants depend upon the chemical environment of soil and that is why the study of soil reaction is so important in the study of soil science. There can be three types of soil reactions:

1. Acidic
2. Alkaline &
3. Neutral

Soil acidity is common in regions where precipitation is high enough to leach appreciable amounts of exchangeable basses from the surface layers of the soils so that the exchange complex is dominated by hydrogen ions. Acid soils, therefore, occur widely in humid regions and effect the growth of plants markedly.

On the contrary, alkaline soils occur when there is a comparatively high degree of base saturation. Salts like carbonates of calcium, magnesium and sodium also give a preponderance of OH^- ions over H^+ ions in the soil solution. When salts of a strong base and weak acid such as sodium carbonate go into solution and hydrolyse, consequently they give rise to alkalinity. The reaction is as follows:

$$Na_2CO_3 \longrightarrow 2Na^+ + CO_3^{2-}$$

$$2Na^+ + CO_3^{2-} + 2HOH \longrightarrow 2Na^+ + 2OH^- + H_2CO_3$$

Since sodium hydroxide dissociates to a greater degree than the carbonic acid, OH^- ions dominate and give rise to alkalinity which may be as high as pH 9 or 10. Alkaline soils are most commonly found in arid and semi-arid regions. In regions where H^+ ions balance OH^- ions neutral soils occur.

Nutritional Importance of Soil pH

Hydrogen ions have considerable influence upon the solubility of nutrients like iron, manganese and zinc. These nutrients become less soluble as the pH increase from 4.5 to 7.5 or 8.0. Moreover, the ease with which these nutrients when already in solution are absorbed and utilized by plants depends upon the pH of the soil solution, e.g., at pH 6.5 or 7.0 ammonium salts are utilized more readily by plants and

[1]Department of Chemistry, Sri Sathya Sai College for Women, (B.U.), Bhopal (India)

in moderate to strongly acidic soils nitrate ions are absorbed with greater ease. Similarly, phosphorus is held with less tenacity in soils al pH around 6.5.

Thus, the soil reaction is correlated with:

1. Exchangeability of calcium and magnesium.
2. Solubility of iron, aluminium and other micronutrients.
3. Availability of phosphorus.
4. Activity of micro-organisms.

1. **Exchangeability of Calcium and Magnesium :** As the exchangeable calcium and magnesium are lost by leaching, acidity develops and in humid regions there is a definite relationship between the soil pH and the amount of calcium and magnesium present in exchangeable form. In arid climate also the same relationship exists except when sodium is absorbed in large amounts.

2. **Solubility of Iron, Aluminium and Other Micronutrients :** Aluminium, iron and other micronutrients except molybdenum are, in fact, so much soluble at low pH value that they may become toxic to plants. At neutral reaction they become deficient and over application of lime renders them totally unavailable. Moreover, excess of calcium hinders the movement of boron in plants in spite of its solubility. It also affects the metabolism of boron.

3. **Availability of Phosphorus :** The kinds of phosphate ions present in the solution depend upon the pH of the soil solution. When the soil in acidic (pH 5.5) $H_2PO_4^-$ ions are dominant, while as the pH value increases HPO_4^- ions are in excess and in distinctly alkaline soils (Ph 7.5) PO_4^{3-} ions are present in excess. Moreover, at low pH values iron, manganese, titanium etc. become more soluble which form insoluble phosphate of iron, manganese, aluminium and titanium and thus render the phosphate ions insoluble. For this reason the pH of the soil which favors a mixture of $H_2PO_4^-$ ions is most commonly preferred. The fixation of phosphorus in the above way is most serious at pH 5.0 or at values below this.

   On the other hand pH values above 7.5 or 8.0 disturb the nutrition of higher plants and micro-organisms in another way. If the pH is above 7.0 and lime is abundant, calcium forms complex calcium phosphates which range in increasing insolubility from oxy-apatite $[3Ca_3(PCO_4)_2CaO]$ to the most insoluble phosphate the flour-apatites $[3Ca_3(PO_4)_2CaF_2]$ which may be formed in the presence of fluorine. Thus, when the pH of the soil is above 8.5 the solubility of both native and added phosphates may be impaired seriously.

   However, soil pH above 7.0 affects in other ways too. If lime is in excess, excess calcium hinders in the translocation of phosphate ions through the cell walls of root hairs. Even if these ions are absorbed once, they may interact with other ions during metabolism and form insoluble phosphates. At pH 6.0 and 7.0, fixation is least and so availability of phosphorus is maximum.

4. **Activity of Micro-organisms :** Generally speaking, bacteria and actinomycetes function better at intermediate and higher ranges of pH values. Fungi, however, work satisfactorily at all pH ranges. Thus, in normal soils, foils, fungi predominate at lower pH values, while at higher pH vaues they may meet a strong competition from bacteria and actinomycetes and leave the field to some degree in their favor.

   In mineral soils, nitrification and nitrogen fixation take place vigorously at pH above 5.5. Decay, aminisation although curtailed at lower pH values still proceed with appreciable intensity as fungi are able to affect the enzymic transfers even at high acidity. Thus, higher plants growing in acidic soils are provided with at least ammoniacal nitrogen.

   It can be concluded that the pH 6.5 to7.5 is most satisfactory for biological regime from the viewpoint of availability of nutrients. There is, however, one exception to the general rule. The exception is micro-organisms oxidizing sulphur to sulphuric acid that are facultative and active in medium to higher pH values and also in markedly acidic conditions.

Buffering of Soil

Before actually understanding the buffering of soils, what the buffer solutions are must be known. Pure water has pH 7.0, but this value is not retained for long as water absorbs carbon dioxide from atmosphere or as a result of the silicates being dissolved from glass containers; the same being true for solutions of single salts.

However, a solution containing a weak acid and a salt of it with a strong base (e.g.. acetic acid and sodium acetate) or a weak base and a salt of it with a strong acid (e.g., ammonium hydroxide and ammonium chloride) has the following properties:

1. It possesses a definite pH.
2. Its pH does not change with time or dilution.
3. It resists a change in pH even when a strong acid or a strong base is added to it, or even if a change occurs it is only slight as compared to the change which would have occurred in the absence of buffers in solutions. Consider the case of a weak acid, i.e., acetic acid mixed with sodium acetate. Acetic acid is a weak acid. Therefore, there will be less hydrogen ions in solution. However, due to a strong electrolyte like sodium acetate there will be an excess of sodium and acetate ions. When a strong acid is added, the hydrogen ions combine with acetate ions to form feebly ionized or unionized acetic acid with the result that no change in the pi 1 value occurs. In the same way, when a strong base is added to it, the base or hydroxyl ions temporarily neutralize the hydrogen ions without any change in the pH value.

Similarly, soils also show a distinct resistance to change in pH values. This property is known as buffering of soil. This is due to the fact that weak acid radicals such as carbonates, bicarbonates and phosphates present in soils behave as buffers. The most important buffering agents are specially those present on the colloidal complex with associated cations. Many weak organic acids are produced in soil due to microbial activity. These provide excellent buffering agents. Briefly it can be said that an average soil is colloidally buffered by humus and clay present in it.

In soils there is always an equilibrium established between the adsorbed hydrogen ions (which give the potential or the reserve acidity) and the hydrogen ions of the soil solution (which gives the active acidity).

$$\underset{\text{(Potential or reserve acidity)}}{\text{Adsorbed } H^+} \leftrightarrow \underset{\text{(Active acidity)}}{\text{Soil Solution } H^+}$$

If lime is added to neutralize the hydrogen ions of the soil solution, some more adsorbed hydrogen ions would give hydrogen ions to the soil solution. Hence, apparently there appears no change in soil pH. However, if some acid is added, some hydrogen ions of soil solution get absorbed and in the same way the pH of the soil solution is not markedly affected.

In general, the higher the exchange capacity of the soil, the greater is the buffering capacity as more hydrogen ions or exchangeable cations must be interchanged to effect a given rise or lowering of the percentage base saturation. In practice, the heavier the texture of the soil and higher the organic matter content, the larger must be the application of lime to effect a given change in soil pH.

Importance of Buffering

Buffering is important because it stabilizes the soil pH. A marked change in pH indicates a radical. Such solutions with reserve acidity or alkalinity are known as buffers which are important specially in reference to availability of plant nutrients. If soil pH values change over a wider range, the higher plants and micro-organisms may suffer.

Thus, buffering is important from the point of view of finding out the quality of amendments required to effect a given change in pH values. In general, the greater the buffering capacity, the larger must be the amount of lime or sulphur required to effect a given change in acidity or alkalinity to the desired point.

Colloidal Control of Soil Reaction

The soil reaction is largely controlled by the colloidal matter present in soil. Thus, soil behaves in a different manner as compared to soil solution as far as reaction is concerned. The following points are worth considering in this context.

1. **Hydrogen ions as against metallic cations:** The control of the soil reaction largely depends on the soil colloidal complex. Colloidal complex of the soil is considered to be a mixture of insoluble acids and their salts. The adsorbed metallic cations and hydrogen ions tend to remain in dynamic equilibrium with the soil solution. Thus the pH of the soil solution largely depends upon the dissociation of these adsorbed cations. The colloidal acid and its calcium (predominant cation) salt dissociate in the following manner:

$$\text{H [Soil]} \leftrightarrow \text{[Soil]} + \underset{\text{(Acidity)}}{H^{+}}$$

$$\text{Ca[Soil]} + 2\text{HOH} \leftrightarrow \underset{\text{H}}{\text{H [Soil]}} + Ca^{++} + \underset{\text{(Alkalinity)}}{2OH^{-}}$$

Thus when a micelle containing adsorbed hydrogen furnishes hydrogen ions to the soil solution, it gives an acid reaction to the soil solution unless it is counterbalanced by the adsorbed bases. Thus, a micelle behaves as a weak acid. In contrast, when metallic cations are adsorbed by these acid radicals, their hydrolysis yields an alkaline reaction because hydroxyl ions dominate over hydrogen ions. However, under natural conditions these reactions occur simultaneously, that is, hydrogen ions and metallic cations are both held by the same micelle. Therefore, the reaction of the soil solution depends upon the relative amounts of adsorbed hydrogen and adsorbed metallic cations, and the acidity or the alkalinity depends upon the domination of hydrogen ions or hydroxyl ions respectively; while a just balance of soil solution yields a neutral soil reaction.

2. **Percentage base saturation:** The pH value also depends upon the percentage base saturation. Percentage base saturation is the percentage of colloidal complex of the soil which is occupied by the exchangeable bases. The relative amounts of adsorbed hydrogen and exchangeable bases of colloidal complex depend upon the percentage base saturation. If the percentage base saturation is 80, four-fifth (80/100) of the exchange capacity is occupied by bases and the remaining one-fifth (20/100) by hydrogen ions. Naturally, a low base saturation will mean acidity and a base saturation approaching 100 will mean neutrality or alkalinity depending upon the nature of the adsorbed bases. Generally, humid region soils dominated by silicate clays and humic acid are acidic if their percentage base saturation is much below 90. When these soils have a percentage base saturation of 90 or more, they are usually neutral or alkaline. The exact pH value depends upon two more factors, i.e., nature of the micelle and the kind of adsorbed bases.

3. **Nature of the Micelle:** The laterite soils present almost the same conditions as the soil whose clay is dominated by silicates, but for the fact that the percentage base saturation falls much lower before giving an acid reaction because of the nature of the micelle radical, which is largely made of hydrous oxides of iron and aluminium in laterite soils. Such clays dissociate to a much lesser extent when dominated by hydrogen ions than silicate clays at a correspondingly low percentage base saturation. Thus, at 50 per cent base saturation a silicate clay must have a pH of 5.2 to 5.8, while that of hydrous oxide clays, at the same percentage base saturation, may be 6.0 to7.0.

   In the case of peat soils the condition is just the opposite. The adsorbed hydrogen ions of the organic complex dissociate quite easily and, therefore, the pH of organic soil is usually lower than that of the humid region mineral soil at the same percentage base saturation. Thus, organic soils have stronger acids and a peat soil at 50 per cent base saturation may give a pH value ranging between 5.2 and 5.8, while at the same percentage base saturation a mineral soil may give higher pH values.

   Different acid silicate clays - kaolinite, montmorillonite, and hydrous mica— dissociate to different degrees; kaolinite the least and montmorillonite the maximum. Similarly, organic colloids also show considerable variety among themselves, however, they have a lower pH value than any of the silicate clays at the same percentage base saturation.

4. **Kind of adsorbed bases:** Another factor which greatly affects pH values of the soil is the kind of adsorbed bases. At a percentage base saturation of 70. the presence of calcium, magnesium, potassium and sodium ions in the ratio of 11:2:1:1 respectively will yield a lower pH than when these ions are in 4:1:2:8 proportion respectively; since in the former case the exchange complex is dominated by calcium, while in the latter case sodium is the dominant ion. Thus, the pH value of soil depends upon three independent factors: the percentage base saturation, the nature of the micelle and the kind of adsorbed bases, and it is not possible to expect a close relationship between the percentage base saturation and pH when soils are compared at random. However, with soils of similar origin, texture and organic matter content there is a rough correlation. However, it should not be expected that two soils of the same pH would seldom posses the same percentage base saturation and vice versa.

Correlation of Higher Plants to Soil Reaction

There are three physiological conditions possible:

1. Conditions presented by a highly acidic soil.
2. Conditions presented by a highly alkaline soil.
3. Those presented by intermediate reaction.

1. **Acid Soil :** A highly acidic soil may have pH 4.5 and present low calcium and magnesium, high solubility of iron, aluminium, manganese, boron, titanium etc., but low availability of nitrogen and phosphorus. Besides these there is a possibility of presence of organic toxins.
2. **Alkaline Soil :** It has a pH of 8.5 and above. It has plenty of active calcium and magnesium, little or no aluminium, and mild yet active humus. Nitrogen is readily available. If the pH increases too much, iron, manganese, copper, zinc, specially phosphorus and boron may not be available in adequate amounts.
3. **Moderately to Slightly Acid Soil :** For average plant growth a moderately to slightly acid soil is quite satisfactory. The chemical and biological agencies of such a soil are well balanced. The nutrient availability and the microbial activity in such a soil are quite satisfactory.

All plants essentially require certain macroelements, which are mainly obtained from soil. These are Nitrogen, Phosphorus and Potassium.

NITROGEN

Origin and Distribution of Nitrogen

The nitrogen contents of surface mineral soils normally range from 0.02 to 0.5 per cent, a value of about 0.15 per cent being representative. A hectare of such a soil would likely contain about 3.3 mg of nitrogen while the air above that hectare of soil would contain nearly 300.000 mg of the element. Obviously, the atmosphere (80%N) is a seemingly limitless source of nitrogen although it is not readily usable by plants in elemental form.

Most of the soil nitrogen is in organic form. Proteins and other organic nitrogen compounds are associated with humus and with some silicate clays, which protect them from rapid microbial breakdown. Only about 2-3 per cent of organic nitrogen is mineralized each year under normal conditions. Ammonium ions tightly bound by clay may account for up to 8 per cent of the nitrogen in surface soils and up to about 40 per cent in subsoils. This clay-fixed nitrogen is only slowly available to plants.

The quantity of nitrogen in the readily available nitrate and ammonium forms is seldom more than 1-2 per cent of the total soil nitrogen, except where large amounts of chemical fertilizers have been applied. This is fortunate because these soluble forms are easily lost from soils through leaching and volatilization. Only enough usable nitrogen is needed to supply the daily requirements of the growing plants.

Reactions of Nitrogen

The worldwide use of nitrogen-containing fertilizers has expanded greatly in recent years. As a result, nitrogen in the soil solution, particularly in the localized soil zones to which the fertilizers have been applied, often is dominated by fertilizer-applied materials. While the ammonium and nitrate ions coming from fertilizes react in a similar way to comparable ions released by microbial breakdown of organic materials, their high concentration in the local zones of application and tendency to acidify the soil deserve special attention.

High Concentration

The localized concentration of anhydrous ammonia (NH_3), ammonium containing salts, and urea (which hydrolyzes to ammonia) stimulates several reactions. The fixation of ammonium ions by clays and organic matter is enhanced. High localized levels of ammonia inhibit the second stage of nitrification, resulting in the undesirable accumulation of nitrite ions. In alkaline soils, a high concentration of ammonium ion can result in the release of some ammonia gas directly to the atmosphere.

The addition of large amounts of nitrogen-containing fertilizers may affect the microbial processes of free fixation and gaseous nitrogen loss. In general, fixation by free-living organisms is depressed by high levels of mineral nitrogen. Gaseous losses, on the other hand, are often encouraged by abundant nitrates. Heavy nitrogen fertilization followed by nitrification thus tends to increase losses of nitrogen and soil.

In most soil solutions, the effects of higher localized concentrations of fertilizer materials on nitrogen transformations are not too serious. Economic considerations often counterbalance the biochemical disadvantage of localized concentration. In any case, it can be assumed that fertilizer nitrogen will react in soils in a manner very similar to that of nitrogen released by biological transformations.

Soil Acidity

Ammonium-containing fertilizers are those that form ammonia upon reacting in the soil can increase soil acidity. The process of nitrification releases hydrogen ions that become adsorbed on the soil colloids. For best crop growth in humid regions, continued and substantial use of acid-forming fertilizers must be accompanied by application of lime.

The nitrate component of fertilizers does not increase soil acidity. In fact, nitrate fertilizers containing metallic cations in the molecule have a slight alkalizing effect.

PHOSPHORUS

Importance of Phosphorus

Phosphorus is essential for plant growth. It is a component of adenosine di phosphate (ADP) and adenosine triphosphate (ATP), the two compounds involved in most significant energy transformations in plants. ATP, synthesized from ADP through both respiration and photosynthesis, contains a high-energy phosphate group that drives most biochemical processes requiring energy. For example, the uptake of some nutrients and their transport within the plant, as well as the synthesis of new molecules, are energy-using processes that ATP helps to implement.

Phosphorus also plays a critical role in the life cycle of plants. It is an essential component of deoxyribonucleic acid (DNA), the seat of genetic inheritance in plants as well as animals, and of the various forms of ribonucleic acid (RNA) needed for protein synthesis. Obviously, phosphorus is essential for numerous metabolic processes.

Among the more significant functions and qualities of plants on which phosphorus has important effects are :

1. Photosynthesis.
2. Nitrogen fixation.
3. Crop maturation: flowering and fruiting, including seed formation.
4. Root development, particularly of the lateral and fibrous rootlets.
5. Strength of straw in cereal crops, thus helping to prevent lodging.
6. Improvement of crop quality, especially of forages and of vegetables.

Phosphorus-fixation Power of Soils

The soils high in clay fix more phosphorus, especially if the clays have primarily Fe&AI oxides and if they are amorphous rather than crystalline.

The high capacity of soil for fixing phosphorus explains why much fertilizer-supplied phosphorus is quick rendered unavailable for crops. Fortunately, over a period of years plants are able to least some of these fixed materials.

Organic Matter, Microbes and Available Phosphorus

Phosphorus held in organic form can be mineralized and immobilized by the same general processes pertinent for nitrogen and sulphur. The following reaction illustrates this point:

Immobilization

$$\underset{\text{Mineralization}}{\text{Organic P forms}} \xrightleftharpoons[\text{microbes}]{\text{microbes}} H_2PO_4 \xleftrightarrow{Fe^{3+},\,Al^{3+},\,Ca^{2+}} \underset{\text{Microbes Fixed phosphates}}{\text{Fe, Al, Ca phosphates}}$$

Soluble phosphorus compounds are released as organic residues and humus are decomposed. The resulting soluble inorganic phosphate ion ($H_2PO_4^-$) is subject to uptake by plants or to fixation into insoluble forms. Should organic residues low in phosphorus but high in other nutrients be added to a soil, rapid microbial activity would take place and available $H_2PO_4^-$ in the soil solution would temporarily disappear, just as was the case for soluble NH_4^+, NO_3^- and SO_4^{2-} ions.

Organic matter influences phosphorus availability in two other ways. First, known sources of organic phosphorus (the nucleic acids) are adsorbed by humic compounds as well as by silicate clays. The adsorptive reaction s probably protects the organic phosphorus from microbial attack. Secondly, specific organic compounds form complexes with iron and aluminium ions and hydrous oxides, thereby preventing these materials from reacting with phosphates. Manure is known to influence the availability of inorganic phosphorus compounds.

POTASSIUM—THE THIRD "FERTILIZER" ELEMENT

The history of fertilizer usage at various fields show that nitrogen and phosphorus received most of the attention when commercial fertilizers first appeared on the market. Although the role played by potassium in plant nutrition has long been known, the importance of potassium fertilization has received full recognition only in comparatively recent years.

The reasons that a widespread deficiency of this element did not develop earlier are at least twofold.

First, the supply of available potassium was so high in many soils that it took many years of cropping for a serious depletion to appear.

Second, even in soils having insufficient potassium for optimum crop yields, production was more drastically limited by the lack of nitrogen and phosphorus. As the use of nitrogen and phosphorus fertilizers expanded, crop yields increased and so did the removal of soil potassium. As a consequence, the drain on soil potassium has been greatly increased. This, coupled with considerable loss by leaching, has raised the demand for potassium in commercial fertilizers.

Effects of Potassium on Plant Growth

Potassium plays many essential roles in plants. It is an activator of dozens of enzymes responsible for such plant processes as energy metabolism, starch synthesis, nitrate reduction, and sugar degradation. Potassium is extremely mobile within the plant and helps to regulate the opening and closing of stomates in the leaves and uptake of water by root cells.

Potassium is essential for photosynthesis, for protein synthesis, for starch formation, and for translocation of sugars. This element is important in grain formation, and is absolutely necessary for tuber development. All root crops generally respond to applications of potassium. As with phosphorus, it may be present in large quantities in the soil and yet exert no harmful effect on the crop.

Potassium increases crop resistance to certain diseases and, by encouraging strong root and stem systems, helps to prevent the undesirable "lodging" of plants that is sometimes caused by excessive nitrogen. Potassium delays maturity, thereby working against undue ripening influences phosphorus can exert. In a general way, potassium exerts a balancing effect on the effects of both nitrogen and phosphorus; consequently it is especially important in a multinutrient fertilizer.

Note that sodium has been found partially to take the place of potassium in the nutrient of certain plants. Where there is a deficiency of potassium, native soil sodium or that added in such fertilizers as sodium nitrate, may be useful.

THE POTASSIUM PROBLEM

Availability of Potassium

In contrast to phosphorus, potassium is found in comparatively high levels in most mineral soils, except those of a sandy nature. In fact, the total quantity of this element is generally greater than that of any other major nutrient element.

The quantity of potassium held in an easily exchangeable condition at any one time often is very small. Most of this element is held rigidly as part of the primary minerals or is fixed in forms that are at best only moderately available to plants. Therefore, the situation in respect to potassium utilization parallels that of phosphorus and nitrogen in at least one way. A very large proportion of all three of these elements in the soil is insoluble and relatively unavailable to growing plants.

Leaching Losses

Unlike the situation with respect to phosphorus, however, much potassium is lost by leaching. Drainage waters from soils receiving liberal fertilizer applications usually have considerable quantities of potassium. A representative humid region soil receiving only moderate rates of fertilizer, the annual loss of potassium by leaching is usually about 35 kg/ha . However, since considerable potassium is adsorbed by soil colloids, leaching losses of this element normally do not result in yield losses except on very sandy soils.

Crop Removal

Potassium removal by crops is high, often being three to four times that of phosphorus and equaling that of nitrogen. The removal of 140-180 kg/ha of potassium by a 60 mg/ha silage corn crop is not at all unusual.

This loss of potassium is made even more critical by the tendency of plants to take up soluble potassium far in excess of their needs if sufficiently large quantities are present. This tendency is termed luxury consumption, because the excess potassium absorbed does nol increase crop yield to any extent.

PRACTICAL IMPLICATIONS IN RESPECT TO POTASSIUM:

Frequency of Application

Frequent light applications of potassium have some advantages over heavier and less frequent ones. Such a conclusion is based on the luxury consumption of potassium by some crops, the ease with which this element is lost from the soil solution by leaching, and the fact that excess potassium is subject to fixation. Although the fixation has definite conserving features, these in most cases tend to be outweighed by the disadvantages of leaching and luxury consumption.

Potassium-Supplying Power of Soil

A second very important suggestion is that full advantage should be taken of the potash-supplying power of soils. The idea that each kilogram of potassium removed by plants or through leaching must be returned in fertilizers may not always be correct. In some soils the large quantities of moderately available forms of potassium already present can be utilized. More often, however, slowly available forms are not found in significant quantities, and supplementary additions are necessary. Moreover, the importance of lime in reducing leaching losses of potassium should not be overlooked as a means of effectively utilizing the power of soils to furnish this element.

Soils of arid zones commonly can supply adequate potassium for many years, even under irrigation. However, continued crop removal can deplete even these soils. Also, deep-rooted plants such as cotton may depend on the subsoil for much of their potassium. Increasing the availability of this element at depths below the plow layer is difficult.

Potassium Losses and Gains

Crop removal of potassium generally exceeds that of the other essential elements, with the possible exception of nitrogen. Annual losses of potassium from plant removal as great as 100 kg/ha or more are not uncommon, particularly if the crop is a legume and is cut several times for hay. As might be expected, therefore, the return of crop residues and manures is average animal manure supplies about 40 kg of potassium, fully equal to the amount of nitrogen thus supplied.

The annual losses of available potassium by leaching and erosion greatly exceed losses of nitrogen and phosphorus. Potassium losses through leaching are generally not as great, however, as the corresponding losses of available calcium and magnesium. In contrast, the total potassium removal by erosion generally exceeds that of any other major nutrient element. Such losses of soil minerals are indeed serious.

Increased Use of Potassium Fertilizers

In the past, potassium in fertilizers was added only to supplement potassium returned in crop residues or released from slowly available forms. While these sources are still very important, fertilizers are used increasingly to supply much of the potassium needed for crop production. This is true especially in cash-crop areas and in regions where sandy soils are prominent. Even in some finer textured soils, the release of potassium from mineral form is too slow to support maximum crop yields. Consequently, increased usage of commercial potassium fertilizers must be expected if yields are to be increased or even maintained.

REFERENCES

1. Adams, Fred(Ed.) Soil Acidity and Liming (1984).
2. G. Nelson Eby, Principle of Environmental Geochemistry, Thomson Brooks, Pacific Grove, Cu (2005).
3. Mehlich, A Influence of adsorbed hydroxyl and sulphate on neutralization of soil acidity "Soil Sc.Soc.Am.Proc. 28:492-496. (1964).
4. Rhodes, J.D and D.L. Corwin monitoring soil salinity, J. Soil & Water Cons, 39:172-175. (1984)
5. Tiessen, H.J.W.B .Stewart, and J.R. Bettany , "Cultivation effects on the amounts and cons, of, C, N,& P in grass land soils Agron. J. 74:831-835. 1982
6. S. E. Manahan, Envtal chemistry, 8th edition, CRC Press, Boca Raton, FI, (2005).

CHAPTER 27

Plastics Waste Management

Dr. Noor A. Khan[1]

1. Introduction

The plastics a marvel of polymer chemistry have become omni present in our daily life through various applications. But indiscriminate use of plastics as well as its reprocessing and disposal plastics waste are posing environmental problems and health hazards beside causing public nuisance.

The quantum of solid waste is ever increasing due to increased in population, development activities, and change in life style and socio economic conditions. Plastics waste is as significant portion of the total municipal solid waste (MSW). It is estimated that approximately 10 thousands tons per day (TDP) of plastic waste is generated i.e. 9% of 1.20 lacs TPD of MSW in the country. The plastic waste constitutes two major category of plastics- (1) Thermoplastics (2) Thermoset plastics. Thermoplastics constitute 80% and thermoset constitutes approximately 20% of the total post consumer plastics waste generated in India. The thermoplastics are recyclable plastics, which include ; polyethene terephthalate (PVC) high density poly ethylene (HDFE), poly propylene (PP), poly styrene (PS) etc. However, thermoset plastics; contain alkyd, epoxy, ester, melamine formaldehyde, poly urethane, metalized and multi layer plastics etc. The environmental hazards due I to mismanage of plastics waste include the following aspects.

(*i*) Uttered plastics spoils beauty of the city and choke drains and make important public places dirty.

(*ii*) Garbage containing plastics, when burnt may cause air pollution by emitting polluting gases.

(*iii*) Garbage mixed with plastics interferes in the waste processing facilities and may air cause problems in land fill operations.

(*iv*) Recycling industries operating in non conforming areas are posing unhygienic problems to environment.

2. Recycled Plastics Manufacture and Usage Rules

Regulation of plastics waste, particularly manufacture and use of recycled plastics carry bags and containers is being regulated in the country as per " Recycled Plastics Manufacture and Usage Rules, 1999 and as amended in 2003. According to these Rules:

- No person shall manufacture, stock, distribute or sell carry bags made of virgin or recycled plastic bags which are less than 8 × 12 inches in size and having thickness less than 20 microns.
- No vendor shall use carry bags/containers made of recycled plastics for storing, carrying, dispensing or packaging of food stuffs;

[1]Principal, Crescent College of Technology, B.U., Bhopal (India)

- Carry bags and containers made of recycled plastics and used for purposes other than storing and packaging food stuffs shall be manufactured using pigments and colorants as per IS 9833.198 entitled "List of pigments and colorants for use in plastics in contact wit food stuffs, pharmaceuticals and drinking water";
- Recycling of plastics shall be undertaken strictly in accordance with the Bureau of Indian Standard specification IS 14534:1998 entitled "The Guidelines for Recycling of Plastics";
- Manufacturers of recycled plastic carry bags having printing facilities shall code/mark carry bags and containers as per Bureau of Indian Standard specification IS 14534: 1998 (The Guidelines for Recycling Plastics);
- No person shall manufacture carry bags or containers irrespective of size or weight unless the occupier of the unit has registered the unit with respective SPCB/PCC prior to the commencement of production;
- The prescribed authority for enforcement of the provisions of these rules related to manufacturing and recycling is SPCB in respect of States and the PCC in Union Territories and for relating to use collection, segregation, transportation and disposal shall be the District Collector/ Deputy Commissioner of the concerned district

3. Options for Plastic Waste Management

A. Recycling of plastics through environmentally sound manner : Recycling of plastics should be carried in such a manner to minimize the pollution during the process and as a result to enhance the efficiency of the process and conserve the energy. Plastics recycling technologies have been divided into four general types - primary, secondary, tertiary and quaternary.

Primary recycling involves processing of a waste /scrap plastics Into a product with characteristics similar to those of original product.

Secondary recycling involves processing of waste/scrap plastics into materials that have characteristics different from those of original plastics product.

Tertiary recycling involves the production of basic chemicals and fuels from plastics waste/scrap as part of the municipal waste stream or as a segregated waste.

Quaternary recycling retrieves the energy content of waste/scrap plastics by burning / incineration. This process is not in use in India.

Steps Involved in the Recycling Process

Selection: The Tecyclers/reprocessors have to select the waste / scrap which are suitable for recycling /reprocessing.

Segregation: The plastics waste shall be segregated as per the Codes 1-7 mentioned in the BIS guidelines (1514534:1998)

Processing: After selection and segregation of the pre-consumer waste (factory waste) shall be directly recycled. The post consumer waste (used plastic waste) shall be washed, shredded, agglomerated, extruded and granulated.

B. Use of Plastic waste in laying of Roads (Polymer Coated Bitumen Road) : The CPCB has undertaken a project in collaboration with Thiagarajar College of Engineering Madurai to evaluate the performance of polymer coated bi roads laid during 2002 2006 in different cities.

The observations are as below:

- The coating of plastics over aggregate improves Impact, Los Angels Abrasion and Crushing Value with the increase in the percentage plastics.

- The extracted bitumen showed almost near value for Marshall stability. The entire road was having good skid resistance and texture values.
- All the stretches in the roods have been found reasonably strong.
- The unevenness index values of these roods are nearly 3000 mm/km which indicate a good surface evenness.
- The plastic tar roads have not developed any potholes, rutting, raveling or edge flaw, even though these roads are more than four years of age.
- Polymer coated aggregate bitumen mix performs well compared polymer modified bitumen mix
- Higher percentage of polymer coating improves the binding strength the mix.
- Foam plastics have better binding values.

C. Plastics waste disposal through Plasma Pyrolysis Technology (PPT) : Plasma Pyrolysis is a state of the art technology, which integrates the thermo chemical properties of plasma with the pyrolysis process. The intense versatile heat generation capabilities of PPT enable it to dispose off all types of plastic wastes including polymeric, biomedical and hazardous waste in a safe and reliable manner.

Plasma Pyrolysis Technology

In plasma pyrolysis, firstly the plastics waste is fed into the primary chamber at 8500C through a feeder. The waste material dissociates into carbon monoxide, hydrogen, methane, higher hydrocarbons etc. Induced draft fan drains the pyrolysis gases as well as plastics waste into the secondary chamber, where these gases are combusted in the presence of excess air. The inflammable gases are ignited with high voltage spark. The secondary chamber temperature is maintained at around 10500 C. The hydrocarbon, carbon monoxide and hydrogen are combusted into safe carbon dioxide and water. The process conditions are maintained so that it eliminates the possibility of formation of toxic dioxins and furans molecules (in case of chlorinated waste). The conversion of organic waste into non toxic gases ($C0_2$, H_2O) is more than 99%. The extreme conditions of Plasma kill stable bacteria such as Bacillus stereothermophilus and Bacillus subtilis immediately. Segregation of the waste is not necessary, as very high temperatures ensure treatment of all types of waste without discrimination.

The CPCB has initiated the study in association with Facilitation Centre for industrial Plasma Technologies (FCIPT), Institute of Plasma Research (IPR) The objectives of the study are to conduct performance study of the PPT on 15 kg/hr prototype demonstration system developed by FCIPT/ IPR for proper disposal of plastics waste and also monitor air quality parameters e.g. suspended particulate matter (SPM), carbon monoxide (CO), hydrocarbons (HC), benzene, dioxins, furans etc. with regards to gaseous emissions. CPCB also proposes to undertake study on safe disposal of plastics waste using higher capacity (approx 50 kg/hr) plasma pyrolysis system as in future and may set up prototype plasma pyrolysis plant on demonstration basis (15 kg/hr waste disposal capacity) at specific locations (hilly and pilgrimage) in consultation with State Government.

D. Conversion of Plastics waste into Liquid Fuel : A reasearch-cum-demostration plant was set up at Nagpur, Maharashtra for conversion of waste plastics into liquid fuel. The process adopted is based on random de-polymerization of waste plastics into liquid fuel in presence o catalyst. The entire process is undertaken in closed reactor vessel followed condensation, if required. Waste plastics while heating upto 2700 C to 3000 C convert into liquid-vapour state, which is collected in condensation chamber the form of liquid fuel while the tarry liquid waste is topped-down from heating reactor vessel. The organic gas is generated which is vented due lock of storage facility. However, the gas can be used in dual fuel dies generator set for generation of electricity. The process includes the steps shown ahead:

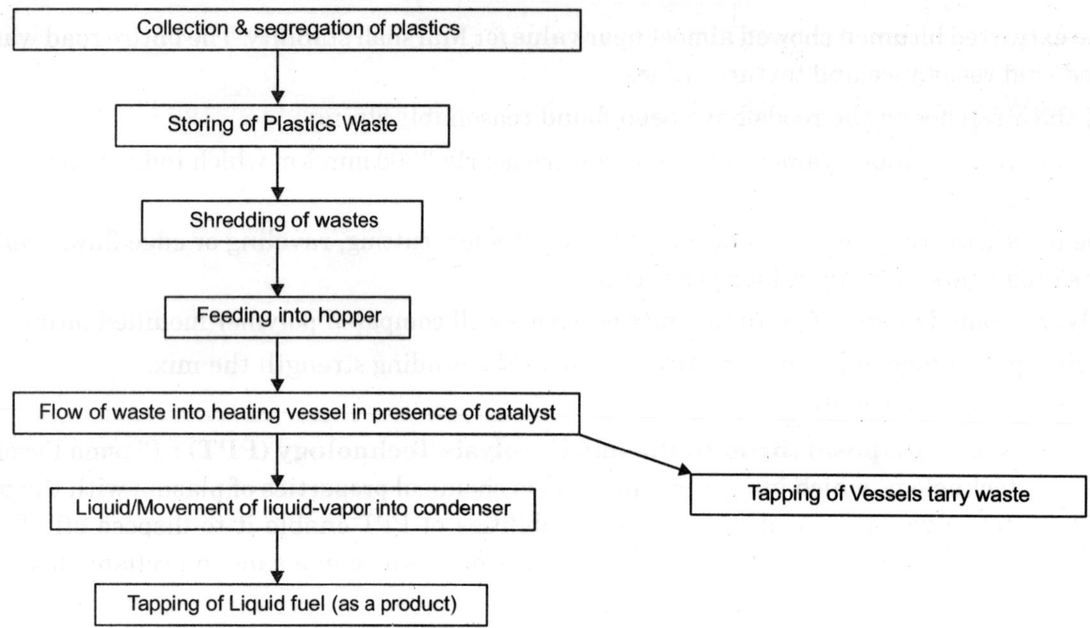

Environment Related Observations during the Process

- There are no liquid industrial effluents and no floor washings as it is a dry process.
- There are no organized stack and process emissions.
- Odour of volatile organics has been experienced in the processing area due to some leakages or lack of proper sealing.
- Absolute conversion of liquid-vapour was not possible into liquid, some portion of gas (about 20%) is connected to the generator. However, the process will be improved in full-scale plant.
- PVC plastics waste is not used and if used, it was less than 1%. In case PVC is used, the chlorine can be converted into hydrochloric acid as a by-product.
- The charcoal (charcoal is formed due to tapping of tarry waste) generated during the process has been analysed and contain heavy metals, poly aromatic hydrocarbon (PAH) which appears to be hazardous in nature. The source of metals in charcoal could be due to the presence of additives in plastics and due to multilayer and laminated plastics.
- Monitoring of process fugitive emissions in the work area as well as emissions from the engines/ diesel generator sets is necessarily required (where this liquid fuel is used) for various parameters such as CO, HCI, Styrene, Benzene, VOCs.

E. Biodegradable Plastics : The environmentally degradable polyolefin films are defined as those materials that contain degradation process of polyolefin article (bag/film/ sheet) under conditions of composting. Often queries are raised regarding biodegradability of plastics but clear-cut answer is not available about the biodegradability of plastics In view of above, CPCB has initiated a study in collaboration with Central Institute of Plastics Engineering and Technology (CIPET) to establish the biodegradability and compostability (e.g. fragmentation rate, degradation role and safety) of polymeric material available in India and abroad. The study will include:

- Inventorisation and assessment of the manufacturing status of biodegradable plastics in India particularly with reference to processing technologies and the environmental issues.
- Establishment of the degradation rate (change in chemical structure, decrease in mechanical strength, fragmentation or weight loss) of the polymeric material plastics material under laboratory scale compos conditions.

- Finding out self-life and its impact on environment (soil, water plastics with reference to colour and additives, once it is disposed off).
- Assessment of effects on foodstuffs with reference to natural colours and additives.

4. Status of Plastics Waste Management in the Country

Central Pollution Control Board (CPCB) is co-ordinating with State Pollution Control Boards (SPCBs) and Pollution Control Committees (PCCs) regard to implementation of Plastics Manufacture and Usage (Amendment) Rules, 2003. The Prescribed Authorities for manufacturing and recycling plastics are SPCBs/ PCCs whereas; Prescribed Authorities for segregation, transport and disposal of plastics waste are District Collectors/Deputy Commissioners. CPCB has compiled information relating to implementation of the said Rules and prepared status report on plastics w management (PWM).

State wise Status of Plastics Manufacturing/Recycling Units (As December, 2006)

Sl. No.	SPCBs/UTs	No. of units	No. of Registration Granted	Comments/Suggestions
1	2	3	4	5
1.	AndhraPradesh	177	177	The Plastics Manufacture and Usage Rules, 1999/2003 are being implemented. Levy of penalties against the violators of recycling Norms are stipulated vide Notification No 25 dated 30.3.2001. The Board is encouraging urban local bodies to segregate waste at source so as to promote recycling. The Andhra Pradesh Pollution Control Board imposed a fine of Rs. 25.000/- each on 16 defaulting units and issued closure orders to 2 units during April 2006 to September 2006.
2.	Andaman & Nicobar Islands	Nil	Nil	Plastics Manufacture & Usage Rules republished by the Administration vide Notification No. 121 dated 5.5.2000. A committee was formed for effective implementations of the rules vide order No. 623 dated 23.7.2004. Administration is organizing regular awareness programmes
3.	Assam	10	Nil	Environmental Awareness programs are being conducted regularly. Bi-lingual publications are brought out by the board regularly. The board is developing Criteria for other plastics products such as ropes, sheets, soap case etc
4.	Arunachal Pradesh	Nil	Nil	The Notification of the Ministry has been circulated to the concerned Department and Deputy Commissioner of the State for compliance by the State Government
5.	Bihar			Advertisements in local daily newspapers for compliance of plastic rules were issued. Inventory is being done. Offices of the State Board have been identified and notified to assist the District Administration for effective implementation of the rules
6.	Chandigarh	9	9if	The UT administration has notified the plastic rules vide Notification no. DC/M A/200 1/187/dated 14.9.2001. Vide notification No. ED/2003/543 dated 16th September, 2003 the Government has restricted the thickness of plastic carry bags to be not below 30 microns and size not less than8x12 inches
7.	Chhatisgarh	32	11	Inventorisation of plastic bag manufacturing units has been done. A committee has been formed to monitor the implementation of compliance. Awareness programmes are being organized from time to time by regional offices of the SPCB through media/ ecoclubs and NGOs etc

...(*Contd.*)

1	2	3	4	5
8.	Delhi	147	147	The Delhi Plastic bags (Manufacture, Sale and Usage and Non-bio-degradable Garbage Control Act, 2001 has been brought out vide DOE/2001 /Rules/451 dated 2-11-2001 to manage plastics waste. Vide notifications No. F.8 (86)/ EA/ Env. /2005(ii)/485, 486 dated 02.06.2005 & F.8 (86)/ EAIEnv./2005(ii)/450 on 25th May, 2006, the Government of Delhi has made the degradable plastic carry bags compulsory in all four and five star hotels, hospitals having bed strength of 100 beds and more and restaurants having sitting capacity of more than 50 seats, all fruits and vegetables outlets of mother diary, all liquor vends and all shopping malls. Public notices in Hindi and English newspapers about the plastic rules as well as management of the plastic wastes issued.
9.	Daman & Diu	-	-	The Notification of the Ministry republished on official gazette
10.	Dadara & nagar Haveli	-	-	The Notification of the Ministry republished on official gazette
11.	Gujarat	200	94	Published public notices in Gujarati and English regarding RPMU rules, 1999 and its amendment in 2003. Gujarat Board has organized meetings at various levels to create awareness on plastics and the awareness programmes are organized periodically. The Board has completed the inventorisation. Government of Gujarat has banned the use of plastic carry bags at religious places i.e. at Ambaji, Dakar, Somnath and Dwarka
12.	Goa	125	17	The State of Goa has notified Goa Non-biodegradable Garbage (Control) Act, 1996. In this Act, the major implementer of this Rule is local authority, which has to provide various places/types of receptacles for deposit of "Non- biodegradable, Biodegradable and Biomedical" garbage/waste and also ensure that owners/occupiers of all lands and buildings abide by the regulations under the above said Act. Notification has been brought out and thickness of plastics carry bags for selling has been raised to 40 microns. Inventory on the units manufacturing carry bags / containers is in progress.
13.	Haryana	—	—	Plastics Manufacture & Usage Rules republished by the State. The SPCB has prepared inventory of 106 units so far and it has been reported that all these units are complying with the rules and they are manufacturing carry bags with more than 20 microns thickness.
14.	Himachal Pradesh	50	15(12)	The State has notified the Himachal Pradesh Non-biodegradable Garbage (Control) Act, 1995. Under this Act, the Government prohibits using colored polythene carry bags manufactured from Recycled Plastics for packaging goods from 1st January 1999. Subsequently, in 2004, vide notification No. STE-A (3)-2/2003 dated' 4.6.2004 the Government imposed a ban on use of plastic carry bags of thickness below 70 microns and size less than 12x18 inches.
15.	Jharkhand	—	—	The compliance to the rules is being periodically monitored by the State Pollution Control Board. Rules disseminated through Public Notices.
16.	J & Kashmir	—	—	Plastics Manufacture & Usage Rules republished by the State Government.

...(Contd.)

1	2	3	4	5
17.	Karnataka	302	Nil	Public Notice issued 30.3.2000 & 7.12.2000. Municipalities are also involved in implementation of plastic rules. Reuse of plastics wastes are used in laying of Roads. Inter-state movement of substandard carry bags/ material etc. is restricted
18.	Kerala	193	10	Wide publicity has been given on the restrictions imposed as per Plastics manufacture and usage rules, 1999. State Government notified vide G.O. (P) N0.264/2003/LSGD dated 1.9.2003 prohibiting the plastic carry bags which are less than 40 microns in the State. Also the Govt. of Kerala has formulated an action plan for plastics waste management within the state.
19.	Lakshadweep	Nil	Nil	The administration imposed ban on packing and carrying of plastic bags for carrying consumer goods. Notification on "Lakshadweep Sanitation Conservancy Byelaws, 1998" was issued vide No. 17/2/96- ST&E dated 17.7.1998 to prohibit the use of non-biodegradable materials for packing and carrying consumer goods. Hoardings are displayed at prominent places indicating, "prohibiting littering of plastic waste". Awareness programmes are periodically conducted for phasing out plastic wastes.
20.	Madhya Pradesh	179	83	The State Government has issued orders dated 4.6.2003 declaring authorities for the implementation of the plastic rules as per the original notification. Board has organized wide publicity campaign regarding the provisions of the rules through various means like Articles in newspapers, workshops, leaflets, pamphlets, rallies exhibitions, TV and radio talks. Inventorisation is complete
21.	Maharashtra	—	—	Maharashtra plastic carry bags (Manufacture & Usage) Rules 2006 notified under Maharashtra Non-biodegradable Garbage Control Ordinance 2006 published. Under these, the Govt. of Maharashtra has banned manufacturing of plastic bags below 50- micron thickness and size of 8x12 inches. Maharashtra State Pollution Control Board (MPCB) has taken actions against units, which are non compliant and also issued show-cause notices, directions and subsequent closures, if required
22.	Mizoram	Nil	Nil	Mass awareness campaign organized through publication, distribution of leaflets / pamphlets, by organizing radio, TV talks, seminars and discussions with NGOs and public.
23.	Meghalaya	1	Nil	The Meghalaya Prohibition of Manufacture, Sale, Use and Throwing of low-density plastic bags Act 2001 notified. As per this Act manufacture, sale, use and throwing of plastic bags less than 40 micron has been prohibited in the State. Authorities are designated by the Government for the proper implementation of the Rules.
24.	Manipur	—	—	Plastics Manufacture & Usage Rules republished by the State Government during June 2004. Monitoring on the compliance of the rules is being carried out by the State Pollution Control Board
25.	Nagaland	4	4	Plastics Manufacture & Usage Rules republished by the State Government vide notification no. GAB-9/26/2003 dated 12.11.2003 through which less than 20- micron poly carry bags are prohibited. The Board is also creating awareness on the eco friendly use of plastics through pamphlets etc.

...(Contd.)

1	2	3	4	5
26.	Orissa	14	2	Plastics Manufacture & Usage Rules republished by the State Government. Advertisement are given in the local newspapers to draw attention of the concerned for compliance of these rules. State level awareness programmes are being carried out regularly. Inventorisation of industries has been completed. In the State of Orissa, the use of plastic carry bags has been banned in the municipality area of Puri and Konark with effect from 01/04/2002
27.	Pondicherry	42	8	Pondicherry administration has republished the plastic rules vide G.O. Ms. No. 16/2003/Envt. Dated 1st December, 2003. Proposing to declare plastic free zone in the town. Also proposed a draft Pondicherry Non-bio degradable Garbage Control Act, 2003. Regular awareness drive is being created.
28.	Punjab	—	—	Usages of poly carry bags for foodstuff banned vide Order 8/21/STE (1)72221 dated 2.11.2000. Inventory of- the plastic manufacturing units completed. Punjab State Council for Science & Technology has introduced a Bill 'Punjab Plastic Carry Bags (Manufacture, Use & Disposal Control) Bill, 2004' which has been cleared by the State Government. This Bill prohibits plastic carry bags below 30 microns & size less than 8x12 inches and has got penalty provisions incorporated in it. Released public notices highlighting the salient features of the Rules in local newspapers for proper compliance. Board has organised several awareness meetings in various districts in the state.
29.	Rajasthan	—	—	Vide Circular No. 8(1) PLG799 1.6.2000 usage of poly carry bags for foodstuff banned.
30.	Sikkim	—	—	Usage of poly carry bags for foodstuff banned vide Sikkim Government Notification GOS/UD & HD/97-2000/6 (83)793 dated 30th March 2001. Regular awareness programmes are being conducted.
31.	Tamil Nadu	588	45	Massive awareness: drive initiated through publications in newspapers, programmes in TV and Radio, hoardings in prominent places and buses. Also a mobile awareness creations van is in operation. Inventory of manufacturing units completed
32.	Tripura«	6	6	The manufacture, sale, distribution and use of virgin and recycled plastic bags and containers are prohibited vide Direction issued by Tripura SPCB dated 1 .9.2003. Board has issued number of advertisements in local newspapers to generate public awareness
33.	Uttar Pradesh			Usage of polythene carry bags for foodstuff banned. Non biodegradable garbage control act notified vide No 2448 (2)/XVII-V-l dated 1- 11-2000
34.	Uttranchal	Nil	Nil	A draft of the Uttaranchal plastic bags (Manufacture, Sale and Usage) and nonbiodegradable garbage (control) act, 2004 has been prepared by the Board and has been sent to Government for notification. A task force has been created to organize mass awareness programmes. Published advertisements in the newspapers on plastic rules. Efficient collection of plastic waste for recycling is being organized.

...(Contd.)

1	2	3	4	5
35.	West Bengal	—	—	West Bengal Government is proposing a non biodegradable garbage control bill. Plastics Rules Notified. The board has issued ban orders on the entry, use sale of plastic carry bags in several heritage / tourist places. The West Bengal Plastic Carry Bags and Garbage Control Bill introduced in the State prohibit manufacture, Storage, transport and use of plastics made of recycled plastics. Thickness of plastic carry bags should not be less than 20 microns and for cups and tumblers are to be with 40 microns thickness. Board has published advertisements in the newspapers & organized awareness campaigns.

5. Strategy for Plastics Waste Management

Commonly littered plastic wastes include; polythene carry bags, plastic wrappings, thermocole packings, plastic plates, cups, spoons, glass, melamine crockery and other non-recyclable plastics waste such as guthka pouches, multilayerpackaging, laminated packagings etc. Estimated quantity of plastic waste is 5-10% of total Municipal Solid Waste (1.2 lakh TPD) generation i.e. 6000 tons per day (TPD). It has been observed that in the present Rules, there are no provisions for the disposal of post consumer plastics waste. With the result, plastics waste is littered as road. The waste often chokes open drains as well as make the land infertile. Considering the ill effects and seriousness on the issue, following strategies are suggested to tackle the menace.

Table 27.2

Issues of Concern	Strategy
Production of sub-standard plastic products.	Regulation of sub-standard plastic products.
Multilayer, laminated and thermoset plastic wastes are not recyclable.	Banning or alternate to non-recyclable plastic packaging.
Improper recycling without environmental consideration.	Improvement in recycling mechanism.
Improper regulatory mechanism.	Stringent action against defaulting units.
Inadequate and unsustainable plastic waste collection and disposal mechanism. (Only 50-60% thermoplastics plastic waste is recycled).	Promotion of alternate options for collection and disposal of plastic waste such as use in road construction, conversion into fuel oil, use in blast furnace/cement kilns, densification, balinge.

The Task Force

To formulate a strategy and an action programme for management of plastics waste, the Ministry of Environment & Forests constituted a Task Force comprising specialists, representatives of industry and civic authorities.

The Task Force in its report (August, 1997) recommended a package of preventive, promotional and mitigative (PPM) measures as also the modus operand! for their implementation. These include guidelines for compliance of environmental safety, specifications for restriction on recycling of poor quality plastics waste, deterrent penalties for littering, industry initiatives and collaboration with the civic authorities for improvement in plastics waste collection system, incentives for development and adoption of appropriate technologies and sustained campaigns for creating public awareness and involvement. Networking of concerned industry Associations, setting up of an Indian Centre for Plastics in Environment (ICPE) and constitution of an implementation and monitoring committee (IMC) were also suggested by the Task

Force for follow-up of the recommendations. The task force has following object:

(*i*) To formulate a strategy and prepare an action programme for management of plastic waste.

(*ii*) To prepare incentives/penalties/levis for checking the growth of plastics packaging waste.

(*iii*) To prepare guide lines for packaging using plastics materials.

On the basis of the Task Force Report, several initiatives have been taken, while some are on the anvil. The Ministry of Environment & Forests has issued a Gazette Notification prohibiting the use of recycled plastics for storing and carrying food staff. To restrict the indiscriminate use of plastic carry bags, the notification also stipulates the minimum thickness for such carry bags. Public awareness programmes have also been launched.

Plastics Waste Industry

The plastics waste industry has diversified its activities over the past 25 years. However, this diversification has not been- accompanied with an appropriate body for plastics waste management in the country. The management of plastics wastes in India presents an interesting and economically feasible solution to the commonly labelled 'menace' of littering plastic wates in public places. The collection of plastic wastes is the source of livelihood for the innumerable 'rag pickers' or waste collectors who are followed by the kabadiwala and waste dealers. In most cases, an entire family is involved in this trade. Plastics waste collection is termed as a 'lucrative' business as against paper, cardboard, glass bottles and metal cans. A typical kabadiwala in Delhi displays the following rate list:

News Papers in English	:	Rs. 4-5/kg
News Papers in Hindi	:	Rs. 3-4/kg
Magazines	:	Rs. 3-3.50/kg
Iron/Loha	:	Rs. 5.50/Kg
Plastics Waste (mixed)	:	Rs. 12-15/kg
Beer Bottles (per bottle)	:	Rs. 2.00

Evidently, the collection of plastics waste is more remunerative vis-a-vis other consumer wastes. Wastes generated from cold drinks/ coffee/ ice-cream cups and catering containers, which are mostly made of polystyrene, fetch anything between Rs. 15 and Rs. 25 per kg. Clear packaging film and polypacks are also attractive plastics waste items that fetch as much. India's plastics wastes recycling industry presently handles over 0.75 million tonne of different types and grades of plastics waste, including around 38000 tonnes of in-house plastics scrap, which together at the recycled stage are valued at around Rs. 2500 crores. With the expected consumption of plastics ranging between 4 to 5 million tonnes by the year 2001, and corresponding growth of packaging applications (flexible and rigid) including PET bottles' and containers, the waste generated would vary between 1 and 2 million tonnes every year.

Field visits to recycling/reprocessing units and waste dealers markets, have brought to light the need for upgradation of the working conditions of operations, as also the recycling technology. Also, it is important to pay some attention to the social status of rag-pickers and waste collectors who contribute towards clearance of plastics waste from public places and thus playa key role in the environmental management of plastics waste.

Packaging is the major application of plastics. Out of 1.88 million tonnes of plastics consumed during 1995-96, over 52 per cent was accounted for packaging applications. This trend is expected to continue. Packaging thus becomes the major source of waste. This includes PE, PVC, Pp, and Multi layer films packaging including around a 30 per cent carry forward of the previous year. This makes .it imperative for the plastics industry to plan its strategy and targets, technologically, socially and environment-ally. This calls for upgradation and diversification of recycling capacity and technology, guidelines' for managing and disciplining plastics waste, maintaining inventory of types, grades and volume of plastics waste generated from various sources, formulating specifications and codes of practices. The need to formulate and issue 'Guideline on Plastics Packaging and Packaging Waste' has been emphasised during various meetings of the Task Force. This has been based on a similar Directive issued by European Union.

To promote increased use of recycled plastics, and upgrading the consumer product applications, there is also a need for undertaking development work which would aim at volume applications, like 'that for the building and construction industry.

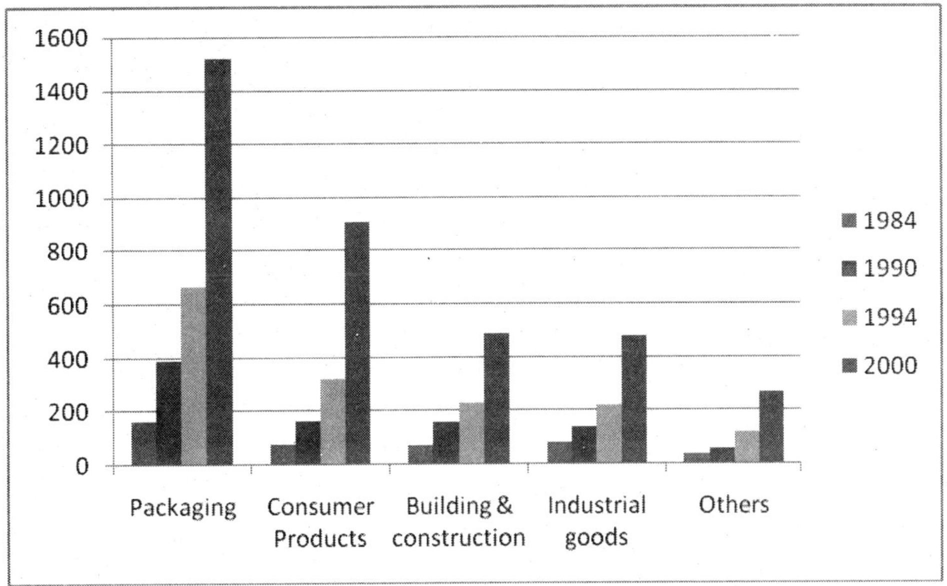

Fig. 27.1. Packaging and Consumer Products Lead the Boom in Plastics Consumption

ENVIRONMENTAL ISSUES

The waste plastics markets all over India are so packed and crowded with waste spilting over all (examples: Delhi- Asia's largest waste plastics market, handling' and trading over 1000 tonnes of waste daily; Jolly Mohala in Bangalore and Dharavi in Bombay), that frequent fires do create environmental problems both for the workers as also for residential areas around. Both the waste dealers and local authorities have to address these issues to work out safety measures. The need has often been expressed to locate such activities in eco-friendly areas.

The visible disposables in plastics' which generate waste and become an environmental eyesore after their useful service, life ends, include the following:

- Plastics packaging/carry bags/bottles/ containers/trash bags;
- Plastics from health and medicare;
- Plastics from the hotel and catering industry; and
- Plastics from air/rail travel

The polythene carrybags have been in the news for their environmental implications, right from cradle to grave. Through various stages of their use, they get disposed via municipal garbage dumps and finally collected by rag pickers to process for recycling. Carry bags of virgin plastics are accepted as user-friendly. The problem arises when plastics are recycled for repeated use. Because these are produced from wastes, these is an apparent consumer resistance.

The basic question for polythene carry bags is whether they should at all be manufactured using recycled materials (100%) and, if so, of what grade —first, second, third or the like. With repeated recycling, not only does the carrying strength of polythene bags deteriorate, but the very appearance gets repulsive (and at times unhygienic). For consumer acceptance, the recycled material of first grade should be used and in respect of the second grade, it should be a blend of 50:50 (recycled: virgin). However, third grade recycled material use should be discouraged for the manufacture of carry bags. To sustain the use of plastics through carry bags, certain gradation and acceptable consumer quality must be insisted: and adhered to by the manufacturers. The recycled polythene bags are normally priced in the range of Rs 45 to . 50 per kg, whereas for the virgin clear/brightly coloured bags, the price per kg is around Rs. 80. What is worse is that a major volume of more-than-once recycled bags remain on ground, and are not collected by the rag pickers as their resale value gets reduced, these together with other food, vegetable oil and detergents packaging (PP, Polyester, multilaye film) when not collected/recycled, become an eye-sore.

In respect of health and medicare, items in plastics, such as disposable syringes, glucose bottles; blood, and uro bags intravenous tubes and catheters, and surgical gloves, though designed for single use and manufactured with appropriate plastics materials, find themselves under attack, when these are carelessly disposed of. Some of these items even return to the market without disinfection. There have been reports of organised picking from garbage dumps around[1] major city hospitals. Such a practice is dangerous and calls for strict action. MoEF has notified rules for management of biomedical wastes which includes plastics wastes.

Plastics items (commonly of PS/PE/PVC/ PET) used in the hotel and catering industry, air and rail travel are prominently identified, and after their use, are seen carefully disposed of through 'Dustbins/ trash bags' placed in the vicinity. These disposed plastics items are quickly cleared by rag pickers/ wastes collectors, and go for recycling at a premium price. Depending upon the scale of operation, a fast-food/ catering establishment generates between 5-75 kg of plastics waste/ day. Whatever may be the merits of disposables in plastics, once their useful service life is over, they are looked down upon as eyesores in the garbage dumps.

PVC mineral water bottles, and PET liquor and mineral water bottles have invaded the market in India, as a replacement for conventional glass bottles. It is reported that around 7000 tonnes of PVC resin per annum are consumed for the manufacture of mineral water bottles and about 70 per cent gets transferred to the waste stream within four days after one time use. Because PVC is of premium grade, the used bottles are efficiently collected back for materials recycling. However, in respect of PET mineral water and liquor bottles, which are currently marketed in India, around 50 percent find themselves in waste stream within a week. A 10,000 tonnes capacity exists for manufacturing and marketing of PET mineral water/liquor bottles and assorted containers. Because of durability and glass-like clarity of PET bottles and containers, a major share of these become a long term asset for the users. However, around 50 percent of current consumption of 6000 tonnes of PET mineral water/liquor bottles used, are available for recycling. Considering the average weight of 27 g per bottle, 3000 tonnes of PET would amount to 115 million numbers of bottles going into waste that largely remains uncollected and unsold. This figure will multiply fourfold by 2001. A system of organised collection of PET bottles waste is required to be encouraged through waste collectors/dealers. Recycling of PET waste is undertaken in Indian by units in Madras, Gajraula, Kanpur and Mumbai. The existing recycling capacity is required to be fully utilised with generation of PET waste from increased use of mineral water/liquor and soft drink bottles. Appropriate product applications are to be identified and promoted in India.

Depending upon the capacity, a passenger airline per trip generates 5-10 kg of plastics waste. This includes PE/PP film, PS cups, PVC/PET bottles. This waste is identified and graded at source and goes for ready recycling.

REGULATION AND LEGISLATION: PRESENT STATUS

Until recently, there has been no definite environmental policy and legislation framed in respect of plastics waste management in India. The HP Non-biodegradable Garbage (control) Act 1995, introduced by the Government of Himachal Pradesh, envisages prohibition of throwing or deposing plastic articles in public places and to facilitate the collection through garbage in identifiable and marked garbage receptacles for non-biodegradables, placed at convenient places, Provisions of this Act, including those of existing laws" for imposing deterrent penalties may be referred to by the local authorities

The Ministry of Environment and Forests has issued the criteria developed by CPCB in association with the Bureau of Indian Standards for labelling 'plastic products' as 'Environment - friendly' under its

'Ecornark' scheme. One of the requirements for plastic products is that the material used for packaging shall be recyclable or biodegradable. Suggestions for recycled plastic products are quoted.

At present, there are no guidelines or codes of practices for collection, sorting and recycling of plastics waste in the absebce of which the conventional practices are adopted and accepted, though need has been voiced to upgrade these, both by the authorities and NGOs. However, while formulating Indian standard specifications for various plastics products, used for critical applications, like plastic piping system, water-storage tanks, packaging for food articles etc., a clause is included which reads "no recycled plastics waste shall be used". An exercise has also been carried out by the Ministry of Environment and Forests in association with the Bureau of Indian Standards to include use of recycled plastics wastes wherever appropriate in the manufacture of plastic products, and this shall be specified accordingly in the relevant Indian Specifications.

The Prevention of Food Adulteration Department of the Government of India, has issued directives of various catering establishments to use only food-grade plastics, while selling or serving food items. Rules have specified use of 'food-grade' plastics, which meet certain essential requirements and are considered safe, when in contact with food. The intention is to preventing possible contamination,' and to avert the danger from use of recycled' plastics. The scheme announced in February, 1995, is being implemented in cooperation with the Bureau of Indian Standards (SIS) which has formulated a series of standards on this subject. The Bureau of Indian Standards Sub-committee PCD 12:17 is charged with formulating guidelines, codes and specifications for recycling of plastics. Two documents viz. 'Guidelines for Recycling of Plastics', and 'Recycled Plastics for Manufacturing of products-Designation' have been finalised by BIS. These two documents, together with the 'Guideline on Plastics Packaging and Packaging Waste' are to be implemented by the industry.

Plastics have become a symbol of our throw-away society. They are non-biodegradable, but recyclable. With the technological advances, plastics recycling is economically feasible for plastics packaging. Recycled materials compete with virgin materials in terms of price and performance. It is an established fact that without reuse, total/absolute diversion of material from waste stream is impossible.

To enhance the demand for recyclable materials, various mechanisms and options are to be assessed. These include: user charge or a tax to ensure that individuals and companies bear the cost of solid waste containing the plastics they produce; government procurement policies, i.e. certain percentage of purchased products be recyclable or made of recycled materials, or price preference on items containing recycled material be encouraged, and finally, through recycling standards (meaning thereby that either the products or packaging be made or recyclable material, i.e. the material must reach a specified recycling site, or products or packaging consist of a certain percentage of recycled material and this should be appropriately labelled on the product. The advantage of recycling standards, if properly designed and applied is that they provide mechanism for coordinating recycling activities, and for establishing broad-based recycling infrastructure.

As an illustrative and useful example for recycling, the one taken by the Irish Business and Employers' Confederation and the Irish Department of the Environment at Dublin Castle, deserves special mention. Repak is the result of a challenge to industry to develop a scheme funded and organised by the industry to recycle packaging waste. It is an excellent example of a self-regulatory approach to implementation of environmental policy. In the United States, competitive nonregulatory recycling systems are responsible for recycling over 25 per cent of the total municipal wastes, with recycling having doubled over the past decade.

Table 27.3. Plastics Waste Management Status In India

	1995-96 Estimates by 2001 (Thousand Tonnes)	
Consumption of Plastics	1889(-%)	4374(-%)
Waste available for Recycling	800(-%)	2000(-%)
Total	2689	6374

(% of plastics waste in MSW 1-4% by wt.)

STRATEGY AND ACTION PROGRAMME

As a result of various meetings and discussions with the industry and experts, and the field visits to waste collection and iii) recycling centres, the Task Force identified the following strategy and action programme.

Target

Evolving integrated plastics waste management policy, with priority for increase in total recovery in terms of materials and energy.

Mission

(*i*) Raising consumer and public awareness, upgrading methodology of waste collection and segregation promoting social and environmental status of waste collectors/rag pickers, who are responsible for collection of plastics waste in India.

(*ii*) Evolving plastics waste management system with appropriate guidelines and directives; and

(*iii*) Promoting upgradation, technically and environmentally, of recycling/ reprocessing systems/ technologies, and end products applications with desired recycled content, and formulating guidelines.

Issues of Concern

Communicational

Social and consumer awareness, to promote proper disposal culture through identified and appropriately located dust bins, both in public places, residential, institutional and industrial areas, including hotels and catering establishments, through audio-visual media, publications/ newsletters/video films/ posters etc./ exhibitions/ seminars/ workshops.

Technical & Environmental

- Plastics packaging, consumption, waste generation, collection and disposal.
- Evolving plastics waste management system;
- Upgradation of materials recycling technology;
- Social and environmental issues relating to working conditions in plastics recycling industry;
- Limits to materials recycling; and »> Technology-based incineration to recover energy.

Industrial

To promote Government-industry interaction, and consumer awareness in respect of plastics waste recycling and demand for recycled content in products through:

(*i*) Systematic applications develop-mental research for promoting end products, their codification/ standardi-zation into critical and non-critical areas:

(*ii*) Industry initiatives and stewardship, by promoting shared producer-user responsibility.

(*iii*) Legislation approach; Incentives/ penalties for checking the growth of plastics packaging waste;

(*iv*) Industry-funded and government supported institutional setup with a view to promoting industry's cause towards plastics waste management through setting up of Indian Centre for plastics in the Environment (ICPE).

INTEGRATED APPROACH FOR PLASTICS WASTE MANAGEMENT

The strategy for effective management of plastics wastes should entail the three Rs: Reduction, reuse and Recycling of wastes. Hence, the action programme suggested by the Task Force includes a package of Preventive, Promotinal and Mitigative (PPM) measures to achieve these objectives. The implementation of the strategy will require active involvement of all sections of the society in which the industry and the civic authorities are the key partners. They have to act in unison to discharge their responsibilities. Public participation and catalytic support from the government are the two important prerequisites for implementation of the strategy.

The action programme for implementation of the strategy covers the following components:

(*i*) Preventive measures: Minimising use of plastics, segregation of wastes and compliance of environmental guidelines

(*ii*) Promotional measures: Improvement in waste collection system and recy-cling technologies.

(*iii*) Mitigative measures: Public awareness programme and penalties for littering, fire protection and safety measures.

Institutional Mechanism

Establishment of a network of concerned Industry Associations, and the Indian Centre for Plastics in the Environment (ICPE), for government-industry interaction.

ACTION PROGRAMME

1. Guidelines on Plastics Packaging : Packaging constitutes 52% of plastics consumption. Accordingly, this issue was addressed by the Task Force and 'Guidelines on Plastics Packaging and Packaging Waste[1] were prepared. Guidelines lay down measures aimed, as the first priority, at preventing the production of packaging waste, and as additional fundamental principles, at reusing, at recycling, and other forms of recovering packaging waste, and hence, at reducing the final disposal of such waste.

2. BIS Guidelines/Specifications : The manufacture of products using recycled plastics should follow appropriate BIS "Guideline for Recycling of Plastics" and Indian Standard " Recycled Plastics for the Manufacturing of Products designation", which have been finalised by Bureau of Indian Standards (BIS).

3. Limits to Recycling : Beyond Type-11 materials: (Post - consumer plastics waste of unknown origin having visible impurities, as per BIS Guideline), recycling of plastics waste should be banned. Alternatively, use of such plastics wastes "(beyond Type-11) should be resorted to for energy recovery. Recycling of multilayer film packaging and plastics wastes beyond Type-11 also be considered for use as composites and volume applications, such as substitutes for wood/ concrete products.

4. Circulation of Dirty Coloured Plastics Carry- bags/ Products : Consumer items, such as toys, water bottles, Kodum, carry bags etc., should not be allowed to use recycled plastics wastes, beyond Type-1 (100%). Instead a blend with virgin plastics be encouraged (50:50), and efforts should be made not to downgrade the quality and performance of end products. Reprocessors using dirty plastics wastes for the manufacture of consumer items will be warned of the environmentally unsound practice. Manufacture of dirty coloured carry-bags with visible contamination and their circulation in the market should be banned

5. Recycling Logistics : The integrated plastics wastes management need the cooperation and participation of plastics industry, local authorities and the consumers: The industry needs to take the lead in supporting pilot collection schemes with the objective of channelising more and more post-consumer plastics wastes for recycling.

6. Consumer Awareness Programme : Social and environmental issues relevant to the plastics industry should be addressed by the industry. For this, it is recommended that a country-wide consumer awareness programme be launched from time to time through media, exhibitions, newsletters, publications, videofilms,posters etc., for the education of common man, environmentalists, government departments, trade associations, educational institutions etc

Applications of Plastics Waste Management

Recycling of PET bottles

Post-consumer PET (number 1) is often sorted into different color fractions. This sorted post-consumer PET waste is crushed, pressed into bales and offered for sale to recycling companies. PET flakes are used as the raw material for a range of products that would otherwise be made of polyester.

PVC recycling

PVC or Vinyl Recycling has historically been difficult to perfect on the industrial scale. But within the last decade several viable methods for recycling or upcycling PVC plastic have been developed.

The most-often recycled plastic, HOPE or number 2, is downcycled into plastic lumber, tables, roadside curbs, benches, truck cargo liners, trash receptacles, stationery (e.g. rulers) and other durable plastic products and is usually in demand.

The white plastic foam peanuts used as packing material are often accepted by shipping, stores for reuse.

In Israel successful trials have shown that plastic films recovered from mixed municipal waste streams can be recycled into useful household products such as buckets.

Similarly, agricultural plastics such as mulch film, drip tape and silage bags are being diverted from the waste stream and successfully recycled into much larger products for industrial applications such as plastic composite railroad ties. Historically, these agricultureal plastics have primarily been either landfilled or burned on - site in the fields of individual farms.

PLASTIC IDENTIFICATION CODE

Resin Identification code

Seven groups of plastic polymers, each with specific properties, are used worldwide for packaging applications (see table below). Each group of plastic polymer can be identified by its Plastic Identification code (PIC) - usually a number or a letter abbreviation. For instance, Low-Density Polythylene can be identified by the number 4 and/ or the letters " LDPE". The PIC appears inside a three-chasing arrow recycling.

The symbol is used to indicate whether the plastic can be recycled into new products.

The PIC was introduced by the Society of the Plastics Industry, Inc. which provides a uniform system for the identification of different polymer types and helps recycling companies to separate different plastics for reprocessing. Manufacturers of plastic products are required to use PIC labels in some countries/ regions and can voluntarily mark their products with the PIC where there are no requirements. Consumers can identify the plastic types based on the codes usually found at the base or at the side of tie plastic products, including food/chemical packaging and containers. The PIC is usually not present on packaging films, as it is not practical to collect and recycle most of this type of waste.

Table 27.4

Plastic Identification Code	Type ofplasticpolymer	Properties	Common packaging Applications
01 PET	Polyethylene Terephthalate (PET, PETE)	Clarity, strength, toughness, barrier to gas and moisture	Soft drink, water andsalad dressing bottles; peanut butter and jam jars
02 PE-HD	High Density Polyethylene (HDPE)	Stiffness, strength, toughness, resistance to moisture, permeability to gas	Water pipes, Hula- Hoop (Children's game) rings, Milk, juice and water bottles; the occasional shampoo/ toiletry bottle.
03 PVC	"Polyvinyl Chloride (PVC)	Versatility, clarity, ease of blending, strength, toughness,	Juice bottles, cling films, PVC piping
04 PE-LD	Low Density Polythylene (LDPE)	Ease of processing, strength, toughness, flexibility, ease of sealing, barrier to moisture.	Frozen food bags: squeezable bottles, e.g. honey, mustard; cling films; flexible container lids.

Plastic Identification Code	Type of plastic polymer	Properties	Common packaging Applications
05 PP	Polypropylene (PP)	Strength, toughness, resistance to heat, chemicals, grease and oil, versatile, barrier to moisture.	Reusable microwaveable ware; kitchenware; yogurt containers; margarine tubes: microwaveable disposable take-away containers: disposable cups and plates.
06 PS	Polystyrene (PS)	Versatility, clarity, easily formed	Egg cartons; packing peanuts; disposable cups, plates, trays and cutlery; disposable take-away containers;
07 O	Other (often polycarbonate or ABS)	Dependent on polymers or combination of polymers	Beverage bottles; baby milk bottles; electronic casing.

Toxic Effects of Plastic Waste

The Major chemicals used to make plastic resins pose serious risks to public health and safety. Many of the chemicals used in large volumes to produce plastics are highly toxic. Some chemicals, like benzene and vinyl chloride are known to cause cancer in humans; many tend to be gases and liquid hydrocarbons, which readily vaporize and pollute the air. Many are flammable and explosive. Even the plastic resins themselves are flammable and have contributed to numerous chemical accidents. The production of plàstic emits substantial amounts of toxic chemicals (eg. ethylene oxide, benzene and xylenes) to air and water. Many of the toxic chemicals released in plastic production can cause cancer and birth defects and damage the nervous system, blood, kidneys and immune systems. These chemicals can also cause serious damage to ecosystems.

Ethylene oxide is used as a sterilant in hospitals. It is also the principle metabolite of ethene, a precursor to polyethylene plastics and other synthetic chemicals. Ethylene oxide can be measured by gas chromatography in air or biological specimens. Ethylene oxide reacts in the body with hemoglobin.

Many food containers for meats, fish, cheeses, yogurt, foam and clear clamshell containers, foam and rigid plates, clear bakery containers, packaging "peanuts," foam packaging, audio cassette housings, CD cases, disposable cutlery, and more are made of polystyrene. J. R. Withey in Environmental Health Perspectives 1976 investigated styrene and vinyl chloride monomer as being similar: "Styrene monomer readily migrates from food contained in it. It makes no difference whether the food or drink is hot or cold, or contains fat or water. It is not inconceivable that the animal body behaves as a 'sink' for styrene monomer until the lipid portion of the animal body either becomes saturated (although death would probably occur prior to this event) or the tissues are equilibrated at the same concentration as the exposure atmosphere."

PVC is used for many products including: flooring, toys, teethers, clothing, raincoats, shoes, building products like windows, siding and roofing, hospital blood bags, IV bags and other medical devices. One of it's major ingredients is chlorine. When chlorine based chemicals are heated in the presence of hydrocarbons they create dioxin, a known carcinogen and endocrine disrupter. All PVC production releases dioxin. Other sources of dioxin are: production and use of chemicals, such as herbicides and wood preservatives, oil refining, burning coal and oil for energy, all car and truck exhaust, cigarette.

Plasticizers are used in PVC that migrate into a blood recipient via the blood bag, IV bag, IV tubing, children's toys are made with PVC.

Anyone who receives blood, is on kidney dialysis or has tubes either inserted in them or has liquid or air transported to their body is at risk. About 85% of medical waste is incinerated, accounting for ten percent of all incineration in the U. S. Approximately five to fifteen percent of medical waste needs to be incinerated to prevent infectious disease. The remaining waste, while not posing any danger from infectious pathogens is very dangerous when burned. It contains high volumes of chlorinated plastics including PVC (also the toxic substances mercury, arsenic, cadmium and lead.)

WHAT YOU CAN DO ABOUT PLASTIC POLLUTION

Plastic bags and bottles, like all forms of plastic, create significant environmental and economic burdens. They consume growing amounts of energy and other natural resources, degrading the environment in numerous ways. In addition to using up fossile fuels and other resources, plastic products create littre, hurt marine life, and threaten the basis of life on earth. We are producing over 25 million tons of plastics per year in the United States, a trivial fraction of which is getting recycled. Here are some steps that you can take to reverse the tide of toxic, non-biodegradable pollution so that it will not overtake our planet.

Table 27.5

Personal Steps	Comments
Take no plastic bags from the grocer's shelf	Put produce in paper, canvas, and other healthy-fiber bags.
Refuse plastic bags at the check-out counter.	If a clerk throws your box of soap into a plastic bag, ask him or her to replace it in one of your bags. Giver the clerk a copy of "Why I Don't Use Plastic Bags," Our experience has been that they appreciate this information.
Don't buy plastic sandwich bags.	Use wax paper bags, cloth napkins, or re-useable sandwhich boxes (e.g., tiffins, described below).
Buy beverages in sustainable containers	Use only glass bottles or cans.
Don't open another plastic water bottle. Take drinking water from the tap.	Bottled water costs over 1000 times more per liter than water from your tap. Buying our most essential nutrient, water, from corporations represents an abdication of community control of the commons. If you have concerns about water safety, investigate a filter system such as Multi-Pure. Better yet, work with your water district to develop stricter standards for water purity.
Buy fresh produce in Mother Nature's wrappers (shell, rind, husk, etc.).	Pre-bagged produce not only use wasteful packaging, but also tends to come from farther away, consuming more of our dwindling oil supplies in transport.
Give up Tupper Ware and related products.	Tiffins (stainless steel food containers) are a long tradition in India. They store food well, have longer lives than Tupper Ware and its look-alikes (you've probably seen the fading, corroding, and chipping that occurs to these plastic containers), are more hygienic, and have a certain panache.
Make a habit of thinking about what comes with each thing that you buy.	Look for and reward earth-friendly packaging choice. e.g. : • Buy greeting cards in paper boxes instead of clear plastic shells• Ask your florist for flowers wrapped in paper, not clear film• Use pens that re-fill instead of land-fill
Make a habit of thinking more in general.	Conscious consumption is not only good for the earth. It's good for you. "Mindfulness," says Thich Nhat Hanh, "is the miracle by which we master and restore ourselves."
Giver away action sheets.	We will give you copies of•"Don't Think of a Plastic Bag!"• "Why I Don't Use Plastic Bags"• Other articles and background sheetsFor copies call Green Sangha at (510) 532–6574.
Ecnocurage stores to change their practices.	Share articles such as those listed above. Ask for a meeting with the manager or owner. We'll join you, or help you prepare for a successful conversation.
Organize a presentation on the hidden costs of plastics.	Members of Green Sangha will be happy to make a presentation for your church, school, civic association, or clur, Call us c/o (415) 459-0176.

Personal Steps	Comments
Study the materials and make a presentation yourself.	Green Sangha is a member of the Campaign Against the Plastics Plague, which provides a slide show and supporting notes at no cost. We offer trainings on how to make presentations in your communities.
Remove plastic from your office or business, and tell your customers why.	Green Sangha will give you sample articles and displays for your restaurant, grocery store, or hotel, explaining to customers the benefits of replacing platic packaging and reducing waste in general.
Showing Support	**Comments**
Get involved hands-on	Help clean up the mess! Across the state, over 290 non-profit and governmental agencies organize volunteer efforts to clean up the coast and prevent pollution. For example, in Marin County: **1. California Coastal Cleanup Day** Happens the third Saturaday of September each year. 9 to noon. Marin County coordinator is Keley Stock (415) 332-3871, or e-mail: Keley. D. Stock@spd02.usace.army.mil **2.** The **Salmon Protection and Watershed Network** (SPAWN) hosts periodic clean-ups. Check their website. Spawunsa-org. or call (415) 488-0370. **3.** The **Cordell Bank Marine Sanctuary** also leads clean-ups at country beaches. Contact Joanne Mohr at (415)561-6625 x307 or jmhor@farallones.org.For a complete listing of agencies in Calfornia, go to : http://www.coastal.ca.gove/publiced/directrory/resdirectory/redindex/html

REFERENCES

1. Anand, S.J.S. 1978. Determination of Mercury, Arsenic and Cadmium in fish by neutron activation. *J. Radio, Anal. Chem.* 44, 101-107.
2. Goldberg, E.D. 1976. *The Wealth of the Oceans,* The UNESCO Press Paris, 172.
3. Lopez, J.N. and G.F. Lee. 1977. Environmental Chemistry of Copper in Torch Lake, Michigan, *Water, Air* and *Soil Pollution*, 8, 373-385.
4. Noggle, G.R. and G.J. Fritz 1986, *Introductory Plant Physiology,* Prentice-Hall of India Pvt. Ltd., New Delhi.
5. Pillai, K.C. 1985. Heavy metals in aquatic environment. In C.K. Varshney (Ed.) 1985. *Water Pollution and Management,* Wiley Eastern Limited, New Delhi.
6. Rai, L.C. and M. Raizada. 1987. Toxicity of nickel and silver ions to *Nostoc muscorum* : Interaction with ascorbic acid, glutathione and sulphur containing amino acids. *Ecotoxiology and Environmental Safety,* 14, 12-20.
7. Rai L.C. and M. Raizada, 1988. Impact of Chromium and lead on *Nostoc muscorum* : Regulation of Toxicity of ascorbic acid, glutathione and sulphur containing amino acids. *Ecotoxiology and Environmental Safety,* 15, 21-32.
8. Srivastava, Alka and V.S. Jaiswal, 1989. Effect of Cadmium on turion formation and germination of *Spirodela polyrrhiza L.J. Plant Physiology*, 134, 385-387.
9. Alloway, B.J. (ed) (1990) *Heavy Metals in Soils,* Blackie and Son.
10. Bowen, H.J.M. (1979) *The Environmental Chemistry of the Elements,* Academic Press, London.
11. Fergusson, J.E. (1990) *The Heavy Elements : Chemistry, Environmental Impact and Health Effects,* Pergamon Press, Oxford.

CHAPTER 28

Bio-medical Waste Management

—Syed Fareed Uddin[1]

Introduction

The management of health care waste is a subject of considerable concern to public health and infection-control specialists, as well as the general public. It is a well known fact that in several types of health care activities, various types of hazardous and contagious materials are generated. Even though the consequences of discarding such waste carelessly are well known, it is only recently that adequate initiatives to manage this waste in a scientific manner are being taken in India.

Unscientific disposal of health care waste may lead to the transmission of communicable diseases such as gastro-enteric infection, respiratory infections, spreading through air waster and direct human contract with blood and infectious body fluids. These could be responsible for transmission of hepatitis B, C, E and AIDS within the community. Health care professionals and the general public are at risk due to the disease spread by improper treatment and disposal of waste. Rag pickers expose themselves to diseases like Hepatitis B, Tetanus, Staphylococci, etc. while handling items like needles, surgical gloves, blood bags etc.

What is Bio-Medical Waste?

Bio-Medical waste is "Any waste which is generated during the diagnosis, treatment or immunisation of human beings or animals or in any research activities pertaining thereto or in the production or testing of biologicals. It includes "any waste which is generated during the diagnosis, treatment or immunisation of human beings or animals or in any research activities pertaining there to or in the production or testing of biologicals, and including categories mentioned in Schedule 1" - as defined by the Bio-Medical Waste (Management and Handling) Rules, 1998.

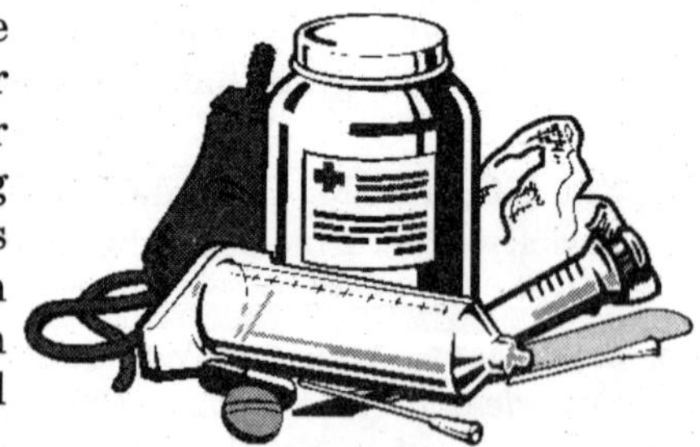

Fig. 28.1

It includes infectious and non-infectious waste. Infectious waste includes pathological waste, cotton, dressing, used needles, syringes, scalpels, blades, glass, etc. and non-infectious waste includes general waste from the kitchen / canteen, packaging material.

What are the Bio-medical Waste Rules?

The Government of India formulated the Bio-Medical Waste (Handling and Management) Rules in 1998 (hereafter referred to as the Bio-Medical Waste Rules) in order to specify procedures that have to be followed in the management and disposal of waste.

[1]Project Operations Manager, JICA/M.P. Reproductive Health Project, Bhopal (India)

The Rules apply to all hospitals, nursing homes etc. in the country.

The Rules apply to all persons who generate, collect, receive, store, transport, treat, dispose or handle Bio Medical Waste in any form.

The Bio-Medical waste (Management and Handling) Rules; 1998 are conferred by section 6,8,25 of the Environment (Protection) Act, 1986 (29 of 1986).

Environment (Protection) Act, 1986

This Act is umbrella legislation providing a single focus in the country for the protection of Environment and seeks to plug the loopholes of earlier registration relating to environment. Several sets of rules relating to various aspects of management of hazardous chemical. Waste, microorganisms, biomedical waste etc.

I. Whoever fails to comply with or contravenes any of the provision of this Act, or the rules made or order or directions issued there under, shall in respect of each such failure of contravention, be punishable with imprisonment for a team may extend to one five year or with fine which may extend to one lakh rupees, or with both, and in case the failure or contravention continues with additional fine which may extend to five thousand rupees for every day during which such failure or contravention continues after the conviction for the first such failure or contravention.

II. If the failure or contravention referred to in sub- section (i) continues period of one year after the date of convocation, the offend shall be punishable with imprisonment for a team which may extend to seven years

The Act on Bio medical waste management dated 27/07/98 puts the time limit to abide by all the specifications required for implementation as 31 Dec, 2002. However, implementation of Bio medical waste management is yet to be commenced in many places. One of the hurdles being posed is the low awareness within the health staff related to this topic.

Table 28.1. Classification of Bio-medical Waste

Category	Description	Treatment & Disposal
1.	Human and Anatomical Waste Human Tissues, Organs, Body parts.	Incineration / Deep Burial.
2.	Animal Waste Animal Tissues, organs, body parts, carcasses, bleeding parts, fluid, blood and experimental animals used in research, waste generated by veterinary hospitals colleges, discharge from hospitals and animal houses.	Incineration / Deep Burial.
3.	Microbiology & Biotechnology Waste from laboratory cultures, stocks or specimens of micro-organisms live or attenuated vaccines, human and animal cell cultures used in research and infectious agents from research and industrial laboratories, wastes from production of biologicals, toxins, dishes and devices used for transfer of cultures.	Incineration Autoclaving/ Microwaving.
4.	Waste Sharps Needles, syringes, scalpels, blades, glass, etc. that may cause punctures and cuts. This includes both used and unused sharps.	Disinfection Chemical disinfection / autoclaving / microwaving and mutilation / shredding.
5.	Discarded medicines and Cytotoxic drugs Waste comprising of outdated, contaminated and discarded medicines.	Incineration / Destruction and drugs disposal in secured landfills.
6.	Solid Waste Items contaminated with blood, and body-fluids including cotton, dressing, soiled plaster casts, lines, bedding, other material contaminated with blood.	Incineration Autoclaving/ Microwaving.
7.	Solid Waste: Items generated from disposable items other than sharps such as tubings, catheters, intravenous sets etc.	Disinfection by chemical treatment Autoclaving / Microwaving and Mutilation / shredding.

8.	Liquid Waste, Waste generated from laboratory and washing, cleaning, housekeeping, and disinfecting activities.	Disinfection by chemical treatment and discharge into drains.
9.	Incineration Ash, Ash from incineration of any Bio Medical Waste.	Disposal in Municipal Landfill.
10.	Chemical Waste Chemicals used in production of biologicals, chemicals used in disinfection, as insecticides, etc.	Chemical treatment and discharge into drains for liquids and secured landfill for solids.

Table 28.2. Colour Coding and Type of Container for Disposal of Bio-medical Wastes

Colour Coding	Type of Container -I Waste Category	Treatment options as per Schedule I
Yellow	Plastic bag Cat. 1, Cat. 2, and Cat. 3, Cat. 6.	Incineration / Deep Burial
Red	Disinfected container/plastic bag Cat. 3, Cat. 6, Cat.7.	Autoclaving/Microwaving/Chemical Treatment
Blue/White translucent	Plastic bag/puncture proof Cat. 4, Cat. 7. Container	Autoclaving/Microwaving/Chemical Treatment and destruction / Shredding
Black	Plastic bag Cat. 5 and Cat. 9 and Cat. 10. (solid)	Disposal in secured landfill

BIOHAZARD Symbol

Bio Hazard Symbol should be put up on all the Coloured Bins for Segregation and the Vehicles being used for transportation of Bio Medical Waste

How do we Deal with the Waste?

Hospital waste is becoming increasingly complex due to changing technologies and increase in the services that the hospitals perform for the community. Management of waste presupposes a scientific approach to the process of waste generation, storage,-transport, treatment and its disposal. It is of utmost importance that bio-medical waste thus generated be managed in an environmentally sound manner, which involves proper understanding of risks associated with the handling of such wastes. Only a cradle-to-grave approach will help in first minimizing, then collecting and finally treating and disposing the waste.

Fig. 28.2. Treatment as per Colour Waste Category Schedule I of the Rules Code

There are a few basic steps that hospitals have to follow in order to deal with Bio-Medical Waste.

The first and the most crucial step being that of *SEGREGATION*. Segregation of waste refers to the basic separation of the different categories of waste generated (as given earlier) at the source of their generation.

The Bio-Medical Waste (Management and Handling) Rules, 1998 have stipulated the following method of segregation on the basis of a simple colour-coded system:

How to Treat Waste?

Each kind of waste requires a different process of treatment depending on the material it is made of as well as the kind of micro organisms and waste it is likely to contain. Given below is a list of the different kinds of waste and the measures that have to be taken for safe handling and disposal of each kind of waste.

Infectious Waste

Bags should be colour coded and should also be properly labelled to avoid confusion while handling or disposing.

Infectious waste should be segregated at the point of generation itself and bins lined with inert material or with inner chambers for bleach should be used.

Usage of a lidded bin will discourage inadvertent use by others and also keep it away from the public.

Personnel involved in the handling of infectious waste should be provided with suitable protective gear. Proper training in managing this waste as well as in handing emergency situations like spillage of the waste should be given to them.

It is easier to handle and transport waste when bags are not completely full. This also reduces the risk of spillage around the bin. The bag has to be sealed at the top before transportation within or outside the hospital.

The recommended method of destruction of infectious waste if autoclaving and microwaving. Incineration is also an option but only for certain kinds of waste. The former however apart from being more environmentally friendly is also more cost effective than incineration.

All Operation Theatre (OT) waste should be sent for Autoclaving.

Disposable Waste

Such items are often single use, disposable products like syringes, IV bottles, sharps, catheters and gloves. These items are often recycled and reused illegally and it is therefore imperative that chemical disinfection be followed for them. They have to be dipped in a chemical disinfectant solution (concentration depends on the potency of the waste - atleast 10gms per litre of water)) for a minimum duration of 30 minutes to 1 hour or autoclaved or microwaved. The bins used for chemical disinfection are a set of bins - one inside the other. The smaller being perforated and easily extractable. This will help ensure that the bleach solution in the outer bin permeates the inner bin containing these waste items and minimises contact with the waste while the waste is being removed.

These items once disinfected have to be cut or mutilated in order to ensure that they are not reused. For instance, the fingers of the gloves should be cut and the IV bottles punctured.

Sharps should be handled with proper protection.

Blood bags should not be handled.

Bleach solution should be changed after every shift.

The plunger and the barrel of the syringe have to be separated before disinfecting it.

Sharps

Sharps have been defined by the Central Pollution Control Board consist of needles, syringes, scalpels, blades, glass and so on. These are all capable of causing punctures, lacerations and cuts.

Sharps need separate attention as the risk of injury and infection in this area is very high. They therefore have to be separated at the point of generation.

Manual bending, breaking or clipping of needles should be avoided as this may cause accidental innoculation. They should be destroyed with a Needle Cutter / Destroyer and then shredded.

Sharps should be placed in a puncture proof container and it has to be marked conspicuously by the Universal Biohazard symbol.

Liquid Waste

Liquid chemical waste has to be neutralised with reagents before disposal.

Liquid pathological waste has to be treated with disinfectant before disposal.

Treatment of Disposal of Segregated Wastes

- The segregated waste form all the bins (except Yellow bin) should be treated before disposal.
- The treated waste should be disposed off according to the categories.

The following table details how to treat the waste generated from each bin before it is disposed.

Table 28.3.

Sections	Black bins	Red bins	Yellow bins	Blue bins/white puncture proof polybag or bin
Observations & LR	Medicine foil, medicine packet, Spoiled medicines	Gauze, Swabs, blood stained cotton, blood stained cloths, sanitary pad	Placenta	Needles, syringes Urobag, Gloves, broken bottles, IV set, IV bottles, Catheters, Blood transfusion bag, mucus extractor, enema set, venflon.
Post natal & other wards	Medicine foil, medicine packet, Spoiled medicines, food packets, plastic carry bags, fruit peels, waste food items, paper cups	Gauze, Swabs, blood stained cotton, blood stained cloths, sanitary pad, baby nappies, pus stained gauze	Umbilical cord shedding	Needles, syringes, Urobag, Blood transfusion bag, broken bottles, IV set, IV bottles, Catheters, mucus extractor, cord clamp, enema set, venflon
Dressing room and Injection room	Medicine foil, medicine packets, Spoiled medicines, paper cups	Gauze, Swabs, blood stained cotton, blood stained clothes, Plaster, pus stained gauze	Tissues	Needless, syringes Gloves, broken bottles

Sections	Black bins	Red bins	Yellow bins	Blue bins/white puncture proof polybag or bin
Laboratory	Laboratory Reagents, paper cups	Gauze, Swabs, blood stained cotton, Blood, urine, stool, sputum sample, used uristiks, pus stained gauge	Biopsy material	Needless, syringes broken bottles, Tubes, pipette, glass slides, lancets
Store	Paper, medicine foils, packing material, expired medicine, paper cups			Broken injection vials, medicine bottles, presterilised disposable items where the packaging is tampered.

Deep Burial Pit

Deep Burial Pit for disposal of Human & Animal waste is the practical solution and has the approval from the Ministry of Environment and Central Pollution Committee for the rural areas and Towns where the population does not exceed 5 lakh and there is no Common Waste Treatment Facility available in the vicinity. The specification of the pit as per the figure given hereunder :-

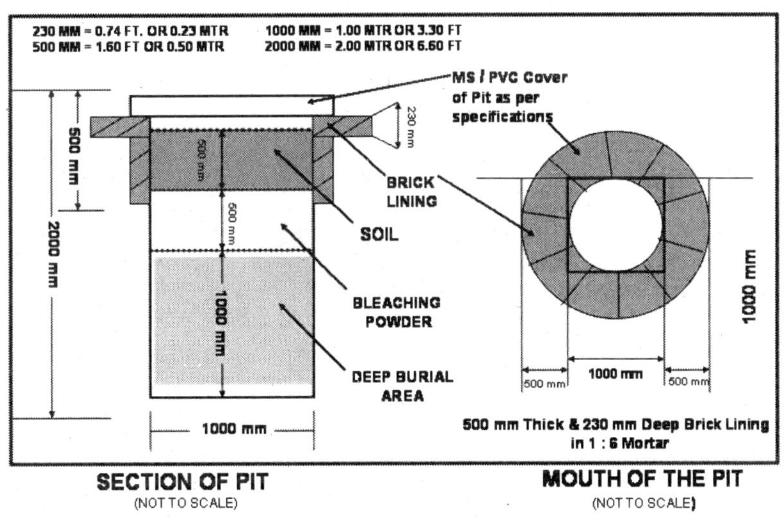

Fig. 28.3

1. A pit or trench should be dug about 2 metre deep. It should be half filled with waste then covered with lime within 50 cm of the surface, before filling the rest of the pit with soil.
2 It must be ensured that animals do not have any access to burial sites. Covers of galvanized iron/ wire meshes may be used.
3. On each occasion, when wastes are added to the pit, a layer of 10 cm of lime powder and soil shall be added to cover the wastes.
4. Burial must be performed under close supervision.
5. The deep burial site should be relatively impermeable and no shallow well should be close to the site.

6. The pits should be distant from habitation, and sited so as to ensure that no contamination occurs of any surface water or ground water. The area should not be prone to flooding or erosion.
7. The location of the deep burial site to be authorised by the prescribed authority.
8. The institution shall maintain a record of all waste for deep burial.

Emergency Situations

Accidents should be avoided and therefore have systems to deal with emergency situations should be developed. One of the most common occurrences while dealing with Bio-Medical Waste is the spillage and leakage of waste.

Spill Protocol is therefore essential for any hospital. This refers to the measures that have to be taken to contain and decontaminate the accident site. This should include the following:

1. The surface containing the spill has to be mopped up with a swab soaked in disinfectant and then the swab should be put in the infectious waste bin.
2. No reagents should be sucked into the pipette with the mouth.
3. Other precautions to be kept in mind are the same as those followed in the rest of the hospital.
4. Laboratories should be well ventilated so as to ensure that personnel do not breathe-in contaminated air. Precautions should also be taken to ensure that the general public does not breathe in this air.

CENTRALISED FACILITY FOR BIO-MEDICAL WASTE:
[Common Waste Treatment Facility (CWTF)]

Waste from Waste from Blood banks Hospitals

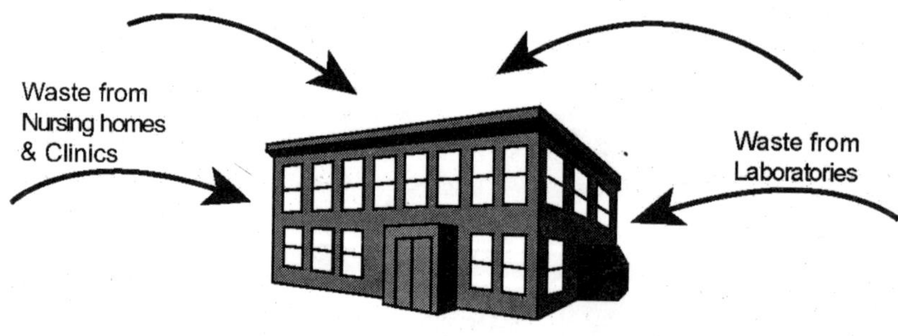

Fig. 28.4

- Small Healthcare Units, like Nursing Homes and Clinics, with minimal investment and space constraints may find it difficult to afford an individual disposal facility for bio-medical waste.
- In such cases these units could overcome this shortcomings by teaming up to establish a centralised facility for disposing off bio-medical waste.
- The Bio-Medical Waste Rules also have provisions for such combined facilities.
- Many municipalities and corporations across the country are proposing to establish such common treatment facilities.

Community's Role in Bio-medical Waste management

While management of bio-medical waste is primarily the responsibility of medical institutions and those who actually generate this waste, the community has a very important role to play in ensuring that the hospital practices the prescribed procedures for treating bio-medical waste.

Patients, who form a part of the community, constantly utilise services of healthcare institutions and hence they also share the responsibility of ensuring that these institutions do not pollute the community.

A community can therefore do the following to ensure a higher level of health:

- Communities should strive to ensure that awareness programmes should be conducted in their areas through residents' associations on the issue.
- Communities should ensure medical practitioners having clinics in their localities do not dispose off their waste in the municipal waste stream, but arrange to send it to the centralised facility.
- Communities should remain vigilant and should promptly report breach in proper bio-medical waste practices by any hospital to the pollution control board.

Patients Role in a Hospital

- Patients can ensure that they dispose off of waste only in the bins provided in the hospital and help keep the premises clean and litter-free.
- Patients should understand the system of Bio -Medical Waste Management followed in the hospital.
- Patients should make sure that they report any irregularities in the hospital to the management.
- Patients with infectious diseases should ensure that they strictly adhere to the procedure suggested by the doctors for disposal of body fluids like sputum.

Fig. 28.5

Procedural Requirements under the Bio-medical Waste Rules

- Hospitals are required to file Form I to the prescribed authority for grant of Authorisation accompanied with the payment of the fee as prescribed by the State or Central Government.
- They are also required to submit a copy of their Annual Reports to the prescribed authority under Form II by January 31st every year.
- These reports shall be sent to the Central Pollution Control Board by the state pollution control board.
- Hospitals are mandatorily required to maintain records and report accidents under form III.

Overview of Technology used for Bio-medical Waste Treatment and Disposal

As bio-medical is a specialised class of waste, that is highly infectious and hazardous, there are specific technologies are required to treat and dispose the waste. The standard technologies being utilised in the country for treatment and disposal of bio-medical waste are:

- Microwave
- Autoclave
- Incineration

Microwave

- The microwave is based on the principle of generation of high frequency waves.
- These waves causes the particles within the waste material to vibrate.
- Generating heat.
- This heat generated from within, kills all pathogens.

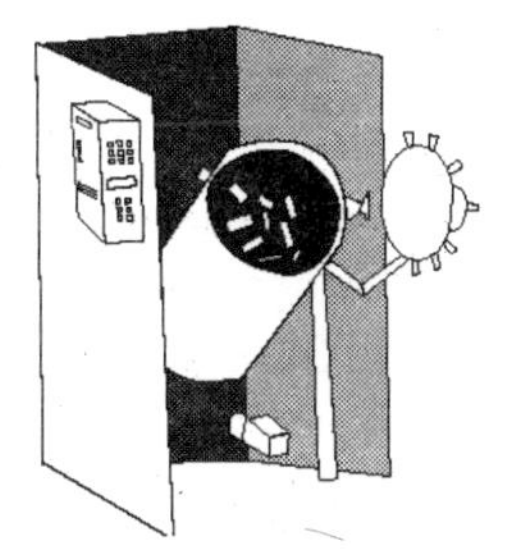

Fig. 28.6

Autoclave

The autoclave operates on the principal of standard pressure cooker

- The process involves using steam at high temperatures.
- The steam generated at high temperature penetrates waste material and kills all the micro organisms.

Incineration

Fig. 28.7

A process that works on the simple principle of burning or combustion is technically called incineration. The incinerator, as the machine is referred to, uses either oil or electricity to power itself. Waste material is fed into the incinerator and is burnt in it. Normally incinerators are operated at temperatures between 300°C to 1100°C based on the volume of waste, the type of incinerator and the type of fuel used. Incinerators used in India are either single chambered or double chambered.

Incineration not only attempts to both kill the pathogens but also destroy the materials in which these reside - most of which are plastic disposables or cellulose rich materials etc. The burning of plastics, especially in unregulated incinerators is extremely hazardous as it creates a new set of chemical toxins, some of which according to current research, are highly toxic even in trace quantities. Some of the *chemical toxins produced by waste incinerators* are:

(*a*) Heavy metals, such as lead, cadmium, arsenic, chromium, nickel and so on, which are compounds that are present in plastics.

(*b*) Acid gases such as sulphur gases, hydrogen chloride and nitrogenous gases, particulate matter

(*c*) Dioxins and furans.

(*d*) Poly-Chlorinated Bi phenyls (PCBs) which, if not trapped in pollution control devices, have grave health effects on humans like endrocrinal problems thus causing disruptions in the human nervous system. If trapped, they become a part of the fly ash which is also very toxic and has to be disposed off carefully. Of these, *dioxins and furans are extremely toxic.* These belong to a family of polycyclic aromatic hydrocarbons compounds which are formed when PVC plastic present in the waste is burned.

What is Dioxin?

Dioxin is the common name for a class of 75 chemicals. Dioxin has no commercial use. It is a toxic waste product formed when waste containing chlorine is burned or when products containing chlorine are manufactured. PVC (polyvinyl chloride) plastic is a major source of the chlorine in medical waste. Commonly used PVC items in health care include medical equipment such as IV bags, gloves, tubing, oxygen tents, mattress covers, packaging and office supplies such as medical binders.

Fig. 28.6

Exposure

When medical facilities burn their waste containing chlorinated plastics like PVC, dioxin will be emitted from the smokestacks of the incinerator. Dioxin particles travel long distances (transboundary). They are highly stable and do not break down and hence travel up the food chain. Ninety percent of human exposure to dioxin occurs through our diets of meat, dairy products, eggs and fish. Dioxin builds up in fatty tissue.

Health Effects

Dioxin is proved to be a human carcinogen by the International Agency for Research on Cancer (IARC). Immune system disruption, reproductive and development Effects and hormone disruption are amongst the major health problems reported from dioxin exposure. (Adapted from the Dioxin Factsheet of Essential Information).

Incinerators are difficult to run : In hospital environments, technologies like incineration fail because untrained hospital staff operate incinerators. Surveys show that most Incinerators (over 85%) are operated at incorrect temperatures, do not destroy the waste completely, are fuel inefficient, and are out of order most of the time. *There is a lot of difference between the theory and practices employed in the operation of Incinerators, making them a high risk method of disposal of waste especially medical waste.*

Sources of Bio-medical Waste

The primary sources of Bio-Medical Waste are - Hospitals, Diagnostic Centres, Laboratories, Blood Banks, Nursing Homes and Clinics and Veterinary Hospitals and Clinics.

Non-infectious waste forms nearly 90% of the waste generated by a hospital.

The remaining 10% comprises of infectious waste and is generated in all the Wards, Operation Theatres, Intensive Care Units, Laboratories and Blood Banks. The waste generated in each of these areas can be categorised as follows:

General ward (Out Patient Department - OPD), Department wards, Intensive Care Unit and Emergency Care : Cotton, dressing, bandages, syringes, needles, IV sets and tubing, blood sets, urine bags all contaminated with blood, pus or other body fluids and other waste like packaging, paper waste and food waste.

Operation Theatres : Pathological waste, Cotton, dressing, instruments, contaminated plastic waste like syringes, tubing, IV sets, Blood sets, contaminated linen, contaminated gloves, caps, masks, hospital gowns used by the patients as well as the staff and doctors.

Laboratories : Contaminated samples, cultures, pipettes, petridishes, tips, test tubes (both plastic and glass), slides.

Blood Banks : Contaminated samples, cultures, pipettes, petridishes, tips, test tubes (both plastic and glass), slides, blood bags, unused blood bags (past the expiry date) and infected blood bags.

Nursing Homes and Clinics : These generate the same kind of waste that hospitals generate but on a smaller scale depending on the facilities provided and the number of beds.

Why Bother about Bio-Medical Waste?

- Such waste is hazardous to health care worker, rag pickers, and municipal workers
- Polyvinyl chloride and disposable medical supplies of plastics can be reused
- Incineration of PVC generates dioxins toxic to humans and environment
- HCW are at heightened risk because of frequent handling and exposures

Occupational Hazards

- Hospital staff and waste handlers in the hospital run the highest risk of contacting disease from following improper Bio Medical Waste Management practices as they are exposed to a multitude of diseases and infection. Therefore, in the interests of the hospital, recognising the importance of Bio Medical Waste Management at the earliest is crucial as apart from contacting disease, hospital staff can also be carriers of disease and infection.
- The current situation prevalent among healthcare waste handlers in hospitals is tantamount to violation of the basic human right of just and favourable conditions of work, in particular the Right to safe and health working conditions but also that of basic Right to Life and security of person.
- The hazardous nature of health care waste may be due to one or more of the following characteristics:
  1. It contains infectious agents.
  2. It is genotoxic (waste with mutagenic, teratoxic or carcinogenic properties). This kind of waste includes - certain cytotoxic drugs (those drugs with the ability to kill or stop the growth of certain living cells), vomit, urine or faeces from patients treated with cytotoxic drugs, chemicals and radioactive materials.

3. It contains toxic or hazardous chemicals or pharmaceuticals.
4. It is radioactive.
5. It contains Sharps.

Infectious waste may contain a wide variety of pathogenic micro-organisms. Pathogens in infectious waste may enter the body through a number of routes: - Through a puncture, abrasion or a cut in the skin. Through the mucous membrane.. By inhalation.. By ingestion.

- There is a particular concern about infection with the Human Immuno-deficiency Virus (HIV) and Hepatitis viruses B and C for which there is strong evidence of transmission via health care waste. These viruses are generally transmitted through injuries from syringe needles contaminated with human blood.
- Different pathogenic micro-organisms have varying abilities to survive in the environment. For example it has been found that the Hepatitis B virus is very persistent in dry air and can survive for several weeks on a surface; it is also resistant to brief exposure to boiling water. It can survive exposure to some antiseptic chemicals and remains viable for up to 10 hours at a temperature of 60° C.
- Equipment that is essential for a worker handling waste: Rubber gloves, shoes or boots, thick trousers and thick long-sleeved shirts. They should also be provided with convenient washing facilities (with warm water and soap) - particularly at the storage and incineration facilities.

  Periodic immunisation of hospital workers and staff against Hepatitis B is an absolute necessity.
- As far as cleaning up of spillages of body fluids or other potentially infectious waste is concerned, workers have to be provided with protection for eyes, respirators and hand tools like a shovel (to avoid direct contact with the waste) in addition to the standard gear mentioned above, in order to avoid any risk of eye injury and inhalation of any toxic fumes or dust.

Mercury in Hospitals

- Mercury is used in hospitals in thermometers, blood pressure instruments, feeding tubes, dilators and batteries, dental applications, fluorescent tubes and in specific laboratory chemicals. Mercury is a highly toxic metal that has the ability to pass all the four main human physiological barriers-skin, blood, brain and placenta. Mercury can cause a variety of diseases on exposure like-bronchitis, muscle tremors, irritability and personality changes. It can also affect the central nervous system in a variety of ways-impaired vision and hearing, paralysis, sleeplessness, emotional instability, developmental defects during foetal development and during childhood. It is particularly dangerous to foetuses, women of child-bearing age, pregnant women and young children.

  It is important to note that toxic heavy metals like mercury produce health effects and symptoms that are not specific and are common to other disorders. Hence, health effects of Mercury maybe wrongly diagnosed or may even escape diagnosis i.e. they may not be attributed to heavy metal poisoning.
- Hospitals and healthcare service providers in western countries are switching to non-mercury alternatives and gradually phasing out mercury in medical applications. Non -mercury substitutes are not only safe but also more accurate than their mercury counterparts.

Duty of Occupier

According the Bio Medical Waste Management & Handling Rules – "It shall be the duty of every occupier of an institution generating bio-medical waste which includes a hospital, nursing home, clinic, dispensary, veterinary institution, animal house, pathological laboratory, blood bank by whatever name called to take all steps to ensure that such waste is handled without any adverse effect to human health and the environment".

A Cause of Concern

The major portion of Bio Medical Waste is generated in Hospitals and Health Care facilities where a good no. of patients visit with a hope of getting cured but if the Bio Medical Waste is not handled or managed properly, not only the patients will get further infected and acquire more grievous diseases than the ones they came to get cured of and even the attendants and the health care providers in these facilities are at equal risk.

Need of the Hour

Bio Medical Waste Management has been "Nobody's business" whereas it should be "Everybody's Concern".

There is need for a good coordination between the Management of Health Care providers and the Enforcement agencies (Pollution Control Boards) and then only the suggestions provided in the Environment Protection Act will get implemented.

There is big gap between enforcers and the implementers. The enforcing agencies should understand their role in a better way and should start facilitating the Health Care Providers to make administrative and financial provisions in their annual Plans. Efforts should also be made towards training the Staff at Hospitals and Clinics on the Bio Medical Waste Management.

There is also a great need to create a public awareness on the subject so that the community at large understands the risks involved.

REFERENCES

1. APHA standard methods for the examination of water and waste water Edited by Andrew D. Eaton, Lenore S. clesceri, Arnold E. Greenberg 19th edition (1995).
2. Arion *et.al., Hospital Solid Waste Management, A Case Study, J. Env. Engg. Div.,* 106 (EE4) August: 741-756 (1980).
3. Biomedical Waste (Management and Handling) Rules, 1998 vide S.O. 630 (E) Ministry of Environment and Forest Notification, dated 20th July (1998).
4. HCOHSA, Report of the results of the Bio-medical waste management survey. Health care occupational health and safety association, Toronto and Ontario Hospital association, Ontario (1985).
5. Mc Gate, A.M. Solid waste incineration and heat recovery at royal jubilees hospital B.C. Toronto (1980).
6. *NEERI,* manual on water and waste water analysis, Nagpur, India (1986).
7. The Environmental (Protection) Act, published in the Gazette vide S.O. 756(E) (1986).
8. Tolerance limits for industrial effluent discharged the inland surface water Bureau of India Standards, IS:2490 (1974).
9. Witts, Information on Science and Technology Application Part-4, Waterfalls Institute of New Delhi (2004).

CHAPTER 29

Biofertilizers : An Ecofriendly Input for Sustainable Agriculture

Dr. Neena Arora[1]

Sustainable agriculture requires the management of resources in a way to fulfill changing human needs without damaging or deteriorating the quality of environment and conserving vital natural resources The strategy for sustaining satisfactory yield levels envisages nutrients balances and efficient nutrient cycling. The nutrient supply from soil can be enhanced by adopting appropriate soil management and conservation practices to reduce losses of nutrient through leaching, erosion and run off. Amelioration of problem soils and improving soil physical conditions insure maximum possible efficiency of applied nutrients.

Soil is one of the most significant ecological factor, which is derived from the transformation of surface rocks. It constitutes an important medium where in numerous plants and animal live. In fact soil of a nation is its most valuable material heritage. The importance of soil may be realized from the statement of Sunder Lal Bahuguna(1987). "The eternal truth that soil and water are the two significant capital of mankind and the natural forests are the mother of rivers and factories for manufacturing soil." The soil provides homes and ideal conditions for living being. Life on earth depends directly on the living soil and the aquatic eco-system of rivers. Without fertile soil and microbial fauna that inhabit it, food would not grow, dead things would not decay and nutrients would not be recycled. An estimate showed that 10% of the fertile soil of the planet has been transformed by human activities from forest into desert, while 25% or more is at risk.

In green revolution use of hybrid seeds, chemical fertilizers and pesticides increased agriculture productivity initially but then a plethora of problem cropped up. Fertilizers over the year not only reduced the average farm yield but also made the crop more susceptible to pests and disease. A gradual increase in the use of chemical inputs in the agriculture forced us to experience the validity of *law of diminishing return,* as in spite of more inputs of chemical fertilizers the yield start falling. Excessive use of nitrogen fertilizer in many rice-and wheat-producing states in comparison to phosphatic and potassic fertilizers has not only deteriorated the soil health but has also impaired the health of human beings and animals. Similarly indiscriminate and excessive use of pesticides produced health hazards in animals and human beings and soil macro/micro flora and fauna. This policy which resulted in the green Revolution in some parts of the world including India is not appropriate/adequate and does not even serve to adresss the problem of maintenance of soil fertility in fragile system. Man depends so much on fertilizer that it is difficult to think of agriculture without their use. Excessive use of these fertilizers by man is one of the

[1]Department of Chemistry, Sri Sathya Sai College for Women, B.U., Bhopal (India)

major source of land as well as water pollution. Throughout the world, soils are increasingly being impacted by mineral nitrogen release resulting from human activities. Some of the adverse effects are :

- When nitrogen fertilizers are used, an inevitable consequence has been higher nitrate levels in water, which can contribute to infant respiratory problems and possioiy 10 ine production of nitrites and the formation of nitrosamine carcinogens.
- Plants grown in high nitrate concentration soil may accumulate nitrate to a level that is harmful for animals. Cereals, grains, many weeds and grass hay contain high nitrate levels when grown in such soils.
- Nitrogen fertilizers can affect the microbial community structure and function. The use of nitrogen fertilizers has led to decrease in filamentous fungal development in a wide variety of soil. The loss of fungi affects the soil structure because fungi link soil particles to stabilize the soil peds. With a weakened and decreased fungal community, mycorrhizal dependent plants become more susceptible to stresses such as drought and toxic metals.
- Excessive and imbalanced use of chemical fertilizers has adversely affected the soil causing decrease in organic carbon, reduction in microbial flora of soil, increasing acidity and alkalinity and hardening of soil.
- Excessive use of N-fertilizer are contaminating water bodies, thus affecting fish fauna and causing health hazards for human beings and animals.
- Eutrification of lakes due to excessive addition of nitrate and phosphate ions in aquatic system.
- 25% to 30% decrease in protein content have been reported in corn, maize, gram and wheat crops when grown in soils fertilized with NPK fertilizer.
- Use of fertilizer produces oversized vegetables and fruits which are more prone to insects and other pests.
- Production of chemical fertilizers adds to the pollution.

Excessive Fertilizer Use can have many Unexpected Global-level Effects

Soil fertility refers to the syndrome of physical, chemical and Biological factors and processes influencing the potential of soil to support crop growth desired by human beings. Nutritional and climatic requirement of crop become as important as the capacity of soil to supply nutrients in determining the production efficiency of agro ecosystems. The capacity of soil to supply nutrients and water depends upon the interaction of a whole range of biotic and abiotic factors operating in the soil system. Soil fertility management, in a broad sense, embraces actions or interventions for meeting the ever increasing crop-based demands of fast multiplying human population.

Soil organic matter and nutrients are important "regulators" of stand productivity. Key processes regulating the sustainability of ecosystems include soil surface hydrology, soil organic matter dynamics and the synchrony between soil processes and plant nutrient demands. Most natural and undisturbed ecosystems have perfect synchrony in nutrient release and uptake by plants.Most of the land management systems suffer from disruption of such synchrony.

In this context, the use of microbial inoculants (biofertilizers, phytostimulators and biopesticides) is considered as the alternate source to meet the nutrient requirement of crops. Knowing the deleterious effects of using only the chemical fertilizers, use of soil microorganisms which can either fix atmospheric nitrogen or solubilize phosphate or stimulate plant growth through synthesis of growth promoting substances will be environmentally benign approach for nutrient management.

Biofertilizer

Biofertilizers are living cell fertilizers or group of microorganism which are able to fix atmospheric nitrogen (N_2) and solubilise other essential nutrient to the plants individually or symbiotically. The term bio means *living;* so bio-fertilizers refer to living, microbial inoculants that are added to the soil. These bio-fertilizers are products consisting of selected and beneficial microorganisms, which are known to improve plant growth through supply of plant nutrients.

Biofertilizers are environment friendly low cost agriculture input playing a significant role in improving nutrient availability to the crop plants. They differ from chemical fertilizers in the sense that the farmers do not directly supply any nutrient to crop plant instead these culture of some specific bacteria and fungi which fix atmospheric nitrogen, enhance the solubility and increased absorption of soil nutrients, stimulating plants growth through normal action or symbiosis or by decomposition of organic residues. On a worldwide basis it is estimated that about 175 million tons of nitrogen per year is added to soil through biological nitrogen fixation (BNF).

In nature there are certain micro organisms and minute plants, which can absorb gaseous nitrogen directly from atmosphere and make it available to plant in solid form. It is called natural nitrogen fixation. Materials containing such organisms are called bio-fertilizers or natural fertilizers.They are multiplied and introduced into root zone of crop plant to supply nitrogen and phosphorus. Nitrogenous biofertilizers harvest atmosphere nitrogen and converts into ammonical form, which in due course is made available to the plants or is released in the soil. Phosphate solubilising micro-organism solubilizes fixed forms of phosphorus already present in the soil and make it available for use of plants. The concept of biofertilizers is to domesticate some of these microorganisms in our agricultural production systems, so that the vast natural reservoir of nitrogen in the atmosphere can be tapped as an additional source to meet our requirements.

TYPES OF BIOFERTILIZERS

On the basis of cells or microorganisms involved in the process, the biofertilizers are of three kind—

(*a*) Bacterial Biofertilizers

(*b*) Blue Green algae / BGA Biofertilizers

(*c*) Mycorhizal /VAM biofertilizers

(a) Bacterial Biofertilizers : Soil contains free living bacteria that are capable of fixing molecular nitrogen into its compound. This process was first recognized in bacterium clostridium pasturianum which is a gram positive, spore forming, rod shaped bacterium. Following bacterial culture are being used as biofertilizers commercialy.

1. Rhizobium : Rhizobium is the most effective biofertilizer world wide among all the biofertilizers known so far. The major environmental factors governing the nitrogen fixation are the type of legume, the effectiveness of the bacteria, inorganic or mineralizable nitrogen content of the soil, the level of available phosphorous & potassium, pH and the presence in usable form of a number of secondary factors have a great influence on nitrogen gain.

Agriculturist have reported that a relatively large group of plants - the legumes are capable of fixing atmospheric nitrogen through a symbiotic association with soil bacteria called- *Rhizobium.* Roots of most of the leguminous plants possess circular out growth or swelling called nodules or tubercles which are formed after infection of the different species of *Rhizobium.*

The leguminous species develop certain definite preference for nodule bacteria and vice-versa. Specificity of *Rhizobium* is based on "cross-inoculation group". The main characteristics of cross inoculation group are—

- Within each cross inoculation group a *Rhizobium* isolated from one legume member of the group would nodulate all other members of that group.
- Rhizobia isolated from one plant in cross inoculation group would not nodulate plants from other group. In accordance with the cross inoculation grouping seven species *of Rhizobium* are recommended as under.

Inoculation Group	**Rhizobium species**
Alfaalfa group	*R. Trifoli*
Clover group	*R. Meliloti*
Bean group	*R. Phaseoli*
Pea group	*R. leguminoserum*
Lupine group	*R. Luipini*
Cowpea group	*R. Species*
Soybean group	*R. Japonicum*

These bacteria have the capacity to fix free atmospheric nitrogen of soil in nodule. The effective nodules are usually larger, a pink in colour due to the presence of red coloured leghaemoglobin. Chemically leghaemoglobin is haem-protein and it contain a haem moiety attached to a peptide chain and represent the globin part of the molecule. The amount of leghaemoglobin in nodules has the direct relationship between amounts of atmospheric nitrogen fixed by legumes. The function of leghaemeglobin is as follows.

- It represents an active site of nitrogen absorption and reduction.
- It acts as a specific electron carrier in nitrogen fixation.
- It regulates the oxygen supply in the nodule.
- It acts as an oxygen carrier.

Rhizobium can fix 40-200 kg of Nitrogen per hectare per year and 40-80 kg.nitrogen is left over in land which is useful for subsequent crop. It increases crop yield by 25-30 per cent.Though responses of Rhizobium inoculation with regards to other crops are rather variable, it may still prove beneficial in certain region. Incoulation is so inexpensive that some farmers have taken to inoculating all pulses and fodder crops such as Barseem and Lucerne with Rhizobium as a routine practice to ensure against crop failure. Depending on the agro climate conditions the fertilizers N equivalent of chickpea, lentil, Pea, pigeon pea, groundnut, green gram and the residual in succeeding crops are given in table 29.1.

Table 29.1. Potential N contribution of N-fixing legumes in Indian Soils

Crops	Fertilizers N equivalent (kg/ha/year)	
	N Fixed	Residual in succeeding crop
Chickpea	20-63	60-70
Green Gram	50-55	30
Ground nut	112-152	60
Lentil	35-100	18-30
Pea	46	20-32
Pigeonpea	68-200	20-49

Subba Rao(1988)

Azotobacter : Azotobacter is also a group of nitrogen—fixing bacteria but unlike rhizobia, they do not form root nodules or associate with leguminous crops. They are free-living nitrogen fixers and can be used for all types of upland crops but cannot survive in wetland conditions. In soils of poor fertility and organic matter, azotobacter need to be regularly applied. In addition to nitrogen-fixation, they also produce beneficial growth substances and beneficial antibiotics that help control root diseases. The azotobacter bacterium is one of the effective fixer of nitrogen and good in phosphate solubilization. This bacteria is more beneficial with Azospirilhum on mixed culture biofertilizers. Experiment shows that Azotobacter biofertilizers are effective in red loamy soil having PH about 8.0 and in the soil having deposition of organic carbon. This biofertilizers is effective in rice, cotton, mulberry and wheat production.

Azospirillum : Like azotobacter, azospirillum species also do not form root nodules or associate with leguminous crops. They are however not free-living and inside plant roots where they fix nitrogen, and can be used in wetland conditions.This group of microorganisms also produce beneficial substances for plant growth, besides fixing atmospheric nitrogen. Azospirillum does well in soils with organic matter and moisture content, and requires a pH level of above 6.0.

Phosphate-solubilising microorganisms : These are a group of bacteria and fungi capable of breaking down insoluble phosphates to make them available to crops. Their importance lies in the fact that barely a third of phosphorous in the soil is actually available to the crop as the rest is insoluble. They require sufficient organic matter in the soil to be of any great benefit.

Phosphorus is also one of the major elements required for plant growth and higher yields. Next to nitrogen, phosphorus is the vital nutrient for plants and microorganisms. This element is necessary for the nodulation by Rhizobium and even to nitrogen fixers, Azolla and EGA. The organic form of phosphorus in soil are compound of phytin, phospholipids, nucleic acid, phosphorylated sugars and coenzymes which come mainly from vegetation and decaying plant residuea. The inorganic forms in the soil include compound of cacium,iron,aluminium. Both the above mentioned organic and inorganic forms of phosphorus in the soil are not available to the plants. Superphosphate is one of the common forms of phosphate fertilizers. Rock phosphate is one of the basic raw materials for phosphatic fertilizers. Direct application of rock phosphate is limited to acidic soil, while in other types of soil the applied phosphate becomes insoluble within a short time. Monocalcium phosphate are converted to dicalcium phosphate which is slowly available to plants. Under such conditions large amount of phosphorus is fixed in the soil which is unavailable to the plants.

A variety of micro-organism shown the ability to solubilize rock phosphate and thus made available to plants. Several bacterial and fungal species like *Pseudomonas, Fusarium, Sclerotium, Bacllus, Micrococcus, Aspergillus,Penicillium, Flavobacterium* possess phosphate solubilizing property. The phosphate solubilising micro-organism bring insoluble phosphate present in the soil into soluble forms by secreting organic acids such as formic, acetic, propionic, lactic, glycolic, fumaric, and succinic acids.

The root fungus association or mycorrhiza has high potential in accumulating phosphorus in the plants. Mixture of charcoal and soil is satisfactory material for these microorganisms in order to prepare commercial inoculants. It is reported that these cultures increase yield upto 200-500 Kg/hectare and thus 30 to 50 kg superphosphate can be saved.These fertilizers are prone to be fixed in the soil.

(b) Blue-green algae : Blue-green algae (BGA) or cyanobacteria are free-living nitrogen-fixing photosynthetic algae that are found in wet and marshy conditions. Blue-green algae are so named for their colour but they may also be purple, brown or red. They are easily prepared on the farm but can be used only for rice cultivation when the field is flooded and do not survive in acidic soils. They also provide extra cellular substances like polysaccharides and carbonic materials. BGA is much effective in saline, sodic & saline sodic soils which have pH about 8.5.

The main BGA fertilizers are Nostoc, Anabina, Calothrix, Phectonema, Hapolasiphon and Tolypothrix etc. This fertilizers can give 10-25 per cent high yield of paddy crop. It is estimated that in general blue green algae supply about 30-40 kg of nitrogen per hectare to paddy crop. Blue green algae, besides fixing nitrogen, carry out photosynthesis and secrete certain growth hormones (Vitamin B12, auxins and ascorboic acid) which are beneficial to rice plants.

Azolla

Azolla is a floating fresh water fern inside which grows the nitrogen fixing BGA Anabaena. It contains 3.4 per cent nitrogen and produces organic matter in soil. This biofertilizer is used for rice cultivation in different countries such as Vietnam, China, Thailand, Phillipines. It is also used in fish culture ponds these days. Application of different doses of Azolla in fish culture ponds shown that a minimum of 25 kgN/ha/yr could be provided through application of 10-12 tonnes of Azolla/ha/year.This can be easily grown in cooler regions. There is need to develop tolerant strains to high temperature salinity, and pests and disease-resistance for its wider adaption.

There are six species of Azolla-A. caroliniana, A. nilotica, A. mexicana, A. filiculoides, A. microphylla and A. pinnata. It grows in ditches and stagnant water. This fern usually forms a green mat over water. The plant has a floating, branched stem, deeply bilobed leaves and true roots which penetrate the body of water. The leaves are arranged alternately on the stem. Each leaf has a dorsal and ventral lobe. The dorsal fleshy lobe is exposed to air and contains chlorophyll. It has an algal symbiont (Anabaena azolla) within the central cavity. Azolla is readily decomposed to NH4 which is available to the rice plants. It has also been observed that N contents of rice receiving A. mexicana are 10 and 35 kg/ha respectively higher than that in controls.

Field trial indicated that rice yields are increased by 0.5-2t/ha due to Azolla appplication. In China, about 18 per cent increase has been observed. In fish culture ponds, application of 25 kg/ha per day Azolla biofertilization has been worked out. Recent studies have revealed potentialities of Azolla as a nitrogenous fertilizer in carp culture ponds.

(c) Mvcorrhizal /VAM (vesicular Arbuscular Mycorrhizal) : The symbiotic association between plant roots and fungal mycelia is termed as mycorrhiza (Fungal roots). These fungi are found associated with majority of agriculture crops. They are ubiquitous in geographic distribution occuring with plants growing in arctic, temperate and tropical regions alike. These fungi are obligate symbionts and have not been cultured on nutrient media.

VAM fungi infect and spread inside the root. They possess special structures known as vesicles and arbuscules. The arbuscules help in the transfer of nutrients from the fungus to the root system and the vesicles, which are saclike structures, store P as phospholipids. VAM have been associated with increased plant growth and with enhanced accumulation of plant nutrients, mainly P, Zn, Cu and S mainly through greater soil exploration by mycorrhizal hyphae. It has also been suggested that VAM stimulate plant growth by physiological effects other than the enhancement of nutrient uptake or by reducing the severity of diseases caused by the soil pathogens. The cereal crops are more benefited with micorrhizal biofertilizers. Both the roots and shoots system of plants shows better growth with this biofertilizers. About 10-22% higher yield can be achieved by the treatment with this biofertilizer.

Table. 29.2 Microorganisms and their activity associated with crops

Organism	Activity	Association if any	Used in crops
Rhizobium (leguminosarum, japonicum, phaseoil, etc.)	N_2-fixation brage crops)	Symbiotic	Legumes (pulses, oilseeds,
Azospirillum	N_2-fixation	Associative	Graminaceous crops like wheat, rice, sugarcane, jowar**
Azotobacter	N_2-fixation	Asymbiotic	Wheat, rice, vegetables
Blue-green algae (Anabaena, Slostoc, Plectonema, etc.)	N_2-fixation	Asymbiotic	Rice
Azolla-Anabaena complex	N_2-fixation	Symbiotic	Rice
Phosphate solubilizing bacteria (Thiobacillus, Bacillus, etc)	Phosphate solubilization	Asymbiotic	Vlany crops
Mycorrhiza (Glomus)	Phosphate solubilization	Associative	Many crops including pulses

Inoculation of Biofertilizers

Bio fertilizers are generally applied to soil, seeds or seedlings, with or withoutuse of some carrier (example-peat, composts or stickers) for the microorganisms. Regardless of methods, the number of cells reaching the soil from commercial products is smaller than the existing numbers of soil or rhizosphere microorganisms. These added cells are unlikely to have a beneficial impact on the plant unless multiplication occurs. In addition, the population of introduced microorganisms will decline and be eliminated in a very short time, often days or weeks. The formulation of inocuia, method of application and storage of the product are all critical to the success of a biological product. Short shelf life, lack of suitable carrier materials, susceptibility to high temperature, problems in transportation and storage of biofertilizer are still need to be solved in order to obtain effective inoculation. Some common inoculation practices are:

(a) Seed inoculation : Seed inoculation uses a specific strain of microbe that can grow in association with plant roots. Soil conditions have to be favorable for the inoculants to perform well. Selected strains of N-fixing Rhizobium bacteria have proven to be effective as seed inoculants for legumes.

The seed treatment can be done with any of two or more bacteria without antagonistic effect. In the case of seed treatment with Rhizobium, Azotobacter, Azospirillum along with PSB, first the seeds must be coated with Rhizobium or Azotobacter or Azospirillum. When each seed has a layer of the aforesaid bacteria then the PSB inoculant has to be treated on the outer layer of the seeds. This method will provide maximum numbers of population of each bacterium to generate better results.

(b) Soil inojculation : In soil inoculation, microbes are added directly to the soil where they have to compete with microbes already living in the soil that are already adapted to local conditions and greatly outnumber the inocuia. Inoculants of mixed cultures of beneficial microorganisms have considerable potential for controlling the soil microbiological equilibrium and providing a more favorable environment for plant growth and protection. Therefore, adequate quality control and a high level of consistency in performance and benefits must be ensured.

Advantages of Bio-fertilizers

The relevance of biofertilizers is increasing rapidly .some of the advantages of using biofertilizers are:

- **Economic and pollution free :** Taking into account the amount of nutrient supplied by biofertilisers are many times cheaper than chemical fertilizers. Production of Biofertilizers does not involve any toxic chemical.
- **Suppliers of micro nutrients :** Biofertilizers not only supply nitrogen and phosphorous but also some micronutrients essential for plant growth. Sometimes yield is limited by micronutrients and application of nitrogenous, phosphatic and potassic fertilizers does not improve yield significantly. In this situation the application of bulky biofetilizers like blue green algae and azolla increases yield due to greater supply of micronutrients.
- **Secretion of growth hormonesi :** Plants also need for their growth and development; some natural complex chemical compounds called hormones. Though growing plants do not themselves synthesize hormones in adequate amounts.Azotobactor blue green algae and azolla have been found to synthesize growth hormones which benefit the main crop. Sometimes ,biofertilizers application gives significant response even if the soil is already rich in plant nutrients. This occurs due to the supply of growth hormones by biofertilizers to the main crop.
- **Supplier of organic matter :** Organic matter is the essential component of the soil. It serves as an inexhaustible source of nutrients and energy for plant as well as for useful micro-organisms. Organic matter has great impact on the physical and chemical properties of the soil. Azolla and BGA produce an average 8-10 tone of biomass per hectare which adds to the organic matter pool of soil.
- **Counteracting negative impact of chemical fertilizers :** When chemical fertilizers are excessively and continuously used for a few years, they may create acidity or alkalinity in the soil and deteriorate the quality of soil. Soil also becomes unresponsive to further use of similar fertilizers. Application of biofertilizers can avoid this problem to a great extent. Besides, large amount of organic matter supplied by the biofertilizers impart tolerance power (buffering capacity) to the soil against acidity or alkalinity. It also withholds metallic elements from entering the plant roots, thereby reducing harmful effects of pesticide thus provides ecological stability to the soil.

In addition, the amount of nutrients provided by them is not enough to adequately meet the total needs of crops for high yields. Therefore, a pragmatic approach more likely to succeed will be to develop a rational and effective combination of biofertilizers and conventional fertilizers for optimum crop yields.

Integrated Use of Chemical, Organic and Biofertilizers

Increased attention is now being paid to developing an Integrated Plant Nutrition System (IPNS) that maintains or enhances soil productivity through balanced use of all sources of nutrients, including chemical fertilizers, organic fertilizers and biofertilizers. The basic concept underlying the IPNS is the adjustment of soil fertility and plant nutrient supply to an optimum level for sustaining desired crop productivity through optimization of the benefits from all possible sources of plant nutrients in an integrated manner.

A field experiment to evaluate the effects of chemical fertilization (CF), organic fertilization (Compost-N and Compost-P), combined use of chemical and organic fertilizer (Compost-P + urea) and integrated use of chemical and organic fertilizer along with biofertilizer ($^1/_2$ (Compost-P + urea) + Biofertilizer) on the growth of different crops have been conducted. It has been reported that the combined use of biofertilizer with compost *(Vi* Compost-P + Biofertilizer) and combined use of biofertilizer, compost and chemical fertilizer [$^1/_2$ (Compost-P + urea) + Biofertilizer] had the advantage of saving compost and/or chemical fertilizer.

Conclusion

Efficient plant nutrition management should ensure both enhanced and sustainable agricultural production and safeguard the environment. Chemical, organic or microbial fertilizer has its advantages and disadvantages in terms of nutrient supply, soil quality and crop growth. Developing a suitable nutrient management system that integrates use of different kinds of fertilizers may be a challenge to reach the goal of sustainable agriculture; however much research is still needed. There is great need to use available fertilizers most efficiently. This will ensure not only clean environment but also higher return by using a somewhat lower fertilizer doses. Fertilizer use efficiency is of high order wiien nutrients are applied in needed quantities in a balanced proportion at appropriate time in a proper form with suitable method in right way to a crop.

In view of the present scenario of the escalating energy cost, to attain sustainable agriculture and for protection of human health there is need to supplement Chemical fertilizers with Biofertilizers.

Biofertilizers are boon for farmers because it helps in increasing the soil fertility and crop productivity. It is only logical that biofertilizers should supplement chemical fertilizers thus leading to an *Era of prosperity and clean environment.*

REFERENCES

1. Alexander, M. (1977): Introduction of soil microbiology, 2nd Ed. (Indian reprint) Wiley eastern Ltd. New Delhi.
2. Bokhtiar, S.M. & Sakurai. K. (2005). Effects of organic manure and chemical fertilizer on soil fertility and productivity of plant and ratoon crops of sugarcane. *Archives of Agronomy and Soil Science.* 51: 325-334.
3. Brown, M.E. & Burlingham, S.K.(1968): Production of plant growth substances by Azotobacter chroococcum, J.Genetics and Microbiol, 53:135-144.
4. Graham, P.H.(2000): Nodule formation in legumes. In *Encyclopedia of microbiology*, 2nd ed., Vol. 3, J. Lederberg,editor-in-chief, 407-417, San Diago, Academic Press.
5. Jensen H.L. (1942) Nitrogen.fixation in leguminous pants. Proc. Linnban society of New South Wales, 65:98-108.
6. Kunda, B.S., & Gaur, A.C. (1984), Rice responses to inoculation with nitrogen fixing and P-solubilizing micro-organics. *Plant and Soil,* 79: 227-234.

7. Mishustin, E.N. & Shilinikova, V.K.; (1969) : Free living nitrogen fixing bacteria of the genus Azotobacter. Soil Biology, Review of Research, UNESCO Publication, 72-124.
8. Okon, Y.; (1985) : Azospirillium as apotential inoculant for agriculture, *Trends in Biotech., 3:* 223-228.
9. Pandey, A. & Kumar, S. (1989): Potential of azotobacter and azospirilla as biofertilizer for upland agriculture: a review, *J.Sci.Ind. Res.,* 48:134-144.
10. Purohit, S.S. & Mathur S.K. (1999) Biotechnology of Nitrogen Fixation and plant productivity in Biotechnology fundamental and Application. Agro Botanica, India, pp. 354-374.
11. Roper, R.M. and Ladha J.K. (1995): Biological N2 Fixation by heterotrophic and phototrophic bacteria in association with straw, *Plant and soil,* 174: 211-224.
12. Sharma, B.K. (1994):. Environmental Chemistry, 5th ed., Goel Pub., Meerut, India.
13. Spaink, H. P. (2000): Root nodulation and infection factors produced by rhizobial bacteria. Annu. Rev. Microbiol., 54:257-288.
14. Subha Rao, N.S. (1982) : Ed. Advances in Agriculture microbiology, Oxford IBH Pub. Co., New Delhi.
15. Subba Rao N.S. (1988) : Ed. Biological Nitrogen fixation-Recent Developments. Oxford and IBH pub. co., New Delhi.
16. Sundara Rao, W.V.B. & Sinha,M. K.(1963) : Phosphate solubilising organism in the soil and rhizosphere. *Ind. J. Agric. Sci.*, 33 : 272-278.
17. Vandenkoornhuyse, P; Balduf, S.L.; Straczek J. & Young, J.P.W. (2000) : Extensive fungal diversity in plant roots, Science, 295 : 2951-2061.
18. Wani, S.P. (1990): Inoculation with associative nitrogen fixing bacteria: role in cereal grain production improvement, *Ind.J. Microbiol., 30 :* 363-393.

CHAPTER 30

Pesticides and Environment

Dr. S.A. iqbal[1] and Dr. M. Ibrahim[2]

Chemistry of Persistent Pesticide

Large quantities of pesticides have been manufactured and distributed in the environment to control plant and animal pests. The environmental stability of many of these compounds is appreciable. However, the toxicity and environmental problems associated with the commercial use of halo-organics that are released in the environment has only recently been appreciated.

Chemistry of Chlorinated Organic Compounds

Organochlorine derivatives are prepared industrially by direct chlorination of hydrocarbons. Chlorination is useful in the preparation of organochlorine compounds, such as 1,3 dichloro-1 propene which are effective in killing nematodes (soil worms). The overall reaction is,

$$CH_2 = CHCH_3 + 2Cl_2 \rightarrow ClCH = CH — CH_2Cl + 2HCl$$

Aromatic compounds are generally more difficult to chlorinate than aliphatic compounds. The aromatic chlorination reaction proceeds by an ionic pathway. Polychlorinated biphenyls are prepared from biphenyl by direct chlorination using ferric chloride as a catalyst. The mechanism involves the initial formation of a chloronium ion that undergoes an electro-philic addition to the aromatic ring system

$$Cl_2 + FeCl_3 \rightleftarrows FeCl_4^- + Cl^+$$

$$Cl^+ + C_6H_6 \longrightarrow [C_6H_6Cl]^+ \longrightarrow C_6H_5Cl + H^+$$

Formation of polychlorinated biphenyls may take place on chlorination of wastes in sewage, disposal plants from textile mills which may be dangerous because it is environmentally stable. Hence chlorination of waste should be avoided.

Organochlorine compound include DDT, BHC, chlordane, Aldrin, Endrin, Dieldrin, Endosulphon etc.

1. DDT : (dichloro diphenyl trichloroethane) : Even though someone created DDT back in the 19th century, only around 1939 Dr. Paul Muller discover that it was effective in killing insects, and shortly after, he won the nobel prize in medicine for this work.

[1]Department of Chemistry, Saifia Science College, Bhopal (India)
[2]H.O.D. Chemistry, Polytechnic Branch, Jamia Millia Islamia, New Delhi (India)

As a pesticide, DDT was first used during-world war II. DDT is a broad spectrum persistent organochlorine pesticide. It has been used against crop predators such as apple, maggots and against the vectors of human diseases such as malaria, yellow fever and plague.

DDT is an organochlorine insecticide used mainly to control mosquito-borne malaria, use on crops has generally been replaced by less persistent insecticides. It is available in several different forms: aerosols, dustable powders, emulsifiable concentrates granules and wettable powders.

Unless otherwise specified, the toxicological, environmental effects and environmental fate and chemistry data presented here refer to the technical product DDT. Technical grade DDT is actually a mixture of three isomers of DDT, principally the P,P′-DDT isomer (Ca.85%) with the O,P′-DDT and O,O′-DDT isomer typically present in much lesser amount.

Physical Properties : The physical appearance of technical product P,P′-DDT is a waxy solid, although in its pure form it consists of colourless crystals.

Synthesis : It is synthesized by the interaction of chlorobenzene and chloral in an acidic medium by a friedel-crafts type reaction. The mechanism of the reaction involves the electrophilic addtion of a carbonium ion species to aromatic ring as shown by the following reaction.

$$CCl_3CHO \overset{H^+}{\rightleftharpoons} CCl_3\overset{+}{C}H - OH$$

Cl—C₆H₄ + $CCl_3\overset{+}{C}HOH$ ⟶ Cl—C₆H₄—CH(CCl₃)(OH) + H^+

Cl—C₆H₄—CH(CCl₃)(OH) ⟶ Cl—C₆H₄—$\overset{+}{C}H$(CCl₃) + H_2O

Cl—C₆H₄—$\overset{+}{C}H$(CCl₃) + C_6H_5Cl ⟶ Cl—C₆H₄—C(H)(CCl₃)—C₆H₄—Cl

DDT

Mechanism of Action of DDT : Insects sprayed with DDT exhibit hyperactivity and convulsions consistent with the action of the DDT on the nervous system. Many theories have been suggested for the toxic effect of DDT, but the exact mechanism is not known with certainty. A plausible theory with experimental basis is that the DDT molecules are of the correct size to be trapped in the pores of the nerve membranes.

2. BHC (Lindane) : The β isomer of BHC is quite persistent in the environment and is frequently reported to occur as a contaminant is human fat and blood. The pure α and β isomers and the mixtures of α andβ isomers are carcinogenic in laboratory mice, producing liver cell tumors following oral administration. It is more rapid in action than DDT.

Cl H H Cl Cl H H Cl Cl H H Cl (C₆H₆Cl₆ ring structure)

3. Endosulphon :

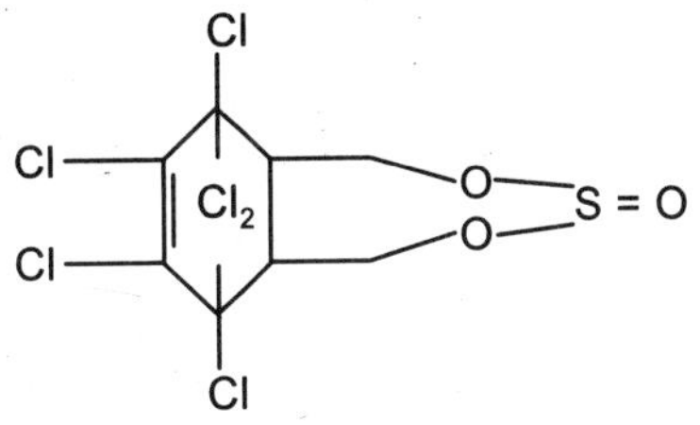

Structure of endosulphan

Endosulphan is a contact and stomach poison, which induces acute neurotoxicity. It is non mutagenic, but evidence for its carcinogenic potential is inconclusive.

4. Chlordane : Chlordane is metabolized to oxychlordane, consequently, oxychlordane is stored in fat and is excreated in the milk of lactating animals. It is a neurotoxicant.

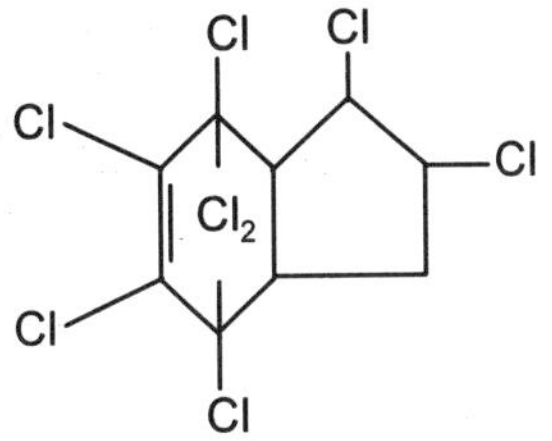

Structure of Chlordane

Environmental Fate

When these non polar pesticidal chemicals reach the aquatic medium, they get absorbed on to soil particles or the suspended matter in water. When representative organochlorines were added to a turbid aqueous suspension, the ratio of adsorbed to free insecticide was 100 : 1 for DDT, 4 : 1 for endosulphon. Most chlorinated hydrocarbons are persistent for more than 18 months, lindane is to persist for a year.

DDT is very highly persistent in the environment, with a reported half life of between 2-15 years. Routes of loss and degradation include runoff, volatilization, photolysis and biodegradation. These processes generally occur only very slowly. Due to extremely low solubility in water, DDT will be retained to a greater degree by soils and soil fractions with higher proportions of soil organic matter. It may accumulate inthe top soil layer in situations where heavy applications are made annually example for apples.

DDT may reach surface waters primarily by runoff, atmospheric transport, drift, or by direct application. The reported half life for DOT in the water environment is 56 days in lake water and 28 days in river water.

Chemistry of Organophosphorus Pesticides

The Chemistry and mechanism of action of the organophosphours insecticides are quite different from those of the orgnochlorine materials. Orgnophosphours compoundmerit consideration because they are being used more frequently in place of the organochlorine compound. The majority of the organophosphorus insecticides are derivatives of phosphoric acid or the sulphur analogues of phosphoric acid. The phosphoric acid esters used in the synthesis of these insecticides are prepared industrially by the reaction of phosphoryl chloride with alcohols. Under suitable conditions, di-ester chlorids may be prepared.

$$Cl{-}P({=}O)(Cl){-}Cl + 2ROH + 2R_3N \longrightarrow RO{-}P({=}O)(OR){-}Cl + 2R_3NHCl$$

Most of the organophosphorus insecticides are triesters of phosphoric or thiophosphoric acid, prepared by the attack of a nucleophile on the phosphochloridate.

Some organophosphorus pesticide are malathion, parathion etc.

1. Malathion : The synthesis of malathion involves a nucleophilic addition of a thioacid to a double bond that is conjugated to a carboxyl group.

$$\underset{\text{(CH}_3\text{O)PSH}}{\overset{\text{S}}{\|}} + \underset{\text{CH}_2\text{COOC}_2\text{H}_5}{\overset{\text{CH}_2\text{COOC}_2\text{H}_5}{|}} \longrightarrow \text{(CH}_3\text{O}_2)_2\overset{\text{S}}{\overset{\|}{\text{P}}}\text{—S—}\overset{\text{CH}_2\text{COOC}_2\text{H}_5}{\overset{|}{\text{C}}}\text{HCOOC}_2\text{H}_2$$

Malathion

Malathion is a non-systemic insecticide and acaricide of low mammalian toxicity. It is extensively used as a home garden insecticide. It is about hundred times less acutely toxic than parathion.

2. Parathion : Parathion is one of the most toxic and widely used of the organophosphorous insecticide. It is metabolized to 4-nitrophenol which is excreted in the urine.

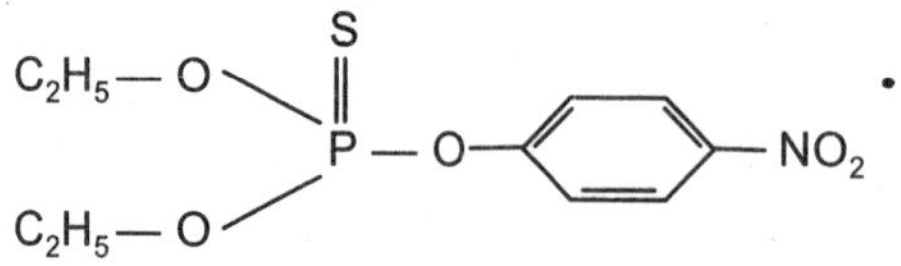

$$(C_2H_5-O)_2\overset{S}{\overset{\|}{P}}-O-C_6H_4-NO_2$$

Structure of parathion.

Environmental Fate

Parathion is very toxic but it neither persists in the soil for longer periods nor does it accumulate in the rood chain. Generally with in a maximum of three months all of it is degraded in the environment.

Malathion is quite stable in neutral and acidic waters. It is readily degraded by aquatic bacteria such as pseudomonas, xanthomonas etc. Parathion is degraded effectively by *Bacillus subtilis.*

Chemistry of Carbamates

Most carbamates are phenyl N-methyl carbamates i.e. carbamate esters derived from substituted phenols. The carbamate insecticides have been used extensively as a replacement for DOT to control the gypsy moth as well as other insects.

Its toxicity to insects is due to its antiacetyl choline sterase action as in organophosphorus insecticide though the mechanism is a little different.

The polarization of carbonyl carbon gives a positive charge to carbon. This leads to protonation of oxygen there by leading to the formation of a reversible complex as a tetrahedral intermediate.

Environmental Fate

Carbamates are readily degraded in the environment. The most common degradation processes are hydrolysis and oxidation. Hydrolytic end products of carbamate insecticides are often more apolar than original carbamates themselves.

Mode of Entry of Pesticides in Environment

Pesticides may enter in aquatic ecosystem either directly, indirectly or unintentionally through the following sources.

1. Rain Water : Pesticides contaminated in vapour phase in air get adsorbed on to the dust particles which ultimately reach soil or water along with the rain water.

Tarrant and Tatton detected residues of BHC, DDT, DDE and Dieldrin in rain water.

2. Spray Drift : It has been estimated that 30 to 40 percent of pesticide used in aerial spray reaches the target while rest fled into air. Water contaminates heavily with pesticides during accidental spray. Wind may also carry the drift to considerable distances while the major fall of pesticides takes place near the site of application polluting water system.

3. Runoff from Agricultural Fields : Generally water soluble pesticides are transported in water system while the insoluble ones get bound to the particulates and carried by water streams. Caro and Taylor have reported that 0.07 per cent of dieldrin applied to the soil persists in run off water. Largest quantities of DOT in water were about 70 mg/L and 440 mg/L. in mud.

4. Industrial Effluents : A huge amount of pesticides are escaped by pesticide producing industries.

5. Domestic Sewage : ICAR (1970) have reported that in Jammu and Kashmir, fish mortality was seen in the ponds of Reasi and in Nehru stream where DOT was used in city drains. DDT concentration in Yamuna water has risen from 0.25 PPb to 0.558 PPb. After the Najafgarh Nallah drains sewage into this river at Wazirabad in Delhi.

Besides these sources other factors, like the amount of suspended particulates, chemical air crafts, accidental spillage, evaporation from soil and plants, distances between the area treated with pesticides and the receiving water, forest cover, time lapse between treatment and rain fall and residues in horticultural products influence the extent of pesticide pollution in water bodies.

Mode of Action of Pesticides

(a) Insecticides

Mode of Action: Insecticides kill the insect pests against which control is desired. Several of these chemicals are toxic to the insects as such, where as others need to be biotransformed to reactive, metabolites, which in turn are responsible for the toxic action.

Chlorinated hydrocarbons are neurotoxicants and they damage the nervous system of the pest organisms. The organophosphorous insecticides and carbamates exert their toxicity by inactivating the vital acetylcholinesterase enzymes. Insecticides such as organochlorine and organophosphorous compouds includes DDT, BHC, parathion, malathion etc.

(b) Herbicides

Herbicides kill the weed plants, by disrupting the photosynthesis or preventing chlorophyll formation. Others act as 'hormones', altering dramatically the growth rate of the weed plants. Interference and distruption of the synthesis of enzymes responsible for various biochemical mechanisms may also be caused by several of these chemicals. Herbicides such as carbamates include carbaryl (sevin), baygon etc. 2,4–D (2,4 dichlorophenoxy acetic acid) and 2,4,5-T(2,4,5-trichlorophenoxy acetic acid) are well known herbicides.

(c) Fungicides

Fungicides can be grouped into protective and eradicative types. Protective fungicides are applied before a fungal infection of the crop takes place. Most of them have an aspecific mode of action and disturb different processes in the cell.

Eradicative or systemic fungicides can penetrate and move within the tissues of the plant. Thus, they may contribute to the protection of the unsprayed freshly growing plant.

Protective fungicides which are important to consider with respect to their possible environmental impact are organomercurials, organotin and copper compounds, dithiocarbamates and captan. Systemic fungicides which are widely used are the benzoimidazoles.

There are some inorganic pesticides such as fungicides, which are copper, nickel, lead and arsenic salts.

ENVIRONMENTAL IMPACTS OF PESTICIDES

Pesticide use in crop production has been suspected of being a major contributor to environmental pollution. The post war period has seen continued increases in both agricultural productivity and pesticide use. There are widespread and growing concerns of pesticide over-use, relating to a number of dimensions such as contamination of ground water, surface water, soils and food and the consequent impacts on wild life and human health.

The use and abuse of pesticides has disturbed the ecological balance between pests and their predators in developed and developing countries. The lesser developed countries still don't use as much pesticide as does the industrialized world. The rate of pesticide use per hectare of land is highest in Japan (17.7 kg/ha). In the developed regions, the pesticide market is deminated by herbicides which tend to have lower acute or immediate toxicity than insecticides.

Acute Pesticide Poisoning

Pesticides are toxic chemicals, by design and as such they represent risks to the users. In developing countries, where users are often illiterate, ill-trained and donot possess appropriate protective devices, the risks are magnified. The poison information centre in NIOH. Ahmedabad reported that op compounds were responsible for the maximum number of poisoning (73%) among all agricultural pesticides. In a study on patients of acute op poisoning (N=190), muscarinic manifestations such as vomiting (96%), nausea (82%), miosis (64%), excessive salivation (61%) and blurred vision (54%), and CNS manifestations such as giddiness (93%), headache (84%), disturbances in consciousness (44%) were the major presenting symptoms.

Assessment of Human Exposure

In human being, the pesticides residue level is an index of exposure, which may be acute, occupational or incidental. In acute exposure, the residue level has a diagnostic potential, and in the occupationally exposed, the residue level merits an insight reflective of industrial exposure. However in general population, the residue level is a measure of the incidental exposure and / or average levels of the persistent pesticides which is mainly through the food chain.

Residues of organochlorine insecticides, especially DDT and HCH have been detected in man and his environent the world over. However, by comparison very high level of these have been reported in human blood, fat and milk samples in India.

Residues in Human Blood

Since blood is the most accessible body fluid for ascertaining residue levels, scientists at the NIOH, Ahemadabad have attempted to provide a database on the residues of DDT and HCH including other cyclodiene derivatives.

Example : Heptachlor, heptachlor epoxide, Aldrin, oxychlordane, HCB and dieldrin in blood samples from general population of Ahemadabad (Rural) area. The total organochlorine insecticide content in all serum samples showed an average of 200.3 PPb with a range of of 58.3 – 321.4 PPb.

Residues in the blood samples of the general population of different cities are given in Table.

Table 30.1. Levels of DDT and HCH content in human blood samples in general population In India.

City	Year	No. of samples	Total DDT(ppm)	Total HCH(ppm)
Lucknow	1980	25	0.02	0.022
Delhi	1982	340	0.71	0.49
Lucknow	1983	48	0.028	0.075
Delhi	1985	50	0.301	-
Ahmedabad (Rural)	1992	31	0.048	0.148
Ahmedabad (Urban)	1997	14	0.032	0.039

Residues in Human Fat

The maximum DOT residues were detected in age group of 20 – 39 years and the higher levels of HCH residues were found in the age group of 40 years.

The wide variation seen may be due to the geographical variations in consumption and use of these chmicals. However, the factors that may influence the storage and bioaccumulation of these chemicals are the compound intensity, efficiency of absorption, species, age, nutritional status and integrity of the organs.

Residues in Human Milk

Monitoring of human milk is important from two standpoints, firstly,] pesticides tend to accumulate in thefat and are relatively easy to isolate anc measure and secondly to evaluate their potential risk to infants, who rely sólely on mother's milk for a substantial period. Residues of these compounds in human milk have been reported from different parts of the world as reviewed by Jenson and from India.Thc levels of these contaminants in human milk samples collected from different cities are given in Table

Table 30.2. Levels of DDT and HCH residues in human milk samples in general population In India.

City	No. of Sample	Whole Milk Basis (ppm)	
		Total DDT	Total HCH
Lucknow	25	0.127	0.107
Ludhiana	75	0.51	0.195
Bangalore	6	0.053	0.014
Calcutta	6	0.114	0.031
Bombay	6	0.224	0.053
Delhi	60	0.344	—
Delhi	60	—	0.38
Ahmedabad	50	0.305	0.224

Residues in Food Commodities and Average Daily Intakes

Pesticide residues in food are of concern.

Hexachlorobenzene (HCB, a fungicide) was identified in water, human milk and human fat samples collected from Faridabad and Delhi. In another study, HCB residues were present in various raw food commodities collected from markets from India.

In a multi-centric study to assess the pesticide residues in selected food commodities collected from different states of the country, DDT residues were found in about 82 per cent of the 2205 samples of bovine milk collected from 12 states. About 37 per cent of the samples contained DDT residues above the tolerance limit of 0.05 mg/kg (whole milk basis). The highest level of DDT residues found was 2.2 mg/kg. The proportion of the samples with residues above the tolerance limit was maximum in Maharashtra (74%) followed by Gujarat (70%), Andhra Pradesh (57%), Himachal Pradesh (56%) and Punjab (51 %). In the remaining states, this proportion was less than 10 per cent.

Measurement of chemicals in the total diet provides the best esti-mates of human exposure and of the potential risk.

Residues in Environmental Samples

The residues of pesticides in air-borne samples collected from Ahmedabad were identified in studies by the NIOH (National Institute of Occupational Health). The levels of BHC and DDT ranged between

2.06-18.96 ng/m^3 and 7.21 – 51.19 ng/m^3 respectively. The maximum levels were seen in summer and the minimum in winter.

In Ahmedabad, the mean levels of DDT and HCH in drinking water samples were 47.4 and 256.9 ng/1 respectively. The residues have also been detected in botn the surface and ground water samples of north eastern districts and khasi hills of India.

Potential Risk in Occupationally Exposed Subjects

The high risk groups exposed to pesticides include the production workers, formulators, sprayers, mixers, loaders and agricultural farm workers. During manufacture and formulation, the possibility of hazards may be more because the processes involved are not risk free.

Workers Exposed to HCH

A study on workers (N=356) in four units manufacturing HCH revealed neurological symptoms (21%) which were related to the intensity of exposure. Significant increase in liver related enzymes (alkaline phosphatase, ornithine carbamoyi-transferase, γ-glutamyl transpeptidase) was also noticed. A remarkably high concentration of HCH residues, especially; of β-HCH was found in the serum samples of all exposed subjects.

Formulators Exposed to Combination of Pesticides

Observations confined to health surveillance in male formulators engaged in production of dust and liquid formulations of various pesticides (malathion, methyl parathion, DDT and lindane) in industrial settings of the unorganised sector revealed a high occurrence of generalized symptoms (headache, nausea, vomiting, fatigue, irritation of skin and eyes) besides psychological neurological, cardiorespiratory and gastrointestinal symptoms coupled with low plasma cholinesterase (chE) activity).

A study on workers associated with the formulations of various oc and op pesticides in small scale industrial units showed depression in serum chE activity, increased level of serum cholesterol, phosphoTipid and generalized toxic symptoms in 73 percent subjects.

Skin diseases in workers handling pesticides were reported in a study involving 117 workers (75 pesticide factory workers and 42 farm workers), fifty five workers had pigmentation on the exposed parts, nine pityriasis versicolor, five chronic urticaria, four dermatitis and five psoriasis, pigmentation and urticaria may be related to skin sensitivity in response to these chemicals.

Health Effects of Methomyl on Sprayers

The magnitude of the toxicity risk involved in the spraying of methomyl, a carbamate insecticide, in field conditions was assessed by the NIOH. Significant changes were noticed in the ECG and the levels of serum LDH and chE activities in the spraymen indicating the cardiotoxic effects of methomyl.

Reproductive Performance in Sprayers

A cytogenic study revealed a significant increase in chromatid breaks and gaps in chromosomes in the peripheral blood in grape garden workers exposed to pesticides.

Ecological Effects of Organochlorine Compound

Bioamplification

DDT has proved to be hazardous to several species of fish and bird because it is not easily metabolised by the animals. Rather, it is known to concentrate in the fatty tissues. Its concentration progressively increases in those species which occupy a higher trophic level in the food chain. This phenomen a is called bioamplification.

As an example, spraying a marsh to control mosquitoes will result in the accumulation of traces of DDT in the cells of microscopic aquatic organisms such as plankton. The DDT then becomes concentrated in the tissue of the fish and shell fish which feed on planktons. This may led in turn to residues in small fish then to larger fish and finally to such fish eating birds as gulls, ospreys, bald eagles, pelicans and falcons. The level of accumulated DDT are often so high that the consequences may be disastrous.

Since one of the greatest potential hazards of the chlorinated hydrocarbon pesticides is that human food may become contaminated with large amounts. Most of the residues found in food arise directly from treatment of the crop, and mainly from the spraying of the growing crop. The higher incidence of BHC and DDT residues in the total environment in India is because of the fact that they constitute about 60% of all pesticidal chemicals consumed in India.

Suggestion and Recommendation

Following measures are being suggested for minimising human ex-posure to pesticides, and to keep the soil environment free from hazardous pesticide.

Intergrated Pest Management—A Viable Alternative

Integrated pest management (IPM), a new concept in the field of crop protection emphasizes the need for simpler and ecologically safer measures for pest control to reduce environmental pollution and other problems caused by excessive and indiscriminate use of the pesticide. The main components of IPM are pest surveillance, use of crop varieties resistant to pest, sound cultural practices, biological control and use of ecofriendly pesticides having less mammalian toxicity. Agenda 21 of the united nation conference on environment and development (UNCED) at Rio de Janeiro in June 1992 identified IPM as one of the requirement for promoting sustainable agriculture and rural development. Considering the harmful impact of pesticides on the environment, the government of India recognised the benefits of IPM and adopted it as the main plank of the plant protection strategy in the overall crop production programme.

Biologically intensive IPM or Biointensive IPM (BIPM) is only a variation of the basic theme of IPM and depends on host plants resistance, biological control and cultural control and use of biorational pesticides, which can be integrated with these. IPM is mainly aimed at developing system based on biological and other non-chemical methods as much as possible.

Cultural measures are associated with crop production to make the environment less favourablefor survival, growth and reproduction of pest species. Use of planting materials such as healthy seeds, seedings is one of the best BIPM practicies. Planting material should be given maximum attention because sometimes disease, pests, insects and nematodes are also carried along with the planting material and the farmer.

Biopesticides

India loses about 20-30 per cent of its crops to pests and disease each year. To counter this loss, pesticides consumption has incrased sharply particularly since the late 1960s, The consumption of pesticides increased from about 100,000 metric tones in 1992-1993.

The widespread use of chemical pesticides has led to a number of problems causing serious concern ex. the damaging impact of these chemicals on the environment and the health hazard due to contamination of food including milk, eggs, meat and fish, fodder and fibre by pesticides residues.

Considering the environmental damage caused by chemical pesticides, and the development of resistance by an increasing number of pests, there is an urgent need for the use of alternative methods of crop protection.

The potential use of biopesticides for this purpose is one of the most appropriate and promising of these methods. The agents employed as biopesticide which include parasites, predators, fungi, bacteria and viruses, are the natural enemies of pests. They are preferred over chemical pesticides as they—

(*a*) do not leave harmful residue;

(*b*) are target specific and do not destroy beneficial organisms; and

(*c*) promote the growth of natural enemies of pests, thus reducing the need for future pesticide application. The use of biopesticides in India is not new. There are a number of instances where biocontrol agents have been successfully employed in India.

Example : The sugarcane scale insect has been controlled with the help of predatory coccinellid beetles in Gujarat, Karnataka.

TNAU-BCIL Biopesticides Technologies

The technology developed by TNAU, which is being offered through BCIL to industry on a commercial basis , is for five biopesticides: These are Trichogramma, Trichoderma, Heliothis NPV spodoptera Litura NPV and chilo Infuscatellus Gv. Trichogramma is a parasite which destroys pest eggs. Trichoderma is a fungus which is effective against root pathogens and is used for the treatment of seeds. The last three biopesticides are natu-rally occurring viruses which can be used for destroying pests.

The technology for the production of these biopesticides has bcci, scaled up and iried at two pilot plants, set up with financial assistance from BCIL. The pesticides produced by these pilot plants have been used on 11,000 hectares each. The pilot plants have demonstrated the feasibility of TNAU technology for the production of these pesticides on a large scale.

These biopesticides are effective against some of the most important pests affecting Indian crops, such as heliothis armigera, spodoptera litura, bollworm, chillo infuscatellus and sugar cane internode borers. some of the important crops affected by these pests include cotton, chickpeas, maize, tomato, groundnut, sorghum, sunflower, tobacco and castor. Trichoderma can be used for seed treatment for protection against various fungal diseases in the case of groundnut, sunflower, sesamum and chickpea.

Table 30.3. Biopesticides and their Target Pests and Crops

Sr. Product	Pests	Crops
1. *Heliothis* NPV	Heliothis armigera	Cotton, Chickpea, Groundnut, Sunflower
2. *Spodoptera litura* NPV	Spodoptera litura	Cotton , Tobacco and oil seeds
3. *Chilo infuscatellugs GV*	Chilo Infuscatellus	Sugarcane
4. *Trichogramma*	Sugarcane internode borer.	Sugarcane
Parasitoid	Boll worms of sorghum cotton,	Sorghum stem borer
5. *Trichoderma*	Root rot, charcoal rot	Groundnut

Importance of Biopesticides

Trichogramma

An importance advantage of Trichogramma is that, being an egg parasite, it eliminates the pest before it has a chance to do any damage Trichogramma is one of the most popular biopesticides and is wiaeiy against lepidopteran insects in many parts of the world.

Trichoderma

It is an antagonistic fungi for controlling variety of seed borne, seedling diseases for plant. Trichoderma is very effective biocontrol agent of plant diseases used world wide since 1930. It is naturally occuring antogonistic fungi. It has been exploited for augmentation in recent years.

Heliothis NPV, Spodoptera NPV

As biopesticides, NPVs (and other microbial pesticides) have a number of important advantages.The most important of these is their specificity to target pests, which makes them harmless to other organisms, including the natural enemies of pests.

Chilo Infuscatellus GV

This virus, which belongs to the group granulosis virus, is similar in its antagonistic characteristics to other baculoviruses, it is extremely target specific. It is active against chilo Infuscatellus (sugar cane borer), which is an important pest of sugar cane. Its effectiveness has also been demonstrated in number of field trials.

Market Potential

As opposition to the use of chemical pesticides mounts on ground of environment and health considerations, the demand for biopesticides i expected to increase considerably in the near future.

According to one estimate, biopesticides are likely to account for about 5 per cent of the total pesticide consumption in India by year 2000.

The government has already set up 122 biocontrol labs/ parasite breeding centres. These efforts are supported by 25 central Integrated pest management centres (CIPMC) which demonstrate the effectiveness of biopesticieds against pest of important crops such as sugar cane, groundnut. cotton, rice, coconut, mct$_O$r gram, tobacco, maize and various vegetables. The activities of these and other similar agencies are expected to increase the popularity and acceptance of these pesticides among the farmers.

REFERENCES

1. Dewan, *A* and saiyed H.N. Acute poisonings due to agricultural pesticides reported to the NIOH poison Information centre. In Proceedings of the WHO workshop on occupational Health problems in agriculture sector. Eds. J.R. Parikh, V.N. Gokani, P.B. Doctor, H.N. Saiyed. National Institute of occupational Health, Ahmedabad. (1998).
2. Agarwal S.B. A Clinical, biochemical, neurobehavioural and socio, psychological study of 190 patients admitted to hospital as a result of acui.u urganophosphorous poisoning (1993).
3. Bhatnagar, V.K. Patel, J. S. Variya, M.R. Venkaiah, K, Shah, M. P. and Kashyap, S. K.Levels of orgaochlorine insecticides in human blood from Ahmedabad (rural), India, (1992)
4. Ramachandran, M., Banerjee, B.D. Gulati, M. Grover, A. Zaidi, S.S. A and Hussain, Q.Z. DDT and HCH residues in the body fat and blood samples from some Delhi hospitals (1984).
5. Pesticide pollution due to chlorinated Insecticides especially DDT in the environment of Man and other livestock in the country. Report of a DSt project. Eds. S. K. Chatterjee, S.K. Kashyap and S.K. Gupta. National Institute of occupational Helath, Ahmedabad (1980).
6. Durham , W.F. Body burden of pesticides in man.
7. Annual Report, National Instutute of Occupational Health, Ahmedabad, (1984-85).
8. Surveillance of food contaminants in India. Report of an ICMR Task force study. Eds. G. S. Toteja, J. Dasgupta, B. N. Saxena and R. I. Kalra. Indian council of Medical research, (1993).
9. Nair, A and Pillai, M.K.K. Monitoring of hexachlorobenzene residues in Delhi and faridabad, India, (1989).
10. Gupta, S.K. Jani, J. P. Saiyad, H.N. and Kashyap S.K. Health hazards in pesticide formulators exposed to a combination of peslicides (1984).
11. A Research Paper by Irani Mukharji and Praveen Kurnar Sharma, A Journal of chemistry and environment.
12. B. K. Sharma, "Environmental chemistry" (1994).
13. Krishnan Kannan, Environmental pollution (1991).
14. P. S. Sindhu, Environmental chemistry.

CHAPTER 31

Polluted & Waste Water Treatment & Recycling

Dr. Vivek Mishra[1]

Waste Water Treatment

Sources and Types of Toxic wattes

Domestic wastes contain animoniacal nitrogen at levels up to 50-60 mg L^{-1} and may contain sulphide at levels up to 50 mg L^{-1}, both of which can cause damage to aquatic life, if discharged. Agricultural wastes may also contain even higher levels of both ammonia and sulphide plus a range of toxic pesticides and herbicides and the drainage from silage may be toxic due to low pH values. The majority of toxic wastes are associated with industry. Here the range of possibilities is too wide. Table 19.1 gives some indication of the main types of toxic industrial waste and the toxins that they contain.

Pre-Treatment of Toxic Wastes

Pre-treatment of industrial wastes prior to discharge to a sewer is widely practised. In some cases this is to comply with the restrictions imposed by Water Authorities and Local Authorities on wastes entering a sewer. There are also a large number of specific toxic substances whose discharge to sewers is controlled.

The main advantages of treatment on site are the possibilities of recovering heat or specific substances in an uncontaminated condition and the economies which might result from treatment at higher temperatures and/or higher concentrations. Where the toxic materials are organic in nature there is often difficulty in treatment due to inhibition of bacteria growth. It is often easier and cheaper to develop the necessary bacterial flora in an *on-site* treatment plant.

This to some extent depends upon the concentration and toxicity of the substances concerned. In some cases dilution of the waste by admixture with sewage reduces the toxic inhibi-tion making it preferable to treat the industrial waste and sewage together. Also many industrial wastes are deficient in some nutrient such as nitrogen or phosphorus. The desirable ratio of BOD: N:P is 100: 5 : 1 and the ratio in domestic wastes is commonly 100 : 18 : 2.5 so that deficiencies in industrial wastes can be balanced.

Even where it is decided not to carry out pre-treatment of the toxic waste by chemical or biological methods it is often useful to install devices to improve the effluent quality by simple physical means. This includes some form of screening 10 reduce solids. Also some form of balancing to reduce variations in concentration, flow, pH etc. and some traps to prevent the escape of oil and grease and some grit arrestors.

[1]Department of Microbiology, Saifia Science College, Bhopal (India)

Table 31.1. Sources of some Common Toxins

Toxin	Sources
Acids—mainly inorganic but some organic causing pH < 6	Acid Manufacture, Battery Manu-facture, Chemical Industry, Steel Industry
Alkalis—causing pH > 9	Brewsry wastes, Food Industries Chemical Industry, Textile Manufacture
Antibiotics	Phan laceutical Industry
Ammoniacal nitrogen	Coke Manufacture, Fertiliser Manufacture, Rubber Industry
Chromium—mainly hexavalent but also less toxic trivalent form	Metal processing, Tanneries
Cyanide	Coke production, Metal Plating
Detergents—mainly anionic but some cationic	Determent Manufacture, Textile Manufacture, Laundries, Food Industry
Herbicides and Pesticides—mostly chlorinated hydro-carbons	Chemical Industry
Metals—mainly copper, cadmium cobalt, lead, nickel, mercury and zinc	Metal processing and plating, Chemical Industry
Phenols	Coke Production, Oil Refining Wood Preserving
Solvents—mostly benzene, acetone, carbon tetrachloride, alcohols	Chemical Industry, Pharmaceuticals

The processes which are used may be classified as physical, chemical and biological. The physical processes are summarised in Table 31.2. Where the toxic wastes contain or are composed of organic materials it may also be necessary to provide some biological treatment especially if the effluent is to be discharged directly into a water way. Many different types of processes are used but the following are the most popular.

(*a*) High-rate filtration using plastic media with very high rates of recircuktion.

(*b*) Activated sludge using contact stabilisation.

Like all biological processes these can suffer from toxicity problems especially where the concentration of toxin is not constant. In general terms it is easier for bacteria and other microorganisms to adap to toxic substances than for organisms like worms, fly larvae et:. For this reason' conventional per-colating filters have not proved successful—the lack of grazing fauna has led to persistent ponding.

Due to a combination of higt organic strength and inhibition from toxic substances it is unusual to obtain complete treatment of toxic industrial wastes by conventional primary and secondary treatment. The effluent from high-rate filters often has a BOD and COD similar to settled sewage and is either suitable for discharge to a sewer or for further biological treatment on site.

Chemical treatment of industrial wastes may be used in addition to and to some extent in place of biological treatment. The aims are somewhat different since biological treatment is mainly a way of oxidising organic matter or a way of converting it into a settleable form. Cheriical treatment can also provide oxidation through chlorine, ozone, etc., but this is used only for oxidising particular com xiunds like cyanide since it is expensive and liable to lead to the production of undesirable chlorinated organics. It is mainly used for pH correction and improving the removal of solids. The commonest chemicals in use are shown in Table 31.3.

Table 31.2. Physical Methods of Pre-Treatment

Process	Aim	Examples
Screening	Removal of coarse solids	Vegetable canneries, Paper Mills
Centrifuging	Concentration of solids	Sludge dewatering in chemical industry
Filtration	Concentration of fine solids	Final polishing and sludge dewatering in chemical and metal processing
Sedimentation	Removal of settleable solids	Separation of inorganic solids in ore extraction, coal and clay production
Flotation	Removal of low specific gravity solids and liquids	Separation of oil, grease and solids in chemical and food industry
Freezing	Concentration of liquids and sludges	Recovery of non-ferrous metals
Solvent Extration	Recovery of valuable materials	Coal Carbonizing and Plas-tics manufacture
Ion Exchange	Separation & Concentration	Metal processing
Reverse Osmosis	Separation of dissolved solids	Desalination of process and wa.:.: water
Adsorption	Concentration and removal of trace impurities	Pesticide manufacture, dye-stuffs removal

Table 31.3. Chemicals used in Industrial Waste Treatment

Chemical	Purpose
Calcium hydroxide	pH adjustment, precipitation of metals and assisting sedimentation
Sodium hydroxide	Used mainly for pH adjustment in place of lime
Sodium carbonate	pH adjustment and precipitation of metals with soluble hydroxide
Carbon dioxide	pH adjustment
Aluminium sulphate	Solids separation
Ferrous sulphate	Solids separation
Chlorine	Oxidation

Primary and Secondary Treatment

If the pre-treatment of toxic industrial wastes is successful, no difficulties should be encountered in subsequent treatment. However, no pre-treatment system is perfect and malfunction will occasionally occur mostly due to variations in the manufacturing process. As a result toxic material together with possible overload of organics and solids may be passed on to the subsequent treatment stages.

The effect of toxic materials on pri jaary sedimentation is insignificant, since this is a purely physical process of sedimentation and flocculation. However the effect of the pri·nary sedimentation on toxic wastes can be very important. Toxic materials in suspension such as partic alate metals are effec-tively removed. Also the flocculant mater al has a great capacity for adsorption and removes the majority of dissolved metals, pesticides and other toxic organics. In one respect this is beneficial since it renders the waste material less inhibitory for biological treatment but it selectively concentrates the toxins in the sludge and may give rise to problems in digestion and in sludge disposal.

The key to successful secondary treatment of wastes containing toxic materials is the adaptation of the microorganisms to the presence of the toxins. Bacteria ard to a lesser extent protozoa show considerable ability to acclimatise to the presence of toxic substances and a great adaptability in degrading new synthetic organics. Metazoa are ratherless adaptable and for this reason activated sludge is senerally better than per-colating filters for treating toxic wastes experience with treating toxic industrial wastes on percolating filters has shown frequent ponding problems due to a lack of activity by the grazing fauna. High-rate filters which utilise hydraulic scouring for film control have been used successfully and the high recirculation ratios help to dilute any incoming toxins. This helps to overcome the other disadvantage of filters which is due to the plug-flow nature of the process. Any shock loads of toxin are not as readily diluted as in a completely mixed reactor.

The activated sludge process is generally preferred for dealing with wastes containing an admixture of toxic materials. In particular the completely mixed version gives immediate dilution of any shock loads. The dangers with activated sludge are that:

(*a*) The toxin may reach concentration which inhibits enzyme activity.

(*b*) Some toxins also affect the bacterial surface and therefore affect settleability.

Nitrification is particularly sensitive to these problems. It is therefore important in treating a toxic waste by either biological process that a microbial population is developed which is acclimatised to the presence of the toxin and in the case of degradable toxins contains sufficient numbers of organisms which can metabolise the toxins. These twin aspects of acclimatisation require great care in the start-up operation and may need a period of several months before successful operation is achieved.

Sludge Treatment and Disposal

Toxins which remain in solution during primary and secondary sedimentation do not appear in the sludges and thus cause no further difficulties. Toxicity in digestion may also occur due to soluble toxins in the treatment of industrial wastes by anaerobic methods. But the main problems are due to toxins which are in settleable form or are readily adsorbed, are selectively concentrated in the sludge and give rise to difficulties in digestion (and in subsequent disposal). The classes of toxins involved are mainly metals, chlorinated hydrocarbons, organic solvents and detergents.

Sewage Treatment

The processes of treating municipal sewage are broadly classified as *primary, secondary,* and *tertiary.* The particular process used in a given situation depends on the volume to be treated, the location of the outfall, the dilution factor, the potentisA hazard to users receiving the water, and in many cases, the cost of the project.

Primary Treatment

The sewage is often disposed of as raw sewage of effluent from primary treatment. Primary treatment consists of removing floating and suspended solids by mechanical means (Fig. 31.1.). More than one-half of the suspended solids can ba removed by primary treatment. First the large solids are screened out and grease and scum are removed. This is followed by sedimentation in a basin to remove the remaining solids, called *primary sludge.* The screens, called trash racks, consist of steel bars about 2 to 4 inches apart. In some cases sand and other coarse material is removed by *grit chambers* to further protect pumps and other equipment from damage. Sometimes the waste water is then run through fine screens. Usually after screening and removal of grit, the waste water is run directly into settling tanks. The settling tanks may have skimming devices, or the removal of scum may be done separately.

The primary sludge is a burdensome problem because it is bulky and must be removed. It contains 94 to 99 percent water. Usually the first step is to remove as much of the water as possible. In some cases the sludge is dried in beds with some of the water being removed by filtration. The residue is disposed of on land. Because the sludge itself can comprise a pollution problem, a better method is to bring about microbial. decom-position in *sludge digestion tanks* before drying. The microbial action occurs under anaerobic conditions, and because this proceeds slowly at low temperatures, the digestion tanks are usually mnintained to 25 to 32°C, at which temperatures the sludge decomposes in 20 to 30 days. The digested sludge will have been reduced to about one-third of its original volume and will be relatively inoffensive.

Primary treatment consists of removing floating and suspended particles by mechanical means (Fig. 31.1a). The other products of primary treatment are gases and the fluid or clarified waste water. The gas is mostly methane, which is usually burned as fuel to provide heat for the digesters and other equipment, the clarified waste water has highly objectionable properties and in most cases is put through a secondary treatment.

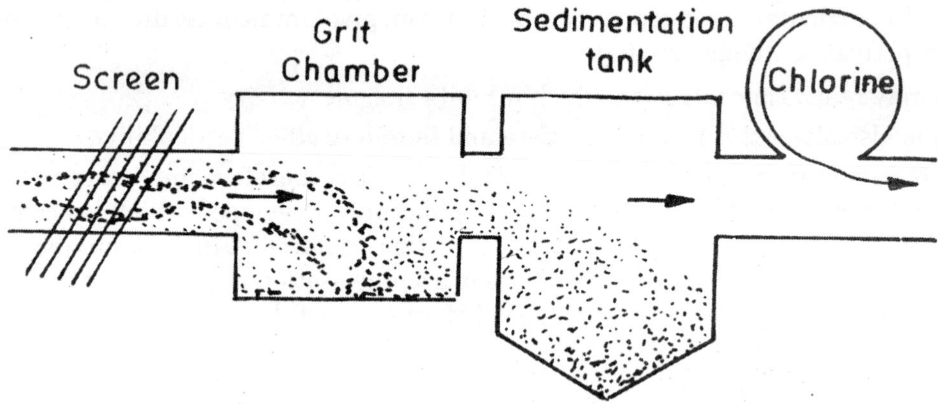

Fig. 31.1(*a*). Primary Sewage Treatment.

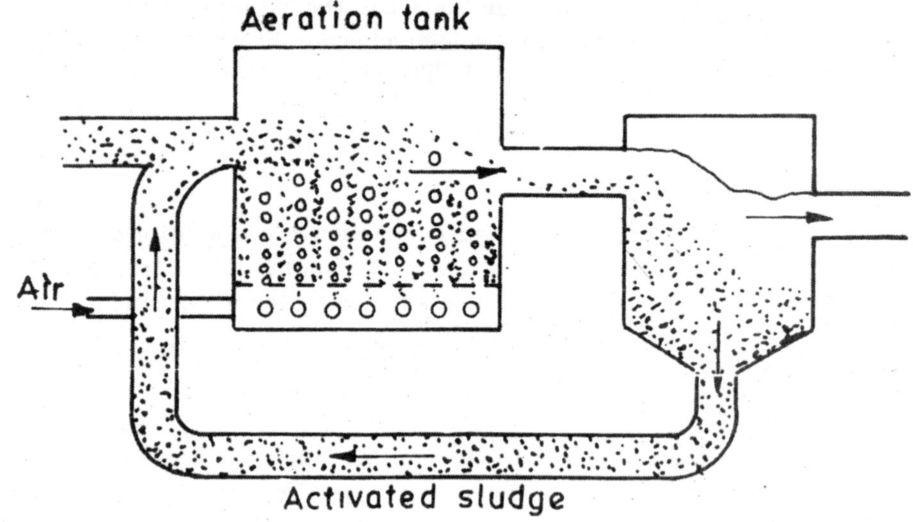

Fig. 31.1(*b*). Secondary Sewage Treatment.

Secondary Treatment

Secondary treatment of waste involves the biological degradation of organic materials by microorganisms under controlled conditions (Fig. 31.1b) The usual method is to bring about the biological oxidation of the organic material under aerobic conditions, in which the waste is aerated to supply oxygen for the microorganisms. The degraded material settles out in secondary settling tanks and is therefore described as being removed by sedimentation. The sediment containing the microbial growths and their by products is called *secondary sludge* or *activated sludge.* The clarified waste water is discharged in the outfalls to rivers, lakes, bays, lagoons, or oceans. Some of the sludge is returned to aeration tanks where it is recycled with the incoming waste. Most of the sludge, however, must be removed, and this operation accounts for one-fourth to one-half of the operation.

Sludge

Both primary and secondary sludges are usually transferred to sludge digestion tanks where decomposition takes place under anaerobicfcponditions. Various methods for water removal are used. Because the removal of the sludge is a major part of sewage disposal, a great deal of effort has gone into drying and finding uses for the product, or locating disposal sites and transporting the sludge there.

A few of the treatment plants heat-dry the activated sludge, usually after some form of mechanical water removal, called *dewatering,* and sell the product as fertilizer. The process is costly. It is less expensive to incinerate the sludge in furnaces or to use it as landfill. Incineration sterilizes the sludge and reduces its volume. However, incineration has some disad-vantages. It creates an air pollution problem and leaves an ash-that must be disposed of. Still, disposing of a small amount of ash is easier than getting rid of a large amount of sludge.

Final Disposal

The methods used for disposal of sludges are usually that are the U ast costly. The choice of disposal site whether on land or sea depends on the proximity of the treatment plant to suitable disposal locations. Dewatered sludges are commonly used as landfill. Liquid sludges are disposed of either on land or in bodies of water. Liquid sludge is used to fertilize or condition agricultural land. However, problems of odour, water pollution, and stimulation of insect and algal growth, as well as other aspects of public health and aesthetic values, must be considered.

Tertiary Treatment

Any treatment to further purify waste water beyond the methods commonly employed for secondary treatment may be called "advanced treatment." The term *tertiary treatment* is often used to refer to certain advanced treatment procedures for treating the effluent coming from secondary treatment. The purpose is the removal of the contaminants of waste water that remain after secondary treatment. These contaminants consist mainly of suspended solids, dissolved organic compounds, and the inorganic plant nutrients nitrogen and phosphorus. A thorough job of cleaning up sewage waste water is difficult and costly.

A number of alternative techniques depending upon the nature of pollutant are available. The methods involve removal of nutrients by autotrophic plants, flocculation of colloidal paniculate matter, use of adsorbants, use of oxidising agents like $KMaO_4$ or O_3, use of *reverse osmosis* by artificial pressure device across semipermeable membrane, or by specific chemical techniques. Chlorination is done to eliminate orthophosphates remaining in water after primary and secondary treatments which can be precipitated by Ca or some other metal ions. Aluminium sulphate, ferric chloride, ferrous (ic) sulphate and calcium hydroxide are good coagulants. Nitrates are removable with the help of *Nitrococcus denitrificans* and *Thiobacillus denitrificans* which denitrify and release nitrogen and water.

REFERENCES

1. Anon, 1970. Drinking Water—Is it Drinkable *Env. Sci. and Tech.* 4 (10).811-813.
2. Eckenfelder, W.W. and D.J. O'Connor. 1961. *Biological Waste Treatment.* Pergamon Press. New York.
3. Sawyer, C.N. and P.L. McCarthy. 1978. *Chemistry for Environmental Engineering.* McGraw Hill, New York.
4. Tearle, K. (Ed.) 1973. *Industrial Pollution Control.* Business Books Ltd. London.
5. Varshney, C.K. (Ed.) 1985. *Water Pollution and Management.* Wiley Eastern Limited, New Delhi.

CHAPTER 32

Natural Resources with Reference to Forest and Wetland

Dr. Rabia Sultan[1]

Every thing available in our environment which can be used to satisfy our needs, provided, it is technologically accessible, economically feasible and culturally acceptable can be termed as "Resource."

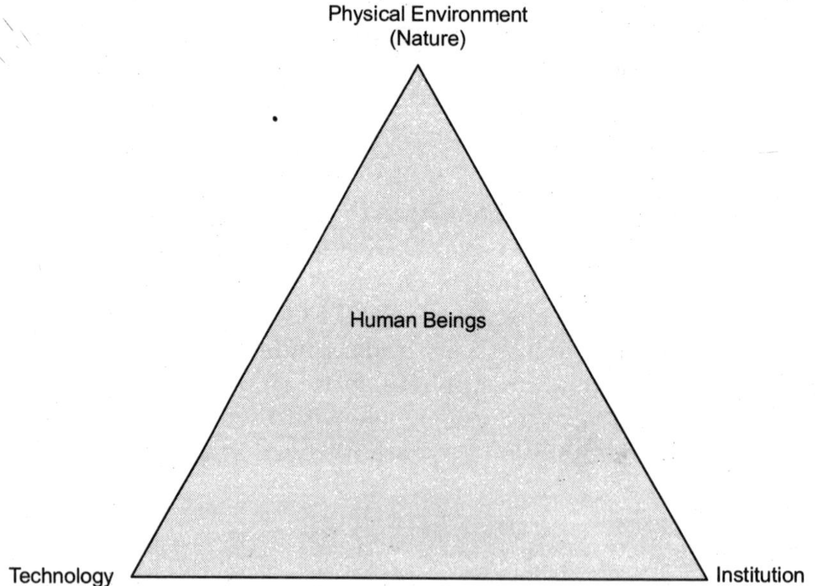

Fig. 32.1. Interdependent relationship between Nature, Technology and Environment.

The process of transformation of things available in our environment involves an inter-dependent relationship between Nature, Technology and Institutions. Human beings interact with nature through technology and create institutions to accelerate their economic-development.

NATURAL RESOURCES WITH REFERENCE TO FOREST AND WETLAND

Resources are vital for human survival as well as for maintaining the quality of life. Resources are free gift of nature. Any material which is required or used to sustain life or livelihood is termed as resource. In other words, resources are all these requirements of organism, population and communities which tend to

[1]Department of Zoology, Saifia Science College, Bhopal 463001 (India)

help in accumulation of energy by their increased availability. Some examples of resources are air for breathing, water for drinking, land for living and growing food, forests for timber and paper, ores for aluminum, copper, iron and other metals and coal and natural gas for producing energy. (Fig. 32.1)

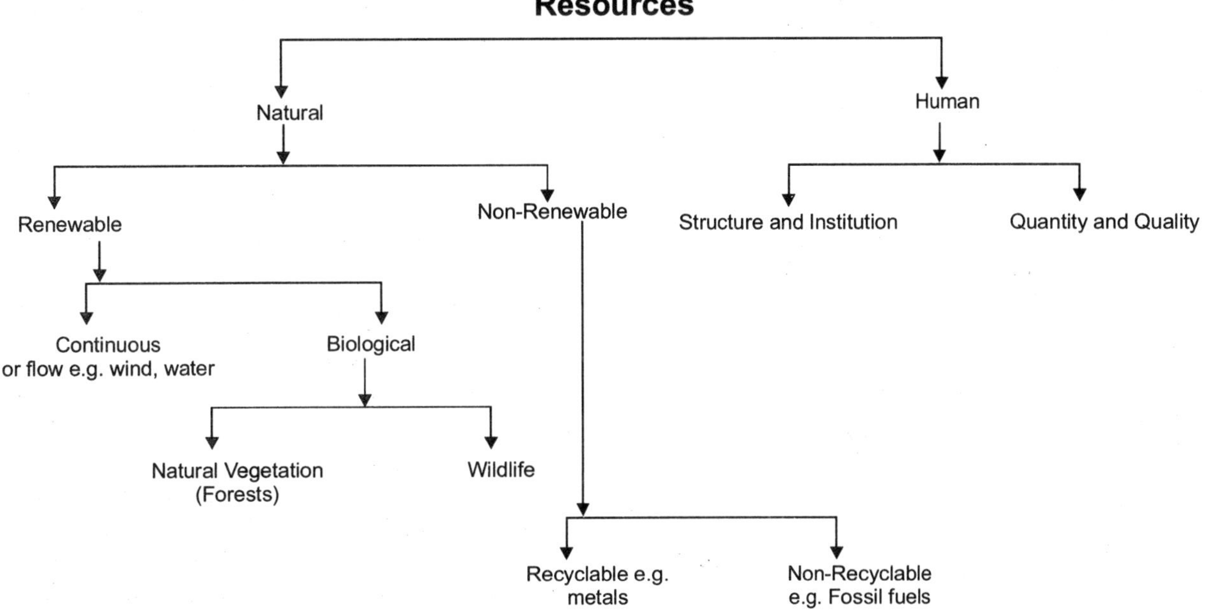

Fig. 32.2. Classification of Resources.

According to Remade (1984) a natural resource is defined as a form of energy of matter which is essential for the functioning of organisms, populations and ecosystems. The basic ecological variables—energy, space, time and diversity are sometimes combined and termed as natural resources. These natural resources maintain ecological balance among themselves. Man is the only organism who has disrupted this natural balance.

On the basis of origin, Natural resources (Fig. 32.2) may be divided into -

1. Biotic Resources.

2. A biotic Resources.

1. Biotic Resources : Biotic resources are the ones which are obtained from the biosphere. Forest and their products, animals, birds and their products, fish and other marine organisms are important examples. Minerals such as coal and petroleum are also included in this category because they were formed from decayed organic matter.

2. Abiotic Resources : A biotic resource comprise of non living things. Example include land, water, air and minerals such as gold, iron, copper, silver etc.

Considering their stage of development natural resources may be referred to in the following ways:

- **Potential Resources :** Potential Resources are those which exist in a region and may be used in the future for example, mineral oil may exist in many parts of India having sedimentary rocks but till the time it is actually drilled out to put into use, it remains a potential resource.
- **Actual Resources :** Actual resources are those which have been surveyed, their quantity and quality determined and are being used in present times. For example the petroleum and the natural gas which is obtained from Bombay High Fields. The development of an actual resource, such as wood processing depends upon the technology available and cost involved. That part of the actual resource which can be developed profitably with available technology is called reserve.

With respect to renewability, natural resources can be categorized as follows:

- Renewable Resources.
- Non Renewable Resources.

- **Renewable Resources :** Renewable resources are those resources which can be regenerated or reproduced, some of the them like sunlight, air, wind etc are continuously available and their quantity is not affected by human consumption. Many renewable resources can be depleted, by human use but may be replenished, thus maintaining a flow. Some of these like agricultural crops take a comparatively longer time, while still others, like forests take even longer.
- **Non Renewable Resources:** Non renewable resources are those resources which can not be regenerated once they are exhausted. Minerals and fossils are included in this category since their rate of formation is extremely slow, they can not be replenished once they get depleted. Out of these, the *metallic* minerals can be reused by recycling them. But coal and petroleum cannot be recycled.

Minerals are important rich source of energy but their distribution is uneven. In some places they may be found in abundance where as at other place it may be in very *less* quantity. For example molybdenum is found in abundance in North America and very scarce in Asia. Asia is rich in tin, tungsten and manganese but these metals are poorly found in North America. World's most precious metal, gold and platinum are found in large quantity in South Africa, but are poor in silver. The reason for such uneven distribution of these metals is unknown. About 200 metals are widely and extensively used in various industries and are a rich source of our economy. This has resulted in forcing a new mineral cycle in the environment, which includes (*a*) mineral extraction (*b*) conversion of minerals into large quantities of metals ceramics and chemicals (*c*) manufacture of medicines, pesticides, fertilizers, utensils and other public utility goods (*d*) their return back to the environment. At every stage of this mineral cycle, energy is consumed, however, this mineral cycle has resulted in progressive deterioration of aquatic, terrestrial and atmospheric environment and this posses a question mark on the future of human civilization and health.

Earth's resources are limited and need very careful utilization. Metallic and non metallic are the important reserve of earth.

Exploitation of Minerals by Mining and Environmental Problems

By mining it means the selective recovery of minerals and materials from the crust of the Earth. The mined materials may be used as such or used for extraction of elements.

In order to recover only small quantity of desired product, mining involves the physical removal of minerals from the crust of the Earth. During their removal, accidents may occur due to human error. Therefore, it becomes essential to provide extensive training to all people who work in mines.

Mining is usually carried out either on surface or deep under the ground. In some of the gold mines in South Africa, mining starts at depths of 1500 m and goes down to 3500 m. During mining, a large quantity of unconsolidated material is produced. This may sometimes cause accidents, thereby causing loss of human lives. Workers do working in such mines at the cost of their health.

All mines face more or less safety problems. However underground mines are more hazardous than surface mines. At the same time, soft-rock mines are more dangerous than hard-rock mines. In underground mines, accidents occur due to fall of rocks on the sidewalls of openings. Accidents may also occur in moving of machineries, water inrushes and explosions due to release of gases (e.g. methane) in the mines. In all mines especially in coal mines, methane is released. Due to this workers may not get enough oxygen to breath and at the same time explosion may occur due to burning of this gas.

In underground mining the weight of the overlaying rock may exceed the strength of down laying rocks thereby resulting in the rock falling explosively a condition known as a rock rust. Research is being carried out to improve design to eliminate or reduce the danger of such occurrence.

The miners working especially in underground mines are more prone to diseases. The miners inhale dust in mines. Thereby causing a number of lung diseases such as black lung, silicosis, asbestosis etc.

Table 32.1. Sources, Uses and Environmental Problems of some important metals.

Metal and its Abundance in the Earth's Crust	Sources and	Properties and Uses	Environmental Problems
1	2	3	4
Aluminum (Al) 8.13%	Bauxite, an impure form of Al_2O_3 Reserves are plentiful.	Only one-third as dense as iron, resists corrosion, good electrical conductor. Used in aircraft, automobiles, beverage cans pots and pans.	Much electrical energy used in production (about 67,000 kilowatt hours per tone of aluminum.) Also some fluorine-containing gases and dusts are produced in manufacture.
Iron (fe) 5.1%	Hematite (Fe_2O_3) and magnetite (Fe_3O_4). The highest grade hematite ores are largely exhausted Magnetite is found in taconite rock, reserves are still vast and extraction technology has improved.	Major metal for all structures and machinery.	Air, water and land pollution occur both at the mine and in the chemical production of iron from the ore. Taconite deposits also contain asbestos.
Copper (Cu) 0.007%	Elemental copper, mined since ancient times, is largely exhausted. Major sources not are copper sulfides, CuS and Cu_2S.	Excellent conductor of heat and electricity. Used for electrical wire, water and steam pipes and cooking utensils.	Acid mine drainage. Smelters tar process the sulfides produce large amounts of sulfur dioxide, SO_2.
Lithium (Li) 0.003%	Widely distributed but in small quantities, in various minerals.	Will be important for the fusion energy program because the 6Li isotope is needed to make tritium.	No serious problems.
Lead (Pb) 0.0016%	Major source in galena, PbS. Reserves are concentrated but not abundant.	Soft, dense metal that is fairly resistant to corrosion and has fairly low melting point (327^{0}C). Used for pipes, slider, electrodes in batteries and pigments in paint. Its use as an antiknock agent in gasoline is declining.	Acid mine drainage. SO_2 is produced in lead smelters. Lead smelters. Lead compounds are cumulative poisons.
Gallium (Ga) 0.0005%	Found as an impurity in ores of zone and aluminum but in small amounts.	Used are solar cells to convert solar energy to electricity. If the solar energy program expands, gallium reserves will be critical.	No special problems.
Mercury (Hg) 0.00005%	Sometimes occurs as the native element in small amounts but the important ore is cinnabar, HgS. Reserves are very limited.	The only metal that is liquid at all ordinary terrestrial temperatures. Used in electrical switches in thermometers and for many special chemical and medicinal purposes.	Mercury compounds are toxic.
Platinum (Pt) 0.00002%	Occurs as the native elements in widely scattered ores.	Unsurpassed as a catalyst for oxidation reactions for which it is used in catalytic converters to reduce pollution from automobile exhausts. Platinum is also widely used as a catalyst in industry.	No special problems.

The explosives used in mining release oxides of nitrogen which are extremely toxic. In some mines, toxic gases such as hydrogen sulphide and carbon monoxide are also produced which are also toxic and cause accidental deaths.

Many mines especially those containing uranium may face radiation problem, thereby causing cancers to miners.

Due to the hazardous nature of mining work all major mining countries have implemented extremely strict legislation as well as regulation governing mine safety. These also cover the quality of the air explosives, lighting, noise, support of mining workers and all other hazards that may be encountered in the mines. In recent years, there happens to be a big change in the design of support system which enjoys great success in reducing the accidents due to fall of ground in British coal mines. These days mining workers are trained in latest technologies to reduce accidents. At the same time the mine managers must have passed examinations in mining safety and law.

Conservation of Mineral Resources

As minerals are exhaustible resources it becomes essential to conserve these resources. The dissipated mineral resources cannot be replaced except overlong geological period which may extend to millions of years. Most of countries including India are exploiting these resources at an accelerated rate to get valuable foreign exchange. However it is better if we export the metals obtained from the minerals.

It is possible to conserve the minerals by improving efficiency in mining technology and benefaction technology. There are many metals such as iron, aluminum, copper, tin, etc. which can be recycled, thereby reducing waste for example scrap iron is being used for iron and steel industry. Scarce materials can be saved by using such substances which are cheaper and can be produced in abundance for example aluminum is extensively used in electrical industry in place of copper.

Forest

Forest is a piece of land that has a dense growth of trees and shrubs more than a third of the earth's 51 million square miles (132 million sq. km) of land is covered by forests or is capable of supporting forests. However, this forest land is unequally distributed among the continents and even within different areas of the same continent.

The forest cover in our country (India)is estimated at 637,293 sq. km. which is 19.39 percent, of the total geographical area (dense forest 11-48 percent, open forest 7.76 percent and mangrove 0.15 percent).

According to the state of forest report (1999) the dense forest cover has increased by 10,098 sq km since 1997 However, this apparent increase in the forest cover in also due to plantation by different agencies. The state of forest report does not differentiate between natural forests and plantations. Therefore, these reports fail to deliver accurate information about loss of natural forests.

Types of Forests

In India much of its forest resources are either owned or managed by the government through forest department or other Government departments. These are classified under the following categories.

(i) **Reserved forests :** More than half of the total forest land has been declared reserved forests. Reserved forests are regarded as the most valuable as far as the conservation of forest resources are concerned.

(ii) **Protected forests :** Almost one third of the total forest area is protected forest, as declared by the forest Department. These forest lands are protected from any further depletion.

(iii) **Unclasped forests :** These are other forests and wastelands belonging to both government and private individuals and communities.

Reserved and protected forests are also referred to as permanent forest estates maintained for the purpose of producing timber and other forest produce and for protective reasons. Madhya Pradesh has the largest area under permanent forests constituting 75 percent of its total forest area. Jammu and Kashmir, Andhra Pradesh, Uttaranchal, Kerala, Tamil Nadu, West Bengal and Maharashtra have large percentages of reserved forests of its total forest area whereas Bihar, Haryana, Punjab, Himachal Pradesh, Orissa and Rajasthan have a bulk of it under protected forests. All North eastern States and part of Gujarat as unclasped forests managed by local communities.

Although the most prominent features of a forest are its trees and shrubs, a forest is much more than such woody vegetation. Forests provide many social economic and environmental benefits. In addition to timber and paper products forests provide wild life habitat and recreational opportunities, prevent soil erosion and flooding, help provide clean air and water and contain tremendous biodiversity. Forests are also an important defense against global climate change, through the process of photosynthesis, forests produce life giving oxygen and consume huge amounts of carbon dioxide, the atmospheric chemical most responsible for global warming by decreasing the amount of carbon dioxide in the atmosphere of forests may reduce the effects of global warming.

Deforestation

However, huge area of the richest forests in our country (India) has been cleared for fuel wood, timber products, agriculture and livestock. These forests are rapidly disappearing. Low productivity of forests is due to the low attention paid towards their development. Per capita forest area is also very low and this needs to be increased considerably. Due to these reasons the forests have played minimum role in the economic development of our country. Proper scientific management, conservation and utilization of forest wealth is likely to increase their resource value. There is need to conserve and preserve the existing forest resources and effects should be made to expand the forests. What would happen if there are no forests at mountain tops? The rivers emerging from them would flood the plains and raise the river beds due to excessive deposition of silt and denudation of mountains affect the land vegetation floods cause heavy loss of crops, property, animal and human life every year and causes reduction of soil fertility due to erosion.

To encourage Forestation and prevent wasteful approach of human being, introduction of proper management methods is necessary. This would check block cutting, deforestation, pest and fire control. A recycling procedure ensuring continuous supply of forest products should be planned on long terms basis, so that the forest resources could last for a larger period. It is very important for our survival for example, we need the modest timber supply annually and this harvested crops should be replaced by an equal amount of timber crop growth. Timber trees reach maturity in 15 to 100 years, so its recycling should be planned accordingly.

Block cutting method is generally adopted in the forest which have even aged trees, of a few species only coniferous forests are cut in this way. Thus a particular area is denoted by cutting down its entire tree population. Such area may be in the centre of interior of the forest and the lost forest area is replaced by reforestation of *adjacent* area of the same size. This would result in a sustained supply of forest products without adversely affecting the actual size of the forest.

Thus the annual felling of forest trees is followed by annual reclamation of the deforested areas.

Timber is an important product of forest trees which is of high economic value. After removing the stems of timber trees which are the main source of timber, stems, leaves, barks, twigs, are left behind as waste. Further wastage occurs in raw mills in the form of raw dust, bark and small pieces of woods. Now with the recent advanced technology all these wastes are being used for industrial and commercial purposes. Very careful and planned utilization of timber resource is the need of hour so that they can be available in future for a longer duration. A proper management of timber forests can only ensure this.

Conservation Measures

The important measures that can be taken for conservation of forests are :

1. Forest fire should be prevented.
2. Care must be taken to eradicate pests by spraying some pesticides to make trees healthy.
3. Resistant varieties of plants should be selected and planted.
4. The trees, if felled for timber and other use should be replaced by planting more trees.
5. The use of fuel wood and wood charcoal should be discouraged. The consumption of fuel wood can be minimized by using biogas plants and Solar Chula has for cooking purposes.
6. The rate of annual deforestation should be followed by greater rate of annual reforestation so that there will be no scarcity.
7. Social forestry programmed should be undertaken on a large scale with active participation of people.
8. A forestation programme should be undertaken in the waste lands.
9. Trees of aesthetic value should be planted along road sides and railway tracks.
10. Regeneration that is renewal of forest crops should be adopted.
11. Reforestation by suitable monoculture should be implemented that is to harvest a single species on short rotation.
12. Unwanted deforestation should be restricted.

Wet Land

The wetland is an area of land whose soil is saturated with moisture either permanently or seasonally such areas may also be covered partially or completely by shallow pools of water wet lands include swamps, marshes and bogs, salt water, fresh water, or brackish.

Wetlands are considered the most biologically diverse of all eco systems dominated by the influence of water but are the transitional zones that occupy an *intermediate* position between dry land and open water. They possess characteristics of both terrestrial and aquatic ecosystems. The habitats range from river, flood plains and rained lakes to estuaries, mangroves, swamps and salt marshes. Though the first sign of civilization are traced back to wetland areas like Nile delta and flood plains of Indus River, but the attention on wet land could be discussed in recent years.

The wetland ecosystem is the only ecosystem type to have its own international convention know as "Ramsar Convention 1971"

Ramsar Convention

The Convention on Wetlands of International Importance, especially as Waterfowl Habitat, or Ramsar Convention, is an international treaty designed to address global concerns regarding wetland loss and degradation. The primary purposes of the treaty are to list wetlands of international importance and to promote their wise use with the ultimate goal of preserving the world's wetlands. Methods include restricting access to the majority portion of wetland areas as well as educating the public to combat the misconception that wetlands are wastelands.

Climate

Temperatures vary greatly depending on the location of the wetland. Many of the world's wetlands are in temperate zones (midway between the North and South Poles and the equator.) In these zones, summers are warm and winters are cold but temperatures are not extreme. However, wetland found in the tropic zone, which is around the equator are always warm. Temperatures in wetlands on the Arabian Peninsula for example can reach 50 °C (122 °F). In northeastern Siberia, which has a polar climate wetland temperatures can be as cold as –50 °C (–58 °F).

Rainfall

The amount of rainfall a wetland receives depends upon its location. Wetlands in Wales, Scotland and Western Ireland receive about 150 cm (59 in) per year. Those in Southeast Asia, where heavy rains occur can, receive up to 500 cm (200 in). In the northern areas of North America wetlands exist where as little as 15 cm (6 in) of rain fall each year.

Under the Ramsar international wetland conservation treaty, wet lands are defined as follows:

Wetlands are areas of marsh, fen, peat land or water, whether natural or artificial, permanent or temporary with water that is static or flowing, fresh, brackish or salt, including areas of marine water the depth of which at low tide does not exceed six meters "Another very broad definition prepared by U.S. Fish and wild life services read as" wet lands are lands transitional between terrestrial and aquatic system where water table is usually at or near surface or the land is covered by shallow water."

There is no single universally accepted definition of wetland because of the wide diversity of wet land types and the difficulty in demarcating the boundary of this ecosystem.

Plant life found in wetlands includes mangroves, water lilies, cattails, sedges, tamarack, black spruce, cypress, gum and many others. Animals include many different amphibians, reptiles, birds and fur bearers. They are generally distinguished from other water bodies or land forms based on their water level and on the types of plants they thrive within them. Specifically wet lands are characterized as having a water table that stands at or near the land surface for a long enough season each year to support aquatic plants but simply, wet lands are lands made up of hydria soil.

The study of wetlands has recently been termed plaudology in some publications.

Conservation

Due to lack of potential financial benefits, wet lands have historically been the victim of large-scale draining efforts for real estate development, or flooding for use as recreational lakes. Wet lands provide a valuable flood control function but building *levees* help replace natural flood controls. Wetlands are very effective at filtering and cleaning water, so to help with the ever increasing challenge of decreasing water pollution, millions of dollars have been invested on water purification, plants and expensive remediation measures. The USA came to understand how biologically productive wet lands are. So the USA passed laws limiting wet lands destruction, and created requirements that if a wet land had to be drained, developers at least had to offset the loss by creating artificial wet lands.

In New Zealand over 90% of the wet lands have been drained since European settlement, predominantly to create farmland, wet land now have a degree of protection under Resource Management Act.

Ecological roles of Wetlands

Wetlands are among the most productive ecosystems in the world, comparable to rainforests and coral reefs. An immense variety of species of microbes, plants, insects, amphibians, reptiles, birds, fish and mammals can be part of wetlands ecosystem. Physical and chemical features such as climate, landscape shape (topology), geology and the movement and abundance of water help to determine the plants and animals that inhabit each wetland. The complex, dynamic relationships among the organisms inhabiting the wetland environment are referred to as food webs.

Wetlands support a rich food web, from microscopic algae and dragonfly larvae to alligators and black bears.

Wetlands can be thought of as "biological supermarkets." They provide great volumes of food that attract many animals' species. These animals use wetlands for part of or all of their life-cycle. Dead plants leave and stems break down in the water to form small particles of organic material called "detritus." This enriched material feeds many small aquatic insects, shellfish, and small fish that are food for larger predatory fish, reptile's amphibians, birds and mammals.

The functions of a wetland and the values of these functions to human society depend on a complex set of relationships between the wetland and the other ecosystems in the *watershed*. A watershed is a geographic area in which water, sediments, and dissolved materials drain from higher elevations to a common low-lying outlet or basin a point on a larger stream, lake, underlying *aquifer*, or *estuary*.

Wetlands play an integral role in the ecology of the watershed. The combination of shallow water, high levels of nutrients, and primary productivity is ideal for the development of organisms that form the base of the food web and feed many species of fish, amphibians, shellfish and insects. Many species of birds and mammals rely on wetlands for food, water, and shelter, especially during migration and breeding.

Wetlands, microbes, plants and wildlife are part of global cycles for water, nitrogen and sulfur. Furthermore scientists are beginning to realize that atmospheric maintenance may be an additional wetlands function. Wetlands store carbon within their plants communities and soil instead of releasing it to the atmosphere as carbon dioxide. Thus wetlands help to moderate global climatic conditions.

Table 32.2. Categories of Wetlands

Type	Characteristics
(A) Inland Fresh Areas	
1. Seasonally flooded basins or flats	Soil covered with water or Waterlogged during variable periods but well drained during much of the growing season.
2. Fresh meadows	Without standing water during growing season: Waterlogged to within a few centimeters of surface.
3. Shallow fresh marshes	Soil waterlogged during growing season: often covered with 15 cm. or more of water.
4. Deep fresh marshes	Soil covered with 15-90 cm of water.
5. Open fresh water	Water less than 3 meters deep. Bordered by emergent vegetation.
6. Shrub swamps	Soil waterlogged: often covered with 15 cm or more of water.
7. Wooded swamps	Soil waterlogged, often covered with 30 cm of water. Along sluggish streams flat uplands shallow lake basins.
8. Bogs	Soil waterlogged; spongy covering by mosses.
(B) Coastal Fresh Areas	
9. Shallow fresh marsh	Soil waterlogged during growing season, at high tides as much as 15 cm of water. On lowland side deep marshes along tidal rivers, sounds deltas.
10. Deep fresh marshes	At high tide covered with 15 cm to 90 cm of water. Along tidal rivers and bays.
11. Open fresh water	Shallow portions of open water along fresh tidal rivers and sounds.
(C) Inland Saline Areas	
12. Saline flats	Flooded after periods of heavy precipitation: waterlogged within few cams of surface during the growing season.
13. Saline marshes	Soil waterlogged during growing seasons: often covered with 60-90 cm of water, shallow lake basin.
14. Open saline water	Permanent areas of shallow saline water. Depth variable.
(D) Coastal Saline Areas	
15. Salt flats	Soil waterlogged during growing season: sites occasionally to fairly regularly cover by high tide. Landward sides or islands within salt meadows and marshes.
16. Salt meadows	Soil waterlogged during growing season. Rarely covered with tidewater: landward side of salt marshes.
17. Irregularly flooded salt marshes	Covered by wind tides at irregular intervals during the growing season. Along shores of nearly enclosed bay sounds etc.
18. Regularly flooded salt marshes	Covered at average high hide tide with 15 cm or more of water, along open ocean and along sounds.
19. Sounds and bays	Portions of salt-water sounds and bays shallow enough to be dike and filled. All water land-ward from average low tide line.
20. Mangrove swamps	Soil covered at average high tide with 19-90 cm water.

Table 32.3. Wetlands of Madhya Pradesh

S.No.	Name of Wetlands	Area (ha)	Features
1.	Yashwant Sagar, Sagar	7200	Created in 1938 primarily to supply to drinking water for the Indore city is a prominent wetland of the region. Has been subjected to pollution due to inflow of agricultural chemicals, Cattle and poultry wastes from the catchment's area. Siltation weed infestation and grazing are the other major cause of deterioration of the lake.
2.	Barma Reservoir, Raisen	7700	The life line of the Singhori wild life sanctuary has been subjected to pollution due to inflow of agricultural chemicals/wastes and silt from its vast catchment's deforestation, grazing etc.
3.	Govindgarh Talab, Rewa	307	Reduction of storage capacity due to siltation, growth of weeds, deterioration of water quality.
4.	Sangram Sagar, Jabalpur	12	Lake having religious importance subjected to *siltation*, inflow of sewage bathing, washing of cloths and cattle, weed infestation, weakening of earthen dam.

Wetlands are not the wastelands. Given the care and consideration it deserves the wetlands will be far from a curse. It will be a boon to improve the ecological balance and environment and at the same time to boost the employment and the economy of the land.

REFERENCES

1. Ajai kumal *et. al.,* (2006). Understanding Wetland. Everyman's Science Vol. XLI, No. 2, June-July 06.
2. A. Raghu vanshi, C. Raghuvanshi (1982). Environment and Pollution. Hindi Granth Acad, Bhopal.
3. Arvind Bhatia, K.S. Kohli, M. Sawroop (2000). Various Dimension of Environmental Biology. Ramesh Book Depot, Jaipur.
4. J. Parikh and K. Parikh (1999). Sustainable wetland, Environmental Governance-2 Indra Gandhi Institute of Development Research, Mumbai.
5. L. M. Cowardin, *et. al.,* (1979). Classification of wetlands and deep water habitats of the United States. U.S. Fish and wild life service, Washington.
6. Omkar Singh Bhadoria (2009). Environmental Studies. Yogi Prakshan, New Delhi.
7. Rajeev Verma and Anees Siddiqui (2007). Environmental Studies. Publ. Devi Ahilya Prakashan, Indore.
8. Protect Water, Protect Life (Feb 2005). Published by Lake Conservation authority of India M.P. Govt.
9. U.P. Singh *et. al.,* (1999). Conservation of water resources. M.P. Govt. Publication.

CHAPTER 33

Solid-Fuels and Environment

Dinesh Agarwal[1] and Vibha Agarwal[2]

Introduction

Fuel is a combustible substance, containing carbon as main constituent, which on proper burning gives large amount of heat, which can be used economically for domestic and industrial purposes. Wood, charcoal, coal, kerosene, petrol, diesel, producer gas, oil gas, etc.

Solid fuels are termed as fuels which are in solid state it includes wood, charcoal, peat, coal, Hexamine fuel tablets, and pellets made from wood , corn, wheat, rye and other grains. Solid-fuel rocket technology also uses solid fuel.

Solid fuels have long been used by humanity to create fire. Coal was the fuel source which enabled the industrial revolution, from firing furnaces, to running steam engines. Wood was also extensively used to run steam locomotives. Both peat and coal are still used in electricity generation today.

TYPES OF SOLID FUELS

Coal

Coal is a readily combustible black or brownish-black sedimentary rock normally occurring in rock strata in layers or veins called coal beds. The harder forms, such as anthracite coal, can be regarded as metamorphic rock because of later exposure to elevated temperature and pressure. It is composed primarily of carbon along with variable quantities of other elements, chiefly sulfur, hydrogen, oxygen and nitrogen.

Coal was formed from layer upon layer of annual plant remains accumulating slowly that were protected from biodegradation by usually acidic covering waters that gave a natural antiseptic effect combating microorganisms and then later mud deposits protecting against oxidization in the widespread shallow seas—mainly during the Carboniferous period—thus trapping atmospheric carbon in the ground in immense peat bags that eventually were covered over and deeply buried by sediments under which they metamorphosed into coal. In this manner, over time, the chemical and physical properties of the plant remains (believed to mainly have been fern-like species antedating more modern plant and tree species) were changed by geological action to create a solid material.

Coal, a fossil fuel, is the largest source of energy for the generation of electricity worldwide, as well as one of the largest worldwide anthropogenic sources of carbon dioxide emissions. Gross carbon dioxide emissions from coal usage are slightly more than those from petroleum and about double the amount from natural gas. Coal is extracted from the ground by mining, either underground or in open pits.

[1]BUIT, Barkatullah University, Bhopal (India)
[2]Department of Chemistry, F.G.I.T.M. Bhopal (India)

Types of Coal

Believed approximate position of the proto-continents toward the end of the Carboniferous period; the light blue represents shallow seas where many of today's coal deposits are found, as opposed to deeper water which gave rise to oil bearing rocks derived from marine species. The ice caps were known to be very large, lowering sea levels extensively by locking up oceanic waters into solid ice, though how large the ice caps became is a matter of debate. The position of most continental foundations in lower latitudes definitely created a series of successive shallow swamplike seas we burn for today's coal sourced electricity.

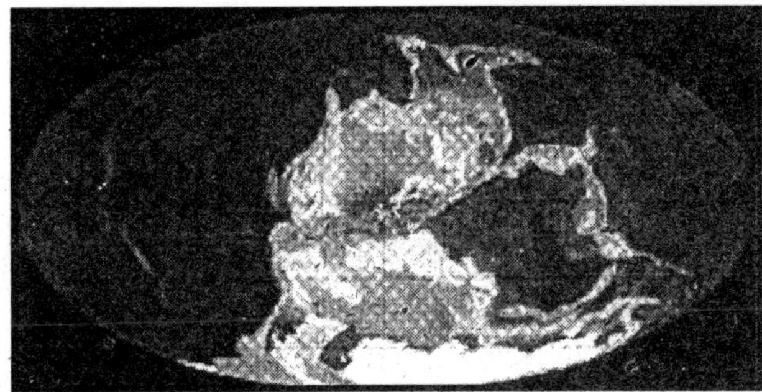

Fig. 33.1

As geological processes apply pressure to dead biotic matter over time, under suitable conditions it is transformed successively into

- Peat, considered to be a precursor of coal, has industrial importance as a fuel in some regions, for example, Ireland and Finland.
- Lignite, also referred to as brown coal, is the lowest rank of coal and used almost exclusively as fuel for electric power generation. Jet is a compact form of lignite that is sometimes polished and has been used as an ornamental stone since the Iron Age.
- Sub-bituminous coal, whose properties range from those of lignite to those of bituminous coal are used primarily as fuel for steam-electric power generation. Additionally, it is an important source of light aromatic hydrocarbons for the chemical synthesis industry.
- Bituminous coal, dense mineral, black but sometimes dark brown, often with well-defined bands of bright and dull material, used primarily as fuel in steam-electric power generation, with substantial quantities also used for heat and power applications in manufacturing and to make coke.
- Anthracite, the highest rank; a harder, glossy, black coal used primarily for residential and commercial space heating. It may be divided further into metamorphically altered bituminous coal and petrified oil, as from the deposits in Pennsylvania.
- Graphite, technically the highest rank, but difficult to ignite and is not so commonly used as fuel: it is mostly used in pencils and, when powdered, as a lubricant.

The Classification of coal is generally based on the content of volatiles. However, the exact classification varies between countries. According to the German classification, coal is classified as follows :

Table 33.1

Name	Volatiles %	C Carbon %	H Hydrogen %	O Oxygen %	S Sulfur %	Heat content kJ/kg
Braunkohle (Lignite)	45-65	60-75	6.0-5.8	34-17	0.5-3	<28470
Flammkohle (Flame coal)	40-45	75-82	6.0-5.8	>9.8	~1	<32870
Gasflammkohle (Gas flame coal)	35-40	82-85	5.8-5.6	9.8-7.3	~1	<33910
Gaskohle (Gas coal)	28-35	85-87.5	5.6-5.0	7.3-4.5	~1	<34960
Fettkohle (Fat coal)	19-28	87.5-89.5	5.0-4.5	4.5-3.2	~1	35380
Esskohle (Forge coal)	14-19	89.5-90.5	4.5-4.0	3.2-2.8	~1	<35380
Magerkohle (Non-baking coal)	10-14	90.5-91.5	4.0-3.75	2.8-3.5	~1	35380
Anthrazit (Anthracite)	7-12	>91.5	<3.75	<2.5	~1	<35300

The middle six grades in the table represent a progressive transition from the English-language sub-bituminous to coal, while the last class is an approximate equivalent to anthracite, but more inclusive (the U.S. anthracite has <6% volatiles).

Cannel coal (sometimes called "candle coal"), is a variety of fine-grained, high-rank coal with a large amount of hydrogen. It consists primarily of "exinite" macerals, now termed "liptinite".

Fig. 33.2 Coal rail cars in Ashtabula, Ohio

Coal as fuel

Coal is primarily used as a solid fuel to produce electricity and heat through combustion. World coal consumption was about 6,743,786,000 short tons in 2006 and is expected to increase 48% to 9.98 billion short tons by 2030.[11] China produced 2.38 billion tons in 2006. India produced about 447.3 million tons in 2006. 68.7% of China's electricity comes from coal. The USA consumes about 14% of the world total, using 90% of it for generation of electricity.

When coal is used for electricity generation, it is usually pulverized and then burned in a furnace with a boiler. The furnace heat converts boiler water to steam, which is then used to spin turbines which turn generators and create electricity. The thermodynamic efficiency of this process has been improved over time. "Standard" steam turbines have topped out with some of the most advanced reaching about 35% thermodynamic efficiency for the entire process, although newer combined cycle plants can reach efficiencies as high as 58%. Increasing the combustion temperature can boost this efficiency even further. Old coal power plants, especially "grandfathered" plants, are significantly less efficient and produce higher levels of waste heat. About 40% of the world's electricity comes from coal, and approximately 49% of the United States electricity comes from coal.

The emergence of the supercritical turbine concept envisions running a boiler at extremely high temperatures and pressures with projected efficiencies of 46%, with further theorized increases in temperature and pressure perhaps resulting in even higher efficiencies.

Other efficient ways to use coal are combined cycle power plants, combined heat and power cogeneration, and an MHD topping cycle.

Approximately 40% of the world electricity production uses coal. The total known deposits recoverable by current technologies, including highly polluting, low energy content types of coal (i.e., lignite, bituminous), is sufficient for many years. However, consumption is increasing and maximal production could be reached within decades.

A more energy-efficient way of using coal for electricity production would be via solid-oxide fuel cells or molten-carbonate fuel cells (or any oxygen ion transport based fuel cells that do not discriminate between fuels, as long as they consume oxygen), which would be able to get 60%-85% combined efficiency (direct electricity + waste heat steam turbine). Currently these fuel cell technologies can only process gaseous fuels, and they are also sensitive to sulfur poisoning, issues which would first have to be worked out before large scale commercial success is possible with coal. As far as gaseous fuels go, one idea is pulverized coal in a gas carrier, such as nitrogen. Another option is coal gasification with water, which may lower fuel cell voltage by introducing oxygen to the fuel side of the electrolyte, but may also greatly simplify carbon sequestration. However, this technology has been criticised as being inefficient, slow, risky and costly, while doing nothing about total emissions from mining, processing and combustion. Another efficient and clean way of coal combustion in a form of coal-water slurry fuel (CWS) was well developed in Russia (since the Soviet Union time). CWS significantly reduces emissions saving the heating value of coal.

Petroleum coke is the solid residue obtained in oil refining, which resembles coke but contains too many impurities to be useful in metallurgical applications.

Refined coal

Refined coal is the product of a coal upgrading technology that removes moisture and certain pollutants from lower-rank coals such as sub-bituminous and lignite (brown) coals. It is one form of several pre-combustion treatments and processes for coal that alter coal's characteristics before it is burned. The goals of pre-combustion coal technologies are to increase efficiency and reduce emissions when the coal is burned. Depending on the situation, pre-combustion technology can be used in place of or as a supplement to post-combustion technologies to control emissions from coal-fueled boilers.

Coal as a Traded Commodity

The price of coal has gone up from around $30 per short ton in 2000 to around $150.00 per short ton as of September 26, 2008. As of October 31, 2008, the price per short ton has declined to $111.50.

In North America, a Central Appalachian coal futures contracts are currently traded on the New York Mercantile Exchange (trading symbol QL). The trading unit is 1,550 short tons (1,410 t) per contract, and is quoted in U.S. dollars and cents per ton. Since coal is the principal fuel for generating electricity in the United States, coal futures contracts provide coal producers and the electric power industry an important tool for hedging and risk management.

In addition to the NYMEX contract, the Inter continental Exchange (ICE) has European (Rotterdam) and South African (Richards Bay) coal futures available for trading. The trading unit for these contracts is 5,000 tonnes (5,500 short tons), and are also quoted in U.S. dollars and cents per ton.

Cultural Usage

Coal is the official state mineral of Kentucky and the official state rock of Utah. Both U.S. states have a historic link to coal mining.

Some cultures uphold that children who misbehave will receive only a lump of coal from Santa Claus for Christmas in their stockings instead of presents.

It is also customary and lucky in Scotland to give coal as a gift on New Year's Day. It happens as part of First-Footing and represents warmth for the year to come.

Charcoal

Charcoal is the blackish residue consisting of impure carbon obtained by removing water and other volatile constituents from animal and vegetation substances. Charcoal is usually produced by slow pyrolysis, the heating of wood, sugar, bone char, or other substances in the absence of oxygen (see pyrolysis, char and biochar). The resulting soft, brittle, lightweight, black, porous material resembles coal and is 85% to 98% carbon with the remainder consisting of volatile chemicals and ash.

Fig. 33.3

The first part of the word is of obscure origin, but the first use of the term "coal" in English was as a reference to charcoal. In this compound term, the prefix "chare-" meant "turn", with the literal meaning being "to turn to coal". The independent use of "char", meaning to scorch, to reduce to carbon, is comparatively recent and is assumed to be a back-formation from the earlier charcoal. It may be a use of the word charren or chum, meaning to turn; i.e. wood changed or turned to coal, or it may be from the French charbon. A person who manufactured charcoal was formerly known as a collier (also as a wood collier). The word "collier" was also used for those who mined or dealt in coal, and for the ships that transported it.

Coke

Fig. 33.4

Coke Burning

Coke is a solid carbonaceous residue derived from low-ash, low-sulfur bituminous coal from which the volatile constituents are driven off by baking in an oven without oxygen at temperatures as high as 1,000 °C (1,832 °F) so that the fixed carbon and residual ash are fused together. Metallurgical coke is used as a fuel and as a reducing agent in smelting iron ore in a blast furnace. The product is too rich in dissolved carbon, and must be treated further to make steel. The coke must be strong enough to resist the weight of overburden in the blast furnace, which is why coking coal is so important in making steel by the conventional route. However, the alternative route to is direct reduced iron, where any carbonaceous fuel can be used to make sponge or pelletised iron. Coke from coal is grey, hard, and porous and has a heating value of 24.8 million Btu/ton (29.6 MJ/kg). Some cokemaking processes produce valuable by-products that include coal tar, ammonia, light oils, and "coal gas".

Hexamine Fuel Tablets

A hexamine fuel tablet is a form of solid fuel in tablet form. The tablets burn smokelessly, have a high energy density, do not liquify while burning and leave no ashes. Invented in Murrhardt, Germany, in 1932, the main component is hexamine.

A number of alternative names are in use, including heat tablet and Esbit. Esbit (which stands for Erich Schumms Brennstoff in Tablettenform or Erich Schumm's Fuel in Tablets) is a genericized trademark as it is used to refer to similar products made by other companies. The tablets are used for cooking by campers, the military and relief organizations. They are often used with disposable metal stoves that are included with field ration packs. Backpackers concerned with ultra light gear tend to buy or make their own much lighter stove. An Esbit beverage-can stove can be made by cutting off the bottom of an aluminum soft drink can, and turning it upside down to support the fuel tablet. A pot can be supported above this with a circle of poultry netting or metal tent pegs. The burning tablets are sensitive to wind, so a simple windscreen should be used; such as a strip of aluminum foil, curved in a circle around the pot and stove. Although not ideal, the fuel tablet can be placed on a rock or on the dirt, with a pot supported above it.

Advantages and Disadvantages

Fuel tablets are simple, ultra-lightweight compared to other stove options, and compact; the entire stove system and fuel can be stored inside a small 850ml cooking pot. However, the heat given off cannot be easily adjusted, so water can be boiled, but cooking requiring simmering is more difficult. Tablets are not a particularly powerful stove fuel, and are sensitive to wind. They are expensive and less widely available than alternatives such as alcohol or petrol.

The solid form is completely safe to handle or even ingest (it is "is not compromised by any known toxicity" according to one study. When burned, however, the chemical oxidation of the fuel yields noxious fumes, requiring foods being cooked to be contained in a receptacle such as a pot or pan, and burned tabs will leave a sticky dark residue on the bottom of pots.

Bagasse

Bagasse is the fibrous residue remaining after sugarcane or sorghum stalks are crushed to extract their juice and is currently used as a renewable resource in the manufacture of pulp and paper products and building materials. Bagasse is often used as a primary fuel source for sugar mills; when burned in quantity, it produces sufficient heat energy to supply all the needs of a typical sugar mill, with energy to spare. To this end, a secondary use for this waste product is in cogeneration, the use of a fuel source to provide both heat energy, used in the mill, and electricity, which is typically sold on to the consumer electricity grid.

Wood Pellets

Wood pellets are used on specially designed stoves. Pellets are typically made from timber waste from sawmills. The wood goes through a fairly lengthy process of transformation before it is finally extruded as hard pellets. In spite of the processing, the fuel is still carbon neutral and so is less harmful to the environment than other fossil fuels. For commercial undertakings, as with any wood fuel, wood pellets do not attract climate change levy tax.

Dung Cakes

Some animal feces, especially those of the camel, bison and cow , is used as fuel when [illegible]ed out Animal dung Cow dung is the waste of bovine animal species.... In many parts of the developing world, cow dung is used as a fertilizer and fuel. But these are mostly used in rural areas.

Economic aspects

Coal liquefaction is one of the backstop technologies that could potentially limit escalation of oil prices and mitigate the effects of transportation energy shortage that will occur under peak oil. This is contingent on liquefaction production capacity becoming large enough to satiate the very large and growing demand for petroleum. Estimates of the cost of producing liquid fuels from coal suggest that domestic U.S. production of fuel from coal becomes cost-competitive with oil priced at around 35 USD per barrel, (break-even cost). With oil prices back at around USD 40 per barrel in the U.S. as of December 15, 2008, liquid coal has currently lost much of its economic allure in that country.

Among commercially mature technologies, advantage for indirect coal liquefaction over direct coal liquefaction are reported by Williams and Larson (2003).

Intensive research and project developments have been implemented from 2001. The World CTL Award is granted to personalities having brought eminent contribution to the understanding and development of Coal liquefaction. The 2009 presentation ceremony will take place in Washington DC (USA) at the World CTL 2009 Conference (25-27 March, 2009).

ENVIRONMENTAL EFFECTS

Aerial photograph of Kingston Fossil Plant coal fly ash slurry spill site taken the day after the event

There are a number of adverse environmental effects of coal mining and burning, specially in power stations.

These effects include:

- Release of carbon dioxide, a greenhouse gas, which causes climate change and global warming according to the IPCC. Coal is the largest contributor to the human-made increase of CO2 in the air.
- Generation of hundred of millions of tons of waste products, including fly ash, bottom ash, flue gas desulfurization sludge, that contain mercury, uranium, thorium, arsenic, and other heavy metals
- Acid rain from high sulfur coal
- Interference with groundwater and water table levels
- Contamination of land and waterways and destruction of homes from fly ash spills such as Kingston Fossil Plant coal fly ash slurry spill
- Impact of water use on flows of rivers and consequential impact on other land-uses
- Dust nuisance
- Subsidence above tunnels, sometimes damaging infrastructured
- Coal-fired power plants without effective fly ash capture are one of the largest sources of human-caused background radiation exposure
- Coal-fired power plants shorten nearly 24,000 lives a year in the United States, including 2,800 from lung cancer.
- Coal-fired power plant releases emissions including mercury, selenium, and arsenic which are harmful to human health and the environment.

Advantages

- They are easy to transport.
- They are convenient to store without any risk of spontaneous explosion.
- Their cost of production is low.
- They posses moderate ignition temperatures.

Disvantages

- Their ash content is high.
- Their large proportions of heat are wasted during combustion. In other words, their thermal efficiency is low.
- They burns with clinker formation.
- Their combustion cannot be controlled easily.
- Their cost of handling is high.
- Their calorific value is lower as compared to that of liquid fuels.
- For complete combustion they require excess of air.
- They cannot be used as 1C engine fuel.

Conclusion

We had seen solid fuel is good in transportation handling and maintenance but bad in when we talk about efficiency heat content and environmental point of view. As solid fuel burns with large amount of smoke accompanied by clinker and soot formation and left behind considerable amount of carbon residue which results in an extra cost for its disposal. Smoke produced by solid fuels contains high amount of carbon dioxide and carbon mono oxide. These gasses are responsible for GREEN HOUSE EFFECT which results in global warming which is one of the major problem world is facing today.The use of some solid fuels (eg. coal) is restricted or prohibited in some urban areas, due to unsafe levels of toxic emissions. The use of other solid fuels such as wood is increasing as heating technology and the availability of good quality fuel improves. In some areas, smokeless coal is often the only solid fuel used. Inspite of above facts due to is cheapness and availability in abundance solid fuels are most widely used fuels in Industries, power plants and in rural areas.

REFRENCES

1. Engineering Chemistry by Jain & Jain
2. Engineering Chemistry by S.S. Dara
3. Wikipedia, Google
4. Images from Google Image Search

CHAPTER 34

Liquid Fuels, Adultration and Environmental Impact

Dr. Mamta Bhattacharya[1]

Introduction

Liquid fuels are those combustible or energy-generating molecules that can be harnessed to create mechanical energy, usually producing kinetic energy they also must take the shape of their container. Most liquid fuels, in widespread use, are or derived from fossil fuels; however, there are several types, such as hydrogen fuel (for automotive uses), which are also categorized as a liquid fuel. It is the fumes of Liquid fuels that are flammable instead of the fluid.

Since the birth of automotives in the 19th century, diesel and gasoline are used as the primary source of energy for the vehicles, though many alternate fuels like CNG, LPG, alcohol, dimethylether, biodiesel, methanol, etc. are emerging in the market. The conventional fuels are basically derived from crude oil, where crude oil is fractioned by continuous distillation into several fractions: Petrol (gasoline), kerosene (kerosene, paraffin oil), gas oil (heavy oil), vacuum gas oil, naphtha, lubricating oil and residue.

1. **Fossil fules** - 1.1 Gasolene, 1.2 Diesel ,1.3 Kerosene
2. **Biodiesel**
3. **Alcohols-3.1** Methanol, 3.2 Ethanol, 3.3 Butanol
4. **Hydrogen**

Fossil fuel : Fossil fuels are also generally liquid fuels. The most notable of these is gasolene. Although unproven, it is generally accepted that they formed from the fossilized remains of dead plants and animals by exposure to heat and pressure in the Earth's crust.

Gasoline : Gasoline is the most widely used liquid fuel. Gasoline, as it is known in United States and Canada, or petrol in India, Britain, Australia, New Zealand, South Africa and many English-speaking countries, is made of hydrocarbon molecules forming aliphatic compounds, or chains of carbons with hydrogen atoms attached. However, many aromatic compounds (carbon chains forming rings) such as benzene are found naturally in gasoline and cause the health risks associated with prolonged exposure to the fuel.

Production of gasoline is achieved by distillation of crude oil. The desirable liquid is separated from the crude oil in refineries. Crude oil is extracted from the ground in several processes, the most commonly seen may be beam pumps. To create gasoline, petroleum must first be removed from crude oil, Gasoline itself is actually not burned, but the fumes it creates ignite, causing the remaining liquid to evaporate.

[1]Department of Chemistry, Sadhu Vaswani College, Bairagarh, Bhopal (India)

Gasoline is extremely volatile and easily combusts, making any leakage extremely dangerous. Gasoline for sale in most countries carries an octane rating. Octane is a measure of the resistance of gasoline to combusting prematurely, known as knocking. The higher the octane rating, the harder it is to burn the fuel, which allows for a higher compression ratio. Engines with a higher compression ratio produce more power (such as in race car engines). However, such engines actually *require* a higher octane fuel.

Diesel : Conventional diesel is similar to gasoline in that it is a mixture of aliphatic hydrocarbons extracted from petroleum. Diesel may cost more or less than gasoline, but generally costs less to produce because the extraction processes used are simpler. After distillation, the diesel fraction is normally processed to reduce the amount of sulfur in the fuel. Sulphur causes corrosion in vehicles, acid rain and higher emissions of soot from the tail pipe.

Kerosene : Kerosene once used in kerosene lamps as an alternative to whale oil, is today mainly used in fuel for jet engines (more technically Avtur, Jet A, Jet A-l, Jet B, JP-4, JP-5, JP-7 or JP-8). One form of the fuel known as RP-1 is burned with liquid oxygen as rocket fuel. These fuel grade kerosenes meet specifications for smoke points and freeze points.Kerosene is sometimes used as an additive in diesel fuel to prevent gelling or waxing in cold temperatures, sometimes it is used in some cars so that it will manage well than a petrolum used in some cars.

Biodiesel : Biodiesel is similar to diesel, but has differences akin to those between petrol and ethanol. For instance, biodiesel has a higher cetane rating (45-60 compared to 45-50 for crude-oil-derived diesel) and it acts as a cleaning agent to get rid of dirt and deposits. This does however depend on locality, economic situation, a host of other factors-and it has been proven to be viable at much lower costs in some countries. Also, it gives about 10% less energy than ordinary diesel.: As with alcohols and petrol engines, taking advantage of biodiesel's high cetane rating potentially overcomes the energy deficit compared to ordinary number 2 diesel.

Alcohols : Generally, the term alcohol refers to ethanol, the first organic chemical produced by humans but any alcohol can be burned as a fuel. Ethanol and methanol are the most common, being sufficiently inexpensive to be useful.

Methanol : Methanol is the lightest and simplest alcohol, produced from the natural gas component methane. Its application is limited due to its toxicity (similar to gasoline). Small amounts are used in some gasolines to increase the octane rating. Methanol-based fuels are used in some race cars and model airplanes.Methanol is also called *methyl alcohol* or *wood alcohol,* the latter because it was formerly produced from the distillation of wood. It is also known by the name *methyl hydrate.*

Ethanol : Ethanol, also known as grain alcohol or ethyl alcohol, is most commonly used in alcoholic beverages. However, it may also be used as a fuel, most often in combination with gasoline. For the most part, it is used in a 9:1 ratio of gasoline to ethanol to reduce the negative environmental effects of gasoline. There is increasing interest in the use of a blend of 85% fuel ethanol blended with 15% gasoline. This fuel blend called E85, has a higher fuel octane than most premium gasolines. When used in a modem Flexible fuel vehicle, it delivers more performance to the gasoline it replaces.

Butanol : Butanol is an alcohol which can be used as a fuel in most gasoline internal combustion engines without engine modification. It is typically a product of the fermentation of biomass by the bacterium *Clostridium acetobutylicum* (also known as the Weizmann organism). This process was first delineated by Chaim Weizmann (1916) for the production of acetone from starch for making cordite, a smokeless gunpowder.

Hydrogen : Liquified hydrogen is the liquid state of element hydrogen. It is a common liquid rocket fuel for rocket applications and can be used as a fuel in an internal combustion engine or fuel cell. Various concept hydrogen vehicles have been lower volumetric energy, the hydrogen volumes needed for combustion are large. Hydrogen was liquefied for the first time by James Dewarin (1898).

General Properties Of the liquid Fuel

This section presents basic information about petroleum-based fuels in general and some of the many physical and chemical properties they share.

Petroleum fuels ignite and burn readily, and produce a great deal of heat and power in relation to their weight. The one composition requirement common to all petroleum fuels is that they consist entirely of hydrocarbon molecules (hydrogen and carbon) except for small amounts of impurities and/or additives.

The refining of crude oil, also known as fractional distillation, produces a range of petroleum compounds that are primarily characterized by their boiling points and molecular weights. Short chain, single ring, or "light" hydrocarbons, are more volatile, less viscous, and have lower boiling points than long chain, multiple ring, or "heavy" hydrocarbons. Denoting the number of carbon atoms (CX) in a hydrocarbon molecule is a way to describe its weight relative to other hydrocarbon molecules. Natural gas is composed primarily of the light hydrocarbons methane (C_1), ethane (C_2), propane (C_3), and butane (C_4). Gasoline typically contains hydrocarbons in the C_4-C_{10} range.

Most petroleum fuels are mixtures of hundreds of different hydrocarbon compounds. The exact number and proportions of these compounds in a particular fuel may vary; therefore, most fuels are formulated to meet general property limits, rather than a specified chemical composition.

For a particular fuel product, governing standards may limit upper and lower percentage composition of certain hydrocarbons as required to meet performance criteria.

Propulsion system demands, such as fluidity, combustion properties, corrosion protection, and impurity limits, are the primary determinants of standardized fuel formulations.

Volatility is a property of fuels that affects its ability to vaporize and form a combustible mixture with air. This property results in a quantifiable characteristic called vapor pressure. Vapor pressure

and volatility are important properties to be considered by designers of fuel storage and delivery systems because they can lead to unwanted evaporative fuel losses or "fugitive" vapors. Lighter fuels such as gasoline and aviation fuel tend to be more volatile and have higher vapor pressures than heavier fuels such as diesel and heating oil at the same temperature and pressure.

Table 34.1

	Gasoline	Diesel	Biodiesel
Chemical Structure	C_4 to C_{12}	C_8 to C_{25}	Methyl esters of C_{12} to C_{22} fatty acids
Cetane Number	N/A	40-55 (a)	48-65 (a)
Pump Octane Number [1]	84-93 (c)	N/A	N/A
Main Fuel Source	Crude Oil	Crude Oil	Fats and oils from sources such as soy beans, waste cooling oil, animal fats, and rapeseed
Energy Content (Lower Heating Value)	116,090 Btu/gal (g)	128,450 Btu/gal (g)	119,550 Btu/gal for B100 (g)
Energy Content (Higher Heating Value)	124,340 Btu/gal (g)	137,380 Btu/gal (g)	127,960 Btu/gal for B100 (g)
Energy comparison (percent of gasoline 100% energy)	100%	100%	B100 has 103% the energy of gasoline or 93% or diesel. B20 has 109% of gasoline or 99% of diesel
Physical State	Liquid	Liquid	Liquid

Common Fuel Impurities

Refined petroleum fuels can contain a variety of undesirable impurities that originate from the crude oil, develop during the refining process, or are introduced during shipment or storage. The most common fuel impurities are discussed below.

Gums are high molecular weight compounds containing hydrogen, carbon, oxygen, and usually sulfur and nitrogen. They are formed when the hydrocarbon molecules in stored fuels are oxidized or polymerized after exposure to air, sunlight, and/or elevated temperatures. When gums precipitate from the fuel, they can clog and form deposits on vital engine components such as filters and injectors, causing mild to severe engine performance problems. Anti-oxidant fuel additives can prevent the formation of gums.

Metals formed during certain refining processes can oxidize and contribute to the formation of filter clogging gums in any type of fuel. This problem is addressed by using a metal deactivator additive.

Microbial contamination occurs after fuels leave the refinery since the refining process sterilizes fuel. Microbes, including algae, bacteria, and fungi feed on the fuel and use the water in the fuel for their oxygen supply. They can multiply and plug fuel filters with an odorous slime. Some of the microbes can also produce corrosive acid byproducts. Minimizing water content and treating with a biocide additive will control microbial growth in fuel.

Sediment is a common contaminant of fuels and usually consists of rust, mineral scale, sand, dirt, and other insoluble impurities. To address this problem, fuels are filtered upon delivery into bulk and operating storage systems to remove as much sediment as possible before the fuel is delivered to the end user.

Sulfur compounds can be corrosive to metals in fuel systems and are controlled by the total sulfur content limits found in the fuel specification.

Water is a very common fuel impurity. Fuel can become contaminated with water during shipping and storage. Water can condense from the fuel itself, may leak into fuel containers from the outside, or it may be present in containers before they are filled with fuel. Water in fuel may also contain other impurities that can cause corrosion problems and damage filters, pumps, and injectors. Water is denser than fuel and can be removed as it collects at the bottom of a storage container.

Common Fuel Additives

Fuel additives are intended to help improve fuel economy, lower maintenance costs, reduce impurities and harmful deposits, reduce exhaust emissions, and improve the overall performance and reliability of the fuel. Different fuels may be formulated with different "packages" of fuel additives. Additives may also be added to fuels during storage or at the time of fueling. Often, the precise chemical composition of many fuel additives and additive packages is proprietary to the manufacturer. Particular combinations and percent content of additives may be specified in a fuel's governing standard. Where additives are approved for use or required by American Society for Testing and Materials (ASTM) standards or military standards, the chemical composition of the additive may be more readily available. Common fuel additives include:

Alkyl lead was a common gasoline additive until the late 1960s used to obtain higher octane ratings and reduce engine "knock." Lead additives have been reduced or entirely phased out of most automotive gasoline formulations due to the environmental hazards associated with lead-containing exhaust emissions. As lead additives have been phased out of gasoline formulations, other oxygenating additives are now used to boost octane ratings and control knock, as well as reduce harmful exhaust emissions. Leaded automotive gasoline typically contained one or more grams per liter (> 1,000 parts per million [ppm]) of alkyl lead. Today, unleaded automotive gasoline contains only a few ppm of lead. Aviation gasoline (Avgas) continues to contain significant concentrations of alkyl lead, typically at levels greater than 1,000 ppm.

Anti-oxidants are primarily used to prevent gum formation in gasolines and aviation fuels.

Biocides may be added to any type of fuel to kill microbes when their growth becomes a recurring problem.

Conductivity additives increase the electrical conductivity of gasolines, aviation, and diesel fuels, thereby reducing the buildup of static charges during mixing, transfer, and shipment.

Corrosion inhibitors protect against corrosion during pipeline transfer and storage of fuels. They have also been found to improve the lubricity, or capacity to reduce friction of fuels. Corrosion inhibitors are used primarily in gasolines, aviation fuels, and diesel fuels.

Detergent additives prevent the buildup of gum deposits in engines and extend fuel injector life. They also help keep fuel filters clean. Detergent additives are primarily found in diesel fuels and automotive gasolines.

Icing inhibitors are used primarily in aviation fuels to prevent the formation of ice crystals from entrapped water in the fuel at freezing temperatures encountered during high altitude flight. Icing inhibitors have also been found to be an effective barrier to microbiological growth. Diethylene glycol monomethyl ether is specified for most military aviation fuels as an icing inhibitor.

Metal deactivators prevent metal contaminants in any type of fuel from oxidizing with hydrocarbons and other compounds to form gums or precipitates.

Oxygenates are oxygen-containing hydrocarbons that are added to automotive gasoline to boost the octane rating, reduce the smog-forming tendencies of exhaust gases, and suppress engine knock. The increased oxygen content promotes more complete combustion, thereby reducing tailpipe emissions. Common oxygenating additives are methyl tertiary butyl ether (MTBE) and ethanol.

Thermal stability additives reduce fuel fouling of critical jet engine components. Thermal stability refers to the ability of the fuel to be used in a system without degradation. Thermal stress results in fuel breakdown that can cause carbon build-up on engine nozzles, afterburner spray assemblies, and manifolds. In some instances, fuel degradation changes the spray pattern in the combuster or afterburner, which leads to damage of engine components, flameouts, and augmentor anomalies.

Gasoline additives increase gasoline's octane rating or act as corrosion inhibitors or lubricators, thus allowing the use of higher compression ratios for greater efficiency and power, however some carry heavy environmental risks. Types of additives include metal deactivators, corrosion inhibitors, oxygenates and antioxidants.

Common forms of Fuel Adulterants

Blending or mixing of adulterants into the base transport fuels exists in various forms and both the type and quantity of adulterants vary from place to place. Moreover, profitability, availability and blendability are the prominent factors governing the choice of adulterants.

Specific Types of Adulteration may be Broadly Classified as Follows:

- Blending relatively small amount of distillate fuels like diesel or kerosene into automotive gasoline.
- Blending variable amount (as much as 30%) of the gasoline boiling range hydrocarbons such as industrial solvents into automotive gasoline.
- Blending small amounts of spent waste industrial solvent such as used lubricants, which would be costly to dispose of an environmentally approved manner-into gasoline and diesel.
- Blending kerosene into diesel, often as much as 20-30 percent.
- Blending small amount of heavier fuel oils into diesel.

There are several petroleum products in our country or abroad, which are close substitutes of petrol (Motor Spirit or MS) and high-speed diesel (HSD), and are available at considerably lower prices. The consequence is that these products are widely used as adulterants. Some of the possible adulterants available in Indian markets are listed in Table 34.2 & 34.3.

Table 34.2 : Potential Adulterants for Gasoline (MS)

Sl.	Solvent/Chemical	Source
1.	Naphtha	Refineries
2.	SBP	BPCL
3.	GAIL Solvents	Bijalpur
4.	GAIL Solvents	Pata
5.	Pentane	GAIL
6.	Cixon	IPCL
7.	Solvent-90	IOC
8.	Hexane	Refineries
9.	Resol	Reliance
10.	Raffinate/Slop	Refineries
11.	C6-C9 Raffinate	Petrochemicals
12.	Naphtha/NGL	GAIL/ONGC
13.	PDS Kerosene	Govt.
14.	Free Kerosene	Parallel Marketers
15.	MTO	Refineries
16.	Pyrolysis Gasoline	Naphtha Crack
17.	Oxygenates	Refineries
18.	Food grd. Hexane	Refineries
19.	Benzene	Koyali
20.	Toluene Koyali	

Source: Report of MoPNG Task Force to examine the use of solvent, raffinate and slop in automobile.

Table 34.3 : Potential Adulterants for Diesel (HSD)

Sl.	Solvent/Chemical	Source
1.	Aromex	Digboi
2.	Lomex	NA
3.	C9 Raffinate	Ptrochmcls
4.	MTO	Refineries
5.	PDS Kerosene	Govt.
6.	Free Kerosene Parallel Markets	

Source: Report of MoPNG Task Force to examine the use of solvent, raffinate and slop in automobile fuel.

Fuel Adulteration and Environmental Effect

Adulteration of transport fuel, which is currently a very flourishing business in our country, can lead to economic losses, increased emissions and deterioration of performance and parts of engines using the adulterated fuels. Some of the effects of adulteration are outlined below:

- Mal-functioning of the engine, failure of components, safety problems etc. The problem gets further magnified for high performance modern engines.
- Increased tailpipe emissions of hydrocarbons (HC), carbon monoxide (CO), oxides of nitrogen (NOx), particulate matter (PM) and can also cause increased emissions of air toxin substances.

- Adulteration of fuel can cause health problems directly in the form of increased tailpipe emissions of harmful & sometimes carcinogenic pollutants. While indirectly in the form of diversion of PDS kerosene to the diesel sector for adulteration, thus prompting the use of biomass as domestic fuel which in turn leads to health problems of various types due to indoor air pollution. It may be noted that all forms of adulteration are not harmful to public health. Some adulterants increase emission of harmful pollutants significantly, whereas others have little or no effect on air quality.
- Significant loss of tax revenue: Various estimates have been made of the extent of financial loss to the national wealth as well as the oil companies as a result of diversion of PDS kerosene, use of off-spec, low value, hydrocarbons mixed with petrol and diesel, evasion of sales tax etc. Although these estimates vary over a wide range, it is safe to assume that the nation is losing at least Rs. 10,000 crores annually as a result of adulteration of fuel.

Adulteration & Emissions

Fuel adulteration causes marked effect on the tailpipe emissions of vehicles, as adulterants alter the chemistry of the base fuel rendering its quality inferior to the required commensurate fuel quality for the vehicles. This in turn affects the combustion dynamics inside the combustion chamber of vehicles increasing the emissions of harmful pollutants significantly. In some cases effects of adulteration are indirect- for example, large scale diversion of rationed kerosene subsidized for household use to the diesel sector for mixing with diesel not only hamper engine performance of diesel vehicles, but also deprives the poor of kerosene which can otherwise be used for cooking and as a consequence of lack of availability of subsidized kerosene force the poor to continue to use biomass which expose them to high levels of indoor pollution.

In general fuel adulteration can increase the tailpipe emissions of hydrocarbons (HC), carbon monoxide (CO), Oxides of nitrogen (NOx) and particulate matter (PM). Adulteration of fuels can also cause emissions of air toxins like benzene and polyaromatic hydrocarbons (PAHs) both well-known carcinogens.

Impacts due to Gasoline Adulteration

Adulterating gasoline with kerosene causes increase in emissions, as kerosene is more difficult to burn than gasoline and this results in higher levels of HC, CO and PM. High sulphur contents of the kerosene can deactivate the catalyst and lower conversion of engine out pollutants. Kerosene addition may also cause fall in octane quality, which can lead to engine knocking. When gasoline is adulterated with diesel fuels, the same effects occurs but usually at lower levels of added diesel fuel. Both diesel and kerosene added to gasoline will increase engine deposit formation.

Gasoline may also be adulterated with gasoline boiling range solvent like toluene, xylene and other aromatics. With the "judicious" adulteration, the gasoline would not exhibit drivability problems in motor vehicles. Larger amounts of toluene and /or mixed with xylene cause some increase in HC, CO, NOx emissions, and significant increase in the level of air toxins -especially benzene - in the tailpipe exhaust. The adulterated gasoline itself could have increased potential human toxicity if frequent skin contact is allowed.

Extremely high levels of toluene (45 % or higher) could cause premature failure of neoprene, styrene butadiene rubber and butyl rubber components in the fuel system. This has caused vehicle fires in some cases, especially in older vehicles.

Adulteration of gasoline by waste industrial solvents is especially problematic as the adulterants are so varied in composition. They will cause increased emissions, may even cause vehicle breakdown. Even low levels of these adulterants can be injurious and costly to vehicle operation.

For gasoline, any adulterant that changes its volatility can effect drivability. High volatility (resulting from addition of light hydrocarbons) in hot weathers can cause vapour lock and stalling. Low volatility in cold weather can cause starting problems and poor warm-up.

Impacts due to Diesel Adulteration

The blending of kerosene with automotive diesel is generally practiced by oil industry worldwide as a means of adjusting the low temperature operability of the fuel. This practice is not harmful or detrimental to tailpipe emissions, provided the resulting fuel continues to meet engine manufacturer's specifications (especially for viscosity and cetane number). However, high-level adulteration of low sulphur diesel fuel with higher-level sulphur kerosene can cause the fuel to exceed the sulphur maximum. The addition of heavier fuel oils to diesel is usually easy to detect because the resultant fuel will be darker than normal. Depending on the nature of these heavier fuel oils and the possible presence of additional PAHs, there could be some increase in both exhaust PM and PAH (Poly Aromatic Hydrocarbons) Commissions.

Approach For Adulteration Detection

A number of analytical techniques are available to detect adulteration. In all cases described below, it is important to have good sampling technique and access to a good petroleum analytical laboratory. For the majority of the tests, accurate data and analysis of original or uncontaminated fuel are also pre- requisite. Some of the approaches for detecting adulteration are outlined below:

***(i)* Full specification tests of the standards :** This may be quite time consuming and many parameters may be well within the requirements even if the fuel is adulterated. In fact fuel standards or specifications are framed to ensure that the fuel corresponds to certain level of quality commensurate to technology requirements of the vehicles. Parameters in the fuel standards may not necessarily stand as checkpoint for any sort of fuel adulteration.

***(ii)* Testing selected parameters :** This asks for testing some critical parameters, which are likely to be affected or altered by adulteration and adversely affect engine performance and emissions and can be evaluated. Lists of such parameters are given below. In general, many of the selected parameters may already be included in the full specification standards. However, dosage of multifunctional additive and cetane improver are intended for following the adulterants by dilution. However, convenient methodology for determination at refinery and outlet levels are yet to be lined up in the country.

Selected Parameters for Gasoline Testing

1. Density
2. Distillation
3. Hydrocarbon Composition

   Aromatic, Vo1%

   Olefms, Vo1%

   Benzene, Vo1%

   Sulphur, ppm
4. Stability

   Existing gum

   Potential gum

5. Octane Number
   Research
   Motor
6. Multifunctional additives-dosage

Selected Parameters for Diesel Testing

1. Flash Point
2. Density
3. Distillation
4. Sulphur
5. Fob/cyclic aromatics (+2 rings)
6. Total sediment
7. Cetane number
8. Cetane index
9. Multifunctional additives-dosage
10. Cetaneimprover-presence,

Dosage : For diesel fuel without cetane improver, cetane index can be utilized. If dosage can be determined, it may be used for detection of adulteration based on depletion from original dosage.

***(iii)* Testing methods :** There can be several alternate approaches, some prominent approaches are:

- Use of conventional manual petroleum testing methods.
- Utilization of automated instruments for conventional petroleum testing. For example- gas chromatographic method is used for simulated distillation. A number of instrumental analysis methods have been developed for establishment of parameters of fuels.

***(iv)* Emerging instruments for fuel surveillance :** Several new instruments are available which claimed to carry out instrumental analysis for the estimation of key parameters of transport fuel. For example the following may be noted.

(*a*) Accurate and comprehensive fuel analyzer. The portable FOxFTIR fuel analyzer is claimed to be ideal for the analysis of commercial fuels.

(*b*) ZeltexZX1O1CC Portable Octane analyzer.

***(v)* Use of marker :** Various markers can be used to identify adulteration, such as kerosene in gasoline. Earlier kerosene was used as major adulterant for adulteration in gasoline and diesel fuel. For detection of adulteration of fuels with kerosene a blue dye and furfural were used. However, where visible dyes have been applied in South Asia, they have not been effective. It is believed that presently the range of adulterants has widened. Various chemical/biochemical markers are available in market e.g. Spectrace marker by Mortan international, petro markers by M/S GFI, Biocode markers by M/S Biocode ltd., etc. The marker is added in trace level with the fuel and whenever the product is to be tested, the marked chemical is detected and measured by specific instruments/ immunoassay. This detection test can also be easily carried out in field with equal accuracy as laboratory tests.

Properties of Markers

- They should be miscible with fuels
- Detectable by simple test procedures

- Difficult to remove from marked fuels
- Non-reactive with other fuel additives
- No interaction with material of construction and fuel impurities; and
- Should be cost effective

Limitations of Marker System

- Relatively a high cost option
- Difficult to maintain a constant dosage at low concentration
- May be leached out by water in the product tanks
- May interact with materials and fuel impurities

Some Important Measures To Control Fuel Adulteration

Some important measures to control fuel adulteration are listed below:

- An important step in tackling fuel adulteration is reducing incentives and opportunities for adulteration. Though it is generally recognized that eliminating pricing differential is the most effective method of controlling adulteration, it will be difficult to eliminate differences among such a wide variety of fuels and solvents meant for different usages.
- Checking adulteration requires a credible monitoring and surveillance system. To ensure that the engine can give the desired performance including low emissions, it is necessary to ensure the fuel quality at the consumer end, which can be achieved by appropriate surveillance programs.
- Any anti-adulteration programme should be backed up by sound financial and legal framework. The fiscal framework should take into account-associated costs like monitoring & testing infrastructure. Policy for imposing severe penalty & exemplary punishment to the adulterators needs to be imbibed into legal framework tb discourage adulteration.
- The manner in which retail fuels are distributed has an important bearing on fuel adulteration. For example, having large numbers of small, independent transport trucks operators moving fuels from terminals to the point of sale creates an environment conducive to adulteration. One effective "market based" approach is the practice in many industrialized countries whereby oil companies market at retail and assume responsibility throughout the supply chain to guarantee fuel quality in order to protect their public image and market share.
- One of the acceptable internationally accepted method for detecting and thereby preventing adulteration of fuels is the use of markers. A number of chemical and biochemical markers are available in the international market. Some of them are dyes, one of which is already being used in India to mark SKO (Superior Kerosene Oil) used for PDS (Public Distribution System).
- The Standard fuel test method being used today when properly executed should be able to give acceptable results. Precision and repeatability could be improved by setting up programmes for cross checking inter-laboratory variability.
- In Industrial countries, practices of adulteration are expected to be less or rare today; in part because public pressure has led most oil companies to take public image seriously and socially responsible behavior is considered as integral part of good business. Thus a culture of "Good Business" needs to be developed within the concerned industry to eradicate adulteration via awareness raising by Government organizations, NGOs, and citizens groups; independent checks by universities and research institutes to "name & shame" those who are not in compliance; efforts by trade associations to identify those retailers that comply in order to "upgrade" the market; international pressure on

large oil companies operating in developing countries; and greater effort by governments to monitor and enforce regulations.

- Use of alternative fuels which are less prone to adulteration, can play a positive role in minimizing adulteration. Thus, promoting use of cleaner fuels like CNG, LPG etc can prove effective in dealing with adulteration.
- Taking & maintaining samples for checking fuel quality is not easy. Finding proper sample containers and not being personally harassed at retail outlets while sampling are just two of the very real operational problems to be resolved

Consumers Front: Anti-Adulteration Tips

Consumers are the sufferers of this malpractice. Any quality conscious consumer has the right to be assured of the quality of the products and if he desires he can get his sample checked for adulteration. Some easy and important checks can be conducted at the retail outlet for MS/HSD: -

Filter Paper Test : For MS. First the mouth of nozzle is cleaned to remove stains. Then, a drop of petrol is put on the filter paper from the nozzle. The petrol dropped on the filter paper is allowed to evaporate for 2 minutes. The petrol should evaporate without leaving any stain on the filter paper. If the colour left on the paper is pinkish, it is the colour of MS and not a stain. Dealers are expected to provide filter paper to customers on demand.

Density test : This is a very simple test for both MS and HSD. This test takes approximately 5 to 10 minutes. Product is taken in a glass jar and then a Hydrometer (separate Hydrometers for MS and HSD) available with the dealer is immersed in the product. A Thermometer is also immersed into the product jar simultaneously without touching the walls of the jar. The readings of Thermometer and Hydrometer are taken. Then, with the help of a conversion chart, the density is converted to 150C and this is compared with the recorded density/reference density, which can be seen from the density register maintained by the dealer. If the variation between the observed density and recorded/ reference density is within + 0.0030, then the product density can be considered to be correct. If the difference is more than + 0.0030, then it indicates possibility of adulteration.

Water contamination checks : For both MS and HSD can be done with the help of a dip rod and water finding paste, available with the dealer.

In case of lubricants : the customer must check the seal of container, date of manufacture and name of the manufacturer. For convenience of 2/3 wheelers, Retail Outlets provide 2-T dispensers/2-T mix dispensing units and also keep tamper proof 2-T pouches.

Green Fuel

Any solid, liquid, or gaseous fuel produced from organic (once living) matter, either directly from plants or indirectly from industrial, commercial, domestic, or agricultural wastes.

There are three main methods for the development of biofuels:

(*i*) The burning of dry organic wastes (such as household refuse, industrial and agricultural wastes, straw, wood, and peat);

(*ii*) The fermentation of wet wastes (such as animal dung) in the absence of oxygen to produce biogas (containing up to 60% methane),

(*iii*) The fermentation of sugar cane or maize to produce alcohol and esters; and energy forestry (producing fast-growing wood for fuel).

Fermentation produces *two main types of biofuels: alcohols and esters.* These could theoretically be used in place of fossil fuels but, because major alterations to engines would be required, biofuels are usually mixed with fossil fuels. The EU allows 5% ethanol, derived from wheat, beet, potatoes, or maize, to be added to fossil fuels. In Brazil ethanol from sugar cane is used in cars run either on ethanol, on gasohol (a blend of petrol and ethanol), or on both ('dual-fuel' engines). Ethanol replaces 40% of the petrol that the country would use for motor transport.

Green fuel, also known as biofuel, is a type of fuel distilled from plants and animal materials, believed by some to be more environmentally friendly than the widely-used fossil fuels that power most of the world. In the desperate search for alternative energy sources, green fuel has evolved as a possible fueling option as the world drains its fossil fuel resources. Detractors suggest that the term "green fuel" is a misnomer, as the processing of crops into biofuel actually creates a considerable amount of pollution that may be just as damaging to the environment as current practices.

In creating basic forms of biofuel, crops are broken down into two types: sugar producing and oil producing. Sugar and starch producing crops, such as sugar cane or corn, are put through a fermentation process to create ethanol. Oil producing plants, like those used in vegetable oils, can be used much like fossil sources of oil; they create diesel that can be burned by cars or further processed to become biodiesel.

Recent technological innovations have created the fields of advanced biofuels, which focus on non-food sources and waster renewal as energy. By converting landfill material, as well as wood and inedible plant parts, into green fuel, we not only cut down on the use of fossil fuels but also effectively recycle enormous amounts of waste. These biofuels help quell the debate on whether growing crops for fuel will result in fewer available food crops.

A new form of fuel can literally called *green,* as it derives from green algae. Algae, often seen growing on bodies of water, is a tiny plant with a rapid growth rate. Its usefulness as fuel derives from the fact that it has an extremely high oil content that can be processed like other oil-producing crops. Many countries are now doing extensive research on algae, which is easy to cultivate and grows extremely quickly. According to some estimates by start-up algae oil companies, one acre of algae can produce 200 times as much oil as one acre of corn.

Some detractors warn against the assumption that green fuel is free from pollution-causing attributes. The processing of sugar and starch plants into ethanol has come under heavy criticism in recent years; not only do these plants take away food-growing space, the fermentation process releases considerable pollution into the air. Moreover, green fuel does not necessarily burn clean, and may emit formaldehyde, ozone, and other carcinogenic substances when used.

It is not yet clear whether the green fuel currently available is the wave of the future or merely an interim step on the journey away from fossil fuel use. Governments around the world are devoting enormous resources to the research of clean, sustainable fuels to replace the pollutant and quickly disappearing oil reserves used today. Green fuel may not be a perfect solution to the problems of oil need and global protection, but it remains an important innovation that may pave the way to a better future.

INDIAN STANDARD

Anhydrous Ethanol for Use in Automotive Fuel—Specification

Scope : This standard prescribes requirements, methods of sampling and test for anhydrous ethanol, which is used either as such or more usually in admixture with petrol and diesel as a fuel for automobile engines.

All standards are subject to revision and parties to agreements based on this standard are encouraged to investigate the possibility of applying the, most recent editions of the standard indicated below:

Table 34.4.

IS No.	Title
2: 1960	Rules for rounding off numerical values *(revised)*
196 : 1966	Atmospheric conditions for testing *(revised)*
264: 1976	Nitric acid *(secondrevision)* 264 : 1993 Hydrochloric acid *(fourth revision)*
323 : 1959	Rectified spirit *(revised)*
1070: 1992	Reagent grade water *(third revision)*
1448	Methods of lest for petroleum and its products:
[P:26] : 1960	Knock characteristics of motor fuels by motor method
[P:27] : 1960	Knock characteristics of motor fuels by research method
2302 : 1989	Tables for alcoholometry by .hydrometer method *(first revision)*
2362 : 1993	Determination of water by Karl Fischer method—Test method; *(second revision)*

Terminology : For the purposes of this standard, the following definitions shall apply.

Ethyl Alcohol (Absolute Alcohol) : Ethyl alcohol (Absolute alcohol) is a clear, colourless and homogeneous liquid, consisting essentially of ethanol admixed with not more than 0.5 percent by volume of water.

Anhydrous Ethanol : Anhydrous ethanol is essentially ethyl alcohol, which is denatured and is meant for use as fuel in automobile engines.

Denaturant : Denaturant is a substance completely miscible in ethyl alcohol and of such a character that while its addition makes the material or any aqueous dilution of it unpleasant and unwholesome for potable purposes, its presence does not render anhydrous ethanol, either as such or blended with petrol or diesel, unsuitable for use in automobile engines.

Standard Atmospheric Conditions : Standard atmospheric conditions for testing shall comprise a relative humidity of 65 ± 2 percent and a temperature of 27 ± 2 °C provided that, in a given series of experiments, the temperature does not vary by more than 1°C.

Requirements

Description : Anhydrous ethanol shall be a clear, colourless and homogeneous liquid, free from matter in suspension.

Denaturant : The denaturant to be admixed with ethyl alcohol and the proportion in which it is to be used shall be as prescribed by law from time to time.

Prohibited Denaturants : Specific mention must be made of some materials that have extremely adverse effects on fuel stability, automotive engines and fuel systems. These materials shall not be used as denaturants, for anhydrous ethanol for use in automobile fuels, under any circumstances. They are as follows: methanol, pyrroles, turpentine, ketones and tars (high-molecular weight pyrolysis products of fossil or non-fossil vegetable matter). Unless a denaturant, such as a higher aliphatic alcohol or ether, is known to have no adverse effect on a gasoline-ethanol blend or on automotive engines or fuel systems, it shall not be used.

Acidity : The material shall be neutral or acidic in reaction to phenolphthalein and when tested as prescribed in Annex D, the acidity, other than due to dissolved carbon dioxide shall not exceed the sacified value.

Packing and Marking

Packing : The material shall be packed in such containers and packages as agreed to between the purchaser and the vendor, subject to the provisions of law in force from time to time.

All containers in which the material is packed shall be dry, clean, free from substances soluble in anhydrous ethanol, and leak-proof.

Necessary safeguards against the risk arising from the storage and handling of large volume of flammable liquids shall be provided and all due precautions shall be taken at all times to prevent accidents or explosions.

Except when they are opened for the purpose of cleaning and rendering them free from alcohol vapour, all empty tanks or other containers shall be kept securely closed unless they have been thoroughly cleansed and freed from alcohol vapour.

Marking

Each container shall be marked legibly and indelibly with the following information:

(*a*) Name of the material;

(*b*) Manufacturer's name;

(*c*) Net, gross and tare weight;

(*d*) Recognized trade-mark, if any;

(*e*) Date of packing;

(*f*) Only for automotive use;

(*g*) Highly flammable; and

(*h*) Hazardous chemical and injurious to health.

BIS Certification Marking

The container may also be marked with the Standard Mark.

The use of the Standard Mark is governed by the provisions of *Bureau of Indian Standard Act,* 1986 and the Rules and Regulations made thereunder. The details of conditions under which the licence for the use of the Standard Mark may be granted to manufacturers or producers may be obtained from the Bureau of Indian Standards.

Quality of Reagents

Unless specified otherwise, pure chemicals and distilled water *(see* IS 1070) shall be employed in tests.

Note: 'Pure chemicals' shall mean chemicals that do not contain impurities, which affect the results of analysis.

1. Scope

This standard prescribes the requirements and the methods of sampling and test for absolute alcohol. The material is intended for use as a raw material, regent and solvent in the chemical and pharmaceutical industries and for the production of power alcohol for which purpose it is partially or completely denatured.

2. Grades

The material shall be of three grades, namely:

(*a*) *Special Crude* to meet special requirements, such as for Defence purposes;

(*b*) *Grade 1* for pharmaceutical and medicinal purposes; and

(*c*) *Grade 2* for general purposes.

3. Requirements

Description : The material shall be clear, colourless and homogeneous liquid free from suspended matter and consisting essentially of ethanol (CH_3CH_2OH).

Table 34.5. Requirements of Anhydrous Ethanol for Use in Automotive Fuel (Clauses 4.3, 4.5, 4.6 and 7.1)

Sl.No.	Characteristic	Requirement	Method of Test, Ref. to Annex
(i)	Relative density at 15.6 /15.6 °C, Max	0.7961	A
(ii)	Ethanol content percent by volume at 15.6/15.6°C, Min (excluding denaturant)	99.50	B
(iii)	Miscibility with water	Miscible	C
(iv)	Alkalinity	Nil	D
(v)	Acidity (as CH_3COOH) Mg/l, Max	30	D
(vi)	Residue on evaporation percent by mass, Max	0.005	E
(vii)	Aldehyde content (as CH_3CHO) Mg/l, Max	60	F
(viii)	Copper, mg/kg, Max	0.1	G
(ix)	Conductivity, μS/m, Max	300	H
(x)	Methyl alcohol, mg/litre, Max	300	J
(xi)	Appearance	Clear and bright	Visual

REFERENCES

1. Central Pollution Control Board Publication, PROBES/78/2000-2001, "Transport Fuel Quality 2005".
2. The World Bank Publication, South Asia Urban Air Quality Management Briefing note No. 7, "Catching Gasoline and Diesel Adulteration".
3. The World Bank Publications, September 2001, Note number 237, "Abuses in Fuel Market".
4. Report from MoPNG on "Steps undertaken to control adulteration of Fuel in India".
5. Website of Anti adulteration cell, www.antiadulterationcell.com.
6. Government of India, "Report of the Expert Committee on Auto Fuel Policy".
7. The World Bank Publications, Pollution Management in Focus. Discussion note number 11, December 2001, "Tranport fuel taxes and urban air quality".
8. Centre for Science & Environment, New Delhi, February 2002, "A Report on the independent inspection of fuel quality at the fuel dispensing stations, oil depots and tank lorries" & its website www.cseindia.org.
9. Government of India, Ministry of Petroleum and Natural Gas, "The Indian Hydrocarbon Sector spreads its wings".
10. Website of The Hindu, www.hinduonnet.com & Business line, www.blonet.com
11. Web Site of The World bank groul, www.rru.worldbank.org.Rapid Response, Topic regulating fuel market for cleaner air.
12. Report by Indian Oil Corporation Limited, R & D Centre, "Monitoring Fuel Adulteration.
13. "http://en.wikipedia.org/wiki/Liquid_fuels"
14. "http://cpcbenvis.nic.in
15. http://en.wikipedia.org/wiki/E85

CHAPTER 35

Environmental-Management & Modeling

Dinesh Agrawal[1] and Vibha Agarwal[2]

Introduction

Environmental management is not, as the phrase could suggest, the management of the environment as such, But rather the management of interaction by the modern human societies with, and impact upon the *environment*. The three main issues that affect managers are those invoking politics (networking), programs (projects), and resources (money, facilities, etc.). The need for environmental management can be viewed from a variety of perspectives. A more common philosophy and impetus behind environmental management is the concept of carrying capacity. Simply put, *carrying capacity* refers to the maximum number of organisms a particular resource can sustain. The concept of carrying capacity, whitest understood by many cultures over history, has its roots in *Malthusian* theory. Environmental management is therefore not the conservation of the environment solely for the environment's sake, but rather the conservation of the environment for humankind's sake.[Citation needed] This element of sustainable exploitation, getting the most out of natural assets, is visible in the EV *Water Framework Directive*.

Environmental management involves the management of all components of the Bio-physical environment, Both living (biotic) and non-living (abiotic). This is due to the interconnected and network of relationships amongst all living species and their habitats. The environment also involves the relationships of the human environment, such as the social, cultural and economic environment with the physical environment.

As with all management functions, effective management tools, standards and systems are required. An environmental management standard or system or protocol attempts to reduce environmental impact as measured by some objective criteria. The ISO 14001 standard is the most widely used standard for environmental *risk management* and is closely aligned to the European *Eco-Management and Audit Scheme (EMAS)*. As a common auditing standard, the *ISO 19011* standard explains how to combine this with *quality management*.

Environmental Management System (EMS)

Environmental management system (EMS) refers to the management of an organisation's environment at programs in a comprehensive, systematic, planned and documented manner. It includes the organisational structure, planning and resources for developing, implementing and maintaining policy for environmental protection.

[1]BUIT. Barkatullah University, Bhopal (India)
[2]Department of Chemistry, GGIIM, Bhopal

An Environmental Management System (EMS):

- Serves as a too to improve environmental performance
- Provides a systematic way of managing an organization's environmental off airs
- Is the aspect of the organization's overall management structure that addresses immediate and long-term impacts of its products, services and processes on the environment
- Gives order and consistency for organizations to address environmental concerns through the allocation of resources, assignment of responsibility and ongoing evaluation of practices, procedures and processes
- Focuses on continual improvement of the system

What is EMS Model?

An EMS follows a Plan-Do-Check-Act Cycle, or PDCA. The diagram shows the process of first developing an environmental policy, planning the 'EMS, and then implementing it. The process also induces checking the system and acting on it. The model is continuous because an (EMS' is a process of continual improvement in which an organization is constantly reviewing and revising the system.

This is a model that can 6e used By a wide range of organizations—from manufacturing facilities to service industries to government agencies.

What are some key elements of an EMS?

- Policy Statement—a statement of the organization's commitment to the environment
- Identification of Significant Environmental Impacts—environmental attributes of products, activities and services and their effects on the environment
- Development of Objectives and targets-environmental goals for the organization
- Implementation-plans to meet objectives and targets
- Training-instruction to ensure employees are aware and capable of fulfilling their environmental responsibilities
- Management Review

What is ISO, ISO 9000, ISO 14000, ISO 14001?

ISO stands for the *International Organization for Standardization*, heated in Geneva, Switzerland. ISO is a non-governmental organization established in 1947. The organization mainly functions to develop voluntary technical standards that aim at making the development, manufacture and supply of goods and services more efficient, safe and clean.

ISO 9000 is a family of *standards* for *quality management systems*. ISO 9000 is maintained by ISO, the *International Organization for Standardization* and is administered By accreditation and certification Bodies. The rules are updated, as the requirements motivate changes over time.

Some of the requirements in ISO 9001:2008 (which is one of the standards in the ISO 9000family) include

- a set of procedures that cover all key processes in the Business;
- monitoring processes to ensure they are effective;
- keeping adequate records;
- checking output for defects, with appropriate and corrective action -where necessary;
- regularly reviewing individual processes and the quality system itself for effectiveness; and
- facilitating continual improvement

A company or organization that has been independently audited and certified to Be in conformance with ISO 9001 may publicly state that it is "ISO 9001 certified" or "ISO 9001 registered". Certification to an ISO 900.7 standard does not guarantee any quality of end products and services; rather, it certifies that formalized Business processes are Being applied.

Although the standards originated in *manufacturing*, they are now employed across several types of organizations. A ";product", in ISO vocabulary, can mean a physical object, services, or software.

ISO 14000 refers to a family of voluntary standards and guidance documents to help organizations address environmental issues. Included in the family are standards for Environmental Management Systems, environmental and EMS auditing, environmental labeling, performance evaluation and life-cycle assessment.

In September 1996, the International Organization for Standardization published the first edition of ISO 14001, the (Environmental Management Systems standard. This is an international voluntary standard describing specific requirements for an EMS. ISO 14001 is a specification standard to which an organization may receive certification or registration. ISO 14001 is considered the foundation document of the entire series. A second edition of ISO 14001 was published in 2004, updating the standard.

Questions may arise when implementing an EMS following the ISO 14001 standard. The V.S. Body that provides input into the standard's development is the V.S. TAG (Technical Advisory Group) to TC 207 (Technical Committee). This same Body has established a formal process to respond to questions that may arise regarding clarification of the ISO 14001 ("the standard"). (Responses will reflect the interpretation of the Standard as intended during the drafting of the Standard and may be found in the "Clarification of Intent of ISO 14001."

How these Standards Developed?

All the ISO standards are developed through a voluntary, consensus-based approach. ISO has different member countries across the globe. Each member country develops its position on the standards and these positions are then negotiated with other member countries. (Draft versions of the standards are sent out for format written comment and each country casts its official vote on the drafts at the appropriate stage of the process. 'Within each country, various types of organizations can and do participate in the process. These organizations include industry, government (federal and state), and other interested parties, like various non-government organizations. For example, EPA and states participated in the development of the ISO 14001 standard and are now evaluating its usefulness through a variety of pilot projects.

What are the 17 requirements of the ISO 14001:2004 standard?

- Environment Policy—develop a statement of the organization's commitment to the environment.
- Environmental Aspects and Impacts—identify environmental attributes of products, activities and services and their effects on the environment.
- Legal and Other Requirements—identify and ensure access to relevant Caws and regulations.
- Objectives and Targets and Environmental Management Program-set environmental goals for the organization and plan actions to achieve objectives and targets.
- Structure and Responsibility - establish rotes and responsibilities within the organization.
- Training, Awareness and Competence - ensure that employees are aware and capable of their environmental responsibilities.
- Communication - develop processes for internal and external communication on environmental management issues.
- EMS (Documentation - maintain information about the EMS and related documents.
- Document Control - ensure effective management of procedures and other documents.
- Operational Control- identify, plan and manage the organization's operations and activities in line with the policy, objectives and targets, and significant aspects.
- Emergency Preparedness and Response - develop procedures for preventing and responding to potential emergencies.
- Monitoring and Measuring - monitor key activities and track performance including periodic compliance evaluation.
- Evaluation of Compliance - develop procedure to periodically evaluate compliance with legal and other requirements.

- Nonconformance and Corrective and Preventive fiction - identify and correct problems and prevent recurrences.
- Records - keep adequate records of EMS performance.
- EMS Audit - periodically verify that the EMS is effective and achieving objectives and targets.
- Management (Review - review the EMS.

Environmental Management Scheme

An environmental management scheme is a mechanism by which landowners and other individuals and bodies responsible for and management can Be incentivized to manage their land in a manner sympathetic to the environment.

SCHEMES BY COUNTRY

United Kingdom

Several schemes are or have Been in operation in the United Kingdom, including:

- Countryside Stewardship
- Environmental Stewards hip
- Environmentally Sensitive Areas Scheme

Currently England operates:

- Entry Level Stewardship
- Higher Level Stewardship
- Organic Entry Level Stewardship
- Higher Level Organic Entry Level stewardship

All of these schemes are administered by Natural England

Scotland

In 2007 Scotland adopted the SRDP (Scottish (Rural Development Programme-a £1.6 billion programme of economic, environmental and social measures designed to develop rural Scotland. Individuals and groups may seek support to help deliver the (government's strategic objectives in rural Scotland.

Males

Until recently the prevailing agri-environment scheme in Wales was Tir Gofal, which means literally 'Land Care'. It was the first scheme in Mates, and indeed in Europe, aimed at promoting whole farm conservation and management. It was different from previous schemes, as it brought farming and conservation into a different level of partnership.

It was recently announced that this scheme has ceased, with a bridging payment scheme until its replacement is launched.

France

In some countries such as *France*, such schemes may be initiated by the central government.

Switzerland

Switzerland has long been a leader in the field. Having started reforming its agricultural policies in 1993 and after a referendum in 1996, since 1998 the country has linked the attribution of farm subsidies with the strict observance of good environmental practice. Before farmers can apply for subsidies, they must obtain certificates of environmental management systems (EMS) proving that they: "make a balanced use of fertilizers; use at (east 7% of their farmland as ecological compensation areas; regularly rotate crops; adopt appropriate measures to protect animals and soil; make limited and targeted use of pesticides."

Environmental Quality Management

Environmental Quality Management, Inc. (EQM) is an environmental engineering and remediation company headquartered in *Cincinnati, Ohio* that has 6een active in providing environmental remediation support in response to terrorist attacks, the space shuttle disaster, superfund site cleanup, hazardous chemical spills and natural disasters.

History

EQ was established in 1990, to provide nationwide environmental consulting, engineering, and remediation/restoration services from its offices throughout the country. In addition to its Cincinnati headquarters, EQM has offices in Chicago, Illinois; Denver, Colorado; Durham, North Carolina; Las Vegas, Nevada; New Orleans, Louisiana; (Pittsburgh, (Pennsylvania; (Portland, Oregon; Roanoke, Virginia; Sacramento, California; San Antonio, Texas; and Seattle, Washington. Recording to the July 07 issue of Engineering News Record's (ENR) listing of The Top 200 Environmental Firms, EQ is currently ranked as the 43rd largest environmental firm in the US; the 17th largest in hazardous waste projects, the 16th largest overall governmental environmental contractor, and the 9th largest "All Environmental" firm in the Us.

The company provides services in the following areas:

Air Quality Services

Emergency Response

Environmental Management

Industrial Hygiene

Steel

Cement

EQM subsidiary companies

The EQM companies currently consist of three companies that are strategically positioned to provide all encompassing environmental and engineering support for their customers, whether they are locator international in scope. The EQM subsidiaries include EQ Engineers, LLC in the Chicago area, and EQ Engineers Slovakia, s.r.o., in Kosice, Slovakia.

World Trade Center Site Monitoring Study

EQM conducted a monitoring study of 'residential buildings located north and southwest of *Ground Zero* to determine the presence of dioxins, PCBs, inorganic metals, and asbestos. ^QM made recommendations to ensure proper cleanup of the asbestos-contaminated dust and to reduce the exposure of cleanup personnel and occupants returning to buildings.

The company assisted in coordination of NIEHS-WETP grantee activities at the WTC Site, assessed the current safety and health status of response personnel wording at the WTC Site, evaluated the current Site safety and health plans or programs and related aspects such as exposure monitoring with respect to worker protections, and performed a preliminary training needs assessment specific to the WTC Site activities.

EQ assisted in the mobilization of response resources, including coordination with the New York City Building and Construction Trades Council and the Construction Employers Association, Bechtel Corporation (the contractor responsible for developing the overall WTC Disaster Site Safety and Health Plan), and other parties with respect to the training programs that could be promptly provided By the grantee organizations.

The assessment of the safety and health status at the Site was Based on on-site observations and analysis of the WTC Disaster Site Worker Injury and illness Surveillance Update Reports issued by the City Health Department.

Anthrax Emergency Response Cleanup

The Hart Senate Office Quitting was closed on October 17, 2001, when aides to Senate Majority Leader Tom Daschle opened an *anthrax*-faced letter in his office. Several other congressional office buildings also showed evidence of anthrax contamination and had to be fumigated.

Responding to these fife-threatening hazards, EQ was on site at Capitol Hill within 24 hours with crews and equipment to develop biocidal treatment of Bacillus Anthracis endospores. EQM led, developed, and trained multiple contractors to deal with personnel equipment, and gasification decontamination.

EQM developed "Sampling for Anthrax" training. EQM also designed, constructed, and manned modular multi-chamber decontamination containment systems in three locations in the Hart Senate Office Building. EQM refined construction units to make them quicker to assemble during daily setup and tear-down. EQM developed handbooks, established Standard Operating Procedures, and taught specialized decontamination procedures.

Space Shuttle Columbia Recovery

The Space *Shuttle Columbia* broke up on re-entry over Texas on February 1, 2003. Within hours of the Columbia disaster, U.S. (Region 6 used its Emergency and Rapid Response Services contract to contact EQM for help in the recovery effort. The company responded immediately by dispatching a crew to the Dallas/Fort Worth area. Before long, EQM and its team had 75 trained response personnel in the field assisting with the recovery effort. The EQM team was involved with tagging and retrieving the debris and transporting it to a secure location.

Due to concerns over hazardous exposures and site disturbances, the company was under pressure to rapidly accomplish and secure recovery. (EQM's staff worked in conjunction with many different local, state, and federal personnel to meet the goal of a quick and complete recovery of the shuttle remains.

This project was conducted over a 500-square-mile (1,300 km^2) area centered around Texas and Louisiana. Chemical hazards encountered included hydrazine and nitrogen tetroxide, in addition to risks of explosive, flammable, corrosive, and reactive materials.

As a result of this important effort, EQM was honored on July 10, 2003 in Washington, D.C at the U.S. EPA's 15th Annual Small & Disadvantaged Business Awards ceremony. The award recognized EQ for its outstanding accomplishments in recovering debris from the Space Shuttle Columbia.

Environmental cleanup response to Hurricanes Katrina and Rita.

EQM provided environmental cleanup and waste management services in eight parishes in Louisiana affected by *Hurricane Katrina* and *Hurricane Rita*. The company was an integral part of the search and rescue operation.

Days after Hurricane Katrina devastated New Orleans and the neighboring Gulf Coast, the Federal Emergency Management Association (FEMA) tasked EQM (through its V.S. EPA Region 6 Emergency & Rapid Response Services contract) to conduct a search and rescue mission, followed by debris removal and cleanup of hazardous waste in the hardest-hit areas.

The company immediately dispatched 120 people (employees and subcontractors) to assist with the effort, fit the height of the rescue mission, the EQM team was wording IS to 18 hours a day and living in a tent city. Sixty Boats were deployed continuously from New Orleans to various parishes in an effort to search for survivors and bring them to shore.

After the search and rescue effort was completed, the company turned its attention to removing debris and hazardous waste. The work involved identifying, labeling, packaging, and arranging transportation and disposal of massive amounts of household waste, manufacturing and refinery waste, and large appliances containing refrigerants.

Institute of Ecology and Environmental Management

The Institute of Ecology and Environmental Management

(IEEM) is the professional body which represents and supports ecologists and environment at managers, mainly in the United Kingdom but increasingly in Ireland and Europe, and the rest of the world.

Established in 1991, IEEM has over 3800 members drawn from local authorities, government agencies, industry, environmental consultancy, teaching/research, and NGOs.

Activities of IEEM

IEEM provides a variety of services to develop the competency and standards of professional ecologists and environmental managers and also to promote ecology and environmental management as a profession.

IEEM is a constituent body of the *Society for the Environment* and the *European Federation of Associations of Environmental Professionals*. IEEM is also a member of the *Europarc Federation, Eurosite*, and the *IUCN-UK Committee*. It is also a signatory of the Countdown 2010 agreement to help save biodiversity.

Environmental Impact Assessment

An environmental impact assessment (EIA) is an assessment of the possible impact—positive or negative— that a proposed project may have on the *environment*: considering natural, social and economic aspects. The purpose of the assessment is to ensure that decision makers consider the ensuing environmental impacts to decide whether to proceed with the project. The *International Association for Impact Assessment* (IAIA) defines an environmental impact assessment as "the process of identifying, predicting, evaluating and mitigating the *biophysical*, social, and other relevant effects of development proposals prior to major decisions being taken and commitments made. After an EIA, the *precautionary* and *polluter pays principles* may be applied to prevent, limit, or require *strict liability* or *insurance* coverage to a project, based on its likely harms. Environmental impact assessments are sometimes controversial.

EV

The *European Union* has established a mix of mandatory and discretionary procedures to assess environmental impacts.5 *European Union Directive* (8S/337/EEC) on Environmental Impact Assessments (known as the EIA Directive)6 was first introduced in 1985 and was amended in 1997. The directive was amended again in 2003, following EV signature of the 1998 *Aarhus Convention*. In 2001, the issue was enlarged to the assessment of plans and programmes by the so called *Strategic Environmental* Assessment (SEA) (Directive (2001/42/EC), which is now in force.5 Under the EV directive, an EIA must provide certain information to comply.7 There are seven key areas that are required:

1. Description of the project
   - Description of actual project and site description
   - Break the project down into its key components, ie construction, operations, decommissioning
   - For each component list all of the sources of environmental disturbance
   - For each component all the inputs and outputs must be listed, eg, *air pollution*, noise, *hydrology*
2. Alternatives that have been considered
   - Examine alternatives that have been considered
   - Example: in a *biomass* power station, will the fuel be sourced locally or nationally?

3. Description of the environment
   - List of all aspects of the environment that may be effected by the development
   - Example: populations, *fauna*, flora, air, soil, water, humans, landscape, cultural heritage
   - This section is best carried out with the help of local experts, eg the RSPB in the UK
4. Description of the significant effects on the environment
   - The word significant is crucial here as the definition can vary
   - 'Significant' needs to be defined
   - The most frequent method used here is use of the *Leopold matrix*
   - The matrix is a tool used in the systematic examination of potential interactions
   - Example: in a windfarm development a significant impact may be collisions with Birds
5. Mitigation
   - This is where EIA is most useful
   - Once section 4 das Been completed it will be obvious where the impacts will be greatest
   - Using this information ways to avoid negative impacts should be developed
   - Best wording with the developer with this section as they know the project best
   - Using the windfarm example again construction could be out of bird nesting seasons
6. Non-technical summary (EIS)
   - The EIA will be in the public domain and be used in the decision making process
   - It is important that the information is available to the public
   - This section is a summary that does not include jargon or complicated diagrams
   - It should be understood by the informed lay-person
7. Lack of know-how/technical difficulties
   - This section is to advise any areas of weakness in knowledge
   - It can be used to focus areas of future research
   - Some developers see the EIA as a starting block for good environmental management

ISO 9000

ISO 9000 is a family of *standards for quality management systems.* ISO 9000 is maintained by ISO, the *International Organization for Standardization* and is administered by accreditation and certification Bodies. The rules are updated, as the requirements motivate changes over time.

Some of the requirements in ISO 9001:2008 (which is one of the standards in the ISO 9000 family) include

- a set of procedures that cover all key processes in the Business;
- monitoring processes to ensure they are effective;
- keeping adequate records;
- checking output for defects, with appropriate and corrective action where necessary;
- regularly reviewing individual processes and the quality system itself for effectiveness; and
- facilitating continual improvement

A company or organization that has been independently audited and certified to Be in conformance with ISO 9001 may publicly state that it is "ISO 9001 certified" or "ISO 9001 registered". Certification to an ISO 9001 standard does not guarantee any quality of end products and services; rather, it certifies that formalized Business processes are being applied.

Although the standards originated in *manufacturing*, they are now employed across several types of organizations. A "product", in ISO vocabullary, can mean a physical object, services, or *software*.

ISO 9000 Series of Standards

ISO 9000 includes the following standards:

- ISO 9000:2005 Quality management systems- fundamentals and vocabulary describes fundamentals of quality management systems, which form the subject of the ISO 9000 family, and defines related terms.
- 9001:2008 Quality management systems - Requirements is

intended for use in any organization regardless of size, type or product (including service). It provides a number of requirements which an organization needs to fulfill to achieve customer satisfaction through consistent products and services which meet customer expectations. It includes a requirement for continual (i.e. planned) improvement of the Quality Management System, for which ISO 9004: 2000 provides many hints.

This is the only implementation for which third-party auditors can grant certification. It should 6e noted that certification is not described as any of the 'needs' of an organization as a driver for using ISO 9001 (see ISO 9001:2000 section 1 'Scope') but does recognize that it may be used for such a purpose (see ISO 9001:2000 section 0.1 'Introduction').

- ISO 9004:2000 Quality management systems - guidelines for performance improvements covers continual improvement. This gives you advice on what you could do to enhance a mature system. This document very specifically states that it is not intended as a guide to implementation.

There are many more standards in the ISO 9001 series (see "List of ISO 9000 standards" from ISO), many of them not even carrying "ISO 900x" numbers. For example, some standards in the 10,000 range are considered part of the 9000 group: ISO 10007:1995 discusses Configuration management, which for most organizations is just one element of a complete management system. ISO notes: "The emphasis on certification tends to overshadow the fact that there is an entire family of ISO 9000 standards... Organizations stand to obtain the greatest value when the standards in the new core series are used in an integrated manner, both with each other and with the other standards making up the ISO 9000family as a whole".

Note that the previous members of the ISO 9000 series 9002 and 9003 have Been integrated into 9001. In most cases, an organization chiming to be "ISO 9000 registered" is referring to ISO 9001.

Contents of ISO 9001

ISO 9001:2008 Quality management systems—Requirements is a document of approximately 30 pages which is available from the national standards organization in each country. Outline contents are as follows:

- Page iv: foreword
- Pages v to vii: Section 0 Introduction
- Pages 1 to 14: Requirements
- Section 1: Scope
- Section 2: Normative Reference
- Section 3: Terms and definitions (specific to ISO 9001, not specified in ISO 9000)
- Pages 2 to 14
- Section 4: Quality Management System
- Section 5: Management Responsibility
- Section 6: Resource Management
- Section 7: Product Realization
- Section 8: Measurement, analysis and improvement

In effect, users need to address all sections 1 to 8, but only 4 to 8 need implementing within a QMS.

- Pages IS to 22: Tables of Correspondence between ISO 9001 and other standards
- Page 23: Bibliography

The standard specifies six compulsory documents:

- Control of Documents (4.2.3)
- Control of Records (4.2.4)
- Internal Audits (8.2.2)
- Control of Nonconforming Product /Service (8.3)
- Corrective fiction (8.5.2)
- Preventive Action (8.5.3)

In addition to these, ISO 9001:2008 requires a Quality Policy and Quality Manual (which may or may not include the above documents).

Summary of ISO 9001:2008 in informal language

- The quality policy is a formal statement from management, closely linked to the business and marketing plan and to customer needs. The quality policy is understood and followed at all levels and by all employees. Each employee needs measurable objectives to work towards.
- Decisions about the quality system are made Based on recorded data and the system is regularly audited and evaluated for conformance and effectiveness.
- Records should show how and where raw materials and products were processed, to allow products and problems to Be traced to the source.
- You need to determine customer requirements and create systems for communicating with customers about product information, inquiries, contracts, orders, feedback and complaints.
- When developing new products, you need to plan the stages of development, with appropriate testing at each stage. You need to test and document whether the product meets design requirements, regulatory requirements and user needs.
- You need to regularly review performance through internal audits and meetings. (Determine whether the quality system is working and what improvements can be made. (Deal with past problems and potential problems. Keep records of these activities and the resulting decisions, and monitor their effectiveness (note: you need a documented procedure for internal audits).
- You need documented procedures for dealing with actual and potential nonconformances (problems involving suppliers or customers, or internal problems). Make sure no one uses bad product, determine what to do with bad product, deal with the root cause of the problem and keep records to use as a tool to improve the system.

Different Versions of ISO 9000

1987 Version

ISO 9000:1987 had the same structure as the UK Standard BS 5750, with three 'models' for quality management systems, the selection of which was based on the scope of activities of the organization:

- ISO 9001:1987 Model for quality assurance in design, development, production, installation, and servicing was for companies and organizations whose activities included the creation of new products.
- ISO 9002: 1987 Model for quality assurance in production, installation, and servicing had Basically the same material as ISO 9001 But without covering the creation of new products.
- ISO 9003: 1987 Model for quality assurance in final inspection and test covered only the final inspection of finished product, with no concern for how the product was produced.

ISO 9000: 1987 was also influenced by easting U.S. and other *Defense Standards* ("MIL SPECS"), and so was well-suited to manufacturing. The emphasis tended to 6e placed on conformance with procedures rather than the overall process of management— which was likely the actual intent.[citation needed]

1994 Version

ISO 9000:1994 emphasized *quality assurance* via preventive actions, instead of just checking final product, and continued to require evidence of compliance with documented procedures. As with the first edition, the down-side was that companies tended to implement its requirements by creating shelf-loads of procedure manuals, and becoming burdened with an ISO Bureaucracy. In some companies, adapting and improving processes could actually be impeded by the quality system.[citation needed]

2000 Version

The Portuguese *ISO 9001* certification image.

ISO 9001:2000 combines the three standards 9001, 9002, and 9003 into one, called 9001. Design and development procedures are required only if a company does in fact engage in the creation of new products. The 2000 version sought to make a radical change in thinking by actually placing the concept of process management front and center ("Process management" was the monitoring and optimizing of a company's tasks and activities, instead of just inspecting the final product). The 2000 version also demands involvement by upper executives, in order to integrate quality into the business system and avoid delegation of quality functions to junior administrators. Another goal is to improve effectiveness via process performance metrics—numerical measurement of the effectiveness of tasks and activities. (Expectations of continual process improvement and tracing customer satisfaction were made explicit.

ISO 9000 standard is continually being revised by standing technical committees and advisory groups, who receive feedback from those professionals who are implementing the standard.[1]

2008 Version

ISO 9001:2008 only introduces clarifications to the existing requirements of ISO 9001:2000 and some changes intended to improve consistency with ISO 14001:2004. There are no new requirements. Explanation of changes in ISO 9001:2008. A quality management system being upgraded just needs to be checked to see if it is following the clarifications introduced in the amended version.[1] Practical, Guide to Implementing ISO 9001:2008

Certification

ISO does not itself certify organizations. Many countries have formed accreditation Bodies to authorize certification Bodies, which audit organizations applying for ISO 9001 compliance certification. Although commonly referred to as ISO 9000:2000 certification, the actual standard to which an organization's quality management can be certified is ISO 9001:2000. Both the accreditation bodies and the certification bodies charge fees for their services. The various accreditation bodies have mutual agreements with each other to ensure that certificates issued by one of the *Accredited Certification Bodies* (CB) are accepted worldwide.

The applying organization is assessed based on an extensive sample of its sites, functions, products, services and processes; a list of problems ("action requests" or "non-compliances") is made known to the management. If there are no major problems on this list, or after it receives a satisfactory improvement plan from the management showing how any problems will be resolved, the certification body will issue an ISO 9001 certificate for each geographical site it has visited.

An ISO certificate is not a once-and-for-all award, but must Be renewed at regular intervals recommended by the certification body, usually around three years. In contrast to the *Capability Maturity Model* there are no grades of competence within ISO 9001.

Auditing

Two types of auditing are required to become registered to the standard: auditing by an external certification Body (external audit) and audits by internal staff trained for this process (*internal audits*). The aim is a continual process of review and assessment, to verify that the system is wording as it's supposed to, find out where it can improve and to correct or prevent problems identified. It is considered healthier for internal auditors to audit outside their usual management line, so as to bring a degree of independence to their judgments.

Under the 1994 standard, the auditing process could be adequately addressed By performing "compliance auditing":

- Tell me what you do (describe the Business process)
- Show me where it says that (reference the procedure manuals)
- Prove that this is what happened (exhibit evidence in documented records)

How this led to preventive actions was not clear.

The 2000 standard uses the process approach. While auditors perform similar functions, they are expected to go beyond mere auditing for rote "compliance" by focusing on risk, status and importance. This means they are expected to make more judgments on what is effective, rather than merely adhering to what is formally prescribed. The difference from the previous standard can be explained thus:

Under the 1994 version, the question was broadly "Are you doing what the manual says you should be doing?", whereas under the 2000 version, the question is more "Will this process help you achieve your stated objectives? Is it a good process or is there a way to do it better?"

The ISO 19011 standard for auditing applies to ISO 9001 besides other management systems like EMS (ISO 14001), FSMS (ISO 22000) etc.

REFERENCES

1. Jo Smith, Pete Smith, "Introduction to Environmental Modelling", August 2009.
2. R. Ryan Dupont, Terry E. Baxter, Louis Theodore, "Environmental Management Problems and Solutions, 1998.
3. Michael V. Russo, "Environmental Managemenr : Readings and Cases", 2nd end., pp. 550-610, Sage 2008.
4. Abbasi, S.A., "Environment Everyone", Discovery Publishing House, New Delhi.
5. Suresh K. Dhameja, "Environmental Engineering and Management", 2nd edn., 2005.
6. S.V.S. Rana, "Essentials of Ecology and Environmental Science, 3rd end., Prentice Hall of India Pvt. Ltd., 2007.
7. Surinder Deswal, Anupama Deswal, "Energy Environment Ethics and Society", 1st edn., Dhanpat Rai and Co. (P) Ltd. 2008-09.

CHAPTER 36

El-Nino Phenomena

Dr. Priya Budhani[1]

The term El Nino—Spanish for "the Christ Child"—was originally used by fishermen to refer to the Pacific Ocean warm currents near the coasts of Peru and Ecuador that appeared periodically around Christmas time and lasted for a few months. Due to those currents, fish were much less abundant than usual. At the present time we use the same name for the large-scale warming of surface waters of the Pacific Ocean every 3-6 years, which usually lasts for 9-12 months, but may continue for up to 18 months, and dramatically affects the weather worldwide.

Graphic by Yiqi Shao

Ei Nino events happen irregularly. Their strength is estimated in surface atmospheric, pressure anomalies and anomalies of land and sea surface temperatures.

The El Nino phenomenon dramatically affects the weather in many parts of the world. It is therefore important to predict its appearance. Various climate models, seasonal forecasting models, ocean-atmosphere coupled models, and statistical models attempt to predict El Nino as a part of interannual climate variability. Predicting EI Nino has been possible only since the 1980s, when the power of computers became sufficient to cover very complicated large-scale ocean-atmosphere interactions.

Historical Observations

EI Nino were observed as early as the 1600s. More systematic study began at the end of the 19th century, when Peruvian geographers noted unusual oceanic and climatic phenomena occurring periodically along the Peru cost. They noticed that eastern Pacific warming was sometimes very strong, In the 1920s The British scientist Sir Gilbert Walker empirically identified that some notable climate anomalies—changes in atmospheric pressure and circulation—happen around the world every few years. He invented the term for those-climate oscillations, "the Southern Oscillation."

Sir Gilbert Walker

While stationed in India studying monsoons, Sir Walker observed pressure differences in the Pacific Ocean. He noticed "a seesaw" of atmospheric pressure measured at two sites: Darwin in Australia, the Indian Ocean, and in Tahiti, an island in the South Pacific, When atmospheric pressure rises at Darwin it falls in Tahiti and vice versa.

[1]Department of Chemistry, Sadhu Vaswani College Bairagarh, Bhopal (India)

In the 1990s. it. was observed that climate anomalies connected with the Southern Oscillation coincided in general with EI Nino occurrences. Around I960., scientists realized that the warming of the eastern Pacific is only a part of the oceanic oscillations that extend westward along the equator, out to the dateline. At about the same time the famous meteorologist Jacob Bjerknes proposed that EI Nino was just the oceanic expression of a large-scale interaction between the ocean and the atmosphere, and that the climate anomaties could be understood as atmospheric **"teleconnections"** spreading from the warm-water regions along the equator in the mid-Pacific.

Professor Jacob Bjerknes

Since approximately1975 scientists have been researching EI Nino and the Southern Oscillation phenomena together. Today we know that EI Nino is a part of an interannual climate oscillation called the EI Nino Southern Oscillation (ENSO) event. EI Nino is a warm phase of ENSO; the cold phase of this event is called La Nina.

El Nino's Impact

The strongest El Nino events of the 20th century occurred in 1982-83 and 1997-98. The effects of 1982-83 included significant storms throughout the southwest United States and one of Australia's worst droughts of the centry.

Drought scene near Narromine, New South Wales. 27 September 1982

According to the World Meteorological Organization, the 1997-'98 El Nino was a major factor in 1997s record high temperatures. The estimated average surface temperature for land and sea worldwide was 0.8°F higher than the 1961-1990 average of 61.7°F. According to the National Oceanic and Atmospheric Administration (NOAA), 1998 has set all-time highs of global land and ocean surface temperatures, above record high levels in 1997. In 1998 the mean temperature wasl.2°F (0.7°C) above the long-term (since 1880) mean of 56.9°F (13.S°C)

The impact of the 1997/8 El Nino has been felt in many parts of the world: Droughts have occurred in the Western Pacific Islands and Indonesia as well as in Mexico and Central America. In Indonesia drought caused uncontrollable forest fires and floods, while warm weather led to a bad fisheries season in Peru, and extreme rainfall and mud slides in southern California. Corals in the Pacific Ocean were bleached by warmer than average water, and shipping through the Panama Canal was restricted by below-average rainfall.

El Nino phenomena dramatically affects the weather throughout the world. Among other weather anomalies, El Nino events are responsible for:

- A shift of thunderstorm activity eastward from Indonesia to the south Pacific, which leads to abnormally dry conditions and severe droughts during both warm and cold seasons in Australia, the Philippines. Indonesia, southeastern Africa and Brazil.
- During the summer season the Indian monsoon is less intensive than normal and therefore it is much less rainy than usual in India.
- Much wetter conditions at the west coast of tropical South America.

El Nino Impacts on the United States, North America and The Atlantic regions include

- Wetter than the normal conditions in tropical latitudes of North America, from Texas to Florida, including more intensive wintertime storms.

- Extreme rainfall and flooding events in California, Oregon and Washington.
- Much milder winters and late autumns in northwestern Canada and Alaska due to pumping or abnormally warm air by mid-latitude low pressure systems.
- Below normal hurricane/tropical cyclone activity in the Atlantic (however, their strength is limited by El Nino).
- Drier than normal North American monsoons, especially for Mexico, Arizona and New Mexico,
- Drier than normal autumns and winters in the U.S. Pacific Northwest.

What Causes El Nino?

The warming of the Pacific occurs as a result of the weakening of trade winds that normally blow westward from South America toward Asia.

Global Wind Patterns: Wind Belts of the General Circulation

The global wind pattern is also known as the "general circulation," and the surface winds of each hemisphere aro divided into three wind belts:

- **Polar Easterlies:** From 60-90 degrees latitude.
- **Prevailing Westerlies:** From 30-60 degrees latitude (aka Westerlies).
- **Tropical Easterlies:** From 0-30 degrees latitude (aka Trade Winds).

The prevalent surface winds across the equatorial Pacific ocean are easterly trade winds. These drag warm surface water away from the coast of Peru and cause colder deep ocean water to come to the surface (so-called "upwelling"). Upwelling causes the thermocline (the zone at the top part of the ocean in which temperature decreases rapidly with depth) to be much shallower in the Eastern Pacific than in the western. Trade winds and the equatorial upwelling maintain warm sea surface temperatures at the western equatorial Pacific and cold surface temperatures in the east. When trade winds weaken, the equatorial upwelling decreases, the thermocline gets deeper, the ocean surface along the coast of South America becomes warmer, and trade winds weaken the, more. This in turn causes surface waters in the eastern Pacific to became even warmer and so on. This mechanism is known as the Bjerknes hypothesis and represents an onset of El Nino.

The questions remain: what stops warming in the eastern Pacific and why do El Nino events last: approximately 12-18 months. The widely accepted (but not unique!) explanation is the delayed oscillator hypothesis.

During the warming event in the eastern Pacific, the thermocline deepens along the equator and rises in the regions about 3 to 8 latitude degrees from the equator. These off equator thermocline anomalies have little effect on the ocean surface temperature, but they propagate westward under the ocean surface as so-called Rpusby waves, with a speed of about 0.8 m/s. When Rossby waves reach the Indonesian archipelago they are reflected back as another type of the ocean underwater waves, namely equatorial Kelvin waves. Because of the deep thermocline in the western Pacific, the arrival of the Rossby signal does not affect surface temperature. Kelvin waves are much faster than Kossby waves and propagate eastward along the equator with an approximately 3 m/s speed as a shallower thermocline anomaly. When Kelvin waves reach the equatorial Eastern Pacific, they move the thermocline (and, therefore, cold deep waters) in this region even closer to the surface, cool the ocean's surface and terminate the warm EI-Nino event.

Detection and Prediction of El Nino

An ENSO observation system has been established over the past 10-15 years. Now scientists can observe the state of upper layers of the tropical Pacific Ocean in real time.

Scientists use a variety of tools and techniques to detect the changes in the state of the Pacific Ocean, and therefore to detect El Nino conditions and the beginning and development of El Nino events. The changes in the tropical Pacific characteristics may be determined by satellites and different kind of buoys. Those observing ocean systems are part of the Tropical Ocean Global Atmosphere program (TOGA).

TOGA consists of scientific research ships, and radiosondes, which form the operational ENSO observing system. Satellites give the information on tropical rainfall, wind, and ocean temperature. Buoys include moored buoys, drifting buoys, and expandable buoys; all these provide the data on upper ocean and sea surface temperatures. Ships observe both the tropical ocean atmosphere and the upper oceanic layer. Radiosondes observe the state of the atmosphere and weather patterns worldwide.

Data from all ocean and atmosphere observing systems are processed by supercomputers. This information is used for both diagnostic and forecasting purposes by different kinds of models. Supercomputers, renitime data transmission via satellites, and modern diagnostic tools allow monitoring of El Nino in real time.

There are two Major Types of El Nino Prediction Models

- Hydrodynamic coupled ocean-atmosphere models. In this type of model, the atmosphere and the ocean are treated as different types of fluids and their behavior is described by a complicated system of differential equations. Solving those equations involve a tremendous amount of computation, and requires the use of the most powerful supercomputers.
- Statistical models: In these models, statistical relationships, derived from the previous El Nino occurrences, are used to predict future El Nino events. These modes are much more computationally efficient than hydrodynamic models. However the period of meteorological and oceanographic observations is too short to produce reliable statistics, and therefore the statistical relationships introduced are usually subjective. The physics of the phenomena is not described, hence the accuracy of the prediction is limited.

There are also hybrid (intermediate) models, where an ocean model is coupled to a statistical atmospheric model. These hybrid models try to combine the computational efficiency of the statistical models with the accuracy of the hydrodynamic, models.

The first successful forecasts of El Nino were made by Mark Cane and Steve Zebiak at Columbia University's Lamont-Doherty Geological Observatory in the US. They developed an intermediate ocean-atmosphere coupled model. The model successfully predicted the onset of the 1986-7 El Nino one year in advance. Since then, the field of El Nino prediction has grown and now there are several models that can predict 61 Nino up to 6-1? months in advance. However El Nino is hard to predict due to its chaotic nature, and its predictability is stili the-subject of debate. Two main factors limit predictability: the effects of high-frequency atmospheric variability and the growth of errors in the initial conditions of numerical models. Initial conditions in ocean-atmospheric models include the array of temperature, pressure, wind, and other data that describe the current state of the ocean and atmosphere. Small errors in initial conditions may cause big changes in the results of a forecast.

Many models have been tested on already known El Nino events of the past, especially on those that took place during the period with relatively reliable meteorological and oceanographic data, that is during the last one and half centuries. Some of those models present retrospective forecasting of El Nino and La NiNinoa events for a period covering more than a century. The results are surprisingly successful, particularly when using information from within six months of the event. All the major climatic fluctuations of the twentieth century are reflected by these models. Obviously the, most difficult aspects to forecast are the intensity of El Nino, the rapidity of its development, and its duration.

As already mentioned, the 1997-98 El Nino was one of the strongest in the 20th century. Many climate models predicted a high possibility of the 1997-98 El Nino event. The National Oceanic and

Atmospheric Administration, for example, successfully predicted it six months in advance. But nobody expected this particular event to be so extremely strong.

In late 2001, ocean and atmosphere conditions were similar to those in 1996. So the scientists suggested a high probability of a new El Nino appearance in 2002. Those forecasting suggestions were right[1]

A moderate intensity El Nino event occurred in the Tropical Pacific in 2002-2003. The classic features of El Nino all were registered: abnormally warm sea surface temperatures of the tropical Pacific, weakening of trade winds, and changes in rainfall variability over the Pacific region. This event, though much less intensive than the 1997-98 El Nino, had a big impact on weather variability worldwide. Its evolution was registered by satellites and through in situ data of the ENSO Observing System.

One cannot underestimate the importance of El Nino forecasting for many regions of the world, especially for those.countries in tropical regions where economic success is based on fisheries, agriculture and food production, all of which depend on weather patterns. Peru is an excellent example of a country that, derives huge economic benefits from El Nino forecasting. Usually a warmer than normal year with moderate and strong El Nino onset is unfavorable for fisheries. When the state of equatorial Pacific is near normal conditions are favorable for agriculture. La NiNinoa conditions (colder than normal ocean surface temperatures) are good for fishing, but not favorable for farmers, bringing them drought and crop failure. El Nino affects the amount of precipitation, so forecasts help to decide when it's better to sow rice (in expected wetter periods), and when to sow cotton (in drier periods).

Therefore ENSO forecasting in Peru presents four possibilities in ENSO phases:

1. near normal conditions
2. a weak El Nino with slightly higher than normal precipitation
3. El Nino conditions with flooding
4. La NiNinoa conditions with colder than normal sea surface temperatures and higher possibility of drought

Predicting El Nino, and especially climatic extremes (floods, droughts, etc.) caused by El Nino, is very important for the United States as well, because this can save the country billions of dollars in damage costs. The U.S. economy is much more complicated than the economy of Peru, so detailed analysis is out of scope of this presentation, but. interested readers will find abundant information in the abovementioned sources.

Numerous models of El Nino still are not as reliable as weather forecasting models, but they are able to represent the main features of the typical El Nino event.

Summary

El Nino is d warm phase of the interannual climate oscillation called El Nino Southern Oscillation (ENSO) event, an example of large-scale ocean-atmosphere interaction, and is characterized by large-scale warming of the surface tropical Pacific Ocean. El Nino events occur every 3-6 years, last 9-12 months, sometimes even up to;.8 months, and have a big impact on world weather.

The major impacts of El Nino are temperature anomalies, changes in precipitation variability, floods and diuugius throughout the world.

El Nino events happen irregularly and are hard to predict. However many numerical climate models predicted the last few El Nino events successfully. El Nino forecasting is becoming more and more reliable with our improving knowledge of the phenomenon's nature, with the help of more and more powerful computers, and with the operational E! Nino Southern Oscillation observation system.

Ei Nine forecasting is especially important for tropical countries where El Nino impacts are the strongest.

CHAPTER 37

Instrumentation in Environmental Analysis

Dr. Subarta Pani[1], Dr. S.A. Iqbal[2] and Ms. Vandana Magarde[3]

Introduction

Environmental science and technology is an area, where scientific observations are used as a basic tool for understanding the critical parameters and their influences on the overall environmental status. Quantification of observation is essentially required to express the measurement of any substance in rational manner. The characterization of pollutants in the environment has been greatly facilitated by the development of a number of instruments and related analytical methods. Some of the most common instruments that are in use in the filed of environmental analysis are discussed in this chapter.

1. SPECTROPHOTOMETER

Spectrophotometer is the instrument that can analyze a wide range of parameters in the liquid medium. All the spectrophotometers basically use the colorimetric methods for the analysis of various parameters. Spectrophotometers work basically on the principal of the absorption of the emitted monochromatic light beam when passed through a liquid medium. The methods employed in the analysis of these parameters are relatively easier, and yield fairly accurate results.

Principle

Colorimetric analysis is based on the Beer's law. Beer's law states "the intensity of a beam of monochromatic light decreases exponentially with the increase in concentration of absorbing substance arithmetically." In all the procedures, the compound to be analyzed is reacted with a chemical or set of chemicals yielding a coloured compound, intensity of which is directly proportional to the concentration except in few exceptional cases like fluoride (SPANDS METHOD).

The intensity of color is measured by passing a monochromatic bean of light, the wavelength of which depends upon the color complex being formed in the analysis. Some part of the light is absorbed by the coloured liquid media that is read as absorbance by the instrument. The absorbance is directly proportional to the intensity of color, which in turn is directly proportional to the concentration of the compound being analyzed. The absorbance is compared with a standard curve & thus the value of absorbance can be converted into concentration in the desired units. There are different types of spectrophotometers (i.e. visible, UV-visible etc.) commonly available for analysis of various compounds. For reference, the application of UV-Visible Spectrophotometer manufactured by Hach is described here.

[1]Lake Conservation Authority of Madhya Pradesh, Bhopal (India)

Hach Dr-4000 UV–vis-spectrophotometer

The Hach DR–4000 UV–vis-Spectrophotometer is one of the most sophisticated spectrophotometers available in the world today. The spectrophotometer covers a wavelength ranging from 190 nm to 2100 nm with a resolution of up to two places of decimals. The instrument is capable of reading the absorbance up to 4 places of decimals ensuring the measurements of certain parameters even up to ppb levels.

Principle Components of the Instruments

1. **Light Source :** The light source in Hach – DR 4000 UV-vis Spectrophotometer are two lamps.
   - **Tungsten lamp :** for visible spectrum
   - **Deuterium lamp :** for UV spectrum
2. **Monochromatic :** Also known as grating, this monochromatic enables to pass light of only desired wavelength to be emitted.
3. **Sample Cell Holder :** This is the compartment where the prepared samples are kept for the analysis.
4. **Recorder/Sensor :** The sensor is a device which senses the light transmitted through the prepared sample. This sensor sends electrical signals the value in either absorbance of any other desired units.

Application

The Hach DR–4000 is a very versatile instrument having a very wide range of application. The instrument comes ready with more than 250 pre – set programs, as Hach programs. The user can also develop own programs as may be required.

Operating manual for Hach Dr – 4000

When the instrument is switched on, the instrument performs a systems check that enables the instrument to ensure proper functioning of all the vital components of the instrument. After the completion of systems checks, the instrument comes in idle mode or the main menu. The desired functions can be chosen from this stage by pressing the soft by corresponding to the respective functions. The following functions can be chosen from the main menu.

Hach Program : This menu consists of more than 250 pre-set analytical programs with regular up gradations and addition of new parameters. The analysis ranges from simple parameters like color, Turbidity to more complex ones like As & TOC. The ready-made chemicals for direct use are available with manufacturer, may be specific firm for the analysis of different parameters. The chemicals provided in a particular quantity in powder pillows, pouches or accu-vac ampoules is added in a determined quantity of samples & the reaction period is provided as per the method. After the reaction time is over, the prepared samples are kept in the cell holder & the values in the desired units can be read from the instrument.

User Program : User Programs can be developed in the instrument & saved for later uses. This function enables the user to use chemicals other than the Hach reagents. Major functions like name of the program (user), wavelength, timers, chemical forms, units, lower limit, upper limit etc. can be chosen. A standard curve can be plotted in the instrument, so that the user program can be used later to derive the values in desired units in the next uses. The program could be set to read either single wavelength or multi – wavelength that can read up to four wavelengths.

Reading Absorbance/ %Transmittance at Single/Multiple /Wavelength – The instrument is capable of reading the absorbance at single or multiple wavelengths (up to 4) which can be achieved by pressing the soft by corresponding to the desired function & choosing the wavelength/wavelengths.

Accessories of the Hach Dr–4000 UV–Vis Spectrophotometer

1. Hachlink Software : This Hachlink software can be installed in a P.C. that enables the control of the instrument using the PC. The data generated by the instrument could be stored in the P.C. and could be obtained as print out, whenever required.

2. Sipper Module : This is a convenient sample cell module that enables to analyze number of samples in a relatively short period. Sipping of the prepared sample is done through intact pump, after which the instrument reads it and after the analysis is over, it is purged out from the module. The sipping period, settling period and the purging period can be adjusted as per the requirement of the user.

3. Carausol Module : This sample cell module can hold six cuvettes or four sample cells and thus analyze 6 or 4 samples at a time.

4. Flow Through Cells : In analysis of certain sensitive parameters it the desirable to minimize the retention of samples in the cuvettes or sample cells in order to prevent contamination of other samples. The flow through cell ensures reading the value when the prepared sample is still flowing through the cell, without any retention and thus preventing any chances of contamination.

Data Storage & Handling : The instrument has a memory that can store data. The instrument can be set to store the data either manually or automatically as desired. The data could be recalled at any time and can also be printed through a printer attachment whenever desired.

2. ATOMIC ABSORPTION SPECTROPHOTOMETER (AAS)

Typical Atomic Absorption Spectrometer has following Components

- Chopper
- Monochromator
- Light source
- Sample cell
- Detector
- Recorder

Light Source

- Hollow Cathode Lamp (Ne or Ar Gas)
- Electrode less Discharge Lamp

Advantages of EDLS	Disadvantages
• Lower Gain • Higher Absorbance • Better Sensitivity • Longer Life Time • Greater Intensity	• Require Warmup • May Require Separate Power Supply

Function of monochromator

- Disperse Polychromatic light into its Various wave lengths
- Allows for the isolation of specific wave lengths

Process in a Flame AA

M·	**$M^+ + e^-$**	**Ionisation**
M^o	M^*	Excitation
MA	$M \cdot + A^o$	Atomization
Solution	**Solid**	**Vaporization**

Burner Head Selection

- 10 cm head – General purpose for Air acetylene operation
- 5 cm head for nitrous oxide operation or short path length air acetylene

Beers law

io = Initial Lamp intensity

i = Final lamp intensity

Absorbance A = log (io/i)

A = a × b × c

Where

A = Absorbance

a = Absorption Coefficient

b = Path length

c = Concentration

'a' & 'b' constant
for a given element and instrument configuration
There fore A = K × C

BEERS LAW VALID AT ONLY LOW ABSORBANCE

Application

Analysis of Heavy Metals in Water, Soil, Blood, Fish, Food material and others.

3. ION SELECTIVE ELECTRODE (MULTIPARMETER KIT - MULTI LAB P5)

Introduction

Multiparmeter kit (MULTI LAB P5) is a German make portable analysis kit. It is having both photometer and ION selective electrode (ISE) facility. All the colorimetric analysis can be done on photometer. Simultaneously with the help of ISE electrodes the respective ions can be analyzed. Instrument can be operated either by direct current or by battery as per requirement.

Ion Selective Electerode (ISE)

Ion selective electrode are the devices which give directly an electrical out put related to the composition of the sample solution. It works on the basis of Nernst equation.

The classical ISE consists of a tube of a good electrical insulator that is closed at its lower end by a sensing membrane. Within the tube there is a filling solution known as internal filling solution containing a fixed concentration of the ion to which the membrane is sensitive.

Working Mechanism

When the ion selective electrode is placed in the sample solution, the inner filling solution is in contact with the inner face of the membrane, while the sample solution contacts the outer surface. If the concentration of the sample solution is greater then that of the filling solution, ions migrate inwards with their associated charges thus, changing the potential across the membrane. Equilibrium potential soon develops which just prevents further ingress of ions. Migration takes place outwards if the sample is less concentrated than the inner solution.

The number of ions transported is so small that the change in sample concentration is infinite small. The magnitude if potential developed across the membrane depends upon the activity of the species being determined in the sample solution. Potential of an ISE electrode can not be measured itself, it can only measured in conjugation with an external electrode whose potential relative to the solution is intended to be constant regardless of composition.

Nernst Equation gives the relation between electrode potential and activity:

$$E = E^0 + \frac{2.303\ RT}{nF} * \log a$$

Where
E = the electrode potential
E^0 = a constant for ISE and reference electrode
R = gas constant
T = absolute temperature
n = ionic change of the analyze ion
F = Faraday constant
a = activity of the analyze ion

Factors Affecting Electrode Measurement

- pH
- Temperature
- Presence of complexing agent

Significance of ISE Electrodes

- Do not bother for turbidity and colour of the sample.
- Small volume of sample can be analyzed.
- Measures species without upsetting chemical equilibrium.
- Measurements are independent of volume.
- Concentration can be read on directly on specific ion meter.
- Electrodes are sensitive down to ppb level.
- Do not require extensive pre –treatment of the sample.
- Relatively rapid response time.

Parameters Analyzed by Multiparameter Kit (Multi Lab P-5)

- PH
- Conductivity
- Fluoride
- Chloride
- Iodide
- Bromide
- Sulphate
- Nitrate
- Ammonical Nitrogen
- Phosphate
- COD
- BOD
- DO
- Copper
- Cyanide
- Lead

Table 37.1. Ion Selective Electrodes in Environmental Analysis

Ion/ Molecule	Types of Membrane	pH Range	Lower limit of detection (moles /lt.)	Principal interference(ions)
Ammonia	GT	7-14	10^{-6}	Amines
Bromide	SS	2-12	5×10^{-6}	I, CN^-, S^-
Calcium	LM	5-10	10^{-5}	Zn, Fe^{+++}, Pb, Cu, Ni
Chlorine	SS	4-5	10^{-7}	Strong oxidants
Cyanide	SS	12-14	10^{-6}	S^-, I^-
Fluoride	SS	5-8	10^{-6}	OH
Iodide	SS	2-12	5×10^{-8}	CN^-, S^-
Nitrate	LM	3-10	5×10^{-5}	I^-, Br^-, NO_2^-

SS - Solid State
GM - Glass membrane
GT - Gas Transfer
LM - Liquid Membrane

Advantages of Multiparameter Kit

- Muliti Lab P-5 is a handy instrument and meant for the analysis at the site.
- Instrument can be operated on direct line as well as on battery.
- Instrument is having the facility of ISE analysis as well as photometric analysis.
- Very easy to handle.
- With a small kit lots of parameters can be analyzed.

Disadvantages of Multiparameter Kit

- Maintenance of electrode.
- Cost of single electrode is very high that is why very much expensive.
- Chemical required for the analysis are supplied by WTW only.

4. GAS CHROMATOGRAPHY(GC)

Introduction

Gas chromatography is basically a partition chromatography where mobile phase is a gas and the stationary phase is a liquid coated on finely divided inert material called 'support'. The stationary phase is packed in column and by varying the temperature, it is possible to vary the partition coefficient of the solute thus effecting the elution rate of the compound on the column. Gas chromatography has more utility in studies of organic component like polynuclear aromatic hydrocarbons and chlorinated hydrocarbons like pesticides viz. DDT etc to their lowest concentration of 10 –12 gms (picogram). Estimation of these compounds is important because many of them are highly toxic and carcinogenic.

Principle

In chromatography, a mixture is separated by adsorption or solution partition between two immiscible phases. One phase is moving and the other stationary. The moving or the mobile phase may be either gas or liquid and the stationary phase can be either liquid or solid.

System Configuration

- A column packed with suitable packing material.
- An injection facility with appropriate gas supply.
- One or more detectors for quantitative analysis.
- An amplifier or recorder for recording the signal from the detector.

Process Involved during Gas Chromatographic Separation

- The sample containing the solute is injected into the heating block where it is immediately vaporized and swept as a plug of vapour by their carries gas stream into the column inlet.
- The solutes are adjusted at the head of the column by the stationary and the carrier gas.
- Each solute will travel at its own rate through the column and subsequently a band corresponding to each solute will form. The bonds will separate to a degree, which is determined by the partition ratio of the solute and the agent of band spreading.
- The solutes are eluted one after the other in increasing order of their partition ratio and enter the detector attached to the column.
- The time of emergence of peak identify the component and the peak reveals the concentration of the component in the mixture.

INSTRUMENTATION

Sample Injection System

Sample must be introduced as a vapour in the smallest possible volume and in minimum amount of time without decomposing or fractionating the sample. Quantity of the sample introduced and the nature of introduction must be reproducible with a high degree of precision. Liquid samples are usually injected by a microsyringe through a self-sealing rubber septum into a metal block that is heated by a controlled resistance heater.

To obtain the desired sample size, a sample splitting device is placed between the injection block and the column.

Chromatographic Columns

The basic type of column is packed or open tabular column. Packed columns are tubes that have been filled with an inert support coated with a non-volatile liquid phase. Open tubular column has an unrestricted hole through the tubing and the separating medium is coated on the wall of the tubing.

Packed column's normally are used in lengths of 0.7 to 2.0 meter & open tubular run anywhere from 30 to 300 meter in length.

Supports

The support in general used for packed column is diatomaceous earth, which has been crushed and calcinised at or which has been mixed with small amount of flux $CaCO_3$ and calcined.

Carrier Gas

Carrier gas is an essential part of the chromatograph that effects all the portions of the process. The choice of carrier gas depends upon the need of particular detector like electron capture detector requires the supply of nitrogen gas. On the other hand flame-ionization detector requires Nitrogen, hydrogen and Air gas cylinders.

DETECTOR

Flame Ionization Detector (FID)

This is the one of the most popular detector widely used because of its high sensitivity. It consists of a small hydrogen flame burning in an excess of air surrounded by electrostatic field. Column effluent enters the burner (flame) base through a Millipore filter and is mixed with the hydrogen entering the burner. Organic compound eluted from the column is burned and during the combustion ionic fragments and free electrons are formed. These electrons are collected, producing an electric current proportional to the rate at which the sample enters the flame. FID responds only to oxidisable carbon atoms and response is proportional to the number of the carbon atom in the sample.

Electron Capture Detector (ECD)

The principle of the ECD is based on the electron absorption by compound having an affinity for free electrons. As nitrogen carrier gas flow through beta particles from the tritium sources (or Nickle-63) ionize the nitrogen molecule and form slow electrons, which migrate to the anode under a fixed potential. These collected elements produce a steady base line current.

The most common application of ECD is the determination of chlorinated particles apart from the halogenated compound they are also selective towards anhydrides, peroxides, conjugated carbonyls, nitrites, nitrates, ozone etc.

Thermal Conductivity Detector (TCD)

This detector works on the principal of wheat stone bridge were four filaments are arranged in an electrical bridge net work. The carrier gas surrounds the filament in the opposite arms of the bridge; the effluent of the column surrounds the other two filaments. The temperature of the filament is determined by the rate of heat loss by conduction through the carrier gas. As components elute from the column, the composition of the gas changes. The resultant change in thermal conductivity produces s change in filament temperature and in turn, a change in the resistance of the filament, causing an electrical output from the wheat stone bridge circuit.

The detection limit is about 5 microgram per milliliter of sample gas concentration or 10 nanogram of sample weight and the range is about 10^6.

Extraction Procedure

Organochlorine pesticides are extracted by liquid-liquid extraction method employing a polar or non-polar organic solvent, cleaned on a suitable adsorbent column and detected by Gas Chromatography against reference.

Chemicals

1. n-Hexane
2. Chloroform
3. Acetone
4. Anhydrous Sodium
5. Glass wool
6. Pestiside standard {EPA}
7. Florisil, Alumina, Silica gel

There is no general extraction and clean up procedure that can handle all pesticides because of different polarities and the different media they will be extracted from. It is recommended to use a comprehensive extraction and clean up method and then simply the procedure as much as possible without limitation in simplify the procedure as much as possible without limitations in regard to recovery and quantification on the gas chromatograph.

For the analysis of organochorine pesticide in water, Hexane is the best-suited solvent for the first stage extraction.

The water sample is thoroughly mixed by shaking it at room temperature for few seconds. One liter of the well-mixed water sample is transferred to a 2-lit separatory funnel.

20 gm anhydrous sodium sulphate {Na_2SO_4} to the water sample in the separatory funnel is added for checking the emulsion formation during extraction in cases of a little dirty, water samples.

50 ml of n-Hexane is added to the separatory funnel and shake the mixture is shaken vigorously for 2-3 minutes with releasing pressure repeatedly in between by opening the stop –corck.

The separatory funnel is put on stand for some time to allow the layers to separate distinctly.

The upper layer containing pesticide residues and hexane is taken in clean 250 ml flask.

The lower layer is taken in 1liter conical flask and the above procedure is repeated two to three times. Again the upper layer is collected in the same 250ml flask.

The sample collected in the flask is now passed through the glass wool column for drying.

The dried hexane is concentrated at 40°C and the final volume is made upto 5ml in a volumetric flask.

After having the sample processed the final concentrated sample is ready for instrumental analysis.

Applications

Application of the gas chromatography are in many fields like petrochemicals, food, biochemical, clinical, herbicides, pesticides, cosmetics, perfumes, protective coatings, plastics beverage, natural fats/oils, rubber, soaps, synthetic detergents etc. and many others.

Petroleum Industry: Gas chromatography has been used in the analysis of crude petroleum products, fractions, gasoline, waxes and LPG etc.

Food Industry: In this industry gas chromatography has been employed to account the flavor, residual solvents in spices and pesticides in food grains.

Biochemical and Clinical field: Application is useful in detecting sugar, steroids etc in blood sample.

Cosmetic and perfumery: Gas chromatography helps in the determination of the composition o various components, quality and their ingredients.

Paint industry: The volatile components of the paint and even non- volatile components after being suitable derivatization can be identified and measured by Gas chromatography.

Plastic industry: This instrument is used in determination of plastics, determination of esters in acrylic co-polymers of styrene monomers in styrene plastics of vinyl acetate in its co-polymers etc.

Soap Industry: Gas chromatography is used in the determination of ethyl alcohol in the liquid detergents and fatty acids in soaps.

Pesticides and Herbicides: One of the important applications of this instrument is in the detection of pesticides from water, soil and plant materials etc. It can be detected in parts per billion.

Drugs and pharmaceuticals: Gas chromatography is used in this field for the quality assessment of the products.

5. ION EXCHANGE CHROMATOGRAPHY

Introduction

Ion exchange may be defined as a reversible reaction in which free mobile ions of a solid called ion exchange are exchanged for different ions of similar charge present in solution.

Principle of Ion Exchange Chromatography

- A water sample is injected into a stream of carbonate bicarbonate eluent and passed through a series of ion exchanges.
- The anions of interest are separated on the basis of their relative affinities for a low capacity, strongly basic anion exchanger (guard and separator columns).
- The separated anions are directed through a hollow fibre cation exchanger membrane (fibre suppressor) or micromembrane suppressor bathed in continuously flowing strongly acid solution (regenerant solution).
- In the suppressor the separated anions are converted to their highly conductive acid forms and the carbonate bicarbonate eluent is converted to weakly conductive carbonic acid.
- The separated anions in their acid forms are measured by conductivity.
- They are identified on the basis of retention time as compared to standards.
- Quantification is by measurement of peak area on peak height.

Fundamentals of Ion Exchange Chromatography

- Ion exchange is a process wherein a solution of an electrolyte is brought into contact with an ion exchange resin and active ions on the resin are replaced by ions (ionic species) of similar charge from the analyte solution.
- A cation exchanger is one in which the active ions exchange materials are cations and the exchange process involves cations.
- An anion exchanger is one in which the active ions exchange materials are anions and the exchange process involves anions.

Cation Exchanger

A cation exchanger is a high molecular weight, cross-linked polymer having sulphonic, carboxylic, pheolic, etc., groups as an integral part of the resin and an equivalent amount of cations. Thus, a cation exchanger is nothing but a polymeric anion to which active cations are attached. In the cation exchangers, the hydrogen ions are mobile and exchangeable with other cations. The anions (–COD⁻, SO_3^- and –O⁻) remain attached to the resin net work.

Reaction 1

When a Cation exchanger is kept in a solution of a salt, some of the H+ ions of the resin enter the solutions and an equivalent amount of the cations of the salt get attached to the resin. The reaction may be represented as follows:

$HnR + nNa^+ = NanR + nH^+$

(Resin)(Solution) (Resin)(Solution)

The resin having sodium ions, produced in the above reaction, can exchange these ions with other cations. The reaction with calcium ions may be represented as follows :

$2NanR + n\ Ca^{2+} = CanR2 + 2n\ Na^+$

Some Commercially Available Cation Ion Exchanger

1. Amberlite IR-120 2. Dowex 3. Zerolite 4. Amberlite-200

Anion Exchanger

An anion exchanger is a polymer having amine or quaternary ammonium groups as integral parts of the resin and an equivalent amount of anions such as Cl^-, SO_4^{2-}, OH^- ions, etc. These anions are mobile and exchangeable. The anions are mobile and exchangeable.

Reaction 2

The anion exchange behaviour of these materials may be represented as follows.

$$2RCln + nSO_4^{2} = R_2\,(SO_4)_n + 2nCl^-$$

(Resin) (Solution) (Resin) (Solution)

$$RCln + nOH^- = R\,(OH)_n + nCl^-$$

(Resin) (Solution) (Resin) (Solution)

Applications of Ion Exchange Chromatography

Separation of the similar ions from one another:

Ion exchange chromatography is being used to separate similar ions from one another because the different ions undergo exchange reactions to different extents.

For instance, a mixture of Li^+, Na^+, and K^+ ions can be separated by passing their solution through a cation exchanger using sulphuric acid (22 mN Sulphuric acid)as mobile phase/as an eluent.

Similarly ion exchange chromatography has been used to separate a mixture containing Cl^-, Br^-, and I-.The mixture solution is passed through a basic anion exchanger. Sodium carbonate and Bi carbonate solution (2.7 mM Na_2CO_3/0.3mM $NaHCO_3$) has been used as an eluent.

Application Note

Application of Ion Chromatograph: Dionex 500

Salient Features

Dionex 500 is the most sophisticated instrument for the detection of all ions including anions and cations of water, soil, animal and plant materials. The instrument is able to detect concentration of ions upto ppt level.

- DIONEX 500, Ion Chromatography includes following basic components
- PUMP
- DETECTOR
- ORGANIZER
- AUTOSAMPLER
- INTEGRATER
- RECORDER/COMPUTER

PREPARATION OF MOBILE PHASE/ELUENT

Chemicals Required for the Analysis of Anions

1. Sodium Carbonate
2. Sodium Bicarbonate

Chemicals Required for the Analysis of Cations

1. Hydrochloric acid
2. Sulphuric Acid
3. Methanosulphonic acid

Operating Procedure

- Loosen the priming purge screw. Prime the system-using syringe.
- After priming, close the screw i.e. priming purge screw.
- Set pump flow rate 0.2 mL/min. & enter & turn on pump.
- Increase flow rate as per given in increasing steps e.g. 0.2, 0.4, 0.6. 0.8,1.0,1.2, 1.4, 1.5 ml/min.
- Bring cursor on SRS position & using select key, set current at 100 mA & enter.
- Turn on detector.
- Observe suppression i.e. bubbles in waste tube.
- Allow the system to stabilize 20-25 min. i.e. upto to when total conductivity displays < 20 US.
- Now load the standard.
- Offset detector - inject - Run - Start on P.C.
- Wait for complete elution of different components/ions.
- Click on results for converting the results in mg/l or in ppm.
- For closing the system.
- Bring cursor on SRS Using select keys off the SRS current & enter.
- Power off the detector.
- Bring cursor on flow rate & reduce flow rate slowly in decreasing order as per given 1.5, 1.4, 1.2, 1.0, 0.8, 0.6,0.4,0.2,0.0 ml/min. After each entry press enter. Backpressure also reduced to zero.
- Turn off pump.
- Instead of mobile phase, connect to DI water (500 ml approximately).

Loose the printing purge screw & allow water to flow over pump heads for 10-15 min. at flow rate of 1 ml/min.

Controlling as40 Automated Sampler using Winchrome Software

Open Winex screen

Table 37.2. Go to schedule files (single click)

Sample name	Method file	Data file
Give sample name (sample kept in Ist vials)	Double Click on Ist line Select the method file according to the application i.e. Anion for anions, Cation for cations.	Double Click on Ist lineGive data file name to which it should save. Give directory i.e. Anion / Cation
Give sample name of the second vial		

After first entry is over-

Click on Edit >Copy

Source
1 to 1

Destination
2 to number of vials kept.
If 5 vials are kept (2 to 5)

Copy Options

1. Remove click (Tick) on sample name
2. Click on sequential Data file name
3. Click OK
   Give appropriate sample names for each files (vial)
   Go to file.Save as, give schedule
   File name
   e.g: Cation.sch
   (Over write an old file)

Your Schedule is Prepared

Go to Run Screen, single click on Run Icon
Go to file >Batch
Options > Click
Select Schedule file which is prepared earlier
Give number of vials to be processed ie 1 to 6
Click OK
Go to Run >Edit Run Time. Keep it 15.00 minute
Go to Operation >Click on' Unattended Mode'
I.e.??? Unattended Mode
Go to File >Batch >Start Batch
Click YES

YOUR OPTION ON WINCHROME IS OVER

Pump Option

1. Take cursor to Direct Control
2. By Select key, select 'Method 0'
   Press Enter

Detector Options

1. Take cursor to direct control
2. By Select key Select 'Method 0'

Autosampler Option

1. Keep bleed = ON
2. Inj Type = Loop
3. Inj Mode = Const
4. Inj/vial = 1

Keep all the vials in proper order

Click on Run>When Ready light glows

Click Load

The system will start and the work is over.

To take Printout of Report

After a Run is over

Go to OPTIMIZE.Go to file >Open data file. Select the data file to be opened.

Go to file>Open Method, Select appropriate Method

Go to Operation >Reprocess>View Report >Click YES

This will give you report concentration

Operational Procedure for Ion Chromatograph

- Switch on cim model
- Double click on winchrome ex
- Single click on run icon
- Open the method file
- File >file operation >anion.met>open
- Sample name
- Go to run >sample name>give the name >enter
- To view report after run is over
- Go to file >view report
- To print report
- Go to print>print report

Calibration Procedure

- Open data file
- Go to edit>components>select the peak with right mouse button taking the mouse on the peak
- Give the component name >amount>and keep the rt window.5
- Click add >ok
- Go to operation >reprocess
- Go to file >file operation >save method file .select the file (e.g. Anion) and save
- Gon to file >file operation >open method file
- Inject your sample.

Preperation of Mobile Phase

- 50% NaOH
- Take 50g NaOH

- Dissolve in water and make it 100 ml
- 50% of the NaOH solution is prepared.
- Take 1.44 ml of the 50% naoh.
- Make it to 1 litre.

Column Washing for Anion Columns

Wash for 40 to 60 minutes with 10X-mobile phase at flow rate 1.5 ml/min

Flush the column with De-ionized water for 15 minutes.

Prepare mobile phase and run the mobile phase. Allow 20-25 minutes to get the column equilibrated.

Suppressor Cleaning (through pump)

Disconnect the Suppressor from the cell

Now connect Eluent Out of Suppressor to (3) No port of injector.

Flush the Suppressor with the De-ionised water for 15 minutes.

Now reconnect the Suppressor in Normal Mode.

Mobile Phase for Anion and Cation

ANION: 2.7 mM Na_2CO_3 + 0.3 mM $NaHCO_3$

Dissolve 296.8 mg Na_2CO_3 and 82.1454 mg $NaHCO_3$ in De-ionised water and make it to 1 litre

CATION: 22mN Sulphuric acid

Dissolve 0.616 ml of concentrated H_2SO_4 in De-ionised water and make the volume upto 1 litre.

Suppressor Cleaning

For cleaning of Cation suppressor run sample (8 ml of 50% NaOH) kept in Eluent Reservoir for 15-20 minutes following above flow diagram. While running the solution connect 3 no port to Eluent out and let the Eluent in go to waste. Don't disturb Regeneration in and Regeneration out flow connection.

STANDARDS PREPARATION

Stock Standards

Table 37.3 : Prepare 1000 ppm standards of each of the following analytes by dissolving the corresponding mass of the salt in 1.000 litre of deionised water.

NaF	2.210g
$NaClO_3.2H_2O$	2.409g
$NaBrO_3$	1.180g
NaCl	1.648g
$NaNO_3$	1.500g
NaBr	1.288g
$NaClO_3$	1.276g
$NaNO_3$	1.271g
KH_2PO_4	1.433g
$NaSO_3$	1.574g
K_2SO_4	1.814g

Mixed Standards

A convenient mixed standards that gives comparable responses for the different analytes can be prepared by diluting the following amounts of the individual standards together to 1000 ppm.Appropriate mixed standard may be prepared from the 1000 ppm standards above.

Table 37.4

Volume of stock standards	Final mixed standard concentration
1 ml of stock	1.0 ppm F^-
5 ml stock	5.0 ppm ClO_2^-
5 ml stock	5.0 ppm Cl^-
6 ml stock	6.0 ppm NO_2^-
10 ml stock	10.0 ppm Br^-
15 ml stock	15.0 ppm ClO_3^-
15 ml stock	15.0 ppm NO_3^-
20 ml stock	20.0 ppm HPO_4–
25 ml stock	25.0 ppm SO_4–

Mobile Phase for Anion and Cation

ANION: 2.7 mM Na_2CO_3 + 0.3 mM $NaHCO_3$

Dissolve 296.8 mg Na_2CO_3 and 82.1454 mg $NaHCO_3$ in De-ionised water and make it to 1 litre.

CATION: 22mN Sulphuric acid

Dissolve 0.616 ml of concentrated H_2SO_4 in De-ionised water and make the volume upto1 litre.

Preparation of Standards (APHA,4-2,1995)

Standard anion solutions, 1000mg/L:

Prepare a series of standard anion solution by weighing the indicated amount of salt dried to a constant weight at 1050C to 1000ml.Store in plastic bottles in a refrigerator; these solutions are stable for atleast 1 month.Verify stability.

Table 37.5

Anion	Salt	Amount g/L
Cl^-	NaCl	1.6485
Br^-	NaBr	1.2876
NO_3^-	$NaNO_3$	1.3707
NO_2^-	$NaNO_2$	1.4998
PO_4^{3-}	KH_2PO_4	1.4330
SO_4^{2-}	K_2SO_4	1.8141

Combined working standard solution, high range

Combine 12 ml of standard solutions 1000mg/L of NO_2, NO_3, HPO_4, and Br, 20 ml of Cl, and 80 ml of SO_2. Dilute to 1000ml and store in a plastic bottle protected from light.Solution contain 12 mg/L of NO_2, NO_3, HPO_4 and Br , 20 mg/L of Cl,and 80 mg/L of SO_4. Prepare fresh daily.

Combined working standard solution, low range

Dilute 25 ml of the high range mixture to 100 ml and store in a plastic bottle protected from light. Solution contains 3mg/L each of NO_2, NO_3 HPO_4, and Br, 5 mg/L Cl, and 20 mg/L of SO_4. Prepare fresh daily.

APPLICATION NOTE

Ion Exchange Chromatography

Aim : To fractionate proteins (Immunoglobulins) by Ion Exchange Chromatography using DEAE-cellulose column.

Principle : Fractionation of immunoglobulins by ion exchange chromatography exploits differences in net charge between various classes of proteins. Depending upon the pH and ionic strength of medium used, different proteins would bind with ion exchange resins with varying strength. At pH values 8.0 and above most of immunoglobulins would have net negative charge. Since an anion exchanger like DEAE cellulose has a fixed positive charge, the negatively charged proteins bind to the matrix. Reducing the strength of binding between the resin and proteins can effect a gradual release of proteins from the column. This is conveniently done by increasing the ionic concentration of the medium. Proteins having comparatively less net negative charge would be eluted first and those with higher negative charges would follow.

Materials

- 1 ml and 10 ml pipettes
- Conical flask
- Beakers
- Latex tubing
- Test tubes
- 3 cm × 45 cm glass tube with sintered glass disc.
- Pinch cork
- Rubber cork
- Capillary tubing
- PBS pH 8.0
- DEAE-Cellulose
- 1 M Sodium chloride
- 1 M Sodium hydroxide
- 1 N Hydrochloric acid
- Dialysis Bag

Magnetic Stirrer and Bar

- Refrigerated centrifuge 5 ml immunoglobulin fraction obtained from ammonium sulphate precipitation

PROCEDURE

Preparation of DEAE-cellulose Column

(*a*) Regeneration of DEAE cellulose

1. Swell 25 gm of DEAE-cellulose in distilled water for 2-3 hours (20-25 mL/g)
2. Remove the supernatant
3. Stir sediment with 500 ml of 1M NaOH for about 30 minutes with intermittent stirring.
4. Let the mixture to stand for 30 min.
5. Remove the supernatant
6. Stir sediment with 500 ml 1M NaCl for about 30 minutes

7. Let the mixture stand for 30 minutes
8. Remove the supernatant
9. Stir the sediment with 500 ml of 1N HCl for 30 minutes
10. Wash the sediment a few times with distilled water and then with buffer to be used for column packing.

(*b*) Packing of a column

1. Mount vertiçally the column on to a stand.
2. Fill the column to about 10 cm from bottom with buffer and close the outlet.
3. Care should be taken to remove air bubbles from glass wool or from below the sintered glass disc and from flow regulating device.
4. Pour the slurry of DEAE-cellulose in the same buffer into column and avoid entrapment of air in the column. It is allowed to settle under gravity until about 5 cm of cellulose is packed.
5. Open the outlet to give average flow rate and more slurry is added as level in column can accommodate until height of packed bed reaches the required level of 30 cm.
6. Adjust the flow rate to 45 ml per hour and washing with the desired buffer should be continued until pH of the effluent is same as that of the buffer used.

Sample Application and Elution

1. Dialyse the sample (Immunoglobulin fraction prepared by ammonium sulphate precipitation) thoroughly against the buffer PBS pH 7.4.
2. Collect the contents of the dialysis bag in a centrifuge tube and centrifuge at 10000 rpm for 10 min. to remove any particulate matter. Collect the supernatant.
3. Allow the level of buffer in column to run down to level of packed DEAE-cellulose and close the outlet.
4. Place a circular disc of filter paper on the cellulose.
5. Apply 4.5 ml of sample carefully on the bed and open outlet till. Let the sample enter the bed.
6. Use few ml of buffer, PBS pH 8.0, to wash down sample from column sides into the bed and subsequently add more buffer for elution.
7. Collect 3ml fractions and measure the absorbance at 280nm.
8. Plot a graph of OD_{280} (Y-axis) against fraction number (X-axis). The first peak is IgG.
9. The remaining proteins attached to the column can be eluted by addition of buffer containing 1M NaCl.
10. The column can be washed with triple distilled water and stored at 4°C for subsequent use. The matrix may also be removed from the column, stored in a beaker and regenerated before use.

6. FLAME PHOTOMETER

Introduction

Flame photometer offers excellent possibilities for the analysis of metallic ions, which can be excited relatively low excitation levels available in the flame. This has been mostly due to the need for the simpler and more accurate methods for determination of ions of alkali metals.

Stages of Sample in Flame Photometer

1. Atomizer : Sample in the form of solution droplets broken into fine particles and wit the force of the compressed air or oxygen is carried to the flame. Flame and fuel gas where the fine particles of the sample are excited and impart characteristic color having certain wavelength.

2. Measurement of the intensity of colors : For measuring the color, photocell and galvanometer or micrometer is used.

Air at a given pressure of 5kg/cm^2 is blown into the atomizer and the section thus produced draws the sample, It is broken into fine particles and is carried out with the compressed air to the flame.

In the mixing chamber, the compressed air meets with the fuel gas at certain pressure and the mixture thus formed is fed into the burner giving out a characteristic flame.

Radiation's from the flame passes through the lens and iris diaphragm through a filter which allows the radiation characteristic of the element and fall on the photocell.

The optical path from the flame to photocell is enclosed in alight tight box so that no stray radiation reaches the photocell. The current development in the photocell is measured by galvanometer.

Preparation of Standard Solutions

Sodium: Dissolve 2.5419 gms of dried Sodium Chloride (NaCl) in distilled water to make 1L of solution. This will give 1000 ppm of Sodium solution. By diluting this solution we can get standard solutions of different lower concentration.

Potassium: Dissolve 1.9064 gms of dried Potassium Chloride (KCl) in distilled water to make 1L of solution. This is 1000ppm of Sodium solution. By diluting this solution we can get standard solutions of different lower concentration.

Advantages of Flame Phot meter

- Instrument is very easy to handle.
- Easy calibration.
- Easily available & cheap LPG gas is required to run the instrument.
- Maintenance is very easy.

Disadvantages of Flame Phot meter

- Every time calibration is required.
- For different metals separate calibration is required.

7. CIRCULAR DICHROISM*

Circular dichroism (CD) is a form of spectroscopy based on the differential absorption of left- and right-handed circularly polarized light. It can be used to help determine the structure of macromolecules (including the secondary structure of proteins and the handedness of DNA). CD was discovered by the French physicist Aimé Cotton in 1896.

Polarized Light

Linearly polarized light is polarized in a certain direction (that is, the magnitude of its electric field vector oscillates only in one plane, similar to a sine wave). In circularly polarized light, the electric field vector has a constant length, but rotates about its propagation direction. Hence it forms a helix in space while propagating. If this is a left-handed helix, the light is referred to as left circularly polarized, and vice versa for a right-handed helix.

The electric field of a light beam causes a linear displacement of charge when interacting with a molecule, whereas the magnetic field of it causes a circulation of charge. These two motions combined result in a helical displacement when light is impinged on a molecule. Since circularly polarized light itself is "chiral", it interacts differently with chiral molecules. That is, one of the two types of circularly polarized light are absorbed to different extents. In a CD experiment, equal amounts of left and right circularly polarized light are radiated into a (chiral) solution. One of the two types is absorbed more than the other one and this wavelength-dependent difference of absorption is measured, yielding the CD spectrum of the sample.

Due to the interaction with the molecule, the electric field vector of the light traces out an elliptical path while propagating.

Application to Biological Molecules

Visible CD spectroscopy is a very powerful technique to study metal–protein interactions and can resolve individual d-d electronic transitions as separate bands. CD spectra in the visible light region are only produced when a metal ion is in a chiral environment, thus, free metal ions in solution are not detected. This has the advantage of only observing the protein-bound metal, so pH dependence and stoichiometries are readily obtained. Optical activity in transition metal ion complexes have been attributed to configurational, conformational and the vicinal effects. Klewpatinond and Viles (2007) have produced a set of empirical rules for predicting the appearance of visible CD spectra for Cu^{2+} and Ni^{2+} square-planar complexes involving histidine and main-chain coordination.

CD gives less specific structural information than X-ray crystallography and protein NMR spectroscopy, for example, which both give atomic resolution data. However, CD spectroscopy is a quick method that does not require large amounts of proteins or extensive data processing. Thus CD can be used to survey a large number of solvent conditions, varying temperature, pH, salinity, and the presence of various cofactors.

CD spectroscopy is usually used to study proteins in solution, and thus it complements methods that study the solid state. This is also a limitation, in that many proteins are embedded in membranes in their native state, and solutions containing membrane structures are often strongly scattering. CD is sometimes measured in thin films.

Experimental Limitations

CD has also been studied in carbohydrates, but with limited success due to the experimental difficulties associated with measurement of CD spectra in the vacuum ultraviolet (VUV) region of the spectrum (100-200 nm), where the corresponding CD bands of unsubstituted carbohydrates lie. Substituted carbohydrates with bands above the VUV region have been successfully measured.

Measurement of CD is also complicated by the fact that typical aqueous buffer systems often absorb in the range where structural features exhibit differential absorption of circularly polarized light. Phosphate, sulfate, carbonate, and acetate buffers are generally incompatible with CD unless made extremely dilute e.g. in the 10-50 mM range. The TRIS buffer system should be completely avoided when performing far-UV CD. Borate and ammonium salts are often used to establish the appropriate pH range for CD experiments. Some experimenters have substituted fluoride for chloride ion because fluoride absorbs less in the far UV, and some have worked in pure water. Another, almost universal, technique is to minimize solvent absorption by using shorter path length cells when working in the far UV, 0.1 mm path lengths are not uncommon in this work.

In addition to measuring in aqueous systems, CD, particularly far-UV CD, can be measured in organic solvents e.g. ethanol, methanol,trifluoroethanol (or TFE). The latter has the advantage to induce structure formation of proteins, inducing beta-sheets in some and alpha helices in others, which they would not show under normal aqueous conditions. Most common organic solvents such as acetonitrile, THF, chloroform, dichloromethane are however, incompatible with far-UV CD.

It may be of interest to note that the protein CD spectra used in secondary structure estimation are related to the Π to Π^* orbital absorptions of the amide bonds linking the amino acids. These absorption bands lie partly in the so-called vacuum ultraviolet (wavelengths less than about 200 nm). The wavelength region of interest is actually inaccessible in air because of the strong absorption of light by oxygen at these wavelengths. In practice these spectra are measured not in vacuum but in an oxygen-free instrument (filled with pure nitrogen gas).

Once oxygen has been eliminated, perhaps the second most important technical factor in working below 200 nm is to design the rest of the optical system to have low losses in this region. Critical in this regard is the use of aluminized mirrors whose coatings have been optimized for low loss in this region of the spectrum.

The usual light source in these instruments is a high pressure, short-arc xenon lamp. Ordinary xenon arc lamps are unsuitable for use in the low UV. Instead specially constructed lamps with envelopes made from high-purity synthetic fused silica must be used. Light from synchrotron sources has a much higher flux at short wavelengths, and has been used to record CD down to 160 nm. Recently the CD spectrometer at the electron storage ring facility ISA at the University of Aarhus in Denmark was used to record solid state CD spectra down to 120 nm.

At the quantum mechanical level, the information content of circular dichroism and optical rotation are identical.

8. ELECTROANALYTICAL METHOD*

Electroanalytical methods are a class of techniques in analytical chemistry which study an analyte by measuring the potential (Volts) and/or current (Amps) in an electrochemical cell containing the analyte. These methods can be broken down into several categories depending on which aspects of the cell are controlled and which are measured. The three main categories are Potentiometry (the difference in electrode potentials is measured), Coulometry (the cell's current is measured over time), and Voltammetry (the cell's current is measured while actively altering the cell's potential).

(*i*) Potentiometry

Potentiometry passively measures the potential of a solution between two electrodes, affecting the solution very little in the process. The potential is then related to the concentration of one or more analytes. The cell structure used is often referred to as an electrode even though it actually contains two electrodes: an indicator electrode and a reference electrode (distinct from the reference electrode used in the three electrode system). Potentiometry usually uses electrodes made selectively sensitive to the ion of interest, such as a fluoride-selective electrode. The most common potentiometric electrode is the glass-membrane electrode used in a pH meter.

(*ii*) Coulometry

Coulometry uses applied current or potential to completely convert an analyte from one oxidation state to another. In these experiments, the total current passed is measured directly or indirectly to determine the number of electrons passed. Knowing the number of electrons passed can indicate the concentration of the analyte or, when the concentration is known, the number of electrons transferred in the redox reaction. Common forms of coulometry include bulk electrolysis, also known as Potentiostatic coulometry or controlled potential coulometry, as well as a variety of coulometric titrations.

(*iii*) Voltammetry

Voltammetry applies a constant and/or varying potential at an electrode's surface and measures the resulting current with a three electrode system. This method can reveal the reduction potential of an analyte and its electrochemical reactivity. This method in practical terms is nondestructive since only a very small amount of the analyte is consumed at the two-dimensional surface of the working and auxiliary electrodes. In practice the analyte solutions is usually disposed of since it is difficult to separate the analyte from the bulk electrolyte and the experiment requires a small amount of analyte. A normal experiment may involve 1-10 mL solution with an analyte concentration between 1-10 mM.

(*iv*) Polarometry

Polarometry is a subclass of voltammetry that uses a dropping mercury electrode as the working electrode. The auxiliary electrode is often the resulting mercury pool. Concern over the toxicity of mercury has caused the use of mercury electrodes to decrease greatly. Alternate electrode materials, such as the noble metals and glassy carbon, are affordable, inert, and easily cleaned.

(*v*) Amperometry

Most of *Amperometry* is now a subclass of voltammetry in which the electrode is held at constant potentials for various lengths of time. The distinction between amperometry and voltammetry is mostly historic. There was a time when it was difficult to switch between "holding" and "scanning" a potential. This function is trivial for modern potentiostats, and today there is little distinction between the techniques which either "hold", "scan", or do both during a single experiment. Yet the terminology still results in confusion, for example, differential pulse voltammetry is also referred to as and differential pulse amperometry. This experiment can be seen as the combination of linear sweep voltammetry and chronoamperometry thus the confusion in which category it should be named.

One advantage that distinguishes amperometry from other forms of voltammetry is that in amperometry, the currents are averaged (or summed) over time. In most of voltammetry, current readings must be considered independently at individual time intervals. The averaging used in amperometry gives these methods greater precision than the many individual readings of (other) voltammetric techniques.

Not all of the experiments which were historically amperometry now fall under the domain of voltammetry. In an amperometric titration, the current is measured, but this would not be considered voltammetry since the entire solution is transformed during the experiment. Amperometric titrations are instead a form of coulometry.

9. ELECTRON PARAMAGNETIC RESONANCE*

Electron paramagnetic resonance (EPR) or *electron spin resonance* (ESR) spectroscopy is a technique for studying chemical species that have one or more unpaired electrons, such as organic and inorganic free radicals or inorganic complexes possessing a transition metal ion.

The basic physical concepts of EPR are analogous to those of nuclear magnetic resonance (NMR), but it is electron spins that are excited instead of spins of atomic nuclei.

Because most stable molecules have all their electrons paired, the EPR technique is less widely used than NMR. However, this limitation to paramagnetic species also means that the EPR technique is one of great specificity, since ordinary chemical solvents and matrices do not give rise to EPR spectra.

EPR was first observed in Kazan State University by a Soviet physicist Yevgeny Zavoisky in 1944, and was developed independently at the same time by Brebis Bleaney at Oxford University.

EPR THEORY

Origin of an EPR Signal

Every electron has a magnetic moment and spin quantum number $s = 1/2$, with magnetic components $m_s = +1/2$ and $m_s = -1/2$. In the presence of an external magnetic field with strength B_0, the electron's magnetic moment aligns itself either parallel ($m_s = -1/2$) or antiparallel ($m_s = +1/2$) to the field, each alignment having a specific energy. The parallel alignment corresponds to the lower energy state, and the separation between it and the upper state is $\Delta E = g_e \mu_B B_0$, where g_e is the electron's so-called g-factor. This equation implies that the splitting of the energy levels is directly proportional to the magnetic field's strength.

An unpaired electron can move between the two energy levels by either absorbing or emitting electromagnetic radiation of energy $\varepsilon = h\nu$ such that the resonance condition, $\varepsilon = \Delta E$, is obeyed. Substituting in $\varepsilon = h\nu$ and $\Delta E = g_e \mu_B B_0$ leads to the fundamental equation of EPR spectroscopy: $h\nu = g_e \mu_B B_0$. Experimentally, this equation permits a large combination of frequency and magnetic field values, but the great majority of EPR measurements are made with microwaves in the 9000 – 10000 MHz (9 – 10 GHz) region, with fields corresponding to about 3500 G (0.35 T).

In principle, EPR spectra can be generated by either varying the photon frequency incident on a sample while holding the magnetic field constant, or doing the reverse. In practice, it is usually the frequency which is kept fixed. A collection of paramagnetic centers, such as free radicals, is exposed to microwaves at a fixed frequency. By increasing an external magnetic field, the gap between the $m_s = +1/2$ and $m_s = -1/2$ energy states is widened until it matches the energy of the microwaves, as represented by the double-arrow in the diagram above. At this point the unpaired electrons can move between their two spin states. Since there typically are more electrons in the lower state, due to the Maxwell-Boltzmann distribution, there is a net absorption of energy, and it is this absorption which is monitored and converted into a spectrum.

As an example of how $h\nu = g_e\mu_B B_0$ can be used, consider the case of a free electron, which has $g_e = 2.0023$, and the simulated spectrum shown at the right in two different forms. For the microwave frequency of 9388.2 MHz, the predicted resonance position is a magnetic field of about $B_0 = h\nu / g_e\mu_B =$ 0.3350 tesla = 3350 gauss, as shown. Note that while two forms of the same spectrum are presented in the figure, most EPR spectra are recorded and published only as first derivatives.

Because of electron-nuclear mass differences, the magnetic moment of an electron is substantially larger than the corresponding quantity for any nucleus, so that a much higher electromagnetic frequency is needed to bring about a spin resonance with an electron than with a nucleus, at identical magnetic field strengths. For example, for the field of 3350 G shown at the right, spin resonance occurs near 9388.2 MHz for an electron compared to only about 14.3 MHz for ^1H nuclei.

EPR Spectral Parameters

In real systems, electrons are normally not solitary, but are associated with one or more atoms. There are several important consequences of this:

1. An unpaired electron can gain or lose angular momentum, which can change the value of its g-factor, causing it to differ from g_e. This is especially significant for chemical systems with transition-metal ions.
2. If an atom with which an unpaired electron is associated has a non-zero nuclear spin, then its magnetic moment will affect the electron. This leads to the phenomenon of hyperfine coupling, analogous to J-coupling in NMR, splitting the EPR resonance signal into doublets, triplets and so forth.
3. Interactions of an unpaired electron with its environment influence the shape of an EPR spectral line. Line shapes can yield information about, for example, rates of chemical reactions.
4. The g-factor and hyperfine coupling in an atom or molecule may not be the same for all orientations of an unpaired electron in an external magnetic field. This anisotropy depends upon the electronic structure of the atom or molecule (e.g., free radical) in question, and so can provide information about the atomic or molecular orbital containing the unpaired electron.

EPR Applications

EPR spectroscopy is used in various branches of science, such as chemistry and physics, for the detection and identification of free radicals and paramagnetic centers. EPR is a sensitive, specific method for studying both radicals formed in chemical reactions and the reactions themselves. For example, when frozen water (solid H_2O) is decomposed by exposure to high-energy radiation, radicals such as H, OH, and HO_2 are produced. Such radicals can be identified and studied by EPR. Organic and inorganic radicals can be detected in electrochemical systems and in materials exposed to UV light. In many cases, the reactions to make the radicals and the subsequent reactions of the radicals are of interest, while in other cases EPR is used to provide information on a radical's geometry and the orbital of the unpaired electron.

Medical and biological applications of EPR also exist. Although radicals are very reactive and so do not normally occur in high concentrations in biology, special reagents have been developed to spin-label molecules of interest. These reagents are particularly useful in biological systems. Specially designed nonreactive radical molecules can attach to specific sites in a biological cell, and EPR spectra can then give information on the environment of these so-called spin-label or spin-probes.

EPR spectroscopy can only be applied to systems in which the balance between radical decay and radical formation keeps the free-radicals concentration above the detection limit of the spectrometer used. This can be a particularly severe problem in studying reactions in liquids. An alternative approach is to slow down reactions by studying samples held at cryogenic temperatures, such as 77 K (liquid nitrogen) or 4.2 K (liquid helium). An example of this work is the study of radical reactions in single crystals of amino acids exposed to x-rays, work that sometimes leads to activation energies and rate constants for radical reactions.

EPR also has been used by archaeologists for the dating of teeth. Radiation damage over long periods of time creates free radicals in tooth enamel, which can then be examined by EPR and, after proper calibration, dated. Alternatively, material extracted from the teeth of people during dental procedures can be used to quantify their cumulative exposure to ionizing radiation. People exposed to radiation from the Chernobyl disaster have been examined by this method.

Radiation-sterilized foods have been examined with EPR spectroscopy, the aim being to develop methods to determine if a particular food sample has been irradiated and to what dose.

High-field-high-frequency Measurements

High-field-high-frequency EPR measurements are sometimes needed to detect subtle spectroscopic details. However, for many years the use of electromagnets to produce the needed fields above 1.5 T was impossible, due principally to limitations of traditional magnet materials. The first multifunctional millimeter EPR spectrometer with a superconducting solenoid was described in the early 1970s by Prof. Y. S. Lebedev's group (Russian Institute of Chemical Physics, Moscow) in collaboration with L. G. Oranski's group (Ukrainian Physics and Technics Institute, Donetsk) which began working in the Institute of Problems of Chemical Physics, Chernogolovka around 1975. Two decades later, a W-band EPR spectrometer was produced as a small commercial line by the German Bruker Company [2], initiating the expansion of W-band EPR techniques into medium-sized academic laboratories. Today there still are only a few scientific centers in the world capable of high-field-high-frequency EPR, among them are the Grenoble High Magnetic Field Laboratory in Grenoble, France, the Physics Department in Freie Universität Berlin, the National High Magnetic Field Laboratory in Tallahassee, USA, the National Center for Advanced ESR Technology (ACERT) at Cornell University in Ithaca, USA, the IFW in Dresden, Germany, and the Institute of Physics of Complex Matter in Lausanne in Switzerland.

EPR spectra of a nitroxide radical as a function of frequency. Note the improvement in resolution from left to right. See this page for an animation of the above figure.

The EPR waveband is stipulated by the frequency or wavelength of a spectrometer's microwave source.

EPR experiments often are conducted at X and, less commonly, Q bands, mainly due to the ready availability of the necessary microwave components (which originally were developed for radar applications). A second reason for widespread X and Q band measurements is that electromagnets can reliably generate fields up to about 1 tesla. However, the low spectral resolution over g-factor at these wavebands limits the study of paramagnetic centers with comparatively low anisotropic magnetic parameters. Measurements at $v > 40$ GHz, in the millimeter wavelength region, offer the following advantages:

1. EPR spectra are simplified due to the reduction of second-order effects at high fields.
2. Increase in orientation selectivity and sensitivity in the investigation of disordered systems.
3. The informativity and precision of pulse methods, e.g., ENDOR also increase at high magnetic fields.
4. Accessibility of spin systems with larger zero-field splitting due to the larger microwave quantum energy hv.

5. The higher spectral resolution over g-factor, which increases with irradiation frequency í and external magnetic field B_0. This is used to investigate the structure, polarity, and dynamics of radical microenvironments in spin-modified organic and biological systems through the spin label and probe method. The figure shows how spectral resolution improves with increasing frequency.
6. Saturation of paramagnetic centers occurs at a comparatively low microwave polarizing field B_1, due to the exponential dependence of the number of excited spins on the radiation frequency v. This effect can be successfully used to study the relaxation and dynamics of paramagnetic centers as well as of superslow motion in the systems under study.
7. The cross-relaxation of paramagnetic centers decreases dramatically at high magnetic fields, making it easier to obtain more-precise and more-complete information about the system under study.[5]

10. ELECTROPHORESIS*

Electrophoresis is the most well-known electro-kinetic phenomenon.

It was discovered by Reuss in 1809. He observed that clay particles dispersed in water migrate under influence of an applied electric field. There are detailed descriptions of Electrophoresis in many books on Colloid and Interface Science. There is an IUPAC Technical Report prepared by a group of most known world experts on the electro-kinetic phenomena. Generally, electrophoresis is the motion of dispersed particles relative to a fluid under the influence of an electric field that is space uniform. Alternatively, similar motion in a space non-uniform electric field is called dielectrophoresis. Electrophoresis occurs because particles dispersed in a fluid almost always carry an electric surface charge. An electric field exerts electrostatic Coulomb force on the particles through these charges. Recent molecular dynamics simulations, though, suggest that surface charge is not always necessary for electrophoresis and that even neutral particles can show electrophoresis due to the specific molecular structure of water at the interface. The electrostatic Coulomb force exerted on a surface charge is reduced by an opposing force which is electrostatic as well. According to double layer theory, all surface charges in fluids are screened by a diffuse layer. This diffuse layer has the same absolute charge value, but with opposite sign from the surface charge. The electric field induces force on the diffuse layer, as well as on the surface charge. The total value of this force equals to the first mentioned force, but it is oppositely directed. However, only part of this force is applied to the particle. It is actually applied to the ions in the diffuse layer. These ions are at some distance from the particle surface. They transfer part of this electrostatic force to the particle surface through viscous stress. This part of the force that is applied to the particle body is called *electrophoretic retardation force*.

There is one more electric force, which is associated with deviation of the double layer from spherical symmetry and surface conductivity due to the excess ions in the diffuse layer. This force is called the *electrophoretic relaxation force*.

All these forces are balanced with hydrodynamic friction, which affects all bodies moving in viscous fluids with low Reynolds number.

The speed of this motion v is proportional to the electric field strength E if the field is not too strong. Using this assumption makes possible the introduction of electrophoretic mobility μ_e as coefficient of proportionality between particle speed and electric field strength:

$$\mu_e = \frac{v}{E}$$

Multiple theories were developed during 20th century for calculating this parameter. Ref. 2 provides an overview.

Theory

The most known and widely used theory of electrophoresis was developed by Smoluchowski in 1903 [10]

$$\mu_e = \frac{\varepsilon\varepsilon_0\zeta}{\eta}$$

where ε is the dielectric constant of the dispersion medium, ε_0 is the permittivity of free space ($C^2 N^{-1} m^{-2}$), η is dynamic viscosity of the dispersion medium (Pa s), and ζ is zeta potential (*i.e.*, the electrokinetic potential of the slipping plane in the double layer).

Smoluchowski theory is very powerful because it works for dispersed particles of any shape and any concentration, when it is valid. Unfortunately it has limitations of its validity. It follows, for instance, from the fact that it does not include Debye length κ^{-1}. However, Debye length must be important for electrophoresis, as follows immediately from the Figure on the right. Increasing thickness of the DL leads to removing point of retardation force further from the particle surface. The thicker DL, the smaller retardation force must be.

Detail theoretical analysis proved that Smoluchowski theory is valid only for sufficiently thin DL, when Debye length is much smaller than particle radius a:

$$\kappa a >> 1$$

This model of "thin Double Layer" offers tremendous simplifications not only for electrophoresis theory but for many other electrokinetic theories. This model is valid for most aqueous systems because the Debye length is only a few nanometers there. It breaks only for nano-colloids in solution with ionic strength close to water

Smoluchowski theory also neglects contribution of surface conductivity. This is expressed in modern theory as condition of small Dukhin number

$$Du << 1$$

Creation of electrophoretic theory with wider range of validity was a purpose of many studies during 20th century.

One of the most known considers an opposite asymptotic case when Debye length is larger than particle radius:

$$\kappa a < 1$$

It is called the "thick Double Layer" model. Corresponding electrophoretic theory was created by Huckel in 1924. It yields the following equation for electrophoretic mobility:

This model can be useful for some nano-colloids and non-polar fluids, where Debye length is much larger.

There are several analytical theories that incorporate surface conductivity and eliminate restriction of the small Dukhin number. Early pioneering work in that direction dates back to Overbeek [12] and Booth.

Modern, rigorous theories that are valid for any Zeta potential and often any κa, stem mostly from the Ukrainian (Dukhin, Shilov and others) and Australian (O'Brien, White, Hunter and others) Schools.

Historically the first one was Dukhin-Semenikhin theory. Similar theory was created 10 years later by O'Brien and Hunter. Assuming *thin Double Layer,* these theories would yield results that are very close to the numerical solution provided by O'Brien and White.

(*i*) Gel electrophoresis

Gel electrophoresis apparatus - An agarose gel is placed in this buffer-filled box and electrical current is applied via the power supply to the rear. The negative terminal is at the far end (black wire), so DNA migrates toward the camera.

Gel electrophoresis is a technique used for the separation of deoxyribonucleic acid, ribonucleic acid, or protein molecules using an electric current applied to a gel matrix. It is usually performed for analytical purposes, but may be used as a preparative technique prior to use of other methods such as mass spectrometry, RFLP, PCR, cloning, DNA sequencing, or Southern blotting for further characterization.

Separation

The term "gel" in this instance refers to the matrix used to contain, then separate the target molecules. In most cases the gel is a crosslinked polymer whose composition and porosity is chosen based on the specific weight and composition of the target to be analyzed. When separating proteins or small nucleic acids (DNA, RNA, or oligonucleotides) the gel is usually composed of different concentrations of acrylamide and a cross-linker, producing different sized mesh networks of polyacrylamide. When separating larger nucleic acids (greater than a few hundred bases), the preferred matrix is purified agarose. In both cases, the gel forms a solid, yet porous matrix. Acrylamide, in contrast to polyacrylamide, is a neurotoxin and must be handled using appropriate safety precautions to avoid poisoning.

"Electrophoresis" refers to the electromotive force (EMF) that is used to move the molecules through the gel matrix. By placing the molecules in wells in the gel and applying an electric current, the molecules will move through the matrix at different rates, usually determined by mass, toward the positive anode if negatively charged or toward the negative cathode if positively charged.

Visualization

Agarose gel prepared for DNA analysis : The first lane contains a DNA ladder for sizing, and the other four lanes show variously-sized DNA fragments that are present in some but not all of the samples.

After the electrophoresis is complete, the molecules in the gel can be stained to make them visible. Ethidium bromide, silver, or coomassie blue dye may be used for this process. Other methods may also be used to visualize the separation of the mixture's components on the gel. If the analyte molecules fluoresce under ultraviolet light, a photograph can be taken of the gel under ultraviolet lighting conditions. If the molecules to be separated contain radioactivity added for visibility, an autoradiogram can be recorded of the gel.

If several mixtures have initially been injected next to each other, they will run parallel in individual lanes. Depending on the number of different molecules, each lane shows separation of the components from the original mixture as one or more distinct bands, one band per component. Incomplete separation of the components can lead to overlapping bands, or to indistinguishable smears representing multiple unresolved components.

Bands in different lanes that end up at the same distance from the top contain molecules that passed through the gel with the same speed, which usually means they are approximately the same size. There are molecular weight size markers available that contain a mixture of molecules of known sizes. If such a marker was run on one lane in the gel parallel to the unknown samples, the bands observed can be compared to those of the unknown in order to determine their size. The distance a band travels is approximately inversely proportional to the logarithm of the size of the molecule.

Applications

Gel electrophoresis is used in forensics, molecular biology, genetics, microbiology and biochemistry. The results can be analyzed quantitatively by visualizing the gel with UV light and a gel imaging device. The image is recorded with a computer operated camera, and the intensity of the band or spot of interest is measured and compared against standard or markers loaded on the same gel. The measurement and analysis are mostly done with specialized software.

Depending on the type of analysis being performed, other techniques are often implemented in conjunction with the results of gel electrophoresis, providing a wide range of field-specific applications.

Nucleic Acids

In the case of nucleic acids, the direction of migration, from negative to positive electrodes, is due to the naturally-occurring negative charge carried by their sugar-phosphate backbone.

Double-stranded DNA fragments naturally behave as long rods, so their migration through the gel is relative to their radius of gyration, or, for non-cyclic fragments, size. Single-stranded DNA or RNA tend to fold up into molecules with complex shapes and migrate through the gel in a complicated manner based on their tertiary structure. Therefore, agents that disrupt the hydrogen bonds, such as sodium hydroxide or formamide, are used to denature the nucleic acids and cause them to behave as long rods again.

Gel electrophoresis of large DNA or RNA is usually done by agarose gel electrophoresis. See the "Chain termination method" page for an example of a polyacrylamide DNA sequencing gel.

Proteins

SDS-PAGE autoradiography : The indicated proteins are present in different concentrations in the two samples.

Proteins, unlike nucleic acids, can have varying charges and complex shapes, therefore they may not migrate into the gel at similar rates, or at all, when placing a negative to positive EMF on the sample. Proteins therefore, are usually denatured in the presence of a detergent such as sodium dodecyl sulfate/ sodium dodecyl phosphate (SDS/SDP) that coats the proteins with a negative charge.[1] Generally, the amount of SDS bound is relative to the size of the protein (usually 1.4g SDS per gram of protein), so that the resulting denatured proteins have an overall negative charge, and all the proteins have a similar charge to mass ratio. Since denatured proteins act like long rods instead of having a complex tertiary shape, the rate at which the resulting SDS coated proteins migrate in the gel is relative only to its size and not its charge or shape.

Proteins are usually analyzed by sodium dodecyl sulfate polyacrylamide gel electrophoresis (SDS-PAGE), by native gel electrophoresis, by quantitative preparative native continuous polyacrylamide gel electrophoresis (QPNC-PAGE), or by 2-D electrophoresis.

(*ii*) Capillary Electrophoresis

Capillary electrophoresis (CE), also known as capillary zone electrophoresis (CZE), can be used to separate ionic species by their charge and frictional forces. In traditional electrophoresis, electrically charged analytes move in a conductive liquid medium under the influence of an electric field. Introduced in the 1960s, the technique of capillary electrophoresis (CE) was designed to separate species based on their size to charge ratio in the interior of a small capillary filled with an electrolyte. While its use has been sporadic, CE offers unparalleled resolution and selectivity allowing for separation of analytes with very little physical difference. Efficiencies of millions of plates are routinely reported. Once thought impossible, separation of large proteins differing in only one amino acid (ie. D-Lysine substituted for L-Lysine) and even an isotopic separation of ^{14}N and ^{15}N ammonium hydroxide have been reported. No other technique has shown such powerful selectivity with the ability for extremely high sensitivity. As few as 6 molecules of a substance have been separated and detected with the help of laser-induced fluorescence (LIF).

Instrumentation

The instrumentation needed to perform capillary electrophoresis is relatively simple. The system's main components are a sample vial, source and destination vials, a capillary, electrodes, a high-voltage power supply, a detector, and a data output and handling device. The source vial, destination vial and capillary are filled with an electrolyte such as an aqueous buffer solution. To introduce the sample, the capillary inlet is placed into a vial containing the sample and then returned to the source vial (sample is introduced into the capillary via capillary action, pressure, or siphoning). The migration of the analytes is then initiated by an electric field that is applied between the source and destination vials and is supplied to the electrodes by the high-voltage power supply. It is important to note that all ions, positive or negative, are pulled through the capillary in the same direction by electroosmotic flow, as will be explained. The analytes separate as they migrate due to their electrophoretic mobility, as will be explained, and are detected near the outlet end of the capillary. The output of the detector is sent to a data output and handling device such as an integrator or computer. The data is then displayed as an electropherogram, which reports detector response as a function of time. Separated chemical compounds appear as peaks with different retention times in an electropherogram.

Detection

Separation by capillary electrophoresis can be detected by several detection devices. The majority of commercial systems use UV or UV-Vis absorbance as their primary mode of detection. In these systems, a section of the capillary itself is used as the detection cell. The use of on-tube detection enables detection of separated analytes with no loss of resolution. In general, capillaries used in capillary electrophoresis are coated with a polymer for increased stability. The portion of the capillary used for UV detection, however, must be optically transparent. Bare capillaries can break relatively easily and, as a result, capillaries with transparent coatings are available to increase the stability of the cell window. The path length of the detection cell in capillary electrophoresis (~ 50 micrometers) is far less than that of a traditional UV cell (~ 1 cm). According to the Beer-Lambert law, the sensitivity of the detector is proportional to the path length of the cell. To improve the sensitivity, the path length can be increased, though this results in a loss of resolution.

Fluorescence detection can also be used in capillary electrophoresis for samples that naturally fluoresce or are chemically modified to contain fluorescent tags. This mode of detection offers high sensitivity and improved selectivity for these samples, but cannot be utilized for samples that do not fluoresce. The set-up for fluorescence detection in a capillary electrophoresis system can be complicated. The method requires that the light beam be focused on the capillary, which can be difficult for many light sources. Laser-induced fluorescence has been used in CE systems with detection limits as low as 10^{-18} to 10^{-21} mol. The sensitivity of the technique is attributed to the high intensity of the incident light and the ability to accurately focus the light on the capillary.

In order to obtain the identity of sample components, capillary electrophoresis can be directly coupled with mass spectrometers or Surface Enhanced Raman Spectroscopy (SERS). In most systems, the capillary outlet is introduced into an ion source that utilizes electrospray ionization (ESI). The resulting ions are then analyzed by the mass spectrometer. This set-up requires volatile buffer solutions, which will affect the range of separation modes that can be employed and the degree of resolution that can be achieved. The measurement and analysis are mostly done with a specialized gel analysis software.

For CE-SERS, capillary electrophoresis eluants can be deposited onto a SERS-active substrate. Analyte retention times can be translated into spatial distance by moving the SERS-active substrate at a constant rate during capillary electrophoresis. This allows the subsequent spectroscopic technique to be applied to specific eluants for identification with high sensitivity. SERS-active substrates can be chosen that do not interfere with the spectrum of the analytes.

Modes of Separation

The separation of compounds by capillary electrophoresis is dependent on the differential migration of analytes in an applied electric field. The electrophoretic migration velocity (u_p) of an analyte toward the electrode of opposite charge is:

$$u_p = \mu_p E$$

where μ_p is the electrophoretic mobility and E is the electric field strength. The electrophoretic mobility is proportional to the ionic charge of a sample and inversely proportional to any frictional forces present in the buffer. When two species in a sample have different charges or experience different frictional forces, they will separate from one another as they migrate through a buffer solution. The frictional forces experienced by an analyte ion depend on the viscosity (η) of the medium and the size and shape of the ion.[3] Accordingly, the electrophoretic mobility of an analyte at a given pH is given by:

$$\mu_p = \frac{z}{6\pi\eta r}$$

where z is the net charge of the analyte and r is the Stokes radius of the analyte. The Stokes radius is given by:

$$r = \frac{k_B T}{6\pi\eta D}$$

where k_B is the Boltzmann constant, and T is the temperature, D is the diffusion coefficient. These equations indicate that the electrophoretic mobility of the analyte is proportional to the charge of the analyte and inversely proportional to its radius. The electrophoretic mobility can be determined experimentally from the migration time and the field strength:

$$\mu_p = \left(\frac{L}{t_r}\right)\left(\frac{L_t}{V}\right)$$

where L is the distance from the inlet to the detection point, t_r is the time required for the analyte to reach the detection point (migration time), V is the applied voltage (field strength), and L_t is the total length of the capillary. Since only charged ions are affected by the electric field, neutral analytes are poorly separated by capillary electrophoresis.

The velocity of migration of an analyte in capillary electrophoresis will also depend upon the rate of electroosmotic flow (EOF) of the buffer solution. In a typical system, the electroosmotic flow is directed toward the negatively charged cathode so that the buffer flows through the capillary from the source vial to the destination vial. Separated by differing electrophoretic mobilities, analytes migrate toward the electrode of opposite charge. As a result, negatively charged analytes are attracted to the positively charged anode, counter to the EOF, while positively charged analytes are attracted to the cathode, in agreement with the EOF.

Their respective electrophoretic and electroosmotic flow mobilities

The velocity of the electroosmotic flow, u_o can be written as:

$$u_o = \varepsilon_o E$$

where μ_o is the electroosmotic mobility, which is defined as:

$$\mu_0 = \frac{\varepsilon\zeta}{\eta}$$

where ζ is the zeta potential of the capillary wall, and ε is the relative permittivity of the buffer solution. Experimentally, the electroosmotic mobility can be determined by measuring the retention time of a neutral analyte.[3] The velocity (u) of an analyte in an electric field can then be defined as:

$$u_p + u_o = (\mu_p + \mu_o)E$$

Since the electroosmotic flow of the buffer solution is generally greater than that of the electrophoretic flow of the analytes, all analytes are carried along with the buffer solution toward the cathode. Even small, triply charged anions can be redirected to the cathode by the relatively powerful EOF of the buffer solution. Negatively charged analytes are retained longer in the capilliary due to their conflicting electrophoretic mobilities.

Electroosmotic flow is observed when an electric field is applied to a solution in a capillary that has fixed charges on its interior wall. Charge is accumulated on the inner surface of a capillary when a buffer solution is placed inside the capillary. In a fused-silica capillary, silanol (Si-OH) groups attached to the interior wall of the capillary are ionized to negatively charged silanoate ($Si\text{-}O^-$) groups at pH values greater than three. The ionization of the capillary wall can be enhanced by first running a basic solution, such as NaOH or KOH through the capillary prior to introducing the buffer solution. Attracted to the negatively charged silanoate groups, the positively charged cations of the buffer solution will form two inner layers of cations (called the diffuse double layer or the electrical double layer) on the capillary wall. The first layer is referred to as the fixed layer because it is held tightly to the silanoate groups. The outer layer, called the mobile layer, is farther from the silanoate groups. The mobile cation layer is pulled in the direction of the negatively charged cathode when an electric field is applied. Since these cations are solvated, the bulk buffer solution migrates with the mobile layer, causing the electroosmotic flow of the buffer solution. Other capillaries including Teflon capillaries also exhibit electroosmotic flow. The EOF of these capillaries is probably the result of adsorption of the electrically charged ions of the buffer onto the capillary walls. The rate of EOF is dependent on the field strength and the charge density of the capillary wall. The wall's charge density is proportional to the pH of the buffer solution. The electroosmotic flow will increase with pH until all of the available silanols lining the wall of the capillary are fully ionized.

Efficiency and Resolution

The number of theoretical plates, or separation efficiency, in capillary electrophoresis is given by:

$$N = \frac{\mu D}{2D_m}$$

where N is the number of theoretical plates, μ is the apparent mobility in the separation medium and D_m is the diffusion coefficient of the analyte. According to this equation, the efficiency of separation is only limited by diffusion and is proportional to the strength of the electric field. The efficiency of capillary electrophoresis separations is typically much higher than the efficiency of other separation techniques like HPLC. Unlike HPLC, in capillary electrophoresis there is no mass transfer between phases. In addition, the flow profile in EOF-driven systems is flat, rather than the rounded laminar flow profile characteristic of the pressure-driven flow in chromatography columns. As a result, EOF does not significantly contribute to band broadening as in pressure-driven chromatography. Capillary electrophoresis separations can have several hundred thousand theoretical plates.

The resolution (R_s) of capillary electrophoresis separations can be written as:

$$R_s = \frac{1}{4}\left(\frac{\Delta\mu_p\sqrt{N}}{\mu_p + \mu_0}\right)$$

According to this equation, maximum resolution is reached when the electrophoretic and electroosmotic mobilities are similar in magnitude and opposite in sign. In addition, it can be seen that high resolution requires lower velocity and, correspondingly, increased analysis time.

(*iii*) Immunoelectrophoresis (IEP)

Immunoelectrophoresis (IEP), also known as gamma globulin electrophoresis, or immunoglobulin electrophoresis, or IEP, was developed by French immunologist Petr Nikolaevich Grabar in the 1950s. It is a method of determining the blood levels of three major immunoglobulins: immunoglobulin M (IgM), immunoglobulin G (IgG), and immunoglobulin A (IgA). It combines electrophoresis and immunodiffusion technique. To run this test, a gel is prepared with alternating wells and slots cut into it. The antigen mixture (usually a serum sample) is placed in the wells and electrophoresis is performed to separate the proteins in the sample. Then an antibody or mixture of antibodies is added to the slots in the gel. The separated antigens and the antibodies migrate toward one another and form precipitin arcs at the region of optimal concentrations.

Immunoelectrophoresis is the electrophoresis of a determined antigen mixture in an agarose gel that allows the separation of different proteins along the gel slide, and then the lateral diffusion in the gel of an immune serum or a monoclonal antibody. Antibodies specific to the antigens form prepicitation arcs seen over a dark background, or by Coomassie blue staining.

This method was very useful to determine the number of antigens recognised against a particular mosaic antigen (crude parasite extracts, for example) but today the Western blot method is preferred because the apparent molecular weight of each antigen can be determined and it is more sensitive, but IEF are still useful when non reducing conditions are needed.

Purpose

Immunoelectrophoresis is a powerful analytical technique with high resolving power as it combines separation of antigens by electrophoresis with immunodiffusion against an antiserum.

The increased resolution is of benefit in the immunological examination of serum proteins. Immunoelectrophoresis aids in the diagnosis and evaluation of the therapeutic response in many disease states affecting the immune system. It is usually requested when a different type of electrophoresis, called a serum protein electrophoresis, has indicated a rise at the immunoglobulin level. Immunoelectrophoresis is also used frequently to diagnose multiple myeloma, a disease affecting the bone marrow.

Precautions

Drugs that may cause increased immunoglobulin levels include therapeutic gamma globulin, hydralazine, isoniazid, phenytoin (Dilantin), procainamide, oral contraceptives, methadone, steroids, and tetanus toxoid and antitoxin.

The laboratory should be notified if the patient has received any vaccinations or immunizations in the six months before the test.

This is mainly because prior immunizations lead to the increased immunoglobulin levels resulting in false positive results.

It should be noted that, because immunoelectrophoresis is not quantitative, it is being replaced by a procedure called immunofixation, which is more sensitive and easier to interpret.

Description

Serum proteins separate in agar gels under the influence of an electric field into albumin, alpha 1, alpha 2, and beta and gamma globulins.

Immunoelectrophoresis is performed by placing serum on a slide containing a gel designed specifically for the test.

An electric current is then passed through the gel, and immunoglobulins, which contain an electric charge, migrate through the gel according to the difference in their individual electric charges.

Antiserum is placed alongside the slide to identify the specific type of immunoglobulin present.

The results are used to identify different disease entities, and to aid in monitoring the course of the disease and the therapeutic response of the patient to such conditions as immune deficiencies, autoimmune disease, chronic infections, chronic viral infections, and intrauterine fetal infections.

There are five classes of antibodies: IgM, IgG, IgA, IgE, and IgD.

IgM is produced upon initial exposure to an antigen.

For example, when a person receives the first tetanus vaccination, anti tetanus antibodies of the IgM class are produced 10 to 14 days later.

IgM is abundant in the blood but is not normally present in organs or tissues.

IgM is primarily responsible for ABO blood grouping and rheumatoid factor, yet is involved in the immunologic reaction to other infections, such as hepatitis.

Since IgM does not cross the placenta, an elevation of this immunoglobulin in the newborn indicates intrauterine infection such as rubella, cytomegalovirus (CMV) or a sexually transmitted disease (STD).

IgG is the most prevalent type of antibody, comprising approximately 75% of the serum immunoglobulins.

IgG is produced upon subsequent exposure to an antigen.

As an example, after receiving a second tetanus shot, or booster, a person produces IgG antibodies in five to seven days.

IgG is present in both the blood and tissues, and is the only antibody to cross the placenta from the mother to the fetus.

Maternal IgG protects the newborn for the first months of life, until the infant's immune system produces its own antibodies.

IgA constitutes approximately 15% of the immunoglobulins within the body.

Although it is found to some degree in the blood, it is present primarily in the secretions of the respiratory and gastrointestinal tract, in saliva, colostrum (the yellowish fluid produced by the breasts during late pregnancy and the first few days after childbirth), and in tears.

IgA plays an important role in defending the body against invasion of germs through the mucous membrane-lined organs.

IgE is the antibody that causes acute allergic reactions; it is measured to detect allergic conditions.

IgD, which constitutes the smallest portion of the immunoglobulins, is rarely evaluated or detected, and its function is not well understood.

Preparation

This test requires a blood sample.

Aftercare

Because this test is ordered when either very low or very high levels of immunoglobulins are suspected, the patient should be alert for any signs of infection after the test, including fever, chills, rash, or skin ulcers. Any bone pain or tenderness should also be immediately reported to the physician.

Risks

Risks for this test are minimal, but may include slight bleeding from the blood-drawing site, fainting or feeling lightheaded after venipuncture, or bruising.

Normal Results

Reference ranges vary from laboratory to laboratory and depend upon the method used. For adults, normal values are usually found within the following ranges (1 mg = approximately 0.000035 oz. and 1dL = approximately 0.33 oz.):

- IgM : 60-290 mg/dL
- IgG : 700-1,800 mg/dL
- IgA : 70-440 mg/dL

Abnormal Results

Increased IgM levels can indicate Waldenström's macroglobulinemia, a malignancy caused by secretion of IgM at high levels by malignant lymphoplasma cells.Increased IgM levels can also indicate chronic infections, such as hepatitis or mononucleosis and autoimmune diseases, like rheumatoid arthritis.

Decreased IgM levels can be indicative of AIDS, immunosuppression caused by certain drugs like steroids or dextran, or leukemia.

Increased levels of IgG can indicate chronic liver disease, autoimmune diseases, hyperimmunization reactions, or certain chronic infections, such as tuberculosis or sarcoidosis.

Decreased levels of IgG can indicate Wiskott-Aldrich syndrome, a genetic deficiency caused by inadequate synthesis of IgG and other immunoglobulins. Decreased IgG can also be seen with AIDS and leukemia.

Increased levels of IgA can indicate chronic liver disease, chronic infections, or inflammatory bowel disease.

Decreased levels of IgA can be found in ataxia, a condition affecting balance and gait, limb or eye movements, speech, and telangiectasia, an increase in the size and number of the small blood vessels in an area of skin, causing redness. Decreased IgA levels are also seen in conditions of low blood protein (hypoproteinemia), and drug immuno-suppression.

(*iv*) Isoelectric Focusing (IEF)

Isoelectric focusing (IEF), also known as electrofocusing, is a technique for separating different molecules by their electric charge differences. It is a type of zone electrophoresis, usually performed in a gel, that takes advantage of the fact that a molecule's charge changes with the pH of its surroundings. A protein that is in a pH region below its isoelectric point (pI) will be positively charged and so will migrate towards the cathode. As it migrates, however, the charge will decrease until the protein reaches the pH region that corresponds to its pI. At this point it has no net charge and so migration ceases. As a result, the proteins become focused into sharp stationary bands with each protein positioned at a point in the pH gradient corresponding to its pI. The technique is capable of extremely high resolution with proteins differing by a single charge being fractionated into separate bands.

Molecules to be focused are distributed over a medium that has a pH gradient (usually created by aliphatic ampholytes). An electric current is passed through the medium, creating a "positive" anode and "negative" cathode end. Negatively charged molecules migrate through the pH gradient in the medium toward the "positive" end while positively charged molecules move toward the "negative" end. As a particle moves towards the pole opposite of its charge it moves through the changing pH gradient until it reaches a point in which the pH of that molecules isoelectric point is reached. At this point the molecule no longer has a net electric charge (due to the protonation or deprotonation of the associated functional groups) and as such will not proceed any further within the gel. The gradient is initially established before adding the particles of interest by first subjecting a solution of small molecules such as polyampholytes with varying pI values to electrophoresis.

The method is applied particularly often in the study of proteins, which separate based on their relative content of acidic and basic residues, whose value is represented by the pI. Proteins are introduced into a Immobilized pH gradient gel composed of polyacrylamide, starch, or agarose where a pH gradient has been established. Gels with large pores are usually used in this process to eliminate any "sieving" effects, or artifacts in the pI caused by differing migration rates for proteins of differing sizes. Isoelectric focusing can resolve proteins that differ in pI value by as little as 0.01.[1] Isoelectric focusing is the first step in two-dimensional gel electrophoresis, in which proteins are first separated by their pI and then further separated by molecular weight through SDS-PAGE.

Principle

Isoelectric focusing (IEF) is anelectrophoretic separationbased on the isoelectric pointsof proteins. The pI is the pointat which the protein has anoverall net charge of zero. Differences of only a few hundredthsof a pH-unit in isoelectricpoints are sufficient to resolve proteins fromeach other. IEF is used as an alternative electrophoresisformat complementing the widelyused SDS-PAGE electrophoresis, which is basedon size of proteins. The combination of bothprinciples is applied in 2D-electrophoresis as outlinedbelow. This technique has becomecrucial for the development of Proteomics.In many applications, closely related proteinshave to be separated. Examples are the differentiationof protein isoforms or enantiomeres. Suchproblems have been successfully solved by isoelectricfocusing (IEF). In contrast to other electrophoretic techniques,pH is not kept constant throughout the wholesystem. Instead, the sample components migrateelectrophoretically in a stationary pH-gradient.Proteins will migrate until they reach the pHpointin the gradient at which the charge of theprotein equals zero (pH = pI) The protein is saidto focus at this point.

Advantages

Optimal resolution in other electrophoreticsystems requires applicationof sample as a narrowzone. In IEF this is not so crucial, even large sample volumes donot lower resolution. Sampleconcentration upon focusingalso results in lower detectionlimits compared with other electrophoresistechniques. Another advantage of this technologyis that separation in IEF does not requiredenaturation of proteins, thus any kind of subsequentinvestigations, such as activity staining (e.g. to find separated enzymes) or antibodydetection, is not hindered.Compared to alternative methods (e.g. PCRbased), IEF offers the following advantages

- efficient
- expressive
- economic (no sophisticated equipment required)
- easy (clear, one-dimensional separation principle)
- fast
- sensitive

11. NUCLEAR MAGNETIC RESONANCE SPECTROSCOPY*

Nuclear magnetic resonance (NMR) is a physical phenomenon based upon the quantum mechanical magnetic properties of an atom's nucleus. NMR also commonly refers to a family of scientific methods that exploit nuclear magnetic resonance to study molecules.

All nuclei that contain odd numbers of protons or neutrons have an intrinsic magnetic moment and angular momentum. The most commonly measured nuclei are hydrogen-1 (the most receptive isotope at natural abundance) and carbon-13, although nuclei from isotopes of many other elements (e.g.^{113}Cd, ^{15}N, ^{14}N ^{19}F, ^{31}P, ^{17}O, ^{29}Si, ^{10}B, ^{11}B, ^{23}Na, ^{35}Cl, ^{195}Pt) can also be observed.

NMR resonant frequencies for a particular substance are directly proportional to the strength of the applied magnetic field, in accordance with the equation for the Larmor precession frequency.

NMR studies magnetic nuclei by aligning them with an applied constant magnetic field and perturbing this alignment using an alternating magnetic field, those fields being orthogonal. The resulting response to the perturbing magnetic field is the phenomenon that is exploited in NMR spectroscopy and magnetic resonance imaging, which use very powerful applied magnetic fields in order to achieve high spectral resolution, details of which are described by the chemical shift and the Zeeman effect.

NMR phenomena are also utilized in low field NMR and Earth's field NMR spectrometers, and some kinds of magnetometers.

Discovery

Nuclear magnetic resonance was first described and measured in molecular beams by Isidor Rabi in 1938. Eight years later, in 1946, Felix Bloch and Edward Mills Purcell refined the technique for use on liquids and solids, for which they shared the Nobel Prize in physics in 1952.

Purcell had worked on the development and application of RADAR during World War II at Massachusetts Institute of Technology's Radiation Laboratory. His work during that project on the production and detection of radiofrequency energy, and on the absorption of such energy by matter, preceded his discovery of NMR.

They noticed that magnetic nuclei, like ^{1}H and ^{31}P, could absorb RF energy when placed in a magnetic field of a strength specific to the identity of the nuclei. When this absorption occurs, the nucleus is described as being in resonance. Different atoms within a molecule resonate at different frequencies at a given field strength. The observation of the resonance frequencies of a molecule allows a user to discover structural information about the molecule.

The development of nuclear magnetic resonance as a technique of analytical chemistry and biochemistry parallels the development of electromagnetic technology and its introduction into civilian use.

THEORY OF NUCLEAR MAGNETIC RESONANCE

Nuclear Spin and Magnets

The elementary particles, neutrons and protons, composing an atomic nucleus, have the intrinsic quantum mechanical property of spin. The overall spin of the nucleus is determined by the spin quantum number I. If the number of both the protons and neutrons in a given isotope are even then I = 0, *i.e.* there is no overall spin; just as electrons pair up in atomic orbitals, so do even numbers of protons and neutrons (which are also spin ½ particles and hence fermions) pair up giving zero overall spin. In other cases, however, the overall spin is non-zero. For example ^{27}Al has an overall spin I = 5/2.

A non-zero spin, I, is associated with a non-zero magnetic moment, μ, via.

$$\mu = \gamma I$$

where the proportionality constant, Y, is the gyromagnetic ratio. It is this magnetic moment that is exploited in NMR.

Electron spin resonance is a related technique which exploits the spin of electrons instead of nuclei. The basic principles are otherwise similar.

Values of Spin Angular Momentum

The angular momentum associated with nuclear spin is quantized. This means both that the magnitude of angular momentum is quantized (*i.e.* I can only take on a restricted range of values), and also that the 'orientation' of the associated angular momentum is quantized. The associated quantum number is known as the magnetic quantum number, m, and can take values from +I to –I in integral steps. Hence for any given nucleus, there is a total of 2I+1 angular momentum states.

The z component of the angular momentum vector, I_z, is therefore:

$$I_z = m\hbar$$

where $\hbar$ is Planck's reduced constant.

The z component of the magnetic moment is simply

$$\mu_z = \gamma I_z = m\gamma\hbar.$$

Spin Behavior in a Magnetic Field

Consider nuclei which have a spin of one-half, like ^1H, ^{13}C or ^{19}F. The nucleus has two possible spin states: $m = ½$ or $m = –½$ (also referred to as up and down or α and β, respectively). The energies of these states are degenerate—that is to say that they are the same. Hence the populations of the two states (*i.e.* number of atoms in the two states) will be approximately equal at thermal equilibrium.

If a nucleus is placed in a magnetic field, however, the interaction between the nuclear magnetic moment and the external magnetic field mean the two states no longer have the same energy. The energy of a magnetic moment μ when in a magnetic field B_0 (the zero subscript is used to distinguish this magnetic field from any other applied field) is given by the negative scalar product of the vectors:

$$E = -B_0 \cdot \mu = -\mu_z B_0$$

where the magnetic field has been oriented along the z axis.

Hence

$$E = -m\hbar\gamma B_0.$$

As a result the different nuclear spin states have different energies in a non-zero magnetic field. In hand-waving terms, we can talk about the two spin states of a spin ½ as being aligned either with or against the magnetic field. If Y is positive (true for most isotopes) then m = ½ is the lower energy state.

The energy difference between the two states is

$$\Delta E = -\hbar\gamma B_0$$

and this difference results in a small population bias toward the lower energy state.

Resonance

Resonant absorption will occur when electromagnetic radiation of the correct frequency to match this energy difference is applied. The energy of a photon is $E = hv$, where v is its frequency. Hence absorption will occur when

$$v = \frac{\Delta E}{h} = \frac{\gamma B_0}{2\pi}.$$

These frequencies typically correspond to the radio frequency range of the electromagnetic spectrum. It is this resonant absorption that is detected in NMR.

Nuclear Shielding

It might appear from the above that all nuclei of the same nuclide (and hence the same Y) would resonate at the same frequency. This is not the case. The most important perturbation of the NMR frequency for applications of NMR is the 'shielding' effect of the surrounding electrons. In general, this electronic shielding reduces the magnetic field at the nucleus (which is what determines the NMR frequency). As a result the energy gap is reduced, and the frequency required to achieve resonance is also reduced. This shift of the

NMR frequency due to the chemical environment is called the chemical shift, and it explains why NMR is a direct probe of chemical structure. If the nucleus is more shielded, then it will be shifted upfield (lower chemical shift) and if it is more de-shielded, then it will be shifted downfield (higher chemical shift).

Unless the local symmetry is particularly high, the shielding effect depends on the orientation of the molecule with respect to the external field. In solid-state NMR, magic angle spinning is required to average out this orientation dependence. This is unnecessary in conventional NMR of molecules in solution since rapid molecular tumbling averages out the anisotropic component of the chemical shift.

Relaxation

The process called population relaxation refers to nuclei that return to the thermodynamic state in the magnet. This process is also called T_1 relaxation, where T_1 refers to the mean time for an individual nucleus to return to its equilibrium state. Once the population is relaxed, it can be probed again, since it is in the initial state.

The precessing nuclei can also fall out of alignment with each other (returning the net magnetization vector to a nonprecessing field) and stop producing a signal. This is called T_2 relaxation. It is possible to be in this state and not have the population difference required to give a net magnetization vector at its thermodynamic state. Because of this, T_1 is always larger (slower) than T_2. This happens because some of the spins were flipped by the pulse and will remain so until they have undergone population relaxation. In practice, the T_2 time is the life time of the observed NMR signal, the free induction decay. In the NMR spectrum, meaning the Fourier transform of the free induction decay, the T_2 time defines the width of the NMR signal. Thus, a nucleus having a large T_2 time gives rise to a sharp signal, whereas nuclei with shorter T_2 times give rise to more broad signals. The length of T_1 and T_2 is closely related to molecular motion.

APPLICATIONS OF NMR

Medicine

The use of nuclear magnetic resonance best known to the general public is in magnetic resonance imaging for medical diagnosis and MR Microscopy in research settings, however, it is also widely used in chemical studies, notably in NMR spectroscopy such as proton NMR, carbon-13 NMR, deuterium NMR and phosphorus-31 NMR. Biochemical information can also be obtained from living tissue (e.g human brain tumours) with the technique known as in vivo magnetic resonance spectroscopy or chemical shift NMR Microscopy.

These studies are possible because nuclei are surrounded by orbiting electrons, which are also spinning charged particles such as magnets and, so, will partially shield the nuclei. The amount of shielding depends on the exact local environment. For example, a hydrogen bonded to an oxygen will be shielded differently than a hydrogen bonded to a carbon atom. In addition, two hydrogen nuclei can interact via a process known as spin-spin coupling, if they are on the same molecule, which will split the lines of the spectra in a recognizable way.

Chemistry

By studying the peaks of nuclear magnetic resonance spectra, skilled chemists can determine the structure of many compounds. It can be a very selective technique, distinguishing among many atoms within a molecule or collection of molecules of the same type but which differ only in terms of their local chemical environment. See the articles on carbon-13 NMR and proton NMR for detailed discussions.

By studying T_2* information a chemist can determine the identity of a compound by comparing the observed nuclear precession frequencies to known frequencies. Further structural data can be elucidated by observing spin-spin coupling, a process by which the precession frequency of a nucleus can be influenced by the magnetization transfer from nearby nuclei. Spin-spin coupling is most commonly observed in NMR involving common isotopes, such as Hydrogen-1 (HNMR).

T_2 information can give information about dynamics and molecular motion.

Because the nuclear magnetic resonance timescale is rather slow, compared to other spectroscopic methods, changing the temperature of a T_2* experiment can also give information about fast reactions, such as the Cope rearrangement or about structural dynamics, such as ring-flipping in cyclohexane. At low enough temperatures, a distinction can be made between the axial and equatorial hydrogens in cyclohexane.

An example of nuclear magnetic resonance being used in the determination of a structure is that of buckminsterfullerene. This now famous form of carbon has 60 carbon atoms forming a sphere. The carbon atoms are all in identical environments and so should see the same internal H field. Unfortunately, buckminsterfullerene contains no hydrogen and so ^{13}C nuclear magnetic resonance has to be used. ^{13}C spectra require longer acquisition times since carbon-13 is not the common isotope of carbon (unlike hydrogen, where ^{1}H is the common isotope).

Non-destructive Testing

Nuclear magnetic resonance is extremely useful for analyzing samples non-destructively. Radio waves and static magnetic fields easily penetrate many types of matter and anything that is not inherently ferromagnetic. For example, various expensive biological samples, such as nucleic acids, including RNA and DNA, or proteins, can be studied using nuclear magnetic resonance for weeks or months before using destructive biochemical experiments. This also makes nuclear magnetic resonance a good choice for analyzing dangerous samples.

Data Acquisition in the Petroleum Industry

Another use for nuclear magnetic resonance is data acquisition in the petroleum industry for petroleum and natural gas exploration and recovery. A borehole is drilled into rock and sedimentary strata into which nuclear magnetic resonance logging equipment is lowered. Nuclear magnetic resonance analysis of these boreholes is used to measure rock porosity, estimate permeability from pore size distribution and identify pore fluids (water, oil and gas). These instruments are typically low field NMR spectrometers.

Process Control

NMR has now entered the arena of real-time process control and process optimization in oil refineries and petrochemical plants. Two different types of NMR analysis are utilized to provide real time analysis of feeds and products in order to control and optimize unit operations. Time-domain NMR (TD-NMR) spectrometers operating at low field (2-20 MHz for ^{1}H) yield free induction decay data that can be used to determine absolute hydrogen content values, rheological information, and component composition. These spectrometers are used in mining, polymer production, cosmetics and food manufacturing as well as coal analysis. High resolution FT-NMR spectrometers operating in the 60 MHz range with shielded permanent magnet systems yield high resolution ^{1}H NMR spectra of refinery and petrochemical streams. The variation observed in these spectra with changing physical and chemical properties is modelled utilizing chemometrics to yield predictions on unknown samples. The prediction results are provided to control systems via analogue or digital outputs from the spectrometer.

Earth's Field NMR

In the Earth's magnetic field, NMR frequencies are in the audio frequency range. EFNMR is typically stimulated by applying a relatively strong dc magnetic field pulse to the sample and, following the pulse, analysing the resulting low frequency alternating magnetic field that occurs in the earth's magnetic field due to free induction decay (FID). These effects are exploited in some types of magnetometers, EFNMR spectrometers, and MRI imagers[2]. Their inexpensive portable nature makes these instruments valuable for field use and for teaching the principles of NMR and MRI.

Magnetometers

Various magnetometers use NMR effects to measure magnetic fields, including proton precession magnetometers (PPM) (also known as proton magnetometers), and Overhauser magnetometers.

Polarography

Polarography is an voltammetric measurement whose response is determined by combined diffusion/ convection mass transport. Polarography is a specific type of measurement that falls into the general category of linear-sweep voltammetry where the electrode potential is altered in a linear fashion from the initial potential to the final potential. As a linear sweep method controlled by convection/diffusion mass transport, the current vs. potential response of a polarographic experiment has the typical sigmoidal shape. What makes polarography different from other linear sweep voltammetry measurements is that polarography makes use of the dropping mercury electrode (DME). Close-up of the DME.

A plot of the current vs. potential in a polarography experiment shows the current oscillations corresponding to the drops of Hg falling from the capillary. If one connected the maximum current of each drop, a sigmoidal shape would result. The limiting current (the plateau on the sigmoid), called the diffusion current because diffusion is the principal contribution to the flux of electroactive material at this point of the Hg drop life, is related to analyte concentration by the Ilkovic equation:

$$i_d = 708nD^{1/2}m^{2/3}t^{1/6}c$$

Where D is the diffusion coefficient of the analyte in the medium (cm^2/s), n is the number of electrons transferred per mole of analyte, m is the mass flow rate of Hg through the capillary (mg/sec), and t is the drop lifetime is seconds, and c is analyte concentration in mol/cm^3.

Limitiations

Heyrovský's Polarograph and DME

There are a number of limitations to the polarography experiment for quantitative analytical measurements. Because the current is continuously measured during the growth of the Hg drop, there is a substantial contribution from capacitive current. As the Hg flows from the capillary end, there is initially a large increase in the surface area. As a consequence, the initial current is dominated by capacitive effects as charging of the rapidly increasing interface occurs. Toward the end of the drop life, there is little change in the surface area which diminishes the contribution of capacitance changes to the total current. At the same time, any redox process which occurs will result in faradaic current that decays approximately as the square root of time (due to the increasing dimensions of the Nernst diffusion layer). The exponential decay of the capacitive current is much more rapid than the decay of the faradaic current; hence, the faradaic current is proportionally larger at the end of the drop life. Unfortunately, this process is complicated by the continuously changing potential that is applied to the working electrode (the Hg drop) throughout the experiment. Because the potential is changing during the drop lifetime (assuming typical experimental parameters of a 2mV/sec scan rate and a 4 sec drop time, the potential can change by 8 mV from the beginning to the end of the drop), the charging of the interface (capacitive current) has a continuous contribution to the total current, even at the end of the drop when the surface area is not rapidly changing. As such, the typical signal to noise of a polarographic experiment allows detection limits of only approximately 10^{-5} or 10^{-6} M. Better discrimination against the capacitive current can be obtained using the pulse polarographic techniques.

Qualitative Information

Qualitative information can also be determined from the half-wave potential of the polarogram (the current vs. potential plot in a polarographic experiment). The value of the half-wave potential is related to the standard potential for the redox reaction being studied.

*Compilation done based on information available in internet.

CHAPTER 38

Noise Pollution: Causes and Health Effect

Ms. Nazia Ahmad Umar[1] and Athar Adil Hashmi[2]

Introduction

Noise, an omen to human existence has now become a silent killer of the same; hence noise against the noise is constraining out. Mankind has always been plagued by a variety of both natural and man-made ills. The manmade plague of the 21st Century is environmental noise from which there is virtually no escape, no matter where we are - in our homes and yards, on our streets, in our cars, at theaters, restaurants, parks, arenas, and in other public places. The word noise comes from the Latin word nausea meaning seasickness. Noise is the most silent pollutant of the environment and can without doubt be called the silent killer and stress booster. Noise pollution (or environmental noise) is displeasing human-, animal- or machine-created sound that disrupts the activity or balance of human or animal life. A common form of noise pollution is from transportation, principally motor vehicles, referred to as environmental noise. Environmental noise consists of all the unwanted sounds in our communities except that which originates in the workplace. Almost everyone has had one experience of being temporarily "deafened" by a loud noise. Noise is a disturbance to the human environment that is escalating at such a high rate that it will become a major threat to the quality of human lives. In the past thirty years, noise in all areas, especially in urban areas, has been increasing rapidly. Environmental noise pollution, a form of air pollution, is a threat to health and well-being. It is more severe and widespread than ever before, and it will continue to increase in magnitude and severity because of population growth, urbanization, and the associated growth in the use of increasingly powerful, varied, and highly mobile sources of noise. Another good reason for its rapid escalation is the sustained growth in highway, rail, and air traffic, which remain major sources of environmental noise. Noise pollution has become an unfortunate fact of life worldwide, despite attempts to regulate it. Second-hand noise is in a way analogous to second-hand smoke, an unwanted airborne pollutant produced by others; it is imposed on us without our consent, often against our wills, and at times, places, and volumes over which we have no control. Sound which pleases the listeners may be termed music and that which causes pain and annoyance—noise. Of course, at times, what is music for some can be noise for others. Typical neighborhood noise comes from premises and installations related to the catering trade (restaurant, cafeterias, discotheques, etc.); from live or recorded music; from playgrounds and car parks; and from domestic animals such as barking dogs. The main indoor sources are ventilation systems, office machines, home appliances and neighbours. There are many countries that have

[2]Department of Chemistry, Jamia Millia Islamia, New Delhi (India)

regulations on community noise originating from road, rail and air traffic, and from construction and industrial plants, but very few have regulations on neighbourhood noise. The reason for this is probably the lack of methods to define and measure it, and the difficulty in controlling it. As compared to many other environmental problems, noise pollution is continually growing, giving rise to an increasing number of complaints from affected individuals. Thus, noise is not simply a local problem, but a global issue that affects everyone and requires urgent and drastic precautionary action in any environmental planning situation.

It is perhaps the first time, when human is dying of environmental pollution m hunt of overnight development and technological advancements. Though, the pollution of any kind is a slow poison, but environmental noise is such a pollution, which is dire enemy of human brain and does not spare even animals and properties too. The noise pollution is in no way lesser hazardous than the air and water pollution. But, comprehensive legislation like water and air pollution is yet to be passed on noise. Although, the sweat music and soneral voice brings joy cordiality relaxations mental peace and makes life worth living but undesired noise brings disturbances to peace mental disorders and hence makes the life miserable. It is heartening that original text of Indian constitution is free from the word environment. In a long series of labour laws in India, the noise pollution is left unattended. It is only 1987, when environmental noise has been recognized, as a pollutant in India but public at large is still unaware about health hazards and remedies of the same. To control noise pollution is a great cause before the whole humanity for its survival. This cause became the subject matter to this treatise.

Man is a social animal; he needs comfort in every sense. People are not able to enjoy their lives completely because of the depletion in the environment and this happens because of the hostility between man and environment. Due to the advancement of science and technology, the problem of noise in recent years has emerged as one of the important pollutants of environment.

Noise pollution is a type of energy pollution in which distracting and irritating sounds are clearly audible and which may result in disturbing any natural process or causes human harm. In other words any unwanted sound may result in noise pollution. These unwanted sounds may adversely affect wild life and human existence. Noise by definition is unwanted sound. What is pleasant to some ears may be extremely unpleasant to others depending upon a number of psychological factors. The sweetest music, if it disturbs a person, who is trying to concentrate or to sleep, is noise for him, just as pneumatic riveting hammer is noise to everyone. In other words any sound may be noise if circumstances cause it to be disturbing.

In recent years, people have become more aware of how pollution is affecting our health and well being. Noise pollution, however, is often over looked in our efforts to reduce pollution. This isn't because it isn't as important, it is simply a fact that many people do not think about noise pollution. It is clear that air pollution causes cancer, and the use of lead has caused serious neurological problems, especially in children. But the effects of noise pollution are not a obvious as these other types of pollution. There is mounting evidence that noise pollution is a serious health problem for both humans and wildlife. But the effects of noise pollution are not as obvious as these other types of pollution. There is no doubt that efforts to reduce noise pollution help to insure a quiet and peaceful existence for everyone and for those who are suffering from the effects of noise pollution, prevention is priceless.

In modern times, and with regard to the increasing number of vehicles circulating in urban road networks, road traffic noise is recognized as a serious health problem and raises several aspects of general public awareness. The evidence that noise interferes negatively with people's daily activities is experienced by almost everybody, and methodical research has brought further irrefutable proves to this fact. This problem is exacerbated when one knows that nowadays the number of vehicles circulating in the urban network of roads is steadily increasing whereas at the same time the number of quiet hours during night time has a tendency to diminish, although at a somehow slower rate. Hence, expressed in terms of social cost, the adverse effects of noise on people in general and of traffic noise in particular result in a reduction of their elements of well being. According to the World Health Organisation, "health is a state of complete physical, mental and social well being.

Ironically, the subject of noise pollution is one of the most silent of all forms of pollutions. But, this little spoken-about form of pollution is nevertheless causing as many health issues as the other forms of pollution do.

Governments have a responsibility for the health of their people which can be fulfilled only by the provision of adequate health and social measures. Noise is in this respect more than just nuisance, and it constitutes a danger that is real to people's health by producing both physical and psychological stress. Between the two decades of the seventies and the nineties, the effectiveness of vehicle noise emission regulations has permitted to reduce progressively and appreciably the maximum permitted noise level of cars. However, the reduction in actual road noise levels does not follow this trend, and consequently smaller achievements are to be expected for the few next coming years. Therefore, noise control programs must include strategies such as land use planning and traffic restraints (Berge, 1994). Several initiatives have been taken by various countries to check the noise level. For example, USA has taken initiative to create sites where human-caused noise pollution will not be tolerated (Geary, 1996). Similarly, the European Union (with more than 250,000 inhabitants) requires that 'noise maps' of big cities aré drawn up by 2002 (New Scientist, 1998). To safeguard against ill effects of noise, the laws of Netherlands do not permit building of houses in areas where 24-hour average noise levels exceed 50 dB. And in Great Britain, the Noise Act empowers the local authorities to confiscate the noisy equipment and fine people who create excess noise at night. Recently, several countries are making amendments in their environmental laws which will make contribution to noise pollution an actionable nuisance.

Causes of Noise Pollution

Environmental pollution in India can broadly be attributed to:

- Rapid industrialization
- Energy production
- Urbanization
- Commercialization
- An increase in the number of motorized vehicles

Vehicles are a major source of pollutants in cities and towns. The concentration of ambient air pollutants in the metro-politan cities of India as well as many of the Indian cities is high enough to cause increased mortality. The rate of generation of solid waste in urban centers has outpaced population growth in recent years with the wastes normally disposed in low-lying areas of the city's outskirts (India: State of the Environment 2001). Urbanization is a process whereby increasing proportions of the population of a region or a country live in urban areas. Urbanization has become a major demographic issue in the 21st Century not only in India but also all over the world. There has always been great academic interest in the Indian urbanization process and a number of scholars have analyzed India's urban experience, particularly in the post independence period (Bose, 1978; MUA, 1988; Mohan, 1996). The level of urbanization in terms of the proportion of urban population to the total is low in India but the urban population in absolute terms is high. Since the first regular census of India was taken in 1881, almost all census reports have commented on the urban growth. During the last three decades in India, the link between urbanization and environment and the threat to the quality of life have emerged as a major issue.

Particularly menacing is the noise pollution that emanates from construction activities that continue throughout the day and well into the night. In Delhi, due to the banning of truck traffic during the day, much unloading of construction equipment and related raw materials takes place throughout the night, and since all kinds of wholesale markets are located adjacent to residential colonies, there is no escape from the crashes and thuds that accompany the manual unloading of bricks, timber, steel and other building materials that are necessary at construction sites. Frequent power outages have made the use of non-insulated generators (as opposed to invertors) ubiquitous. These can be particularly annoying and especially unsettling at night.

To make matters worse, ordinary citizens (particularly in Punjab and the Northern Plains) compound the problem with their penchant for raucous wedding celebrations complete with marching bands that weave through city streets blasting away pop tunes all through the night. The practice of bursting fire-crackers on festivals and other holidays is also common as is the tendency to amplify all manner of religious celebrations at odd hours of the night.

Although the trend towards the loudspeaker "jagran" began with some Sikh communities during the Khalistan years, Hindu, Jain and Muslim communities have now adopted this tortuous custom with a vengeance. In doing so, they have all revealed an appalling lack of concern of the right of the general population to a good night's sleep. Whereas the Muezzin in the mosque will use the loudspeaker only for a limited time to signify the start of prayers, all-night sessions are not at all unusual in periodic Hindu or Jain religious gatherings. In some Punjab Gurudwaras, the use of loudspeakers is a daily occurrence, and in some towns, miked Sangeet sessions can start at 3 AM in the morning.

Recent Supreme Court rulings have acknowledged the gravity of the problem, but noise prevention statutes do not yet offer provisions for hefty fines or other serious punishment, even for chronic and repeated offenders. As it is, law-enforcement agencies are often reluctant to pursue noise-prevention statutes because of cultural, religious or other identification with the violators. But even when police officers are diligent and sincere in their efforts to act against citizen complaints regarding noise violations, the penalties are too small to prevent such violations from recurring. Unless police-officers are mandated to collect substantial fines, or make on-the-spot arrests of unrepentant violators, hapless citizens bombarded by high-decibel assaults are unlikely to find relief through the existing statutes.

The last century was the noisiest in the history of the world, the fossil fuel era of noise. The good news is that the greener we get, the quieter we'll also get. Electric cars and lawn equipment, for instance, make less noise, as do more fuel-efficient hybrid vehicles. Improved technology can also provide some measure of solace. Fire engines and police cars could replace those eardrum-busting sirens with models that better aim the sound in one direction, limiting spillover noise. And you can turn clown the volume inside by replacing noisy household appliances with quieter, energy-conserving models, Not all sound is bad. but too much of the wrong sounds harm your health. What many people don't know is that everyday items such as lawn mowers and kitchen blenders can emit noise at hazardous levels.

Noise pollution (regulation and control) rules, 2000 was enacted to take care of the depletion in the environment due to excessive noise. These are rules regarding the noise levels that should be maintained in certain areas. The noise pollution rules came up with a silence zone i.e. an area comprising not less than 100 meters around hospitals, educational institutions, court, religious places or any other area which is declared as such by the competent authority. According to the noise pollution rules authorities could take action and initiate prosecution against people who do not follow the rules. After the commencement of these rules also one can see many cases where people just honk without any reason creating chaos on the road and other places which comes under the purview of silence zone. These people while honking and using loudspeakers does not realize the difficulties that they cause to others and to themselves. One can even lose hearing; stress levels can go high and even mental instability. Honking unnecessarily while driving has become a trend these days and one can hear it very well at the traffic signals.

The problem of noise in recent years has become a serious threat to personal liberties, enjoyment of which is very essential for the quality of life in any civilized society. Noise, besides disturbing the normal working of the people in day time, may also disturb their sleep during night. Therefore noise has become a growing menace for the life and liberties of the people guaranteed under the constitution. In Maneka Gandhi V. Union of India, the Supreme Court pointed out that the expression 'personal liberty' does not mean only liberty of the persons but also liberty or the rights attached to tne oersoi-Further in Francis Coralie V. Union territory of Delhi, Justice Bhagwati while stressing the quality of life and its enjoyment within the purview of Article 21 has rightly said:

"The right to life enshrined in article 21 cannot be restricted to mere animal existence. It means something much more than just physical survival. The right to life includes the right to live with human dignity and all that goes along with it".

The questionability as to how far the violation of liberties essential for life caused by environmental pollution lies within the scope of article 21 has been discussed by the High Court of Andhra Pradesh in T.Damodar Rao V. S.O Municipal Corporation . The court observed that the enjoyment of life and its attainment and fulfillment guaranteed by Article 21 of the Constitution embraces the protection and preservation of natures gifts without which life cannot be enjoyed and in that case slow poisoning by the polluted atmosphere caused by environmental pollution was regarded as violation of article 21 of the constitution, If the environment is polluted no one can enjoy life fully because the chances of being affected by various diseases are higher and it can also deprive the person of proper sleep, food, peaceful living etc.

Rights to sleep, food, recreation, peaceful living and conversation etc are such basic liberties without which the enjoyment of life with all human dignity is not possible. If these are disturbed by noise, their violation would certainly lie within article 21 of the constitution especially in those cases where the license for the use of such sources of noise has been granted directly by the state administration. This happens often during state festivals and elections. In such circumstances, the state government should take up the responsibility if it fails to control the manner of use of such sources of noise, which ultimately results into the violation of personal freedoms besides causing a problem of environmental pollution through noise. The authorities which include police and other officials should consider the violation of these rules as those against right to life and should strictly impose fine on people who violate the rules and this one step by the state and central government would curb the problem of noise pollution in the country.

No doubt, these noise standards put forth through the Noise pollution (regulation and rules) 2000 and various judgments of the Supreme Court has been helpful in controlling the problem of noise to a great extent, however to control the rapid growth of noise in the country and to keep the environment noise free for the people the need for effective implementation of the laws and awareness amongst the masses are necessary. The environmental protection rules should be followed by all, honking and usage of loud speakers should be banned. Many countries have banned honking and it has proved to be very effective way of preventing noise pollution and most importantly people must know how to live without disturbing others and to protect the environment we live in i.e. there must be 'Individual social responsibility' and only this kind of responsibility can protect our environment from depletion and would allow us to work for the benefit of the environment which plays an important role in our life. Unless and until the masses take up the social responsibility of protecting the environment from all kind of depletion, no laws will be fruitful.

Most of the people inhabiting metropolitan cities or big towns and those working in factories are susceptible to the adverse effects of noise. Characteristically, it affects the rich and the poor alike. The problem of noise pollution is less in small towns and villages. But, those residing in villages/ towns along the national/ state highways or close to railway tracks do bear the burnt of excessive noise. Indiscriminate use of horn by the vehicles and widespread use of loud speakers in Indian social and religious ceremonies cause several health hazards to the urban inhabitants. It may cause deafness, nervous breakdown, mental disorder, heart troubles and high blood pressure, head-aches, dizziness, inefficiency and insomnia (Bhargawa, 2001). The noise level and exposure area depends on its source and its strength. Road noise, especially at some distance from the road can be described as a steady state noise that does not fluctuate much. But, rail and aircraft noise are acoustically characterized by high noise levels of relative short duration. Noise from industrial installations, construction sites and fixed recreation facilities radiates from a point source and the shape of the exposure area is generally a circle. The noise from various sources may either be steady for a long period or fluctuate over a specified period considerably. Road traffic is a key source of noise in big cities.

The speed and exhaust system determines the noise released by road traffic. The contact between tyres and the road surface is dominant source of noise at speeds above 60km/h for light vehicles. In future, tyre to surface noise is likely to become an important issue to be addressed in noise abatement strategies. In urban areas, fast acceleration and re-starting the engine in traffic could result in emissions up to 15dB higher than the normal levels of emission resulting from smooth driving.

Another major source of noise is public address system used by temples, mosques etc. Indian Constitution under Article 25-28 guarantees freedom of religion to all persons. But, this freedom of religion is not an absolute one. Freedom of religion is subject to public order, health and morality.

In a recent decision, Supreme Court held that no religion prescribes that the prayers should be performed by using loudspeaker or by beating drums. Further, it was held that if religious people make use of such equipment, it should not affect the right of other people. The High Court of Tamil Nadu allowed a petition filed by the Welfare Association of KKR Nagar (Chennai) against Church, and directed the respondent that the noise level should not exceed the permissible decibels. Thus, the State can put a restriction on an institution for maintaining public health. Since the noise disturbs the living and leaves a bad impact on the health of the people, restriction imposed by a state on the noise level does not amount to violation of fundamental right.

The analysis table below indicates that a very large proportion of respondents in each age group is being affected by noise emanating from the loudspeakers. The range is 71% to 86% with overall %age of 83%. However, % of such people in age group of 20-40 years is marginally lower. Similar is the situation with automobiles. Majority of respondents across different age groups feel that automobile noise affects their activities.

Sources of noise pollution according to **different** age groups:

Table 38.1 : Age groups

Source of Noise	Upto 20	20-40	40-60	Above 60	Total
Loud speaker	29 (85)	35 (71)	37 (86)	20 (83)	121 (83)
Automobiles	23 (67)	31 (63)	32 (74)	18(75)	104(75)
Neighborhoods	17 (50)	23 (47)	20 (46)	13 (54)	73 (54)
Religious functions	21 (62)	29 (59)	25 (58)	14(58)	89 (58)
Total respondents	34 (100)	49 (100)	43(100)	24(100)	150(100)

Figure in parenthesis are percentages

Respondents (54% across various age groups) acknowledge adverse effect of noise generated by neighborhoods. An almost equal proportion of respondents (58%) across different age-groups claim that noise originating from religious functions affects them.

In general, apart from the loudspeakers and automobiles, religious functions, as well neighborhoods act as significant sources of noise pollution. Thus, metropolitan cities are becoming a victim of a new and very silent class of pollution i.e. noise.

Effects of Noise Pollution

Noise pollution is a highly sensitive social issue. It has its effect on all living things. The problem of noise pollution carries its effect on human beings, animals and birds equally. Hearing loss, in fact, is the most obvious medical consequence of noise pollution, but it is hardly the only one, explains environmental psychologist Arline Bronzaft. In her research, Bronzaft Ibimd that constant noise exposure can impede children's learning ability and cognitive development. And beyond all that, periodically, "you've got to take a break from sound," says Bronzaft. "We need some serenity in our lives."

Noise intensity is measured in decibel units. The decibel scale is logarithmic; each 10-decibel increase represents a tenfold increase in noise intensity. Human perception of loudness also conforms to a logarithmic scale; a 10-decibel increase is perceived as roughly a doubling of loudness. Thus, 30 decibels is 10 times more intense than 20 decibels and sounds twice as loud; 40 decibels is 100 times more intense than 20 and sounds 4 times as loud; 80 decibels is 1 million times more intense than 20 and sounds 64 times as loud. Distance diminishes the effective decibel level reaching the ear. Thus, moderate auto traffic at a distance of 100 ft (30 m) rates about 50 decibels. To a driver with a car window open or a pedestrian on the sidewalk, the same traffic rates about 70 decibels; that is, it sounds 4 times louder. At a distance of 2,000 ft (600 m), the noise of a jet takeoff reaches about 110 decibels—approximately the same as an automobile horn only 3 ft (1 m) away.

Studies about the noise pollution reveals that common ailments caused by noise amongst fully grown human beings are

hearing loss,

insomnia,

blood pressure,

cardiovascular,

digestive problems,

hypertension,

headache,

annoyance and irritation.

Noise pollution related ailments found in children are: dizziness and neurophysiologic reactions. The problem of increasing heart rates, blood pressure in animals .and disturbance in breeding system has been noticed as important ailments.

The studies on noise pollution also prove that normal tolerance of noise in human beings lies between 40-50dB and exposure to noise of more than 90dB may result in permanent hearing loss. Taking these harmful ill effects of noise pollution into consideration, the Noise pollution (regulation and control) rules (2000),was passed to keep a control on the noise levels.

Subjected to 45 decibels of noise, the average person cannot sleep. At 120 decibels the ear registers pain, but hearing damage begins at a much lower level, about 85 decibels. The duration of the exposure is also important. There is evidence that among young Americans hearing sensitivity is decreasing year by year because of exposure to noise, including excessively amplified music. Apart from hearing loss, such noise can cause lack of sleep, irritability, heartburn, indigestion, ulcers, high blood pressure, and possibly heart disease. One burst of noise, as from a passing truck, is known to alter endocrine, neurological, and cardiovascular functions in many individuals; prolonged or frequent exposure to such noise tends to make the physiological disturbances chronic. In addition, noise-induced stress creates severe tension in daily living and contributes to mental illness.

The World Health Organization is one of the bodies which recognizes the special place of low frequency noise as an environmental problem. Its publication on Community

Noise (Berglund et al., 2000) makes a number of references to low frequency noise, some of which are as follows:

- *"It should be noted that low frequency noise, for example, from ventilation systems can disturb rest and sleep even at low sound levels".*
- *"For noise with a large proportion of low frequency sounds a still lower guideline* (than 30dBA) *is recommended".*
- *" When prominent low frequency components are present, noise measures based on A-weighting are inappropriate".*

- *"Since A-weighting underestimates the sound pressure level of noise with low frequency components, a better assessment of health effects would be to use C-weighting".*
- *"It should be noted that a large proportion of low frequency components in a noise may increase considerably the adverse effects on health".*
- *"The evidence on low frequency noise is sufficiently strong to warrant immediate concern.*

"Low frequency noise causes extreme distress to a number of people who are sensitive to its effects. Such sensitivity may be a result of heightened sensory response within the whole or part of the auditory range or may be acquired. The noise levels are often low, occurring in the region of the hearing threshold, where there are considerable individual differences. There is still much to be done to gain a fuller understanding of low level, low frequency noise, its effects, assessment and management. Survey papers of low frequency noise and its occurrence include (Backteman et al., 1983a; Backteman et al., 1983b; Backteman et al., 1984a; Backteman et al., 1984b; Berglund et al., 1996; Broner, 1978a; Hood and Leventhall, 1971).

The auditory effects are described as buzzing, grinding, humming, ringing, roaring, rushing, screeching, squeaking, vibrating, whirring, and whistling. Bodily sensations of tingling, numbness or vibrations sometimes accompany the sounds. There is a parallel between these descriptors and complainants of low frequency noises, especially for those whose experience is worse when trying to sleep.

Noise is usually the term attributed to mean any kind of irritating and obtrusive sound, and noise pollution is one of the environmental concerns to many of us. A consideration of the differences between individuals and the inclusion of acoustical and non-acoustical factors makes it however difficult to assess the effects of noise on people. Noise is in general the term by which it is designated any undesirable sound. A more technical and objective distinction between an acoustical signal in its wider sense and noise stems from the fact that the former has particular characteristics both in time and frequency, and which are usually lacking in the case of noise. However, there still is some ambiguity when reference is made to traffic noise because of the obvious fact that although random, traffic noise has its distinguished spectral and temporal uniformities. In broad terms, these effects can be classified into three main categories: psychological (attitudinal), social (behavioral) and physiological effects. Noise is thought to evoke physiological responses which are characteristic for stress and this has led several researchers to consider the hypothesis that long-term noise exposure contributes to the genesis of serious diseases. However, the controversy of the results of epidemiological studies permits to keep this hypothesis only under the assumption of a particular noise sensitivity presented by the affected population. However, it is a well established fact that across measurement techniques and cultures, noise-reaction relationships show a remarkable similarity.

The auditory effects of noise on people have been quite well-known for some decades ago but, becoming a relatively accessible personal need, cars are invading the urban landscape increasingly, contributing thereby to a higher level of noise pollution than any other man-powered engine. Therefore, most of today's research on noise control is focused on noise from transportation with special emphasis on that of urban traffic. One should note that until the late seventies or so, active research on the hazardous effects of traffic noise on people focused mostly on auditory-related topics, without however neglecting the non-auditory health dimensions of the environmental noise is a subject of continuous and increasing concern to people as annoyance and sleep disturbance are the direct and most important of its health effects

A person busy with some activity may react to noise either immediately, or in a more discrete manner in the form of annoyance. Several publications have reported the efforts of researchers for developing models to help in predicting the degree of annoyance by people as a function of exposure to different noise sources. It has been found that a strong relationship exists between activity interference and the probability of annoyance.

The fact that noise is a serious environmental pollutant urged governments, especially in the industrialized world, to set organizations and commissions acting for the regulation of people's exposition to noise (Bryan and Tempest, 1973; Fidell, 1996; Pettersson, 1997; Saadu et al., 1996; SOU, 1974). However, predicting a community's reaction to urban traffic noise is not easily made based only on simple quantitative measures. The major aim of collecting data over the various physical characteristics of urban noise from different parts of the world during the early seventies was to assess the real size of the problem, and to possibly process suitable descriptors for the subjective judgment of noise exposure by means of simple objective measures. This mission, as pointed out earlier, is not easy to fulfill when individual differences come into play. Several attempts have thus been made with the goal of correlating between subjective annoyance and noise exposure (Langdon and Scholes, 1968; Matschat et al., 1977; Osada, 1991) and after reviewing numerous surveys, Schultz was the first to be able to synthesis a single curve for the relationship between community reaction and noise exposure.

Though people seem to adjust to noise by ignoring it, the ear, in fact, is always operative by transmitting signals to our nervous system stimulating therefore reactions to noise from our bodies. The fact that irritability is a very apparent reaction to noise has made of legislators to often consider public annoyance as the basis of noise control programs. The annoyance that humans feel when faced with noise is the most common outward symptom of stress building up inside them. Therefore these symptoms may be considered as indications for possible more serious health problems.

Among the direct and most obvious effects of noise on awake subjects, annoyance is in general the one which interferes negatively with the individual's speech communication, on concentration ability and consequently on the performance of tasks. Variables influencing the subjective judgment of a situation where prevails noise from traffic, like the weight of running vehicles, traffic fluidity and time of the day, may also contribute to the degree of annoyance and to causing sleep disturbances during the night with resulting after effects the following day. However, some recent reviews suggest that despite different measurement techniques, and which are used within different cultures, the relationships noise-reactions show in general some similarities. Factors like noise sensitivity and attitude towards the noisy situation have presumably some potential effect on modifying noise reactions and account for more variation within data surveys than do noise exposure parameters.

The set of personal characteristics may also cause some annoyance but not necessarily physiological reactions. Consequently, annoyance may lead to some obvious change in one's actions, e.g. shutting the window to isolate the source of noise (or that people with less tolerance to noise tend to move away to quieter areas), or it may engender less evident effects in the form of emotional reactions. Annoyance is in general defined as a feeling of displeasure that is tied to a cause that is believed to affect negatively an individual or a group of individuals. Hence, it follows that judging the degree of noisiness caused by a change in sound level contains two components, the first of which is cognitive and is concerned with the expectations for the sound to meet some characteristics for an ideal environment. The other component is pure emotional, and is related to the change in mood of the affected person as caused by the exposure to the noise event.

Exposure to noise pollution exceeding 75 decibels for more than eight hours daily for a long period of time can cause loss of hearing. The hazards increase with the intensity of the noise and the period of exposure. The sound produced by a bursting cracker, exceeding 150dB, can cause a ringing sensation called 'tinnitus' and can impair hearing permanently. In general about 1 percent of the population suffers from noise-induced pollution. Nagi et al. (1993) found that the noise level produced by household equipment and appliances sometimes reaches up to 97 dB which is more than double the acceptable (45dB) noise level. This excessive noise could carry several ill-effects viz. annoyance, speech interference, sleep disturbance, mental stress, headache, and lack of concentration. Workers exposed to high noise levels have a higher incidence of circulatory problems, cardiac diseases, hypertension, peptic ulcers, and neurosensory and motor impairment. The adverse effects of noise have not even spared the birds (Robins, sparrows, wrens and blackbirds).Those living near busy roads could not hear each other and thus unable to contact for propagation (Deutche Presse-Agentur, 2003). We can visualize from the table below that noise interferes with communication, disturbs the sleep and reduces the efficiency of individuals under its umbrella. Majority of sample respondents exposed to noise pollution report occurrence of annoyance and hearing problem. As many as 35% reported the deafness and almost equal number reported mental breakdown. The survey data shows that the effect of noise is not similar among various age groups. Generally, growing age bears the burnt of excessive noise pollution. For example, the rising proportion of sample respondents in higher age groups acknowledges depression, sleeplessness and deafening effect. A very large proportion of respondents feel that noise interferes with inter-personal communication and causes annoyance. Extreme effects (i.e. mental breakdown and deafness) are acknowledged by one third of survey population. However, there is a much higher incidence of deafness effects on old people (above 60 years of age). Further, a general perusal of table shows that psychosomatic (e.g. depression, sleep) and physiological (deafness) disorders are acknowledged by a smaller proportion of respondents. For elderly patients suffering from heart disease (or other serious or chronic ailments), or students preparing for exams, or for any other citizen with sensitive ears, exposure to such continuous loud noise, especially when it leads to chronic sleep-deprivation or physical/emotional tension can be extremely traumatic. But as research undertaken in Europe, Australia and the US shows quite decisively, even those who may seem indifferent to environmental noise may develop physiological symptoms that may in the long run affect their overall health and well-being. This aspect needs to be particularly emphasized, and consciousness on this issue needs to be raised across all sections of the Indian population.

Although a number of PILs (Public Interest Litigations) have been filed by concerned citizens and lawyers, there has not been enough attention and support from the mainstream media, educators, community health providers and other influential public figures. Politicians have not been particularly unhelpful, preferring to pander to obscurantist religious forces, or cater to selfish commercial or private interests. Print and television media especially need to do more in raising awareness concerning the growing menace of noise pollution, and the general insensitivity to this growing health hazard.

As India's towns and cities become more densely packed, the problem of unwanted noise is likely to increase both during the day and at night. While recent pronouncements by the Supreme Court are an important victory for noise-control activists, it is but a very small step forward. Much more needs to be done if India's laws and implementation are to match what has been achieved in the EU and Australia in recent years.

Table 38.2. Effect of noise on different age groups

Effect of noise Age Groups

	0-20	20-40	40-60	Above 60	Total
Effect on hearing	23 (68)	28 (57)	34 (79)	22 (92)	107 (71)
Interfere with communication	33 (97)	47 (96)	41 (95)	20 (83)	141 (94)
Cause annoyance	25 (73)	38 (78)	35 (81)	18 (75)	116(77)
Disturb sleep	33 (68)	46 (94)	41 (95)	22 (92)	132 (88)
Result in deafness	9 (26)	15(31)	15 (35)	13 (54)	52(35)
Mental breakdown	8(23)	17(35)	17 (40)	(32) 50	8 (32)
Total	34 (100)	49(100)	43 (100)	24(100)	150(100)

Figures in parentheses are percentages to the total of the respective columns

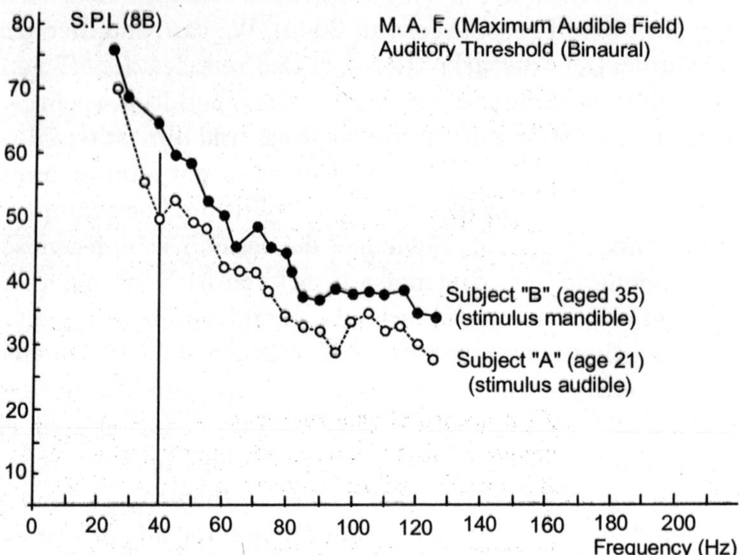

Fig. 38.1. Individual thresholds snowing regions of enhanced sensitivity.

The adverse health effects of noise pollution can be summarized as follows: Effects on the Ear

- *Deafness*
  - o Temporary Deafness: This Persists for about 24 hours after exposure to loud noise.
  - o Permanent Deafness: Repeated or continuous exposure to noise of around 100 dB results in permanent hearing loss. Even single exposure to noise of 160 dB can lead to rupture of ear drum and permanent deafness.

In cases of long term exposure to moderately loud noise, the onset and progress of noise induced deafness is very gradual and by the time the individual is already somewhat deaf, he/she many not be aware of the deafness until the deafness starts affecting the person's ability to hear normal conversation, telephone rings and doorbells etc.

- *Auditory Fatigue* : Noise of 90 dB causes buzzing and whistling in the ears.

Effects on other systems

- ***Decreased Work Efficiency*** : with increasing noise, efficiency of work decreases because of disturbed concentration, annoyance and early onset of fatigue.
- ***Increased Intracrcmial Pressure*** : (Fluid Pressure of the Cerebro Spinal Fluid, the fluid present inside the cavities of brain and between brain and skull)

This leads to Headache, Nausea and Giddiness.

- *Increased Blood Pressure* : Noise can very effectively raise the Blood Pressure of even a normal person.
- *Increased Heart Rate, Respiration rate and Sweating.*

Diminished Night Vision, Colour Perception and visual disturbances.

Thus it is high time that the Indian people and Indian law enforcement and legislative authorities also recognize the gravity of this problem, and begin to pass and enforce laws that ameliorate this growing menace of the modern industrial age.

REFERENCE

1. Bhargawa, Goapl: Development of India's Urban and Regional Planning in 21st Century. Gian Publishing House, New Delhi, pp. 115-116 (2001).
2. Birgitta, Berglund and Lindvall, Homas: A Draft Document of Community Noise. Who Environmental Health Criteria 12, World Health Organization, Geneva (1995).
3. Bond, Michael: Plagued by noise. New Scientist, November 16: 14-15 (1996), Business Line, New Delhi, Oct. 8(2002).
4. Charlotte Clark and Stephen A. Stansfeld International Journal of Comparative Psychology, 2007, 20, 145-158.
5. Deutche, Presse-Agentur: Noisy cities make them dumb. Business Line, May 10 (2003). Dr Geoff Leventhall, A Review of Published Research on Low Frequency Noise and its Effects.
6. Different Noise Sources: a Review of Research" Proceeding of Inter-Noise 83, Vol. II, p. 993-996.
7. Djamel Ouis, Noise and Health Journal
8. Gjestland T., 1987 "Assessment of Annoyance from Road Traffic Noise" Journal of Sound and Vibration 112, p. 369-375.
9. Geary, James: Saving the sounds of silence. New Scientist, 13 April: 45 (1996).
10. Griffriths I. D. and Langdon F. J., 1968 "Subjective Response to Road Traffic Noise" Journal of Sound and Vibration 8, p. 16-32.
11. Kiernan, Vincent: Noise pollution robs kids of languages skills. New Scientist. May 10:5 (1997).
12. Kapoor, B.S. and Singh, K.: 'Noise'-the insidious killer. The Tribune, Nov. 25 (1995).
13. Murli Krishana, R.V. and Murthy, K.P. Vithal: Noise pollution due to traffic in Vishakhapatanam. Indian Journal of Ecology, 10(2): 188-193 (1983)
14. Nagi, G.K., Dhillon, M.K., and Dhaliwal, G. S.: Noise Pollution. Coomonwealth Publishers, New Delhi, p. 5 (1999).
15. Nagi, G., Dhillon, M.K., Bansal, A. S. and Dhaliwal, G.S.: Extend of noise pollution from household equipment and appliances. Indian Journal of Ecology, 20(2): 152-156 (1993).
16. Narendra Singh and S. C. Davar J. Hum. Ecol., (16) (3): 181-187 (2004).
17. P.S. Jaiswal and Nistha Jaiswal–Environmental Law, Second End. 2003, p.327
18. Shetye, R.P., Kapoor, R. K. and T.N. Mahadeven, T. N. : The noise festivals: can we not change? Scavanger. April, pp. 3-8 (1981).
19. Singh, D.P. and Mahajan, C. M.: Noise pollution: Its effect and control. In: G.K. Nagi, M.K. Dhillon and G. S. Dhaliwal (Eds.): Noise Pollution. Coomenwealth Publishers, New Delhi, p. 22 (1990).
20. Singh, P.: Noise pollution. Every Man's Science. 25(1&2): 231-35 (1984).
21. The Times of India. New Delhi. Aug. 7(2001).
22. The Environment (Protection) Act. Universal Law Publishing Co. Ltd, New Delhi (1986).
23. Watts G.R. and Nelson P. M., 1993 "The Relationship Between Vehicle Noise Measures and Perceived Noisiness" Journal of Sound and Vibration 164, p. 425-444.

CHAPTER 39

Social Impact Assessment of Environment

Dr. (Smt.) Sunita Kataria[1]

On the basis of structure, the environment may be divided into two basic types e.g. physical (abiatic) and biotic environment. Physical environment includes atmosphere, lithosphere and hydrosphere and mountain, plateau, plain, Lake River, maritime, glacier, desert and coastal environments. It may be pointed out that of all the organisms man is the most skilled and civilized and therefore his social organization is most systematic. As physical man is one of the organisms of population or biological community and requires physical environment like air, water and food, other biological populations and release wastes into the ecosystem. Social man established social institution form social organizations, formulates laws, principles and policies to safeguard his existence, interest and social and economic man derives and utilizes resources from the physical and biotic environments with his skills and technologies. The economic man equipped with superior technology continued and still continues to exploit the natural resources not withstanding the rebounding repercussions on his own existence. Environmentalist has alarmed the economic man against the devastating impact of unplanned, unscientific exploitation of nature. The philosophy behind this movement by Bhat and Sunderlal Bahuguna is called wilderness ethic, it has intrinsic scientific value because natural or wild areas least affected by human impact, ecosystem where functioning of different components has observed and studied properly, can know the level of degradation and determined the ecological balance. Every Nation has attempted to protect some National Parks and wild life sanctuaries with the help of Sate Governments.

Environment and society are closely related to and are independent. Different social groups and social structures like industrial, agricultural, political, cultural, religious, aesthetic, etc. have evolved and developed during various stages of development of human civilization and these social structures represent mans accumulated cultural resources primarily based on natural environment. Many more questions may be raised in terms of relative value of resources spent on the restoration of quality of environment and pollution abatement programmes and immediate material gains to the society occuring from such programmes. Without doubt the society is now awakened towards environmental problems and the public concern about the quality of environment has reached the motional peak but can this tempo be sustained for long time?

Based on the above framework of cycle of public interest in the environment, downs is of the opinion that the public interest in environmental quality and ecological balance is presently passing through mid-way of the aforesaid cycle and is expected to diminish in coming future.Marxism preaches to organize society's control over the rapacious exploitation of natural resources and to develop harmony between man and nature. The emphasis on rational exploitation of natural resources and ecological balance is evident in the Constitution of the former U.S.S.R. The inherent socialist ideology of public ownership of natural resources is explicit expression of effective nature conservance. In other words, the deep involvement

[1]Prof. (Sociology) Govt. Geetanjali Girls P.G. College, Bhopal (India)

of society in production and consumption processes injects a sense of belonging to natural wealth and therefore the public becomes conscious about the uses and misuses of natural resources and possible danger emanating there from.

The natural resources in the form of matter and energy are of vital significance for the successful survival of all types of life on the planet earth in general and for human being in particular. In fact, all aspects of human society (social, cultural, political and economic) depend on resources. Therefore, the meaning, classification, assessment and evaluation, uses and abuses, conservation and management of all sorts of resources either natural or cultural, renewable or non-renewable, are very significant aspects of environmental geography.

The resources are fundamental base for the economic growth and development of human society but their withdrawals from the nature, mode of their uses by human being and their disposal have enormous adverse effect on the environment. In fact, according to deterministic perspectives of man environment relationships, man is subordinate to natural environment as all aspects of human life viz. physical (health and comfort), social, economic, political, ethical and aesthetic etc. not only depend but are dominantly controlled by physical environment.

W.M. Davis, in 1903 and 1906 clearly demonstrate that human activities, racial characteristics and cultural elements are related to greater extent to the environment. Deterministic/environmentalist approach was fully organized on scientific plane by E. Huntington. His 'Civilization and Climate' (1915). 'the human and habitat(1927), Season of Birth (1938) etc. clearly demonstrate the influences of physical environment on man. Darwinian concepts of natural selection, adaptation and survival of the fittest. (C.C. Park, 1980 p.113) is fully reflected in Ratzels social Darwinists concept of geography as the study of means relationship to his environment which had held away – whether it was with emphasis on the role of human choice (as in the possibilist tradition of Frech geography) Ellen Churchil Semple in U.S.A. and A.J. Herbertson in U.K.) R.D. Dikshit, 1985, p.69) (*ii*) Application of deductive approaches to scientific enquiry and (*iii*) Acceptance and application of Newton concept of cause-effect relationships.

Possibility Approach

Right from the very inception of the school of environmental determination there was dissenting voice raised by those who believed that no doubt of physical environment influences man and his activities but there is ample scope for man to change the environment so much so that it becomes suitable for man and his society.

Febrver, man can never entirely rid themselves, whatever they do, of the hold their environment has on them. Talking this into consideration they utilize their geographical circumstances more or less according to what they are and take advantage of their geographical possibilities. Influence and influence by more moderate terms response or adjustment. G. Tantham while bridging the gap between environmental determinisms and possibilism maintained that the maxiın should not be conquest of nor submission to, but co-operation with nature.

Economic Approach

This approach is based on the basic ideology of man's mastery over environment and continued economic and industrial expansion through the application of modern technologies. The basic thesis of the growth (affluence) school is that because economic growth is required for political, social and economic stability, the quality of environment normally assumes lower priority in formulating planning proposals and in long term planning be cause the deterioration of the environment is generally protected and socially less oblique than a deterioration in the economy (C.C. Park, 1980). W. Zelinsky (1966), believes in mans ability to solve environmental problems arising out of continued economic growth and industrial expansion. It may be pointed out that this extreme concept of economic determinism led to rapacious exploitation of natural resources in the western developed countries and thus created most of the environmental and ecological problems of global dimension.

Ecological Approach

It may be pointed out following C.C. Park (1980) that man environment relationships and debate should be viewed taking into account the multidimensional aspects of environmental problems which are the result of complex series of several factors viz. physical, economic, social political ethical etc but any positive approach adopted for the study of man environment relationships must take into account the fact that there should be harmony and not hostility between man and environment.

Environment and Man

The environment affects man through (*i*) biophysical limitations (*ii*) behavioral controls and (*iii*) resource availability. Weather and climate affect human well being and health. The study of reactions of human body to changes in the atmospheric environment is knows as human biometeorology which lays emphasis on to establish how much of the overall biological variability is the result of changes in weather, climate and season (J.E. Hobbs, 1980). According to M.Bates (1966) three levels of climatic environment affect human behaviour viz. (*i*) microclimate (which represents weather conditions surrounding an individual organism) (*ii*) ecological climate or ecolimate (represents weather elements of the habitual of the organism, in the case of man the habitat may be his house and working places like factory, office, mine, agricultural farms, pasture or forest).

The most significant aspect of the environment in influencing human activity is the availability of resources. The richness or poorness, quality and quantity and above all availability of renewable and non-renewable resources decide the type of human activities, economic viability, social organizations, political stability, international relations etc. very rich reserve of mineral oil in the Middle East is the main reason for political instability in the Gulf Countries. The inter state disputes sharing river waters are because of location of rivers in more than one country.

Man-environment relationship is becomes clear that purely natural relationship between physical primitive man and natural environment during prehistoric period has changed to hostile relationship between technological man and the environment at present. This substantial change and shift in the nature and magnitude of mans interactions with the natural environmental problems of serious consequences because the changes effected by man in the environment have become unadjustable by the inbuilt self regulatory mechanism of the natural environmental system/ ecosystem.

Though man became successful in transforming the natural environmental resources in his way but the nature was still supreme and master and man continued to be guided by physical environment. Extreme concept of the western world, advanced technological man led to reckless and indiscriminate rapacious exploitation of natural resources for industrial expansion and urban growth which have altogether created most of the present day environmental and ecological problems of global dimension.

The impact of modern technological man on natural environment are varied and highly complex as the transformation or modification of one natural condition and process leads to a series of changes in the biotic and abiotic components of biospheric ecosystems.

The man's impacts on environment fall into two broad categories viz. (*i*) Director or International Impacts and (*ii*) Indirect or Unintentional Impacts. Direct or intentional impacts are intentional impacts are preplanned and premeditated because man is aware of the consequences. Such changes include land use changes (clearing of forests and burning of grass land for crop cultivation, felling of trees for commercial purposes, changes in cropping patterns in relation to new farming techniques, new high yielding seeds, irrigational facilities etc.), constructions and excavations (constructions of dams, reservoirs and canals, diversion and manipulation of river channel, construction of embankments and dykes to protect the area from floods, construction of roads and bridges, increase in urbanization, mining, drilling of mineral oils, withdrawal of groundwater etc.), agricultural practices (mechanization of agriculture, use of chemical fertilizers, pesticides and insecticides) weather modification programmes (cloud seeding to induce precipitation, dispersal and clearing of clouds and fogs, checking of hailstorm etc.), nuclear programmes etc. it is significant to point out that the effect of such anthropogenic changes in the natural environment are noticeable within short period and may continue to affect the environment for long time but these effects are reversible because both before and after studies (which are possible) may enable the man to set the adverse effects right to certain extent if so intended by making suitable changes in the initial programmes. For example, withdrawal of groundwater for drinking water and irrigational purposes is a general practice is almost all the countries but some times the impact is so enormous that it becomes disastrous and pounds back on man and society. The example of Brooklyn (kings country, New York, U.S.A.) is sufficient to demonstrate environmental impact of groundwater withdrawal. The pumping of water from beneath the ground surface of Brookyn city for urban dwellers resulted in cavity of 5-milidiameter reaching to a depth of 35 feet below sea level by 1936 (drilling continued for he first three decades of the 20th century). The water table dropped considerably due to withdrawal of groundwater at the rate of 75 million gallons per day on the one hand and poor replenishment of groundwater from natural sources. This resulted into the formation of big cavity beneath the city, consequently saline sea water leaked into the cavity and wells became contaminated due to salty water which forced the city authorities to close down these contaminated wells. A few rechargc wells were constructed and used water was allowed to return to the groundwater through these recharge wells. Excessive withdrawal of groundwater also results in lands subsidence. The cases of land subsidence due to withdrawal of groundwater have occurred in several localities and cities all over the world e.g. Houston City (Tesax, U.S.A. 0.3 to 1.0 m) Mexico city (ground subsidence from 4 to 7m between 1891 to 1959) Venice (Italy) etc. Ground subsidence also occurs due to mining activities.

Man changes the river regime and ecology through flood control measures, reservoirs, construction of dykes and flood wall to restrict the water into river valleys, flood diversion systems and stream canalization (strengthening, shortening, widening depending of river channels to prevent seasonal overbank flooding).

The indirect impacts of man on the environment are not premeditated and planned and these arise from those human activities which are directed to accelerate the pace of economic growth, especially industrial development.

The release of toxic elements into the ecosystem through their uses as insecticides, fertilizers etc. changed the food chains and food webs (e.g. introduction of D.D.T), similarly, the release of industrial wastes into stagnant water, rivers and seas contaminates water and causes several diseases and deaths of organism and thus disturbs ecological balance (e.g. washing and dumping of tailings or waste sledges from factories, release and concentration of specks of asbestos, release of mercury in its toxic methyl form, leakage of crude oil from oil tankers, release of lead, mixing of different quantities of dissolved inorganic matter etc.

Urbanization, industrial expansion and land use changes very often change weather and climate though in long term perspective. Economic activities of man are capable of affecting the heat balance of the earth and its atmosphere which in turn transforms weather and climate at regional and global scales. In fact man changes the atmospheric conditions through (i) changes in the natural gaseous composition of the atmosphere mainly in the lower part, (ii) changes in the water vapour content of the troposphere and the stratosphere through direct (cloud seeding) and indirect means (deforestation) (iii) changes and alteration of land surfaces (deforestation, mining, urbanization etc.), (iv) introduction of aerosol in the lower atmosphere, (v) release of additional heat in the atmosphere (from urban and industrial sources) etc.

The overall negative effects of increase in human population will be adverse changes in physical conditions, biological community and depletion and destruction of natural resources.

Environment and Sustainable Development

Following are some of the methods suggested for sustained development without depletion of natural resources and ecodegradation.

1. Development of battery driven vehicles
2. The use of biogas plants so that agricultural and animal wastes can be utilized for the production of energy.

3. The use of wind and tidal power should be explored more extensively
4. exploitation of sea and its resources
5. Cultivation of more fuel wood trees and shrubs in the various areas of the land where they are not growing.

Conservation and Protection of Environment and the Environmental Laws

Conservation is the judicious use of natural resources and aim at improving the quality of human life. Both economic development and conservation of nature are necessary for the improvement of human lot. Conservation is different from preservation to the extent that preservation means maintaining of nature as it is or as it was before the intervention of human beings or natural forces. While conservation means rational use of the earth's resources to achieve the highest possible quality of life for mankind.

Since excessive use of resources by increasing human population has depleted and degraded natural resources and adversely affected the environment it has became important to conserve and protect out environment. Some of the practices that are helpful in protecting the environment are follows.

1. Planting trees, restoring forests and grass covers.
2. Rotation of crop.
3. Intensive cropping and use of proper drainage canal.
4. Judicious use of fertilizers i.e., use of biofertilizers in place of chemical fertilizers.
5. Sewage treatment.
6. Recycling of materials like paper, glass, plastics etc.
7. Proper disposal of solid waste.
8. Making biogas from biodegradable waste.
9. Establishment of National Parks, Sanctuaries and Forest Reserve etc.
10. Harvesting of rain water.

Some of the laws which have been passed in our country for conservation and protection of the environment are given below:

1. The Motor Vehicle Act, 1938, amended in 1988
2. Prevention of Cruelty to Animals Act, 1960
3. Wildlife (Protection) Act, 1972 and amended in 1991
4. Water (Prevention and Control of Population) Act, 1974
5. Forest Conservation Act, 1980
6. Air Prevention and Control of Pollution Act, 1981
7. Environment (Protection) Act, 1986

The IUCN (International Union for Conservation of Nature and Natural Resources) was set up in 1948. this was sponsored by UNESCO. The 10th general assembly of IUCN was held in New Delhi in 1969 and the year 1972 was declared as the International Year of Nature Conservation.

SAVING PAPER IS SAVING TREES

SAVE ELECTRICITY

Electricity is the most widely used lighting, cooling and heating energy at homes, schools and work places. With some care and awareness a great amount of electricity can be saved without any sacrifice to your comforts.

SAVE WATER

Water is one of our vital natural resources. The water you get while you open a tap has gone through several processes of purification, storage and supply costing money. In many places in our country water is not easily available. We must, therefore, use water wisely and without waste.

Reduce Vehicular Pollution

Vehicular exhaust is one of the major air pollutants. Cars produce large amounts of carbon dioxide which is a major contributor to "Greenhouse Effect". Vehicles also emit nitrogen oxides, carbon monoxide and hydro-carbons which contribute to acid rain and smog.

Measures Suggested to Control Population Explosion The continuing population explosion in our country calls for some soul searching. The government is aware of the magnitude of the problem and considers the alarming population growth as the biggest challenge facing the nation as well as the government. Opposed to the "Kerala Model" is the "Chinese model" of coercion in population control, looking at the near catastrophic state of our population scenario, thinkers support this 'coercive model' as the only effective solution of the population problem.

Economic Development

Economic development may prove to be the best contraceptive. We have to go for quick population control at any cost on sheer economic principles of supply and demand. To balance any economic equation, we can either increase the supply which depends on both financial and material resources, or reduce the demand which depends on the number of people asking for varied services and commodities.

Role of NGOs

Success of any programme depends on its acceptance by the people. Unless the community is fully involved in the programme and it considers it to be its own programme. It may not be possible to achieve the desired results. This can be achieved in a better way by the non government organizations as these have very intimate relations with the people. Their role in removing deep rooted beliefs favouring large families and male children, improving female literacy, raising age at marriage of girls, essential newborn care, birth spacing, etc can be very significant. In addition to its important role in checking the population explosion, family planning will help to improve the general status of women. A woman who has a large number of children to support and who goes through repeated deliveries spends more time as a mother and a wife and is confined to the four walls of her homes. She cannot play any role in the community and the society, unless she is able to limit her family to a reasonable size. Family planning will improve not only family welfare but will also contribute to social prosperity and individual happiness.

Thinning of Ozone Layer

Ozone is a form of oxygen with three atoms instead of the normal two, and forms a fragile shield in the stratosphere absorbing the suns ultra violet radiation, away from the earth's surface.

Chlorofluorocarbons (CFCs) are substances that are inert, inexpensive, stable, non-inflammable, non-toxic, easy to store and produce. The population explosion is one of the causes of the increase of CFCs due to use of air conditioners, insulation, solvents and fire-killers. In 1986, three US agencies, viz., National Aeronautics and Space Administration (NASA), National Oceanic and Atmospheric Administration (NOAA) and Federal Aviation Administration (FAA) started reporting on ground and satellite-based data on ozone depletion in stratosphere by the use of chemicals like halon 1301, methyl the chloroform, carbon tetrachloride, solvents-FC-113, aerosols, foams and refrigeration CFC-11, and air conditioning CFC-12 that are depleting the ozone layer 4,5,8,12 and 26 percent, respectively, on global scale. The increase in ozone layer depletion will invite "hole" in the ozone layer, which appears every spring over the Antarctica.

Action on the ozone layer. Some of the important international actions taken on ozone protection are listed below:

1977	UNEP coordinating committee on the ozone layer established
1985	Vienna Convention of the protection of ozone layer
1987	Montreal Protocol on substances that deplete the ozone layer
1988	Entry into force of the Vienna Convention

1989	Entry into force of Montreal Protocol
1990	Montreal Protocol amended in London. The amendment required signatory governments to regulate consumption and production of CFCs.
Aug. 1992	Entry into force of the amendment
Nov. 1992	At Copenhagen, parties agreed to accelerate; phase out schedule, to Jan.1996

Green House Effect

The phenomenon commonly known as 'greenhouse effect' occurs due to emission of certain pollutants in the air which after the heating of the atmosphere causes the average global temperature to rise. There has emerged a growing consensus in the scientific community that the anthropogenic greenhouse gas emission of carbon dioxide, nitrous oxide, CFCs, methane and low level ozone threatens to disrupt human societies and natural ecosystems.

The developed world accounted for nearly 80 percent of the total industrial carbon dioxide emissions and the rest of 20 percent originated in the developing world. On the other hand, 98 percent CFCs are produced in developed and two percent in developing world. With the current trend of emissions a WHO study speculates that the combined effects of the six most important greenhouse gases could by the early years of 2030s commit the global warming equivalent to doubling of pre-industrial concentrations of carbon dioxide. The scientists are of the opinion that the earth will be hotter by 2° to 4°c by the middle of the 21st century.

EL Nino

EL Nino is a periodic rise in temperature of the Eastern Pacific, around Peru ad Galapagos Islands. There was a strong EL Nino in 1982 and 1983 and a lesser version of the phenomenon was observed in late 1990s.

Scientists have been trying to come up with conclusive answers on the extent of damage El Nino would cause on the entire world. It has occurred twice in the last 15 years.

Acid Rain

Acid rain is a matter of great global concern and has become on of the major environmental problems. Sulphur dioxide reacts in air to form sulphur trioxide, sulphur trioxide reacts rapidly with atmosphere moisture to produce sulfuric acid.

Sulphur trioxide + Water → sulfuric acid

Acid rain is an increasing problem in industrial areas of America, Europe and Asia.

Desertification

The UN Conference on Desertification (UNCOD) has defined desertification as "the diminution or destruction of the biological potential of the land, which can ultimately lead to desert-like conditions". According to the UN Environmental Programme (UNEP), about 1900 million people are threatened by desertification, which affects more than 6.1 billion hectares, about 35 percent of the earth's land area, each of the 16 years that have passed since the UN Conference on Desertification (UNCOD) in 1977 has seen approximately 6 million hectares of previously productive land rendered infertile.

Deforestation

Forests are not merely a natural resource to supplement multiple human requirements but are a very vital aspect of the environmental system. The forest plays a pivotal role in balancing the ecosystem. Growing industrialization, urbanization and ruthless exploitation of forests has created chaotic conditions and severe phyto-geographical and environmental imbalances. The use of forest has been linked with marking from times immemorial basically for food, shelter and clothing, but population exodus and stresses on forest resources created a condition where natural growth of forests and their maintenance is incapable of coping with the demand.

Hazardous Waste Trade

In the late 1980s, the media published a spate of news stories about the dumping of hazardous wastes in the developing world. In the wake of these incidents, hazardous waste trade moved near the top of the international agenda. UNEP accelerated talks addressing the regulation of this trade, and the world's nations debated the issue. These negotiations resulted in the 1989 Basle Convention, which established a prior informed consent system. The convention, adopted in Marh 1989, entered into force in May 1992. it was strengthened in March 1994, when the contracting parties agreed to a complete ban on shipping of hazardous waste.

Biodiversity

Biodiversity or biological diversity refers to the variety of life on earth. The term encompasses three categories genetic diversity, species diversity and ecosystem diversity. Genetic diversity describes the variation of genes within a species, species diversity describes the variety of species within region, and ecosystem diversity refers to the number and distribution of ecosystems. The increasing interest in biodiversity is a result of concerns regarding species extinction, depletion of genetic diversity, and disruption to the atmosphere, water supplies, fisheries, and forests.

The Declaration of UN Conference on the Human Environment and the Action Plant for the Human Environment addressed the importance of protecting forests, which among other things, are important habitats for wildlife, it recommended that wildlife be monitored to assess the impact of pollutants and that the economic value of wildlife be assessed. In the intervening period between the Stockholm and Rio Conferences, the body of legislation concerning biodiversity continued to grow. Relevant major international legislations included the convention concerning the Protection of the World Cultural and Natural Heritage (1972), the Convention on International Trade in Endangered Species of Wild Fanua and Flora (1973), and the Convention on Conservation of Migratory Species of Wild Animals (1979). Other conventions of biodiversity include May 1992 at Nairobi, June 1992 at UNCED, while in December 1993 Convention on Biological Diversity enters into force.

Adverse Effects of Insecticides and Pesticides

The excessive use of chemical pesticides and insecticides has resulted in degradation of environment which has become a threat to human health. Throughout the ages, farmers have had difficulty in protecting their crops from small animals and disease organisms.

The modern pesticide era started in about 1940 when a chemical called DDT was discovered to be a potent insecticide, DDT was cheaper and more effective than the previously known chemicals and led to dramatic early successes. But this miracle chemical had fallen from grace and on 1 January, 1973, all inter-state sale and transport of DDT was banned in USA, ban was also imposed on several other pesticides like aldrin, dieldrin, etc. most of the insecticides are broad spectrum poisons and their use has led to disruption of ecosystems.

Pollution of the Oceans

Oil spilling in oceans is a major problem. Between 1969 and 1983 there were more than 500 tanker accidents that involved oil spills. Altogether, more than one million tons of oil was release. During Gulf war a huge amount of oil spilled in nearby seas and caused great damage to the ocean life. Apart from accidents, seepage of oil is also common. Sometimes thousand tons of oil spills out of the tanker by large leakage as happened recently (October 1997) near Singapore. Even tanker captains often clean the oily holds of their ships in ocean waters.

Environmental Warfare

Using weather as a weapon and manipulating the environment for military purpose is common and has come to focus in the Gulf War. The United States was the first to use weather modification as a weapon in Indo-China region from 1967-72. Artificial triggering of earthquakes and tidal waves, modifying hurricanes or hailstorms to change their course, creating artificial fog to impede enemy traffic, etc.

Environmental Planning in India

The following are the important Acts:

1. Damodar Valley Corporation (Prevention of Pollution of Water) Regulation Act, 1948
2. River Board Act, 1956
3. Water Reservation and Control of Pollution Act, 1974 and 1977
4. Atomic Energy Act, 1972
5. Radiation Protection Rules, 1971
6. Wildlife Protection Act, 1972
7. Factories Amendment Act, 1987
8. Environmental Protection Act, 1986
9. Central Motor Vehicles Rules, 1989
10. Public Liability Insurance Act, 1991

Above all, in 1977, a major step in this direction has been taken in the form of 42nd Constitution Amendment. Accordingly Article 48-A imposes a duty on the state 'to protect and improve the environment and to safeguard the forests and wild life of the country". Article 51A(g) imposes a duty, on citizens of India "to protect and improve the natural environment including forests, lakes, rivers and wildlife and to have compassion for living creatures".

In India, there is a need for sustainable growth, i.e., growth which is not harmful to environment. The eco-friendly development can solve the problem of environmental degradation. Efforts are going on for the environmental management.

Global Environmental Problems and International Co-operation

The most significant global environmental problem faced by the world community is related to global environmental charges (GEC) greenhouse gases at alarming increasing rate, result of global warming would be climatic changes at local, regional and global levels. The international communities are scared of catastrophic adverse effects of future climatic change on different spheres of man and nature e.g. delectation and sea level changes, submergences of island nationals and major coastal lowlands, atmospheric dynamics including evaporation and precipitation, global radiation balance, photosynthesis and ecological productivity, plant and animal community, human health and wealth and many more. The major sources of global environmental problems have been identified e.g. rapid rate of industrialization and urbanization, population growth at alarming rate, advancement in productive technology, major land use changes etc. and efforts are afoot for tackling this alarming problem of global warming leading to climatic changes.

Trend of concentration of green-house gases and aerosols during 1980-89

1. The concentration of atmospheric carbon dioxide increased at the rate of 1.5 ppmv per year (i.e. increase if 0.4% or 3.2 billion tones of carbon per day) due to human economic activities. This increase is equivalent to approximately 50 percent of anthropogenic emission during 1980-89.
2. The rate of increase of atmospheric methane declined but again registered increase in 1993. the atmospheric methane increased at an average rate of 13 ppmv per year (0.8% or 37 million tones).
3. Nitrous oxided registered an increase of 0.75 pp,v (0.25% or 3.7 million tones of nitrogen)per year.
4. The rates of increase of atmospheric concentrations of several ozone-depleting hydrocarbons have fallen, demonstrating the impact of Montreal Protocol and as amendments and adjustments.
5. The trends of increase in troposphere ozone during 1980-89 were variable and uncertain.

Table 39.1

Trend of concentration of green-house gases and aerosols during 1980-89

1. The concentration of atmospheric major greenhouse gases (GHG) viz. carbon dioxide, methane and nitrous oxide (N_2O) has increased by 30, 145 and 15 per cent respectively, largely due to burning of fossil fuel, land use changes and agriculture.
2. Many greenhouse gases like CO_2 and N_2O remain in the atmosphere for centuries, hence it takes a long time to reverse their effect on climate.
3. If the emissions of carbon dioxide at 1994 level are stabilized, the atmospheric concentration of carbon dioxide would be doubled from the pre-industrial level of atmospheric carbon dioxide by the year 2000.
4. Global mean surface air temperature has increased between 0.3 and 0.6°C since the late 19th century. The global sea level has risen by 10-25 cm over past 1000 years (upto 1994).

Table 39.2 : Knowledge for Development World Development Report, 1998-99 and hence it becomes very difficult for comparative analysis and understanding of real picture

Country	Total (Million Tonnes)		Percapita emission (Tonnes)	
	1980	1995	1980	1995
World	13385.7	22702.2	3.4(W)	4.0(W)
U.S.A.	4515.3	5468.6	19.9	20.8
China	1476.8	3192.5	1.5	2.7
Canada	420.9	435.7	17.1	14.7
India	347.3	908.7	0.5	1.0
Australia	202.8	289.8	13.8	16.0
Regions				
East Asia & Africa	1832.7	4140.0	1.4	2.5
Europe and Central Africa	886.9	3722.0	—	7.9
Latin America and Caribbean Middle	850.5	1219.8	2.4	2.6
East and				
North Africa	500.5	982.9	2.9	3.9
South Africa	392.4	1024.1	0.4	0.8
Sub-Saharan				
Africa	350.5	477.1	0.9	0.8

Table 39.3 : Forest Area and Deforestation

	Country 1990	Forest area 1980-99	
	(000 km²)	(000 km²)	Percentage Change
World	69,595	133.4	0.3
Russian Federation	7,681	15.5	0.2
Brazil	5,611	36.7	0.6
Canada	4,533	–47.1	0.6
United States	2,960	3.2	0.1
Australia	1,456	0	0
China	1,247	8.8	0.7
India	517	3.4	0.6
Region			
Sub-Saharan Africa	522	40.7	0.7
East Asia & Pacific	3,986	43.5	1.0
South Asia	658	5.5	0.8
Middle East and North Africa	466	–1.4	–0.3
Latin America and Caribbean	9,786	74.8	0.7

First Earth Summit (Rio Summit)

United Nations Conference on Environment and Development (UNCED), better known as Earth Summit or simply Rio Summit was organized from June 3 to 14, 1992 in Rio De Janeiro city of Brazil under the aegis of United Nations for the protection of the earth and its environment, maintenance of ecological balance and to enrich biodiversity. The representatives of 178 developed and developing countries attended the conference. The primary objectives of the conference were to arrive at commonly acceptable agreements and their implementation to tackle the problems of global warming, depletion of ozone layer and ozone hole, deforestation, biodiversity, weather and climate change, acid rain, sustainable development etc. the following were five important agenda of the conference. (1) rise in global temperature (global warming), (2) forest protection, (3) biodiversity, (4) agenda 21, and (5) Rio declaration.

Second Earth Summit

The second earth summit was held from June 23 to 27, 1997 in New York City of the U.S.A. in order to evaluate the progress and implementation of proposals and Agenda 21, which were agreed during the First Earth Summit, organized in 1992 in Rio De Janeiro city of Brazil. Representatives of 170 countries with 70 heads of government attended the second earth summit. This summit is also known as *Plus-5 Summit* because this summit was organized after 5 years from the second earth summit (Rio Summit) and the programmes and action plans accepted during Rio Summit were discussed and reviewed but ultimately no concrete and fruitful results could be achieved because no agreement cuddle made on any agenda.

Kyoto Protocol

A summit to reduce global warming was held from December 1 to 10, 1997 in Kyoto city of Japan and an agreement to this effect was also signed. The representatives of 149 countries attended this summit. This agreement is popularly known as *Kyoto Protocol of Kyoto Thermal Treaty*. The following are the main items of this historic agreement:

(1) A proposal of 30 percent cut in the emission of carbon dioxide by 2008-12 A.D. was presented by the island nations on the fear that the temperature is estimated to raise 2°C to 3.5°C at the present rate of global warming but the proposal was strongly opposed by the developed and industrialized countries. Ultimately an agreement on 5.2 percent cut from 1990 level of carbon emission could be signed. 11 industrialized countries would implement this cut in carbon emission. It may be pointed out that the U.S.A. European community and Japan agreed to curtail 7 percent, 8 percent and 6 percent emission of carbon dioxide respectively but the developing countries did not agree for any cut in carbon emission.

(2) This is termed as *carbon trading or hot air trading*. As a consequence of Kyoto Protocol Russia and Japan have struck such deal between them. Japan found it difficult to implement the quota of 6 percent cut in carbon emission from 1990 level. On the other hand, Russia can meet its target of zero percent rises in carbon dioxide because of its economic recession. Under the agreement Japanese companies would invest 120 Russian power plants and industries to cut greenhouse emission. These reductions of Russian Power Plants and Industries to cut greenhouse emission.

(3) The Kyoto Protocol and agreement would automatically be invalidated if at least 60 countries of conference of Parties (COP) do not endorse and implement its provisions and resolutions.

(4) A Clean Development Fund (CDP) would be established which would be funded by the lines realized from countries which flout the protocol.

Chapter 40

Environmental Audit : Need and Practice

Mohammed Akram Khan

Introduction

The growing problem of increasing human population and rapid growth of industrialization has created a lot of environmental overburden on the earth. There is now the need of time and conscience effort by the society to save the nation and the world from the menace of the pollution. People from every walk of life have become aware about the adverse impacts of air, water, noise and soil pollution. Governments, all over the world, have formulated laws and regulations to correct and cure the violations arising out of bad environmental practices.

We are familiar with the term 'Audit' wherein financial accounts and records are examined. 'Environmental Audit' refers to verification and assessment of environmental measures adopted by an organization. It is a management tool, which simply inspects the environmental management activities performed by the industries or organizations and makes them aware of new cleaner technologies. It is an approach comprising of systematically documented, periodic and objective valuation of performance of the organization to protect environment with the aim of,

- facilitating management control of environmental practices
- assessing compliance with company's policies including observance of the existing regulatory requirements
- facilitating professional competence

Objectives of Environmental Audit

Before we look into the objectives of environment audit, it will be helpful to know the brief history of environment audit. British Petroleum international group first used the concept of 'Environmental Audit' in the 1972, when a perfumery and petrochemical complex was subjected to a stringent study. In Europe, Ciba-Geigy Company was the first corporation to adopt Environmental Audit in the 1981. The United States Environmental Protection Agency published their environmental policy in the year 1986 followed by International Chamber of Commerce booklet on environmental auditing in year 1988.

The following are the main objectives of Environmental Audit:

1. Monitoring, measuring and evaluating the performance of a company, assessing the probable risks/hazards associated with the malfunctioning of the pollution control equipments.

[1]Advanced Materials & Processes Research Institute, (AMPRI) Bhopal (India)

2. Assessing the scale of optimum utilization of the resources and evaluating the relative position and orientation of the company at national and international level.
3. Improving the energy consumption and suggest for using alternative energy for conservation of energy resources.
4. Determining the mass balance of various materials used and the efficiency of machineries and processes.
5. Identifying the areas of water usage, utilization of wastewater and determine the characteristics of wastewater and their effects on the surrounding environment.
6. Determining the polluted air emission, their sources, quantities and characteristics.
7. Identifying the categories of solid wastes, hazardous wastes and their sources, quantities and characteristics.
8. Help in minimizing the wastes through modern cleaner technologies.
9. Identifying and documenting the environmental compliance status.
10. Help in producing environmentally educated and technically sound personnel by regular environmental auditing.
11. Protecting the corporation from the probable liabilities.

Before auditing a particular industry or organization, the third party or the auditors design a set of plans and programme. It is known as *'Auditing Protocol'*. It gives step-by-step guide to the auditor on how evidence is to be collected. It includes outlines of total audit work, and its time period. The audit work can be done systematically and efficiently with the help of environmental auditing programme. It helps in the proper utilization of natural resources and as a whole it improves environmental quality.

Environmental Audit Scheme and Components

Environmental Audit Scheme is a management tool involving a systematic, objective, documented and periodic evaluation of Environmental Management System, it mainly focused on the following:

- Minimization of the wastes
- Assessing compliance with regulatory requirements.
- Prevention and control of effect of pollutants.
- Establishing relationship between technical professionals, industries, State Pollution Control Board, other public authorities and industrial associations. Environmental Audit Scheme has the following three components:

(*a*) State Pollution Control Board : The State Pollution Control Board should have an`active participation in environmental audit for effective implementation of the audit scheme. *The roles of State Pollution Control Boards are:*

- Preparation of format of audit report incorporating information on all the aspects of the environmental protection.
- Selection of internal auditors to be engaged by industries to prepare audit report.
- Accepting / receiving the audit report prepared by internal auditors.
- Forwarding the audit reports to appropriate agency for evaluation and verification.
- Initiating action if any, based on auditor's comments.

(*b*) Internal Auditor : Environment has a large domain of study and includes subjects from different aspect. For this reason, there must be a team of auditors consisting of experts from various backgrounds and experience having

- Degrees in the disciplines of chemical / civil / environmental engineering
- Degrees in microbiology /biochemistry / environmental science / allied disciplines. The team must be recognized by the State Pollution Control Board and provided with well-equipped laboratory

facilities for analysis of water and air samples. Internal audit report is to be prepared as per the format prescribed by State Pollution Control Board.

(*c*) External Auditor : The external auditor team should be constituted with component institutions / agencies as approved by State Pollution Control Board. The team of external auditors examines the audit reports submitted and verifies the facts given therein. After verification, external auditors send their comments to State Pollution Control Board for further action.

Environmental audit scheme may have some limitations. However, its ultimate goal would be achieved and become fruitful only when the audit scheme is willingly accepted by the industries.

Principal Areas of Environmental Audit

The concept of Environmental Audit has wide scope covering the following areas:

(*a*) Material audit

(*b*) Energy audit

(*c*) Water audit

(*d*) Health and safety audit

(*e*) Environmental quality audit

(*f*) Waste audit

(*g*) Engineering audit

(*h*) Compliance audit

(*i*) Management audit

(*j*) Liability definition audit

(*a*) Material Audit: Material audit examines the use of different raw materials or natural resources for production, quantity use per unit, cost per unit, process wise consumption, process wise wastage etc. and available information pertaining to material usage. Data for the last few years (at least 3 to 4 years) should be collected and analyzed. Other environmental factors like conservation of raw material, scientific storage and reuse of wastage material are taken into consideration in material audit. A comparison of data with similar industries should be made to gather more information pertaining to material usage.

(*b*) Energy Audit: Energy audit examines consumption of various forms of energy in different processes in an industry or organization. Energy Audit aims at minimization, elimination of avoidable losses of valuable energy and their conservation. Without input of energy, we cannot get any type of processed product from an industry. Thus, conservation of energy is very vital as it saves money, minimizes operation costs and gives more profit. It also partly reduces environmental problems.

(*c*) Water Audit: Water audit conducts consumption of water at different sources. It aims at evaluation of raw water intake, performance evaluation of different wastewater treatment plants, cost involved in water treatment and reutilization of wastewater. It should aim at concept of zero discharge. Data analysis of last few years should be reviewed and taken into account.

(*d*) Health and Safety Audit: As the trained workers and employees are the wealth of the industries, due care should be taken for health and safety of the employees. Health and safety audit aims at examining existing health and safety measures in the industry or organization. The risk and incidence of occupational health hazards and accident statistics of the organization should be examined carefully and an on-site audit must be carried out. Physical management of hazardous wastes, toxic pollutants and fire prevention measures etc. should also be evaluated.

(*e*) Environmental Quality Audit: A thorough examination of the status of existing air quality within the industrial complex, noise levels at different sites, functioning of effluent treatment plant, contamination of soil and review of the greenbelt development programme undertaken should be covered under the environmental quality status.

***(f)* Waste Audit:** Waste Audit covers a number of steps like examining the waste generation sites and identification of the type of waste etc. It is also to be analyzed that whether waste can be minimized or eliminated or recycled or ceased, whether it can be a raw material for another organization, whether the particular industry has the recycling facility or not and whether it disposes wastes with or without segregating different types of wastes. Accordingly, suitable strategies should be recommended for waste minimization.

***(g)* Engineering Audit:** Engineering aspects like efficiency of different parts and machineries, engineering methods and techniques must be examined. Accordingly, preferable and suitable processes are recommended.

***(h)* Compliance Audit:** The different aspects of audit that are required to be carried out as per the regulations, procedures and policies of that particular industry are known as Compliance Audit. Each activity of an industry has certain limits. Some set of rules and regulations in every aspect must be obeyed for running an industry smoothly. There are different types of regulations like Environment Protection Act, 1986, Hazardous Waste (Management and Handling) Rule 1989, Hazardous Waste (Management and Handling) Amendment Rules 2002, Water Act 1974, Air (Prevention and Control of Pollution) Act 1981 and Municipal Solid Waste (Management and Handling Rules), 2000 which contain important procedures for environmental management.

***(i)* Management Audit:** A management audit determines whether an adequate compliance management system is established, implemented and used correctly to integrate environmental compliance into every day operating procedures. Such an audit examines socio-cultural management and operational elements, which include internal policies, human resources training programme, budgeting and planning, and monitoring, reporting system and information management system. Some of the broad activities are done through management audit like responsibility of the owners' or other technical persons, adequate system of authority, division of duties, training-of personnel, documentation and internal verification etc

***(j)* Liabilities Definition Audit:** Another type of environmental audit is liabilities definition audit. Such an audit identifies the environmental problems that could reduce the value of a property or expose the buyer to any liability.

Process of Environmental Audit

Despite the variety of audit programme objectives, some elements are common to most audit programmes Each audit programme involves having a team of individuals conduct field assessment, gather information, analyze information and make judgments about the facility, environment compliance status and audit report findings. While not all audit programmes are designed precisely as outlined below, each makes some provision for including these steps in their audit process. What does vary among companies is the time and effort devoted to pre-audit and post-audit activities.

The environmental audit approach has three different processes as given below:

(i) Pre-audit activities

(ii) On-site activities

(iii) Post-audit activities

Pre-Audit Activities: Pre-audit activities include various preparatory works such as:

(a) Collection of preliminary information

(i) Location of the industry with surrounding land use,

(ii) Climatic conditions,

(iii) Products manufactured

(iv) Raw materials used,

(v) rganizational set up and company's policies for environmental management.

(b) Formation of the audit team and their composition from different fields.

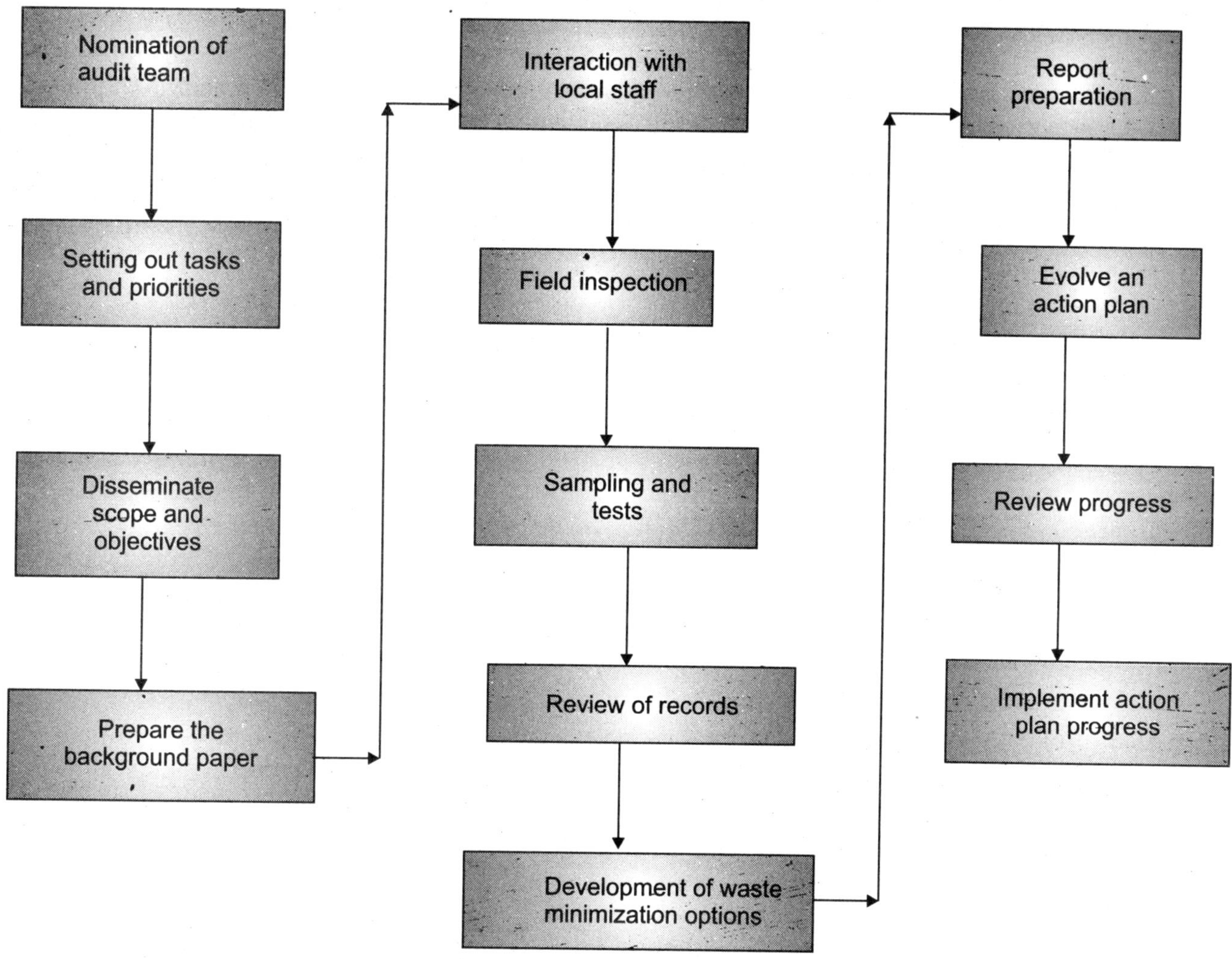

Fig. 40.1. Schematic of Environmental Audit Process

(*c*) Imparting education of environmental audit to every person of the industry.

(*d*) Careful formulation of time frame for the audit so that all the activities are accomplished within a stipulated time period.

(*e*) Preparation of necessary questionnaire to collect and analyze relevant information before the commencement of the audit.

On-site Activities: The main purpose of on-site activities is to review documents maintained at the facility, to interview the employees at site and to inspect thoroughly all relevant operations conducted at the facility.

The activities at the site include the following:

(*i*) Deriving material balance

(*ii*) Identifying wastes flow line

(*iii*) Monitoring of waste characteristics

(*iv*) Evaluation of the performance of process and pollution control equipment/system

(*iv*) Assessing environmental quality

(*v*) Reviewing the safety measures

(*vi*) Reviewing the operating and administrative records and documents

(*vii*) Holding discussions with the Manager and finally preparing the draft report.

Post-Audit Activities: Post-audit activities include discussions among the auditors about the present status of a particular industry after site visit, preparation of an audit report based on the information obtained with suitable suggestions for environmental management. A typical Environmental Audit report contains general information about the industry i.e. name, size, manufacturing details etc. and about the environmental management performance of that industry. Different steps are:

1. Issue of draft report followed by its correction
2. Issue of final report
3. Distribution of final report and highlighting the requirements for action plan
4. Implementation of action plan based on audit findings in the final report
5. Conducting follow-up to the corrective action

Benefits of Environmental Audit

An environment audit programme if designed and implemented carefully can enhance industries performance with respect to environment aspects thereby improving the material balance and energy efficiency profile of production. This will also give clues on how to conserve natural resources and minimize generation of waste to attain sustainable development on one hand and provide direct and indirect economic benefits to industry and stimulate national economy on the other. A properly implemented environmental audit plan provides a wide range of benefits for an organization as well as environment. The benefits of environmental audit are:

- Provides a framework for measuring and managing environmental performance.
- Identifies potential ways of cost savings, which can be achieved through reduction in raw material consumption by ways of waste minimization, recycle and reuse and reduction of pollution load.
- Increases awareness of environmental policies and responsibilities.
- Helps to understand technical capabilities and environmental status ofcompany.
- Helps in timely warning, management of potential future problems and plan for emergencies.
- Helps in interchange of information between different units of the same industry and even on a wide scale between similar industries.
- Enhances production as well as reputation of the company.
- Helps in boosting confidence and co-operation amongst the labors of the company
- Produces trained personnel and thus production capability increases.
- Brings awareness of environmental policies between staff and generates an information base for disaster management.

As a whole, environmental audit plays an important role in minimizing the environmental problems locally, regionally, nationally and internationally.

Environmental Audit Status in India

In India, recognizing the importance of environmental audit and procedures, it was first notified under the Environmental Protection Act, 1986 by the Ministry of Environment and Forest by issuing a Gazette Notification (No. GSR 329 (E)) on 13th March 1992, through which submission of the environmental audit reports have been made mandatory. The industries are now required to submit their audit reports to the concerned State Pollution Control Boards on or before 15th May of every year beginning from 1993. The Environment Audit report has now been renamed as Environment Statement as is being followed by all the industries for compliance.

Acknowledgement

The author is extremely grateful to the Director, Advanced Materials and Processes

Research Institute (CSIR), Bhopal for his kind support and permission for the compilation of Chapter on Environmental Audit. The support rendered by all the colleagues and friends in this endeavor is also deeply acknowledged.

REFERENCES

1. Shreerup Goswami and Prabhasini Pati, Environmental Auditing and Its Countenance, Everyman's Science, Vol. XLIII No. 1, April - May(2008).
2. A. Mehrotra, Pramila Maini, B. Chakradhar, Ashok Maini and A.B. Harapanahalli, A - Z of Environmental Audit, Society of Forest & Environmental Managers, Bhopal.
3. Y. Anjaneyulu, Environmental Impact Assessment Methodologies, (Ed. Sastry C. A.), 254-281, BS Publications, Giriraj Lane, Sultan Bazar, Hyderabad, (2002).
4. M. C. Dash, Ecology, Chemistry and Management of Environmental Pollution, 276-277, Macmillan India Ltd. New Delhi (2004).
5. S. C. Santra, Environmental Science 649-660, New Central Book Agency Pvt. Ltd, Kolkata,(2004).
6. M. Selvam, The need for an Environmental Audit, In : Environmental Accounting and Reporting (Ed. Pramanik A. K.), 82-85, Deep and Deep Publications Pvt. Ltd. Rajouri Garden, New Delhi(2003).
7. A. K. Pramanik, Environmental Audit and Indian Scenario, In: Environmental Accounting and Reporting (Ed. Pramanik A. K.), 261-271, Deep and Deep Publications Pvt. Ltd. Rajouri Garden, New Delhi, (2003).
8. N. K. Uberoi, Environmental Management, 2nd edition, 231-332, EXCEL Books, New Delhi (2003).

CHAPTER 41

Environmental Laws and Legal Aspects

Dr. S.A. Iqbal[1] & M.Tawkir Sheikh[2]

A. ENVIRONMENTAL LAWS

Environmental law is a law that was passed in the 1970, and is known as the National Environment policy Act. This law was passed along with another environment known as the Environment Quality improvement Act. The purpose of these law was to protect the environment from public or private actions that could cause damage.

The mission of NEPA (National Env. Policy Act.) is to form policies and make assessments that are all geared towards the protection of the environment. The Act is responsible for providing the awareness and the means to eliminate danger to the environment and eco-system and establishing goals and laws that will minimize on the risk of environmental harm.

These are a number of federal environmental statutes that make up environmantal law. These are all aimed environment, and include.

1. The clean Air Act.
2. The clean water Act.
3. The pollution prevention Act.
4. The Toxic Substances control Act.
5. The national environmental policy Act.
6. The occupational safety and Health Act.
7. The federal Insecticide, fungicide and Rodenticide Act.
8. Emergency planning and community Right to know Act.

There are many statutes and regulations that come under the umbrella of environmental law, and these are all targeted at the reduction of danger to the environment and an increase public health and awareness. These law are usually long and complicated for anyone not usually involved in the field of environmental law. These are laws and regulations in place to cover literally every environmental issue of potential problems.

Environmental law is a combinations of three things the first is the statute that is passed through (Congress). The second is the list of regulations imposed by MEPA in relation to that particular statute.

[1]Department of Chemistry, Saifia Science College, (B.U.) Bhopal (India)

[2]A.C.S. College, Tukum, Chandrapur-442401 (M.S.) (India)

And the third is the legal interpretation of these regulations by the federal courts. In the event of an environmental issue or problems.

Prior to the 1970 Act, studies were carried out with regards to environmental issues but there were no mechanisms in place to actually do anything about these problems. However, as the environmental issues becomes worse and even the public realised the dangers since it was introduced, their is still much room for improvement. As more complex pollutants threatens to heart our eco-system, it is likely that this law will have to be vastly improved and amended in order to continue to provide protection to the environment and to public health.

1. Clean Air Act

The air that we breathe is essential to maintain life, and low quality, polluted air can pose a serious risk on both the environment and public health. The aim of the clean Air Act is to monitor safety level of airborne pollutants, and to initiate regulations and controls to protect the air from dangerous levels of pollution. The clean air Act program is overseen by the environmental protection Agency responsibility to monitor air quality standards as well as implement the necessary regulations to minimize air pollution.

The original clean Air Act of 1977 was designed to reduce the levels and risk of air pollution, such as ozone / carbon monoxide and particulate matter. The Act was amended in 1990 in order to make it more comprehensive and more protective. However, even now their are millions of people living in areas that have higher contaminations levels than is deemed standards.

There are many contributory factors, to air pollution such as:-

(*a*) Chemical plants.

(*b*) Car and motoring emissions.

(*c*) Paints.

(*d*) Gas Stations.

(*e*) Industrial Production and waste.

As industry continues to grew and more and more cars take to the road, the jobs of the clean Air Act will considerations are going to have to be taken into accounts in federal statutes in order to maintain safe levels of air pollution.

2. Clean Water Act

As the name might suggest the clean water Act is dedicated to the defence and protection of the nations water. This act is designed to ensure that the water retains the correct balance in terms of its physical, biological and chemical make up permits must be applied for under this act for any source that could be a potential pollutant of the water. This program is overseen by the environmental protection Agency.

Originally enacted in 1972, this act was originally known as the federal water pollution control Act Amendments of 1972, it became known simply as the clean water Act following amendments in 1977. The clean water Act set rules and regulations to protect the water from pollutants and contaminations as well as setting water quality standards.

Over the years, additional amendments have been made to clean water Act in the hope of continually improving the protection provided through this Act. Changes have also been made to its way the Act is funded and rather than the old municipal grants scheme it is now funded through the clean water state revolving fund.

3. Resource Conservation and Recovery Act

The mission of its resources conservation and Recovery Act is to provide effective regulations and means to comprehensively manage solid and hazardous waste from generation to disposal. These regulations must be adhered to by any person a organization that hazardous waste including the production, transportation, storage or disposal of the waste. The Resource Conservation and Recovery Act program is overseen by Environmental protection Agency.

This Act was enacted in 1976 and was designed to oversee the life cycle of solid and hazardous waste as well as providing the means for its safe disposal. The act has been amended as a number of occasions to broden the scope of protection provided and to help to made the Act more comprehensive and efficient.

The operating principles of the Resource Conservation and Recovery Act are :

(*a*) To protect public health.

(*b*) To protect the environment.

(*c*) To reduce or eliminate hazardous waste generation.

(*d*) To conserve natural resources.

(*e*) To conserve energy.

Hazardous waste under the Resource conversation and Recovery Act must be solid waste and the RCRA's definition of this type of waste is :

"Garbage, refuse, sludge from a waste treatment plant, water supply treatment plant & air pollution control facility, and other discarded material including solid liquid, semisolid or contained gaseous material agricultural operations and from community activities.

4. Environmental Response and Liability Act

The purpose of this Act is to provide a fast and safe solution to the clearance of contaminated areas as well as to identify the responsible parties and recoup the costs of the clearance from those parties. Anyone found to have disposed of hazardous waste at the contaminated site would be deemed as a responsible party. The environmental response and Liability Act program is overseen by the environmental protection Agency.

The Environmental Response and Liability Act gives the environmental protection Agency the power to force responsible parties to found the clean up of the contaminated site or get the responsible parties to carry out the clearance if deemed safe to do so. However, if no responsible party is identified & is unable to fund the clearance through bankruptcy are a trust fund is in place, which can be used to fund these clean-ups.

This Act was introduced in 1980, and upon its enactment introduced a tax on chemical industries and petroleum industries. This act also provided the authority for direct federal response to the threat of contamination that could pose a risk to public health and environmental health.

5. I suspect my Ground Water & Property may be Contaminated

If you have been using the water for drinking, you should contact your doctor for checkup and further advice, as you may have been affected by the contamination seeking legal advice is also a good idea, as this can prepare you in the event that the contaminated water has caused any harm and you wish to seek compensation.

If you feel that your property has been affected by contamination in any way, make sure you contact the environmental agency closest to you. They will need to carry out investigations and work towards rectifying the problem. You may also be entitled to seek damages depending upon the source of contamination, which is legal assistance will come in useful.

B. LEGAL ASPECTS

It has been covered under the following heads.

1. Constitutional Measures

India is the first country in the world which has provided for constitutional safe guards for the protection and preservation of the environment. In the constitution of India, specific provisions of environment have been in corporated by the constitution (42 amendment) Act, 1976. Now, it is an obligatory duty of the state and every citizen to protect and improve the environment.

The Directive principles of state policy contain specific provisions enunciating the state commitment for protecting the environment.

The state shall endeavor to protect and improve the environment and to safeguard forests and wildlife of the country.

Further more, duties of the citizens towards environment are contained in Article 51-A(g). This Article Says - "It shall be the duty of every citizen of India to protect and improve the natural environment including forests lakes, rivers and wildlife and to have compassion for living creatures.

2. Legislative Measures

The constitution provisions are implemented through environmental protection laws of the country, India has a large body of laws and regulations governing the environment. These includes laws on acted by central and state Governments as well as an increasing body of judicial decisions affecting industrial activities that generates pollution. Further there are more than 200 statutes that have a bearing on environmental matters in India. However the major legal provisions made in the last twenty years are summarised below :

(*a*) The wild life (protection) Act 1972 amended in 1983, 1986 and 1991.

(*b*) The water (prevention and control of pollution) Act 1974, amended in 1988.

(*c*) The Water cess Act 1977 amended in 1991.

(*d*) The forest (conservation) Act 1980, amended in 1988.

(*e*) The Air (prevention & control of pollution) Act 1981, amended in 1987.

(*f*) The environment (protection) Act 1986.

(*g*) The motor vehicle Act 1938, Amended in 1988.

(*h*) The Hazardous waste (Management & Handling) Rules 1989, amended in 2000.

(*i*) The manufacture, storage and import of Hazardous chemicals Rules 1989, amended in 2000.

(*j*) The manufacture, use, import, export and storage of Hazardous Micro-organisms or cells Rules 1989.

(*k*) The public liability insurance Act 1992.

(*l*) A notification of Environmental Statement 1993.

(*m*) A notification on environmental clearance 1994.

(*n*) A notification on environmental clearance 1994.

(*o*) The National environmental Tribunal Act 1995.

The existing laws and regulations on environmental pollution which are administered by the ministry of environment and forests are :

(*i*) The water (prevention and control of pollution) Act 1974 (amended in 1978 and 1988).

(*ii*) The water (prevention and control of pollution) cess Act 1977.

(*iii*) The Air (prevention and control of pollution) Act 1981 amended in 1987.

(*iv*) The environment (protection) Act 1986.

(*v*) The public liability insurance act 1991.

(*vi*) The national environment Tribunal Act 1995.

I. The Water (Prevention and Control of Pollution) Act 1974 (As Amended in 1978 and 1988)

The Water act is a comprehensive legislation providing for the prevention and control of water pollution and for maintaining or restoring the wholesomeness of water streams & wells. Act provides for the establishment of the central pollution control board of the centre and state pollution control boards in the respective states :

(*a*) The functions of the central board at the national levels are to :

(*i*) Advice the central government on matters relating go prevention and control of water pollution,

(*ii*) Provide technical assistance and guidance to the state boards,

(*iii*) Co-ordinate the activities of the State Baord and resolve disputes among them,

(*iv*) Carry out and sponsor research and investigation in the problems of water pollution,

(*v*) Set the standards for streams and wells,

(*vi*) Create environmental awareness, and

(*vii*) To act as state board for the Union Territories.

(*b*) State Board has executive and territorial functions which include :

(*i*) Planning for prevention, control abatment of pollution of streams and wells.

(*ii*) Inspection of sewage or industrial effluent including municipal wastewater treatment plants for the treatment of sewage or trade effluents,

(*iii*) Advice the State Govt. on matters relating to water pollution.

(*iv*) Setting standards for the sewage and industrial effluents discharges. Their is a provision of Joint boards for two or more contiguous - In states. In case of dispute between the state boards, the central board has authority to arbitrate.

Important provisions in the water (prevention and control of pollution) Act, 1974 (As amended in 1978 and 1988) are :

(*a*) Pollution control board (PCB) has the right :

- to obtain any information regarding the construction, installation or operation of an industrial establishment & treatment and disposal system.
- to take samples of treat effluent for the purpose of analysis in the prescribed manner.
- to enter and inspect any industrial establishment, record, register document or any other material object.
- to prohibit use of stream or sewer stand for disposal system without prior consent of the PCB.

(*b*) Restriction on establishment and the operation of any industry process or any treatment and disposal system without prior consent of the PCB.

(*c*) PCB's to grant consent within four months after the date of receipt of the application complete in all respects.

(*d*) PCB's right to refuse or withdraw consent for discharge of effluents.

(*e*) Industry to comply with the condition stipulated in the consent.

(*f*) Industry to appeal to the appellate Authority, in case of grievances against the order passed by the PCB regarding grant, refusal of withdraw of the consent within the specified time in the prescribed manner.

(*g*) Industry to furnish information to the PCB and other specified agency(ies) in case of discharge of poisonous, noxious or polluting matter into a stream, sewer & land, occurred & likely to occur resulting in pollution due to an accident & any other unforeseen event.

(*h*) PCB's right to issue order restraining or prohibiting an industry from discharging any poisonous, noxious & polluting matter in case of emergencies, warranting immediate action.

(*i*) PCB's have power to make an application to the court for restraining likely disposal of polluting matter in a stream & on land.

(*j*) Bar of Jurisdiction in Civil Court in respect of any matter purview of the appellate Authority constituted under the Act and no grant of infunction in respect of any action taken & proposed in pursuance of the Act.

(*k*) Bar of filing of any suit & legal proceedings against the Govt. or Board officials, for action taken in good faith in pursuance of the Act.

(*l*) PCB's to make inquiries, in the prescribed manner, for grant of consent for discharge of effluents.

(*m*) PCB's power to issue directions for the closure, prohibition & regulation of any industry, operation or process or the stoppage or regulation of supply of electricity, water or any other services to industry in the prescribed manner.

(*n*) Industry to comply with the directions of the PCB within the specified time.

(*o*) PCB's to maintain a consent register containing particulars of the consent issued and to provide access to industry at all reasonable hours.

II. The Water (Prevention and Control of Pollution) Cess Act, 1977 (Amended in 1991)

The water cess Act provides for the levy of a cess on water consumed by persons carrying on specified industries given in schedule-I of the Act and also local authorities entrusted with the duty of supplying water under the laws by or under which they are constituted at the rates specified in schedule II of the Act.

The cess is levied and collected by the state Govt. concerned and credited to the consolidated fund of India. An industry which Install and operates its effluent treatment plant is entitle to a rebate of 25% on the cess payable. The cess has been introduced mainly to augment the resources of the central and state pollution control boards.

III. The Air (Prevention and Control of Pollution) Act, 1981 (Amended in 1987)

The Act provides for the setting up of Central / State Boards of prevention and control of Air pollution, however, section 4 of the Act stipulates that in any state in which the water (prevention and control of pollution) Act 1974 is in force and the state Govt. has constituted a state pollution control board that state board shall be deemed to be the state board for the prevention and control of air pollution for union. Territories the central pollution control board is empowered to perform the functions of a state pollution control board under the Act. The state Governments in consultation with their respective state board. Se empowered to declare air pollution control areas. As per the provisions of the Air Act no person can establish & operate any industrial plant in an air pollution control area without obtaining the consent from the concerned state board.

IV. The Environment (Protection) Act 1986

The provisions under this Act are :

(*a*) To take all necessary measures for protecting the quality of environment.

(*b*) Plan and execute a nation wide progamme for the prevention, Control and abatement of environmental pollution.

(*c*) Lay down standards for discharge of environmental pollutants.

(*d*) Empower any persons to enter, inspect take samples and test.

(*e*) Appoint or recognise environment analyst.

(*f*) Establish or recognise environmental laboratories.

(*g*) Lay down standards for the quality of environment.

(*h*) Restrict areas in which any industries, operations, processes may not be carried out subject to certain safeguards.

(*i*) Constitute an authority or authorities for exercising its power.

(*j*) Issue directions to any person officer or authority including the power to direct closure, prohibition & regulation of any industry, operation or process or stoppage or regulation of supply of electricity, water of any other service.

(*k*) The Central Government is empowered to take action under the provision of the Environment (Protection) Act 1986, Powers, under section 5 of the environment (protection) Act 1986 have been degated of the central government to states and Union Territories.

V. Environment Statements

All those on an industry, operation or process requiring consent under water (prevention and control of pollution) Act 1974 (6 of 1974) and or under Air (Prevention and Control of pollution Act 1981 (84 of 1981) and authority under the Hazardous waste (Management and Handling) Rules 1989, are required to submit the environmental statement in prescribed form - V, for the financial year ending 31st march to the concerned. State pollution control boards / pollution control committees in the Union Territories on & before 30th Sept. every year.

VI. Hazardous Waste (Management and Handling) Rules, 1989

The Hazardous wastes (Management and Handling) Rules, 1989, Provide for an effective inventory and controlled handling and disposal of hazardous wastes, under the rules 18, categories, of hazardous waste are identified along with their regulatory quantity Industries generating any of these waste beyond. The regulatory quality are required to seek authorization from the concerned state pollution control board for its temporary storage in the premises and their disposal possibility, of common treatment facilities including landfill are envisaged. The operator of such facility is also required to obtain authorization from the Board. The Boards a expected to specify condition on safe handling and disposal of the waste in authorization. Treatment of the waste at the premises before disposal could also be specified. Import of hazardous waste for processing has to be got approved by the Central Govt.

VII. Manufacture, Storage & Import & Hazardous Chemical Rules 1989

The principal objective of the regulation is prevention of major accidents arising from industrial activity the limitation of the effects of such accidents both of human and environment and the hormonization of the various control measures and the agencies to prevent and limit major accidents. The industrial activities covered by the regulation are defined in terms of process and storage methods involving specified hazardous chemicals.

An important feature of the regulation is that the storage of hazardous chemicals not associated with the process is treated differently from those coming under process use for which a different list of Hazardous chemicals and their manufacture and storage procedures applies under the provisions isolated storage / cover sites are to be separate tank farms or warehouses. The central pollution Control Board and the state pollution control board, as the case may be are the enforcement agency for these storage.

VIII. Safety Reports

A safety report is required to be prepared as par rule 10 in this Act. It involves identification of the nature and use of hazardous chemicals at the installation. The report will also give account of arrangements for safe operation of an installation including Control of any serious deviation that could lead to a major accident and for emergency preparedness at the site. The report will identify that type and the relative likelihood of consequences for any major accident that might occur. It will also demonstrate that the manufacturer or the occupier has identified the major potential accidents from the activity and has provided appropriate Controls.

IX. The Public Liability Insurance Act, 1991

This is an Act to provide for liability insurance for the purpose of providing immediate relief to the persona affected by accidents occurring while handling hazardous substances. The Act casts on the person, who has control over handling any hazardous substance, the liability to give the reliefs specified in the Act to all the victims of any accidents which occurs while handling such substance. It would be the duty of every owner to take necessary insurance policies to discharge to discharge his liabilities.

X. National Environmental Tribunal Act, 1995

This is an Act to provide for strict liability for damages arising out of any accidents occurring while handling any hazardous substance and for the establishment of a national environment tribunal for effective and expeditious disposal of cases arising from such accident. This was enacted with a to view to giving relief and compensation for damages to persons, property and the environment and for matters connected these with or incidental thereto.

XI. Regulatory Standards

Standards for effluent and emissions from industries have been notified and the industries have been directed to adopt which programs leading to compliance with these standards on a time bound basis. The central and the state government are playing a more active role in enforcing these environmental standards. Many polluting units in the country face shifting / closure orders from the courts. It is to be noted that with increasing awareness in environment related issues in the country, the public is becoming more active in high lighting polluting industries and there is an increasing number of public interest litigations in the court.

International Environmental Laws

International environmental law is the body of international law that concerns the protection of the global environment.

Originally associated with the principle that states must not permit the use of their territory in such a way as to injure the territory of other states, international environmental law has since been expanded by a pletora of legally-binding international agreements. These encompass wide variety of issue-areas, from terrestrial, marine and atmospheric pollution through to wildlife and biodiversity protection.

The key constitutional moments in the development of international environmental law are:

- the 1972 United Nations Convention on the Human Environment (UNCHE), held in Stockholm, Sweden;
- the 1987 Brundtland Report, *Our Common Future,* which coined the phrase 'sustainable development';
- the 1992 United Nations Conference on Environment and Development (UNCED), held in Rio de Janeiro, Brazil

The 1972 United Nations Conference on the Human Environment focused on the 'human' environment. The conference issued the Declaration on the Human Environment, a statement containing 26 principles and 109 recommendations (now referred to as the Stockholm Declaration). The creation of an environmental agency was also approved, now known as UNEP. In addition, there was the adoption of a Stockholm Action Program. There were no legally binding outcomes resulting from the Stockholm Conference. Principle 21 of the Declaration was a restatement of law already in existence since Roman times, namely that of 'good neighbourliness'. The Action Plan was never successfully followed by any country.

The 1992 conference (also known as the Earth Summit) led to the adoption of several important legally binding environmental treaties, being the 1992 United Nations Framework Convention on

Climate Change and the 1992 Convention on Biological Diversity. In addition to these, the parties adopted a 'soft law' (non-binding agreements) *Declaration on Environment and Devlopement* which *reaffirmed the Stocholm Declaration and provided 27princples guiding environment* and *development* (now referred to as the Rio Declaration. Another influential soft law document that the parties adopted was Agenda 21, a guide to implementation of the treaties agreed to at the Summit and a guide as to the principles of sustainable development. Agenda 21 also established the United Nations *Commission on Sustainable Development* (CSD) *and the Global Environment Facility (GEF)*. Finally, the non-legal, non-binding Forest Principles were formed *at the Earth Summit.*

A further meeting was helfin 2002, known as the World Summit on SustainableDevelopment (WSSD), held in ohannesburg, South Africa Notable is the absence from its title of the *word 'environment'*. Although this meeting was held to mar the tenth anniversary of the Earth Summit, it is considered by many environmentalists and environmental lawyers to have less than successful in environmental terms It *attained only limited progress towards* stricter global regulation of human impacts on the natural environment.

Nonetheless the WSSD brought a renewed emphasis on the synergies between combating poverty and improving the environement.

Sources of International Environmental Law

International environmental law derives its content from four main sources:

- International agreements (also called treaties, conventions, international legal instruments, pacts, protocols, covenants)
- Customary international law
- General principles of law
- Other/new sources (e.g., court decisions (case-law), resolutions, declarations, doctrine, recommendations given by world organisations etc.)

International Agreements

International environmental agreements can be bilateral, regional or multilateral in nature. The multilateral environmental agreements are frequently referred to as MEAs for short and have become far more common in recent decades. Treaty law is known as a traditional source of law.

The majority of the conventions relating to international environmental law are specific; that means that they deal directly with environmental issues. There are some general treaties with one or two clauses referring to environmental issues but these are rarer. There are about 1000 environmental law treaties in existence today; no other area of law has generated such a large body of conventions on a specific topic.

Protocols

Protocols are like mini-agreements that "hang off" the main treaty. They exist in many areas of international law but are especially useful in the environment field, whee they can be used to update scientific knowledge. They also permit countries to reach agreement on a framework agreement that would otherwise be contentious, by allowing the details to be left to a later date for determination. Protocols are generally much easier to generate than a treaty and they can enter into force very quickly. *The most widely-known protocol in international environmental law is the Kyoto Protocol.*

Customary International Law

Customary international law is important in international environmental law. These are the norms and rules that countries follow as a matter of custom and they are so prevalent that they bind all states in the world. When a principle becomes customary law is not clear cut and many arguments are put forward by states not wishing to be bound.

Examples of customary international law relevant to the environment include:

- the duty to warn other states promptly about emergencies of an environmental nature and environmental damages to which another state or states may be exposed
- Principle 21 of the Stockholm Declaration ('good neighbourliness' or sic utere)

Judicial decisions

International environmental law also includes the opinions of international courts and tribunals. While there are few and they have limited authority, the decisions carry much weight with legal commentators and are quite influential on the development of international environmental law.

The courts include: the International Court of Justice (ICJ); the international Tribunal for the Law of the Sea (ITLOS); the European Court of Justice; regional treaty tribunals. Arguably the World Trade Organisation's Dispute Settlement Board (DSB) is getting a say on environmental law also.

Organizing Principles

International environmental law is heavily influenced by a collection of organising principles.

As with international law, the chief guiding principle is that of soysreijjnty^which means that a country (state) has full power in its own territory to do as it pleases (subjectlo international laws it has agreed to). All other international environmental law principles evolved with this principle in the background and to varying degrees have either supported it or modified it to some extent.

Some of the organising principles of international environmental law include:

- the precautionary principle
- the polluter pays principle
- the principle of sustainable development (Brundtland Report, WSSD) - integration of environmental protection and economic development
- environmental procedural rights
- common but differentiated responsibilities
- intergenerational and intragenerational equity
- common concern of humankind
- common heritage
- partnership (WSSD)
- requirement to conduct a comprehensive environmental impact assessment

Key International Environmental Law Cases

Important cases have included:

- the Trail Smelter Arbitration, 33 AJIL (1939)
- the various nuclear weapons testing cases such as between New Zealand and France before the International Court of Justice;
- Gabcikovo-Nagyramos Dam Case, 1CJ Rep (1997)

Training of International Environmental Lawyers

International environmental lawyers often receive specialized training in the form of an LL.M. degree after having a first law degree – often in another country from where they got their first law degree.

Notable programs include:

- the Lewis and Clark College (http://law.lclark.edu/dept/elaw/)
- the LL.M. in Environmental Law Program at the University of Oregon (http://www.law.uoregon.edu/org/llm/)
- the George Washington University Law School (http://www.law.gwu.edu/Academics/Master+of+Laws+Degree/LL.M.+in+Environmental+Law.htm)
- UNITAR (http://www.unitar.org/elp/)
- the Centre for International Environmental Law (http://www.ciel.org/)

International Environmental Law Groups

Groups active in the area include:

- the Environmental Law Alliance Worldwide(E-LAW) (http://www.elaw.org/)
- the Centre for International Environmental Law (http://www.ciel.org/)
- the Wildlife Interest Group, American Society of International Law (http://www.internationalwildlifelaw.org/)
- Earth Rights International (http://www.earthrights.org/legal/)

International Environmental Law Research Portals

Here are some of the most useful research portals for international environmental law:

- Globelaw (http://www.globelaw.com/)
- ASIL Electronic Guide (http://www.asil.org/resource/envl.htm)
- Envirolink(http://www.envirolink.org/resource.html?itemid=200209240628400.579026&catid=5)
- International environmental law research tutorial (Georgetown University (http://www.ll.georgetown.edU/tutorials/intl/l Oa_ias.html)
- Hastings Law Library Research Guide International Environmental Law (http://library.uchastings.edu/library/Legal%20Research/Research%20Guides/intlenvironbib.htm)
- University of Washington Law School International Environmental Law Research Guide (http://lib.law. washington.edu/_archive/intenv.htm)
- UNEP Department of Environmental Law (http://www.unep.org/delc)

See Also

- ICEAC (International Court of Environmental Arbitration and Conciliation)
- Environmental tariff
- International Environmental Law Research Centre (IELRC)
- List of environmental agreements

External links

- Infography about International Environment Law (http://www.infography.com/content/439331236499.html)

  Retrieved from "http://en.wikipedia.org/wiki/International_environmental_law"

  Categories: International environmental law
- This page was last modified on 9 October 2009 at 00:23.
-

Envoronmental Policies and Laws (India)

Policies

1972	—	Conference on Human Environemnt at Stockholm.
	—	Creation of National Committee on Environmental Planning and Co-ordination (NCEPC) in the Department of Science and Technology, an apex advisory body on all matters relating to environmental protection and improvement.

1976		Forty-second Amendment Act–transfer of forests and wildlife from the state list to the conceurrent list of the consitution: Central Govt. empowered to act directly in managing India's forests and wildlife.
1980		Creation of the department of Environment: Subsequently separate Ministry of Environment and Forests.

Central Laws of Environment

1974	—	Water (Prevention and Control of Pollution) Act for restoration and maintenance of wholesomeness and cleanliness in our national aquatic resources by prevention and control of pollution.
	—	Central and State Pollution Control Boards established to advise the Central Govt. on water pollution issues, co-ordinate the activities of the State Pollution Control Boards, sponsor research projects related to water pollution and develop a comprehensive plan for prevention and control of water pollution.
1981		Air (Prevention and Control of Pollution) Act for prevention, control and abatement of air pollution and preserving the air quality in the country.
1986		The Environment Protection Act (EPA) for protection and improvement of the human environment and prevention of hazards to human beings, plants, animals and property. (Act 1 introduced in the wake of the Bhopal disaster).
		EPA gave the Central Govt. the authority to issue written orders including orders to close, prohibit or regulate any industry, operation or process or to stop or regulate the supply of electricity, water or any other services to these industries.

Environmental Impact Assessment (EIA)

That basic objective of environmental impact assessment or EIA is to identify, predict and evaluate the likely economic, environmental and social impact of developmental activities and taking necessary steps as a remedy as part of the overall environmental management plan (EMP).

It is the Govt. policy that any industrial project (particularly major industry) has to obtain EIA clearance from the Ministry of Environment before approval by the Planning Commission.

Environmental Audit (1992)

Recently the Government of India promulgated an amendment to the EPA rules (1992). Every person carrying on any industry, operation or process requiring consent under the Water Act (1974) or Air Act (1981) or hazardous waste rules (1989) issued under EPA (1986) shall submit an environmental audit report for each financial year ending 31st March in a prescribed form to the concerned state pollution control board. This form covers information on quantity of raw material consumed including water per unit of product, total production, quantity of pollutants generated in terms of solid, liquid and gaseous substances and hazardous wastes as well as the extent of recycling and reuse.